ADVANCED CONTENT DELIVERY, STREAMING, AND CLOUD SERVICES

WILEY SERIES ON PARALLEL AND DISTRIBUTED COMPUTING

Series Editor: Albert Y. Zomaya

A complete list of titles in this series appears at the end of this volume.

ADVANCED CONTENT DELIVERY, STREAMING, AND CLOUD SERVICES

Edited by

Mukaddim Pathan
Telstra Corporation Ltd., Australia

Ramesh K. Sitaraman
University of Massachusetts, Amherst and
Akamai Technologies, USA

Dom Robinson
id3as-company Ltd., UK

WILEY

Cover Image: iStockphoto © nadla
Cover Design: Wiley

Published by John Wiley & Sons, Inc., Hoboken, New Jersey.
Published simultaneously in Canada.

Library of Congress Cataloging-in-Publication Data:

Advanced content delivery, streaming, and cloud services / editors, Mukaddim Pathan, Ramesh K. Sitaraman, Dom Robinson.
pages cm
Includes index.
ISBN 978-1-118-57521-5 (hardback)
1. Cloud computing. 2. Computer networks. I. Pathan, Mukaddim. II. Sitaraman, Ramesh Kumar, 1964- III. Robinson, Dom.
QA76.585.A377 2014
004.67′82–dc23

2014005235

10 9 8 7 6 5 4 3 2 1

To my wife Ziyuan for her inspiration, love, and support. This book would not have been completed, if she did not single-handedly take care of everything, while I was too busy in writing and compilation!—Mukaddim

To my wife Vidya for her love and support. And to our lovely children Anu and Siddu for reminding me that despite our best efforts the Internet still isn't fast enough for them!—Ramesh

To my wife Mariana and our wonderful kids Sofia and Zac—I am sure you will enjoy this book as a bedtime reading. And to my parents (that funny "computer thing" you bought me as a kid came in handy!)—Dom

CONTENTS

PREFACE

The ever-evolving nature of the Internet brings new challenges in managing and delivering content to end-users. Content Delivery Networks (CDNs) improve Web access and streaming performance, in terms of response time and system throughput, while delivering content to Internet end-users through multiple, geographically distributed edge servers. The CDN industry, that is, content delivery, consumption, and monetization, has been undergoing rapid changes. The multidimensional surge in content delivery from end-users has led to an explosion of new content, formats, and an exponential increase in the size and complexity of the digital content supply chain. These changes have been accelerated by economic downturn in that the content providers are under increasing pressure to reduce costs while increasing revenue.

The main value proposition for CDN services has shifted over time. Initially, the focus was on improving end-user-perceived experience by decreasing response time, especially when the customer website experiences unexpected traffic surges. Nowadays, CDN services are treated by content providers as a way to use a shared infrastructure to handle their peak capacity requirements, thus allowing reduced investment cost in their own hosting infrastructure. Moreover, recent trends in CDNs indicate a large paradigm shift toward a utility computing model, which allows customers to exploit advanced content delivery services, hosted on commodity hardware, without having to build a dedicated infrastructure.

From a market perspective, historically buyers based the bulk of demand and spending on "core" CDN products that facilitate the delivery of Web-based content services. Over the last few years, offering from video streaming and value-added services (VASs) peaked as the most demandable CDN products. They formed the basis of most of the present-day CDNs' offering, while strong demand for the basic CDN services still continues. Market research shows that on average, buyers reported 43% of total CDN spending on core products, such as caching and content delivery, while 57% spending on VAS-based products.

While satisfying the market demands, CDN providers are more and more focusing on higher margin, VAS offering in order to gain (or stabilize) overall profit margins. These VASs include mobile data acceleration, content protection, content management, application acceleration, mobile data delivery, and cloud-based storage. While these products currently have a reasonable market penetration, they represent even more substantial near-term growth opportunities.

In addition to the emergence of innovative CDN models, such as managed CDNs, licensed CDNs, and federated CDNs, Telco/operator CDNs are evolving into major market share holders. Telcos/operators around the world have started building CDN platform, technology, and support to aid content consumption, delivery, and rich media experience by end-users. The geographic expansion of Web-based content continues to grow and drive global CDN business requirements. Many CDN players have started with a regional focus and then expanded to offer services in new regions. It is expected that these trends in the CDN industry will continue, as the definition and scope of a CDN gets broader.

1.1 OVERVIEW AND SCOPE OF THE BOOK

The book entitled *Advanced Content Delivery, Streaming, and Cloud Services* presents fundamental and trendy CDN technologies, with a comprehensive coverage of evolution, current landscape, and future roadmap. The book builds on academic and industrial research and developments, and case studies that are being carried out at different organizations around the world. In addition, the book identifies potential research directions and technologies that will drive future innovations. This book is aimed at a large audience including systems architects, practitioners, product developers, and researchers. It can be used as a reference/textbook for graduate students and a roadmap for academicians, who are starting to research in the field of content delivery. We expect the readers to have at least the basic knowledge about Web technologies and the Internet. In particular, readers should be knowledgeable about Web caching, replication, Internet-based services and applications, and basic networking.

Upon reading this, book readers will perceive the following benefits:

1. Learn the state of the art in research and development on content management, delivery, and streaming technologies.
2. Obtain a future roadmap by learning open research issues.
3. Gather the background knowledge to tackle key problems, whose solutions will enhance the evolution of next-generation content networks.
4. Use the book as a valuable reference and/or textbook.

1.2 ORGANIZATION OF THE BOOK

This book is organized into three parts, namely, Part I: CDN and Media Streaming Basics; Part II: CDN Performance Management and Optimization; and Part III: Case Studies and Next-Generation CDNs. Specifically, the topics of the book are the following:

- *CDN*—Infrastructure, architecture, and technology for web content delivery, content management services, and media streaming.

- *Adaptive Bitrate Streaming (ABR).* Techniques for multimedia streaming over computer networks using the HTTP protocol.
- *Cloud-Based Content Delivery.* Integration of cloud computing with traditional CDN model for content and Web application delivery.
- *Wide Area Network (WAN) Optimization.* Optimization algorithms to increase data transfer efficiency in an end-to-end delivery path across WANs.
- *Mobile Acceleration Service.* Optimizing content and video streams to mobile devices to meet dynamic and personalized content needs of mobile users.
- *Transparent Caching.* Carriers network caching technology to control over what content to cache, when to cache, and how fast to accelerate the content delivery.
- *Request-Routing Techniques.* Known and advanced algorithms for redirecting end-user requests, such as DNS-based routing, anycasting, and content-based routing.
- *CDN Performance, Availability, and Reliability.* SSL processing, network-based personal video recorder (PVR), and measurement techniques.
- *Next-Generation CDNs.* Overview of managed/licensed CDN, Telco/carrier CDNs, P2P CDN, and federated CDNs.
- *CDN Case Studies.* Overview of operational infrastructure and services from the major CDNs.
- *CDN Business Intelligence.* Coverage of the CDN market trends, ongoing planning, and management.

Part I of the book focuses on the basic ideas, techniques, and current practices related to content delivery and media streaming. Chapter 1 by Pathan presents an overview of CDNs, operational models, and use cases. It covers recent market and technology trends, as well as identifies a few research issues in the CDN domain. Robinson, in Chapters 2 and 3, provides a comprehensive description of the live media streaming ecosystem and demonstrates the practical configuration of live streaming using a few tools. In Chapter 4, Haßlinger identifies key properties of caching and content delivery in broadband access network, and describes how efficiency can be achieved by configuration and performance tuning. Alzoubi et al. in Chapter 5 present mechanisms and algorithms to effectively redirect end-user requests in a CDN platform. This chapter demonstrates the applicability of IP anycasting for request redirection. Basics of content delivery to cloud-based home ecosystem is covered in Chapter 6 by Cruz et al., highlighting key challenges, industry practices, and recent trends. In Chapter 7, Narayanan et al. describe the challenges in delivering video in mobile networks and present various adaptation techniques for mobile video streaming.

Part II of the book provides a coverage of CDN performance measurement techniques, tools, reporting, and analytics. In Chapter 8, Siglin covers CDN analytics tools and explores a variety of analytic practices and their implications in practical context, including new methods for analyzing adaptive bitrate (ABR) streaming technology. Mathematical modeling to optimize CDN services, such as video on demand (VoD) content delivery, is covered in Chapter 9 by Bektaş and Ercetin. It makes the reader aware of fundamental optimization problems arising in content delivery and the ways of effectively solving these problems. Molina et al. in Chapter 10 present a basic analytical model to analyze the basic and advanced properties of a CDN. Zhanikeev in Chapter 11 describes a method for cloud-based multisource streaming and compares its performance over traditional methods. In Chapter 12, Islam and Grégoire discuss on the intersection of CDN and cloud computing by exposing a number of trade-offs on the deployment of multimedia processing functions inside the cloud and identify relevant performance factors. In Chapter 13, Yoshida describes the performance of a dynamic streaming CDN, comprising techniques for dynamic network reorganization, and load distribution and balancing to realize dynamicity, as well as techniques for stream segmentation and reconstruction, and QoS assurance. Cesario et al. in Chapter 14 present the analysis of mining streaming data in a CDN, improving efficiency and effectiveness of a CDN architecture. A hybrid multidomain architecture is described that solves the problem of computing frequent items and frequent itemsets from distributed data streams. In Chapter 15, Davies and Pathan cover the capacity planning process that is instrumental for the ongoing operation of a deployed CDN infrastructure. It includes a practical application and workflow of the CDN capacity planning process.

Part III, the final part of the book, consists of a handful of representative case studies on present- and next-generation CDNs. In Chapter 16, Sitaraman et al. discuss different network overlays that are crucial for meeting the needs for Internet-based services. Architecture and techniques of representative overlays are discussed, along with their practical usage and implications. Chapter 17 by Pai provides coverage of a variety of next-generation CDNs and presents a case study of CoBlitz, a research CDN that became a commercial licensed CDN. In Chapter 18, Talyansky et al. describe the challenges of content delivery in China, by drawing on experience from ChinaCache, a carrier-neutral CDN. A brief coverage of content-aware network services offered by ChinaCache is provided, along with future trends of content delivery within China. Chapter 19, by Czyrnek et al., presents a case study of a high definition (HD) interactive TV platform, called PlatonTV. In addition to describing the PlatonTV architecture, different aspects of content delivery such as content ingest, content distribution, and management within the CDN are discussed. In Chapter 20, Srebrny et al. present CacheCast—a link layer caching system for single-source, multiple destination data transfers. In this case study, CacheCast architecture, operational methodology, and deployment details are presented. Sourlas et al. in Chapter 21 present a generic three-phase framework for content replication in information centric networks (ICNs). Algorithms supporting efficient replication in ICN are discussed and performance benefits are demonstrated. Chapter 22 by Fortino et al. describes content delivery techniques in vehicular ad hoc networks (VANets). A content broadcasting methodology is presented, which improves content transfer time and delivery efficiency in the radio network. Finally, in Chapter 23, Kilanioti et al. discuss approaches to leverage information from online social networks (OSNs) for rich media content delivery in CDNs. Future research directions in this area, along with a few commercial implications for CDNs, are also discussed.

Mukaddim Pathan
Telstra Corporation Ltd., Australia

Ramesh K. Sitaraman
University of Massachusetts, Amherst and Akamai Technologies, USA

Dom Robinson
id3as-company Ltd., UK

ACKNOWLEDGMENTS

This book came into light because of the direct and indirect involvement of many researchers, academics, and industry practitioners. We acknowledge and thank the contributing authors; research institutions; and companies whose papers, reports, articles, notes, websites, and study materials have been referred to in this book. We are thankful to Professor Albert Zomaya, editor of the Wiley Series on Parallel and Distributed Computing, for his support in accepting the book proposal and guiding us through Wiley's publication process. We express our gratitude to Simone Taylor, Director, Editorial Development of John Wiley & Sons, Inc., and Wiley's publication staff, for handling the book project and ensuring a timely publication.

All chapters were peer reviewed, and authors have updated their chapters addressing the review comments. Prior technical sources are acknowledged citing them at appropriate places in the book. In case of any errors, we would like to receive feedback so that it could be taken into consideration in the next edition.

We hope that this book will serve as a valuable text for students especially at graduate level and a reference for researchers and practitioners working in the content delivery domain.

Mukaddim, Ramesh, and Dom

CONTRIBUTORS

Hussein A. Alzoubi Case Western Reserve University, Cleveland, OH, USA

Tolga Bektaş University of Southampton, Highfield, Southampton, UK

Carlos T. Calafate Universitat Politècnica de València, Valencia, Spain

Jaime Calvo Universidad de Salamanca, Escuela Politecnica Superior de Zamora, Zamora, Spain

Juan C. Cano Universitat Politècnica de València, Valencia, Spain

Eugenio Cesario ICAR-CNR, Rende (CS), Italy

Tiago Cruz Faculdade de Ciências e Tecnologia da, Universidade de Coimbra, Coimbra, Portugal

Mirosław Czyrnek Poznan Supercomputing and Networking Center, Poznań, Poland

Phil Davie Telstra Corporation Limited, Melbourne, Victoria, Australia

Ozgur Ercetin Sabancı University, İstanbul, Turkey

Manuel Esteve Universitat Politecnica de Valencia, Valencia, Spain

Paris Flegkas University of Thessaly, Oktovriou, Volos, Greece

Giancarlo Fortino University of Calabria, Rende (CS), Italy

Chryssis Georgiou Department of Computer Science, University of Cyprus, Nicosia, Cyprus

Vera Goebel University of Oslo, Oslo, Norway

Jean-Charles Grégoire INRS-EMT, Montréal, QC, Canada

Gerhard Haßlinger Deutsche Telekom Technik, Darmstadt, Germany

Salekul Islam United International University, Dhaka, Bangladesh

Manish Jain Akamai Technologies, Inc., Cambridge, MA, USA

Jędrzej Jajor Poznan Supercomputing and Networking Center, Poznań, Poland

Jerzy Jamroży Poznan Supercomputing and Networking Center, Poznań, Poland

Mangesh Kasbekar Akamai Technologies, Inc., Cambridge, MA, USA

Dimitrios Katsaros University of Thessaly, Oktovriou, Volos, Greece

Anuj Kaul Nokia Siemens Networks, Mountain View, CA, USA

Irene Kilanioti Department of Computer Science, University of Cyprus, Nicosia, Cyprus

Ewa Kuśmierek Poznan Supercomputing and Networking Center, Poznań, Poland

Seungjoon Lee AT&T Labs—Research, Florham Park, NJ, USA

Woody Lichtenstein Akamai Technologies, Inc., Cambridge, MA, USA

Pietro Manzoni Universitat Politècnica de València, Valencia, Spain

Carlo Mastroianni ICAR-CNR, Rende (CS), Italy

Andreas Mauthe InfoLab 21, Lancaster University, Lancaster, UK

Cezary Mazurek Poznan Supercomputing and Networking Center, Poznań, Poland

Benjamin Molina Universitat Politecnica de Valencia, Valencia, Spain

Edmundo Monteiro Faculdade de Ciências e Tecnologia da, Universidade de Coimbra, Coimbra, Portugal

Ram Lakshmi Narayanan Nokia Siemens Networks, Mountain View, CA, USA

Vivek S. Pai Princeton University, Princeton, NJ, USA

Carlos E. Palau Universitat Politecnica de Valencia, Valencia, Spain

George Pallis Department of Computer Science, University of Cyprus, Nicosia, Cyprus

Mukaddim Pathan Telstra Corporation Ltd., Melbourne, Victoria, Australia

Thomas Plagemann University of Oslo, Oslo, Norway

Michael Rabinovich Case Western Reserve University, Cleveland, OH, USA

Dom Robinson id3as-company Ltd., Rottingdean, Brighton, Sussex, UK

Mili Shah Nokia Siemens Networks, Mountain View, CA, USA

Timothy Siglin Braintrust Digital, Inc., Harriman, TN, USA

Paulo Simões Faculdade de Ciências e Tecnologia da, Universidade de Coimbra, Coimbra, Portugal

Ramesh K. Sitaraman University of Massachusetts, Amherst, and Akamai Technologies, MA, USA

Dag H.L. Sørbø University of Oslo, Oslo, Norway

Vasilis Sourlas University of Thessaly, Oktovriou, Volos, Greece

Oliver Spatscheck AT&T Labs—Research, Florham Park, NJ, USA

Piotr Srebrny University of Oslo, Oslo, Norway

Maciej Stroiński Poznan Supercomputing and Networking Center, Poznań, Poland

Domenico Talia ICAR-CNR, Rende (CS), Italy; DIMES, University of Calabria, Rende (CS), Italy

Michael Talyansky ChinaCache, Sunnyvale, CA, USA

Leandros Tassiulas University of Thessaly, Oktovriou, Volos, Greece

Alexei Tumarkin ChinaCache, Sunnyvale, CA, USA

Kobus Van Der Merwe University of Utah, Salt Lake City, UT, USA

Jan Węglarz Poznan Supercomputing and Networking Center, Poznań, Poland

Hunter Xu ChinaCache, Beijing, China

Yinghua Ye Nokia Siemens Networks, Mountain View, CA, USA

Norihiko Yoshida Information Technology Center, Saitama University, Saitama, Japan

Ken Zhang ChinaCache, Beijing, China

Marat Zhanikeev Kyushu Institute of Technology, Iizuka, Japan

PART I

CDN AND MEDIA STREAMING BASICS

1

CLOUD-BASED CONTENT DELIVERY AND STREAMING

Mukaddim Pathan

Telstra Corporation Ltd., Melbourne, Victoria, Australia

1.1 INTRODUCTION

Over the last decade, end-users have been increasingly using the Internet to access not only typical websites but also high definition (HD) video and rich media content. While accessing the Web, end-users expect high bandwidth, improved performance, and low latency for the delivered content. End-users' requirements of high quality, consistent, dependable, responsive viewer experience can be characterized by faster loading Web pages, quick channel changes and downloads, fast-start video, and quality of experience (QoE) in mobile devices. Similarly, content providers require an efficient content delivery mechanism to increase growth and scale, reliability and performance, and engagement and reach, while decreasing cost, risk, and network load.

Content delivery networks (CDNs) [1–4] improve websites, streaming, and download performance of Internet content by end-users while reducing the cost to serve for content providers. A CDN is a collaborative collection of network elements spanning the Internet, where content is replicated over mirrored Web servers (i.e., point of presence (PoP), edge or replica servers), located at the *edge* of the Internet service providers' (ISPs') networks to which end-users are connected. As shown in Figure 1.1, content is served into the CDN once, then content is delivered to end-users from the edge servers

Advanced Content Delivery, Streaming, and Cloud Services, First Edition.
Edited by Mukaddim Pathan, Ramesh K. Sitaraman, and Dom Robinson.

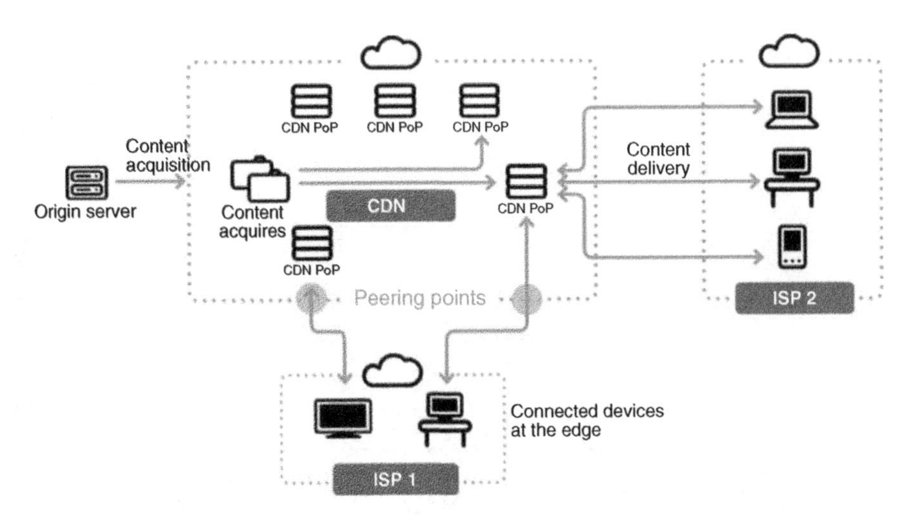

Figure 1.1 Abstract view of a CDN.

rather than serving each individual request directly from the content provider's origin server. Nowadays, CDNs are a prerequisite for delivering quality online experience for live, linear, and on-demand delivery of website and media content. They often leverage cloud (compute and storage) infrastructure and services to provide scalability, resiliency, availability, and performance.

The reason why a CDN may be used varies depending on the particular enterprise—providing live streaming coverage of major events, distributing training videos to employees, providing fast and efficient software downloads to customers, or enhancing the performance of an e-commerce website. A few of these practical use cases are the following:

- *Accelerated Web Performance*. CDNs help to improve the delivery of website content (static and dynamic) so that websites perform better, load faster, and generate more revenue for content providers. CDNs not only cache static content at the edge but also handle dynamic, transactional content from e-commerce providers, online auction sites, by accelerating the data transfer to provide an improved experience to end-users.
- *Software Updates and Downloads*. CDNs enhance the automatic or on-demand delivery of software or file downloads, including software patches. For example, consumers can fast download the latest release of operating system or word processing software releases online instead of ordering and purchasing in-store.
- *Rich Media Content Streaming*. CDNs help deliver rich media content, that is, interactive digital media such as audio and video files in different encoding formats (HyperText Transfer Protocol (HTTP) adaptive streaming with Microsoft Smooth, Adobe HDS, and Apple HLS [5,6]), to specialized streaming clients and devices of end-users. Such streaming can be live or on-demand. Live content

streaming provides webcasts of corporate announcements, investor briefings, and online coverage of events such as the Olympics. On-demand streaming is performed with stored/archived content, based on end-user request to view it.

- *IPTV Use Case.* CDNs are used for Internet Pay TV (IPTV), such as catch-up TV service for recent programs of broadcasters, as well as online live video channels. This type of service features delivery of most media types, digital rights management (DRM), multiple bitrate streaming, and distribution of video content to multiple regions. IPTV services over CDN apply both linear content delivery and on-demand content delivery.
- *Managing and Delivering User-Generated Content.* CDNs are often used to facilitate the hosting and distribution of user-generated content. YouTube is the prime example where a CDN is used to deliver content uploaded by end-users. In addition, CDNs open the way for new applications such as Enterprise TV (e.g., Enterprise YouTube where corporate staff can upload and access video content) for interactions, training, and knowledge sharing.
- *End-to-End Online Video.* CDNs at its fullest use are often employed as the glue in integrating online video and content management platform to create an end-to-end delivery chain, offering a real differentiation through high quality user experience. Such end-to-end video delivery is composed of content ingest over Internet Protocol (IP) or satellite; content management and repurposing using specialized software, corporate storage, and hosting; content delivery using a CDN; and analytics-based management of IP-connected devices of end-users.

This chapter sets the context for the book, by providing an overview of CDN technologies, use cases, trends, and research directions in this domain. It presents how CDN is positioned with respect to the video ecosystem, followed by operational details of a CDN for several use cases. Coverage of recent technology trends in the CDN domain is provided to assist practitioners in this field; in addition, a few research issues are listed to demonstrate an R&D pathway to CDN researchers. While introductory concepts on many CDN terminologies, concepts, and technologies are covered in this chapter, readers should refer to later chapters in the book for detailed information and discussion on relevant CDN technologies.

1.2 CDN OVERVIEW

In addition to the growing popularity of news, sports, and e-commerce website content, accelerating demand for online HD video content, coupled with improvements in broadband technologies, and the increased availability of media-rich content are the key drivers behind using a CDN by content providers. At the most fundamental level, a CDN is about the efficient movement of digital content (video, software, audio, and files) across the Internet middle mile on a massive and ever-increasing scale. Let us consider a specific example (Figure 1.2) on how a CDN can be useful for large-scale video streaming. Good video QoE requires high network bandwidth and low network loading

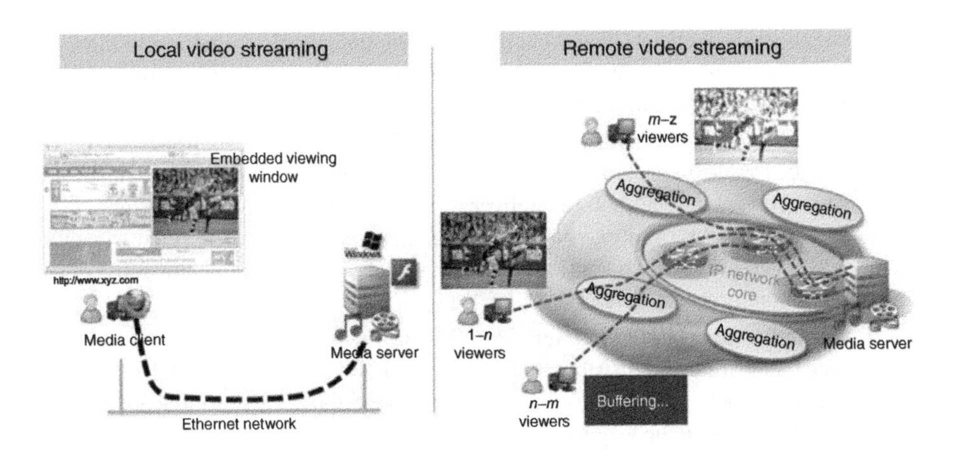

Figure 1.2 Rationale behind CDNs: video streaming example.

to avoid contention. Moreover, the streaming server needs to be close to the end-user, characterized as local video streaming. When significant network infrastructure (i.e., Internet middle mile) is present between end-users and the originating media server, in the remote video streaming scenario, satisfactory performance may be obtained at a small scale. Nevertheless, as demand grows, end-user viewing experience gets increasingly worse (e.g., video buffering, jitter). Centralized media server suffers performance and reliability problems, and significant cost is incurred to deliver video files across large network distances. A CDN can be used to tackle such scenario where it helps to minimize the cost of video content delivery, ensures that network resources are utilized efficiently, and optimizes end-user experience.

Without a CDN, bandwidth is overused as each request for the same content is retransmitted over and over. Moreover, a "brute force best effort" solution means that media content is not optimized (linear streaming vs video on demand (VoD)) for optimal edge delivery, as required for real-time video streaming. With a CDN, bandwidth usage of the content provider is optimized, as if a single end-user has requested unique content only once. No additional investment is required by the content provider to increase bandwidth capacity, as media content is packaged for delivery by the CDN infrastructure.

1.2.1 CDN Types

The CDN market is highly dynamic, comprising a myriad of players, with different offering and targeted market segments. Table 1.1 lists the different CDN types.

1.2.2 Market and Product Segments

A CDN is composed of content distribution and management, content routing, content edge delivery, content switching, intelligent network services, origin server

TABLE 1.1 Types of CDNs

CDN Types	Description
Pure-play CDN	CDNs that provide over-the-top (OTT) delivery of video and audio content without the ISPs being involved in the control and distribution of the content itself. Pure-play CDNs deliver content over ISPs' network (e.g., Akamai) or own infrastructure (e.g., Limelight Networks, Level 3).
Carrier/Telco CDN	Broadband providers and Telcos, for example, Verizon, Telstra, AT&T, that provide content delivery as a means to reduce the demands on the network backbone and reduce infrastructure investment, using hardware and software from vendors, for example, Cisco, Juniper, or Alcatel-Lucent.
Managed CDN	Pure-play CDNs can help carriers to build and manage the CDN component of the carrier's network with their professional services group, for example, Limelight Deploy. This approach leverages the expertise, infrastructure, and software of a pure-play CDN.
Licensed CDN	Pure-play CDNs can also provide CDN software for integration, testing, and deployment on the carrier's infrastructure. Although integration assistance is initially available from the pure-play CDN provider, licensed CDN is managed by the network operator itself. For example, EdgeCast, Highwinds, and Akamai Aura provide licensed CDN product.
Federated CDN	Multiple CDNs can interconnect their networks and compete more directly against pure-play CDNs. It is interesting for a content provider willing to deliver its content to the aggregated audience of the federation [7,8]. Cisco and Highwinds are working toward federated CDNs.

integration, self-service portal, professional services, and core networking. From a product perspective, a CDN provider can bundle content management, storage, and customer-onboarding (portal) services, in addition to the core CDN functionalities and value-added services (VAS). Figure 1.3 shows an example product portfolio of a CDN, showing the five constituents. Although earlier CDNs have been feature-rich in core functionalities, recent market trends indicate that traditional CDN services (e.g., caching, routing, delivery) are running the risk of being commoditized and sold mainly on price. Therefore, CDN providers are adding new services on top of their basic offering to generate revenue from running the business. There is a growing demand of VAS (e.g., website/application acceleration, Front-End Optimization (FEO), application security, rights protection, and analytics) and ancillary offering (e.g., origin storage, encoding/transcoding, content management, and ingest). It may also be required by a CDN to offer personalized service and custom solutions for individual enterprise customer needs. Details on the representative value-added and ancillary CDN services can be found in the later chapters of the book.

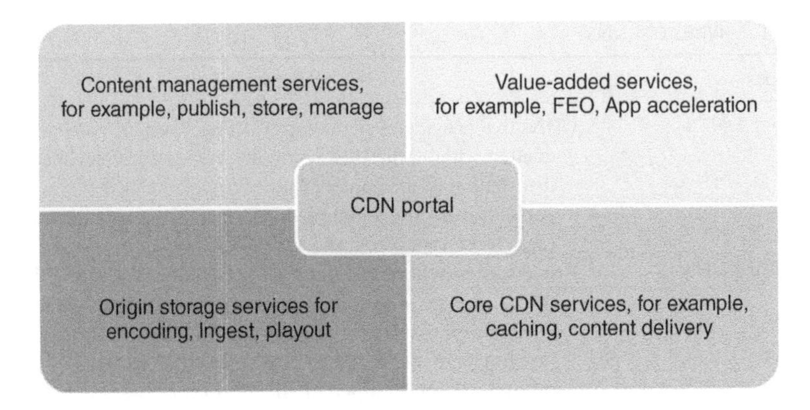

Figure 1.3 Example product portfolio of a CDN.

There are a number of driving forces in the CDN marketplace that influence the amount of revenue a CDN provider can make [9,10]. A representative list of these market drivers is provided in Table 1.2.

From a market perspective, a CDN provider offers its services to the following market segments:

- *E-commerce and Consumer Products/Retail.* CDNs offer whole-site and dynamic content delivery services to website owners. A large portion of such websites is composed of applications and dynamic content, such as e-commerce/online retailers (eBay, Amazon), auction website (graysonline), and consumer products website (Walmart, JC Penny). These websites are heavily used during sales and special events (e.g., Thanksgiving, Christmas), requiring a CDN to tackle the heavy traffic demand for them.
- *Media and Entertainment.* CDNs offer progressive download, linear, VoD, and live streaming services to digital media companies such as content broadcasters (CNN, BBC), Internet-based publishing (Fairfax Media), and experimental digital media (DXARTS, ACMI). These services are predominantly based on video content, with limited website static object caching. Users access the VoD or live streaming in mobile devices, set-top box, and online. CDNs help to deliver the video content at scale as well as tackle bursty traffic [11] for special events, such as the Cricket/Football World Cup and the Olympics.
- *Hospitality, Travel, and Leisure.* This market segment features websites and services used by end-users for accommodation booking, holiday booking, air ticket purchase, and so on. Similarly, in e-commerce websites, the websites Zuji, Expedia, and HotelClub have embedded applications and dynamic content, but differ in terms of back-end search and computation functionalities. This market segment is aided by CDNs by taking care of the high traffic to websites, as well as application and database query execution at scale, during airfare/accommodation sale times.

TABLE 1.2 CDN Market Drivers and Impacts

Market Drivers	Impact
Uptime/availability of service	The most crucial market driver to set up a CDN's reputation as a reliable provider (and hereby to attract revenue-generating customers) is its service availability.
Throughput performance	CDN customers are increasingly looking at network throughput performance, as the key driver for choosing a CDN.
Network performance: first byte delivery	A CDN's network performance, in terms of first byte delivery, provides the very first impression of the service provided by it and also gives the customer confidence on its chosen CDN.
Price	Over the past years, CDN pricing has reached a commoditized level. CDNs are now focusing on value-added services (VAS) to generate revenue. There are six underlying cost models that appear frequently, namely, Gbps billing, GB transfer billing, GB storage billing, pay-as-you-go, monthly commit, and capped (percentile-based pricing).
Customer service	The success of the relationship with the service provider is defined by the moment the service fails and when it is too complex to initialize. When things go wrong, communication barriers only add to the frustration. Therefore, the customer service staff should have fluency to communicate in the language spoken by regional/global customer.
Reporting/analytics	For mission critical CDN services, real-time monitoring, "heartbeat" reporting, analytics, and live alerts are to be provided proactively, in order to cater for failure resolution.
Customer portal/online account management	An online account management system allows faster customer onboarding and allows the customer to track the service activation process, as well as ongoing account and/or service management.
Provider's financial stability	The CDN marketplace is very dynamic and has experienced number of acquisition and/or mergers, as well as close operations. A CDN provider has to demonstrate that it is making stable revenue from the market to gain customers' trust.
Sales process/ease of doing business	A customer always expects to be professionally guided through the sales process by the sales executive. The sales process is a significant opportunity to impress and secure the customer.
Range of products	A CDN is often assessed by customers based on its product portfolio and the VAS it provides. A set of innovative and trendy product feature set characterizes a CDN's revenue potential.

- *Banking, Financial Services, and Insurance.* CDNs help this market segment by delivering a large amount of static website content, as well as dynamic content and applications, such as calender, currency exchange, financial projections, and loan calculators. Traffic profiles to these websites remain stable around the year, with high peaks only during important financial times, for example, end of the financial year and beginning of the year.
- *Public Sector/Government and Enterprise.* CDNs also help the Government (such as the White House website) to deliver heavy websites of secure and static content. Traffic profiles to these websites are constant, featuring limited flash crowds, except during the online release of important legislations or Government changes.

1.2.3 Video Ecosystem and CDN

Increased level of Internet activity, the popularity of online video, the use of rich media for marketing and advertising, and an increase in the types and volumes of Internet-connected devices (smart phones, iPhone, iPad, tablets, connected smart TVs, and gaming consoles) are driving significant growth in the volume of online data being consumed. Toward this end, a number of technologies and evolving business model, centered on video and changing consumer lifestyle, are propelling the rise of online video and fundamentally transforming media distribution, TV, advertising, and content delivery methods.

Video content by far is the significant driver for the increased utilization and relevance of CDNs. In fact, over 48% of global CDN revenue is associated with video content delivery [12]. When video and online content consumption is broadly distributed, the QoE of end-users becomes significantly important. The further the content has to travel, the greater the latency in the content delivery and the poorer is the end-user experience. A CDN with its rich technical features, capabilities, and VAS is at the core of the video ecosystem, enhancing the end-user experience by accelerating the content delivery method (download/streaming speed) across geographically distributed end-users.

Figure 1.4 shows the position of a CDN with respect to the end-to-end video ecosystem supply chain. Video content is ingested from the source, packaged into the right container format by encoding platform, and is indexed in the storage by the content management system. CDN as the delivery vehicle distributes the content among the edge servers for delivery to IP-connected devices. User profiling through reporting and analytics ensures that further revenue generation opportunity is created through ad insertion, based on end-user behavioral profile. Such advertisements are shown as embedded in the video streams, on Web pages, or within the electronic program guide (EPG) of set-top boxes.

1.3 WORKINGS OF A CDN

As discussed earlier, a CDN delivers static, dynamic, and streaming content. On the basis of operational functionalities, a CDN can be described from four perspectives, namely,

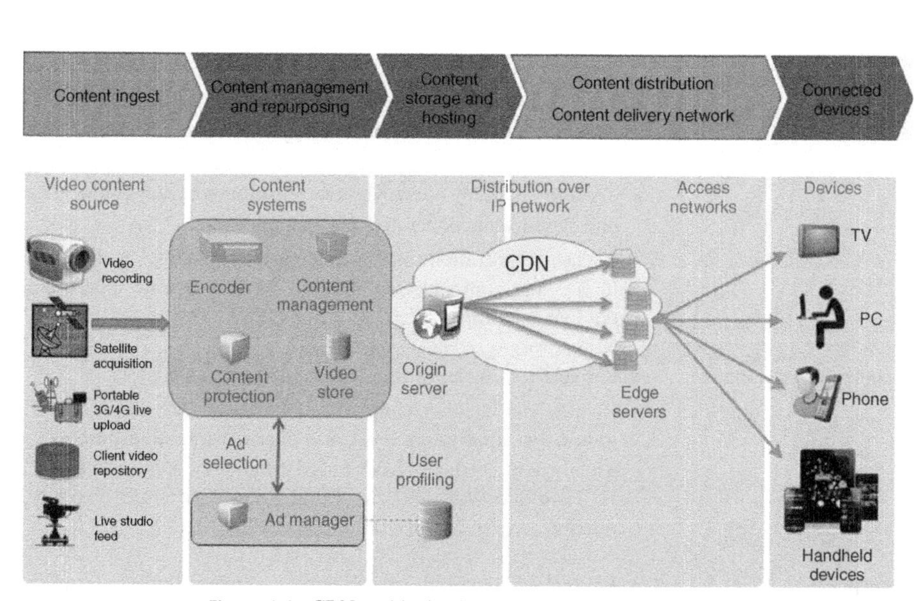

Figure 1.4 CDN positioning in a representative video ecosystem.

TABLE 1.3 CDN Functional Attributes

Aspects	Impact
CDN composition	• CDN organization—overlay and network approach. • Servers—origin and replication server. • Relationship—client-to-edge-to-origin, network element-to-caching proxy, and interproxy. • Interaction protocols—network elements and intercache interactions. • Content/service types—static, dynamic, streaming content and value-added services.
Content distribution and management	• Content selection and delivery—full and/or partial site: empirical, popularity, object, and cluster-based. • Replica placement—single or multi-ISP approach using K-center, greedy, hotspot, topology-informed, tree-based, and scalable replica placement. • Content sourcing—cooperative push-based, non-cooperative pull-based, and cooperative pull-based. • Cache organization and management—caching techniques: intra- and intercluster caching; cache update: periodic, on-demand, propagation, and invalidation.
Request routing	• Request-routing algorithms—adaptive and nonadaptive. • Request-routing mechanisms—DNS-based: NS and CNAME redirection, application level, and IP anycasting; transport layer: usually combined with DNS-based techniques; application layer: HTTP redirection, URL rewriting using centralized directory, distributed hash table (DHT), or document routing.
Performance measurement	• Internal and external measurement—network statistics acquisition: network probing, traffic monitoring, surrogate feedback to measure performance metrics such as geographical proximity, latency, packet loss, average bandwidth, downtime, startup time, frame rate, server load, and simulation-based performance measurement.

CDN composition, content distribution and management, request routing, and performance measurement [13]. Table 1.3 summarizes the functional attributes of a CDN.

CDN functional attributes are closely related to each other. The structure of a CDN varies depending on the content/services it provides to its customers. Within a CDN structure, a set of edge servers is used to build the content delivery component, some combinations of relationships and mechanisms are used for redirecting end-user requests to an edge server and interaction protocols are used for communications between CDN elements. Content distribution and management are strategically vital in a CDN for efficient content delivery and overall performance. Content distribution includes content selection and delivery based on the type and frequency of specific end-user requests; placement of edge servers to strategic positions so that they are close to end-users; and content sourcing to decide which methodology to follow to acquired content. Content distribution and management are largely dependent on the techniques for cache organization (i.e., caching techniques, cache maintenance, and cache update). Request-routing mechanisms are used to select the optimal edge server in response to a given request. Finally, performance measurement of a CDN is done to evaluate its ability to serve the customers with the desired content and/or services.

In the following, we briefly discuss the operational methodology of five distinct types of content delivery.

1.3.1 Static Content Delivery

The basic use case for a CDN is the delivery of static content, for example, static HTML pages, embedded images, PDF documents, files, and software patches. Figure 1.5 shows the interaction flows for website-embedded static content delivery.

End-user requests content by specifying its unique resource locator (URL) in the Web browser. User's request traverses the transport and routing infrastructure over the Internet middle mile, eventually reaching the origin server. When the origin server receives a request, it makes a decision to provide only the basic content (e.g., index page

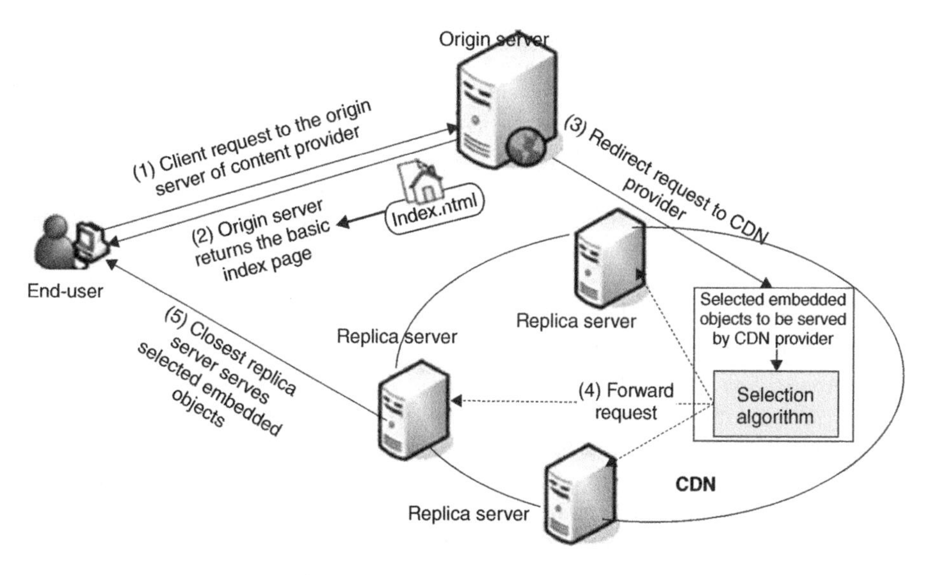

Figure 1.5 Static content delivery.

of the website). To serve the high bandwidth demanding and frequently asked content (e.g., embedded objects—images, logos, navigation bar, and banner advertisements), user's request is redirected to the CDN infrastructure. The CDN performs acquisition of the content from the origin and distributes the content among geographically replication edge servers. Using a selection algorithm (often proprietary), an edge server is selected that is the "closest" to the end-user to serve the requested embedded objects. Selected edge server serves the user's request and caches it for subsequent request servicing.

1.3.2 Dynamic Content Delivery

Dynamic content is generated on demand using Web applications. Dynamic content changes based on a contextual situation, such as user profile, and interaction characteristics. For example, a website that shows the current time is actually showing dynamic content: the information (the numbers shown) changes based on the situation (the current time). Types of dynamic content include animations, changing text, content generated using server-side scripts (ASP, ColdFusion, Perl, PHP, Ruby, WebDNA) or client-side scripts (JavaScript, ActionScript), and DHTML.

Dynamic content delivery is aided with a multitier logical architecture of a Web system (Figure 1.6), which separates the HTTP interface, the application (and business) logic, the data repository, and, when existing, the user-related information for authentication and content personalization.

When a CDN is employed to deliver dynamic content, each layer of the Web system is replicated. In the following, we describe the workings of each layer.

- *Front-End Layer*. It receives HTTP connection requests from end-users, serves static content from the file system, and represents an interface toward the application logic. It employs techniques such as *fragment-based*, that is, storing content segments, or *sequential* caching, that is, storing partial stream segment to reduce buffering [14]. Fragment-based caching commonly makes use of Edge Side Includes (ESI) [4,15], an XML-based markup language that enables to distinguish cacheable (static) content and noncacheable (dynamic) content. Edge servers perform identification and markup of page fragments for assembly of the dynamic Web page content before delivering it to end-users. Alternatively, a CDN can use FEO techniques to reduce the number of page resources required to download a given page and make the browser process the page faster. We elaborate on FEO techniques in Section 1.4.
- *Application Layer*. It handles business logic and computes information to respond with dynamic content. Content generation often requires interactions with the back-end and user profile layers. This layer improves the delivery of dynamically generated content by offloading the task from the origin, using either client-side or server-side scripting or a combination of the two. *Client-side scripting* refers to the Web applications that are executed at the client side by the end-user's Web browser. Client-side scripts are often embedded within an HTML or XML document, but they may also be contained in a separate file, which is referenced by the document that uses it. Upon request, the necessary script files

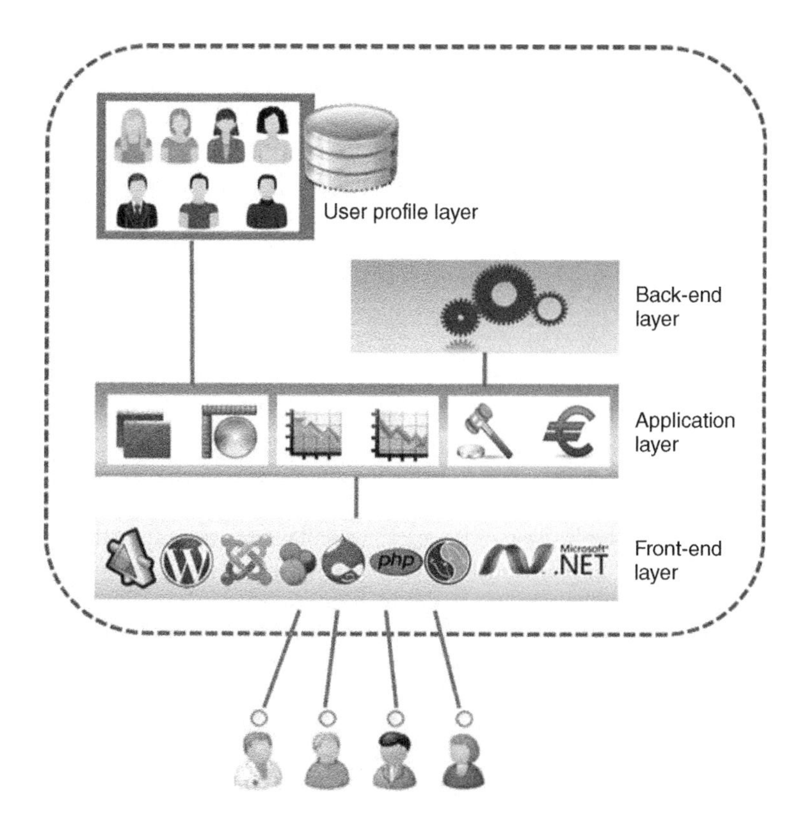

Figure 1.6 Logical layers of a Web system for dynamic content delivery.

are sent to the user's computer by the CDN. The user's browser executes the script, and then displays the document, including any visible output from the script. JavaScript and ActionScript used for DHTML and Flash technologies are examples of client-side scripting. *Server-side scripting* refers to applications running on the edge server of a CDN to change the content on various Web pages. Such Web pages are often created with the help of server-side languages such as ASP, Perl, PHL, Ruby, and WebDNA. Server-side scripts are embedded into the HTML source code, which results in a client's request to the website being handled by a script running on the edge server before it responds to the client's request. Server-side scripting is usually used to provide an interface and limit access to proprietary databases or other data sources. These scripts may assemble client characteristics for use in customizing the response based on those characteristics, user requirements, and access rights. *Combination of client- and server-side scripting* is also common. Ajax is a Web development technique for dynamically interchanging content that sends a request to the edge server for data. The edge server returns the requested data, which is then formatted by a client-side script. This technique reduces load time because the client does not request the entire Web page, only the content that will change is transmitted. Google Maps is an example that uses Ajax techniques. Table 1.4 summarizes the client-side and server-side scripting languages and dynamic content types.

- *Back-End Layer*. It manages the main information repository of a Web-based service. It typically consists of a database server and storage of critical information to generate dynamic content. The layer provides additional scalability to the management of application data. Replication of the back-end data can be full or partial. The partial replication of data can be obtained by exploiting a caching mechanism of the most popular queries to the data storage (*content-blind caching*) or by actively replicating portions of the back-end data, selected on the basis of usage patterns, network, and infrastructure status (content-aware caching) [16,17].
- *User Profile Layer*. It stores information on the user preferences and context. This information is accessed for dynamic content generation to provide personalized content. Replication of the user profile layer across edge servers involves technique to partition and distributes user information based on access patterns. Some techniques support user mobility during consecutive sessions, across multiple edge servers, by allowing data to follow the user. User profile can be updated manually through a Web form or by the Web-based services automatically on the basis of the user behavior.

1.3.3 HTTPS Content Delivery

HTTPS is the protocol for accessing a secure Web server when authentication and encrypted communication are possible. Technically, it is not a protocol in and of itself; rather, it is the result of simply layering the HTTP on top of the Secure Sockets Layer/Transport Layer Security (SSL/TLS) protocol, thus adding the security capabilities of SSL/TLS to standard HTTP communications [18,19]. Web browsers know how to trust HTTPS websites based on pre-installed certificates from Certificate Authorities (CA), for example, VeriSign, GeoTrust, and Symantec.

HTTPS- and SSL-based delivery ensures an end-to-end trust tree between the origin server hosting the website/assets and the user/browser connection. This end-user to server interaction can traverse many networks and systems, and its data have a risk of man in the middle attacks. Building a widely accepted HTTPS feature in a CDN based on industry standard, end-to-end SSL, certificate integration and key exchange hierarchy allows a CDN to attract customers from the Government, as well as from banking sectors.

CDNs are used to deliver HTTPS content as well as to offload processing of SSL encryption/decryption from the origin. HTTPS content delivery is of major importance to e-commerce retailers, banking industry, and financial institutions that need this feature for dynamic content delivery, transactional traffic, and SSL acceleration. Owing to its computational requirements, SSL decryption can impact IT Web infrastructures and can cause performance and high availability risks. CDNs can offer SSL offloading to a certain degree (by scaling of the computational infrastructure).

TABLE 1.4 Scripting Technologies Used at the Application Layer

Client-Side Scripting		Server-Side Scripting		Combined Client and Server-Side Scripting
Language	Type	Language	Type	Technology
JavaScript using client-side scripting	*.js	ASP	*.asp	• Ajax: JavaScript: jQuery, MooTools
				• Java: Apache Wicket, BackBase
				• C++: Wt
Visual Basic Script	*.vbs	ActiveVFP	*.avfp	• .NET: ASP.NET AJAX, Web.Ajax
ActiveX Scripting	*.axs	ASP.NET	*.aspx	• Perl: Catalyst
TCL	*.tcl	C via CGI	*.c, *.csp	• PHP: Phery, CJAX framework
		ColdFusion Markup Language	*.cfm	• Python: Pyjamas
		Java via JavaServer Pages	*.jsp	• Ruby: Ruby on Rails framework
		JavaScript using server-side JavaScript	*.ssjs, *.js	• Groovy: ZKGrails
		Lua	*.lp, *.op	• Scala: Lift framework
		Perl CGI	*.cgi, *.ipl, *.pl	
		PHP	*.php	
		Python (via Django framework)	*.py	
		Ruby (Ruby on Rails framework)	*.rb, *.rbw	
		SMX	*.smx	
		Lasso	*.lasso	
		WebDNA	*.dna, *.tpl	
		Progress WebSpeed	*.r, *.w	

In order to provide HTTPS content delivery capability in a CDN, the following elements need to be integrated with the CDN platform.

- *Certificate*. Certificates are digital identification documents that allow both CDN edge servers and end-users to authenticate each other. A certificate file usually has a .crt extension. Server certificates contain information about the website owner and the organization that issued the certificate (such as VeriSign, GeoTrust, or Thawte), while client certificates contain information about the user and the organization that signed the certificate. A Certificate Signing Request (CSR) is generated each time a certificate is created. It has a .csr extension. Once the CSR file is signed by the CA, a new certificate is made and can be used for HTTPS websites.
- *Session Key*. The end-user and edge server use the session key to encrypt data. It is created by the end-user via the server's public key.
- *Public Key*. The end-user encrypts a session key with the server's public key. It does not exist as a file but is produced when a certificate and private key are created.
- *Private Key*. The server's private key decrypts the client's session. The private key has a .key extension and is part of the public–private key pair.

The basic form of HTTPS content delivery is the capability to securely deliver static object, music streaming, and progressive downloads. By enabling this feature, a CDN is able to support content delivery for organizations that require SSL security on non-dynamic and cacheable transactions. For SSL-encrypted HTTPS requests, the CDN is expected to perform full SSL processing from setting up the SSL session to encrypting all the outbound data. CDN edge servers process all the SSL requests, and the origin server is not burdened with this processing. Requests that a CDN does not cache, such as stock quotes and bank account balances, are sent to the origin server for retrieval. CDN retrieves this information and, without caching it, encrypts the data and sends it on to the requesting browser.

1.3.4 Live Streaming

A major use case for CDNs is to provide live streaming coverage of sporting, cultural, and corporate events, made available for viewing on IP-connected devices. Live streaming events take place over a defined time interval with defined start and end times, for example, a football game or a car racing event. Live streaming can be performed in different content delivery formats. Most recently, HTTP-based adaptive bitrate (ABR) streaming technologies, such as Microsoft Smooth, Apple HLS, and Adobe HDS, have been heavily used in the CDN industry for live streaming (Chapters 2 and 3 provide further details).

ABR streaming is a multibitrate video streaming technology that allows a media player client to dynamically adjust the quality of the video stream, depending on the prevailing network conditions experienced during the streaming session. To ensure the best possible viewer experience, an end-user's media player client is able to select a streaming bitrate that best suits the network conditions at the time, and can dynamically change the bitrate during the session if network conditions improve, or deteriorate, without terminating or rebuffering the video presentation.

In order to describe the configuration and operational details of live streaming, let us consider a playback experience using the Microsoft Smooth technology. MS Smooth streaming allows CDNs and content providers to deliver streaming video content over HTTP/IIS to Silverlight clients in an adaptive manner. This delivery mechanism uses an HTTP compliant Web server, a manifest file (XML playlist) to index the video/audio fragments, and Silverlight as the media player client to decode the fragments.

Silverlight is the Microsoft media player client used to connect to the manifest file that holds and manages the indexed information of the encoded fragments. Silverlight then uses its built-in heuristics and actively monitors the available network and CPU resources every 2 s and makes the request for the most adequate data segment of video and audio. Figure 1.7 shows an example setup of live streaming over CDN using MS Smooth.

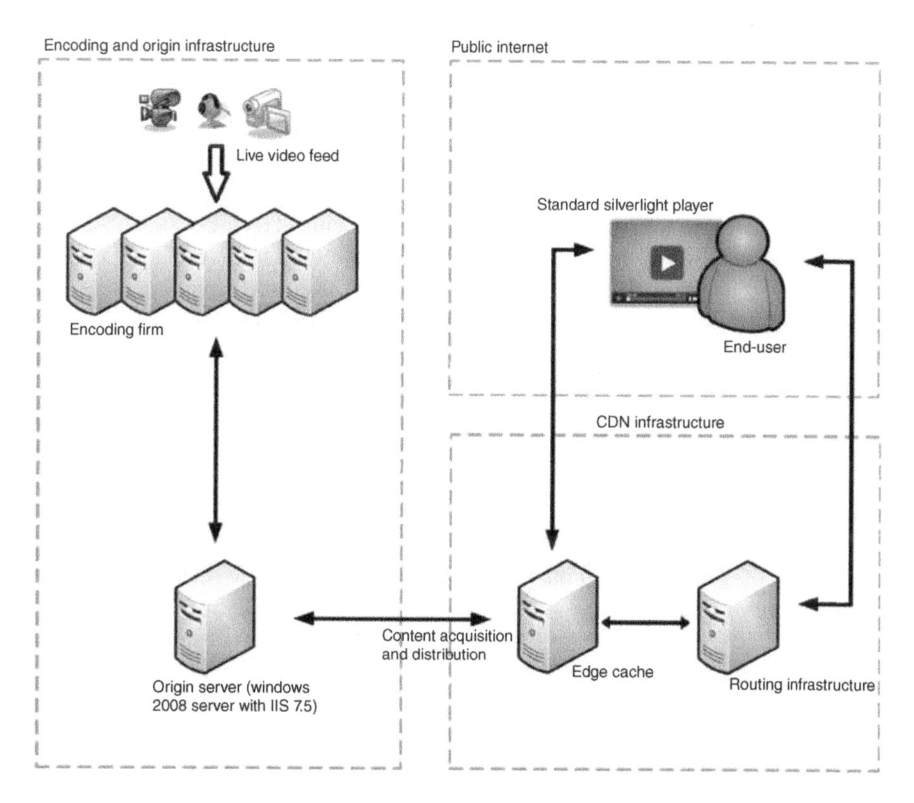

Figure 1.7 Live streaming example.

The encoder takes a live feed and converts it to video/audio where it is segmented into different qualities to allow the media player client to select the most appropriate segment to retrieve the content seamlessly. The Smooth manifest file (XML) is created at the encoder level and is sent to the origin server (in this example, a Windows 2008 server with IIS 7.5) by the encoder along with the media fragments. Other delivery impacting factors such as output caching rules, DRM, advanced logging, and HTTP response headers are also controlled at the encoding and origin infrastructure layer. Many of the end-to-end delivery behavior and performance outcomes can be configured at the origin. It is important that origin configuration follows best practice to avoid negative impact on the CDN and end-user experience during live streaming.

The CDN acquires live streaming content from the origin and distributes it among the edge servers for delivery. The purpose of the manifest file is to enable the origin server and eventually the CDN, to interpret the incoming live stream. It also assigns appropriate semantic meaning to the tracks in the stream, creates necessary parameters to decode the video/audio data, and makes appropriate heuristic decisions. This XML Smooth manifest file ultimately is the asset that the media player client connects to. The routing infrastructure plays a key role in selecting the optimal edge server to deliver the live stream to end-users. The media player client used by the end-user can be a Smooth streaming player, built using Microsoft Media Platform Framework, utilizing Silverlight technology.

1.3.5 On-Demand Streaming

On-demand streaming allows end-users to select and watch/listen to video or audio content on demand, for example, VoD, audio and video on demand (AVoD), and subscription video on demand (SVoD) services. IPTV technology is often used to bring VoD services to televisions and personal computers. Television VoD systems stream content either through a set-top box, a computer, or other device, offering viewing in real time, or download to a device such as a computer, personal video recorder (PVR), or portable media player for viewing at anytime.

The majority of cable and telco-based television providers offer both VoD streaming, including pay-per-view, and free content, whereby an end-user buys or selects a movie or television program. User can begin to play the video content on the television set almost instantaneously, or download to a digital video recorder (DVR) rented from the provider, or download onto a PC for viewing in the future. IPTV is an increasingly popular form of VoD service. It enables the broadcast of multiple linear channels with an EPG, enabling the viewer to look ahead for program information. This presents a digital TV-like experience on any IP-connected device. Figure 1.8 shows an example IPTV service using a CDN.

On-demand streaming at scale using a CDN can be obtained in the following way. Content is first acquired from the video source and ingested into the encoding infrastructure, which transcodes it into digital assets. Cloud-based encoding can be used to enable content providers to cost-effectively transcode large libraries of content to Web format in a very timely manner. The content is wrapped with an appropriate DRM, and metadata is assigned to the content for inclusion in the content catalog. Content is

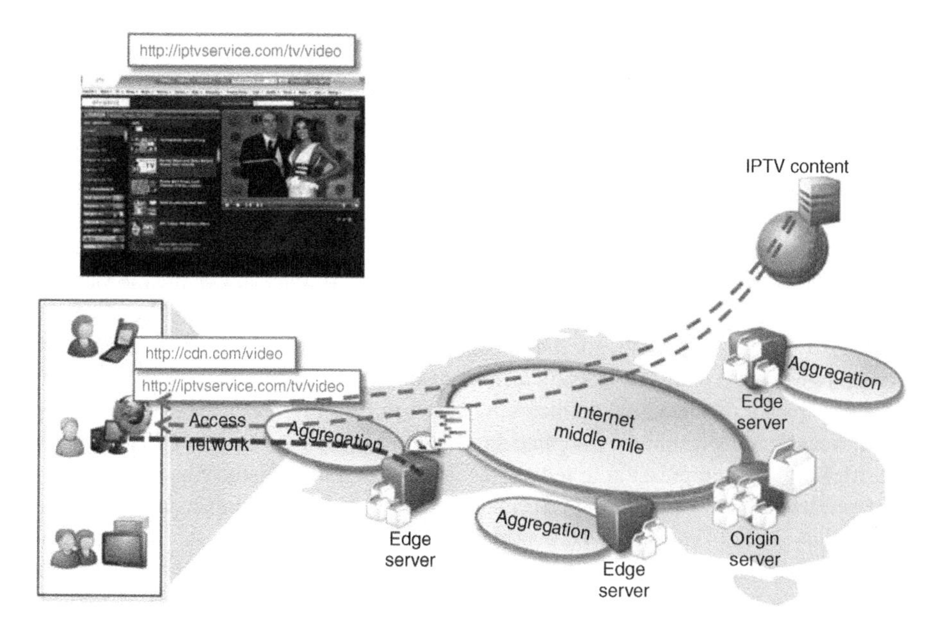

Figure 1.8 On-demand streaming—IPTV.

stored in a common repository at the origin server. The CDN acquires the content from the origin, assigns a CDN-specific URL, and stages the asset among the edge servers. Request redirection is configured to allow the content delivery at scale from the optimal server. End-user selects the desired content either through a Web browser or by selecting a channel/program in the EPG. Upon user request, the CDN provides the previously configured HTML redirecting to the requested content from the optimal edge server.

CDNs allow on-demand streaming distribution to IP-connected devices from a single management platform with Web-based interfaces. When CDNs are used for IPTV services, it provides cable TV-like experience to end-users without an expensive distribution network (cable or satellite). Similarly for catch-up TV services, CDNs allow publishing to multiple devices from a single content store.

1.4 CDN TRENDS

The CDN marketplace consists of a handful of pure-play providers to maintain strong growth rates, as they simultaneously scale up current offerings and diversify their customer bases as well as product offerings. However, over the past few years, the industry has seen growth in telco/operator-CDN market. Rapid innovation and emerging competitors are directly contending for core CDN services as well as looking to meet niche demands for VAS. In this section, we highlight a few major technology and feature trends in CDNs.

1.4.1 Dynamic Content and Application Acceleration

End-users often complain that the website they are trying to access is too slow. This is mostly due to the dynamic content embedded in the website and the applications running over the network that are causing most of the delays. To tackle this, many present-day CDNs employ a suite of acceleration techniques that act at the higher layers of the protocols stack, from Transmission Control Protocol (TCP) to HTTP, to improve performance through the optimization of protocols and the applications themselves. As for instance, legacy TCP/IP slows down data transfer because of the way Web and application servers interact with the client. It is ill-suited to address the sophisticated delivery requirements of highly dynamic Web application and content. The optimization of TCP reduces the overhead associated with TCP session management on servers and improves the capacity and performance of the actual application, which in turn results in improved response times. Acceleration techniques applied in CDNs can often leverage cached data and industry standard compression to reduce the amount of data transferred.

Optimized Web application performance drives greater usage of externally facing applications such as extranets; B2B portals; and Web channels for sales, fulfillment, product updates, and support. Drivers behind dynamic content and application acceleration in CDNs are of the following:

- Dynamic applications, estimated up to 75% of the today's Web applications, require a technique employed to improve the response time.
- Web services, used for strategic business-to-business (B2B) applications, require improvement of the business process and availability in order to avoid high abandonment and poor end-user satisfaction.
- Whole-site delivery requires a way to ensure the scalability and performance guarantees for planned and unplanned peak usage.

Techniques used to achieve dynamic content and application acceleration in CDNs perform optimizations at three discrete layers—(i) routing, (ii) transport, and (iii) application [20,21]. They work in combination to provide optimal application response time. Some of the core technologies behind dynamic content and application acceleration in CDNs are listed in Table 1.5.

1.4.2 Front-End Optimization

CDNs employ FEO to reduce the number of page resources required to download a given page and make the browser process the page faster. In contrast to edge caching and application acceleration, FEO does not bring content/application closer. Alternatively, it facilitates faster content rendering by optimizing the client-side delivery of website resources. The technological features of FEO can be categorized into three broad classes: (i) request reduction, (ii) byte reduction, and (iii) faster page rendering [22,23]. Table 1.6 lists the techniques used by FEO.

A few CDN providers use application delivery controller (ADC) capabilities along with FEO to further accelerate page rendering and website content delivery [23]. Some of

TABLE 1.5 Dynamic Content and Application Acceleration Techniques

Category	Techniques
Routing	• Optimized route and path selection • Global traffic management across multiple data centers • Real-time route optimization to reduce round trip time (RTT) and Internet availability • Latency and packet-loss optimization of live streams • Server load balancing, level 4–7 switching, and packet redundancy
Transport	• High performance transport protocol with secure SSL termination, optimized window sizing, long-lived persistent HTTP connections, pipelining of requests, and intelligent retransmission • TCP optimization including transport flow optimization, selective acknowledgement, adaptive buffering, and congestion avoidance • Hierarchical queuing and scheduling • Optimized routing and path selection
Applications	• Hierarchical edge caching • Parsing of HTML pages and prefetching of embedded content • Live and on-demand streaming and HTTP downloading • Content-aware compression • Stream splitting • Content prepositioning
Data transfer efficiency	• Client-aware compression • Application-aware compression • Header compression • Delta caching • Object and metadata caching

these ADC capabilities include hardware acceleration (SSL termination/offload, layer 7 content switching), application optimization (TCP multiplexing, TCP optimization, and SPDY protocol optimization), and network optimization (traffic rate shaping, network health checking, and WAN optimization).

1.4.3 Mobile Content Acceleration

With the advent of innovative cellular (3G/4G) and wireless (Wi-Fi) services, mobile devices are capable of accessing the Internet and other data services at a speed comparable to traditional wired services [24]. CDNs use mobile content acceleration techniques to optimize service delivery to handheld devices [42]. It helps manage all mobile Internet traffic to ensure reliable and high speed mobile data services and quality of service (QoS)

TABLE 1.6 FEO Techniques

Category	Feature	Description
Request reduction	File versioning	File versioning ensures content invalidation and efficient caching by giving each file a unique name. When a website asset is modified, the old content is not referenced anymore.
	Asset consolidation	It includes website assets of similar type into a single package, which only requires a single request for delivery.
	Inlining/landing page consolidation	Inlining is the technique of embedding small website objects directly into the pages or resources that reference them.
	On-demand image loading	On-demand image loading allows a page to only load the images that are visible within the current viewport. As a user scrolls down, new images are loaded on demand.
	Browser cache optimization	Under this technique, traffic patterns are analyzed to identify objects that should be "seeded" into the browser cache so that subsequent requests can be served from local browser cache.
	Reduce DNS lookups	DNS has a cost. It typically takes 20–120 ms for DNS to lookup the IP address for a given hostname. FEO caches DNS lookups for better performance.
Byte reduction	Payload reduction	Text-based mime types are highly compressible. FEO uses compression algorithms and reduces the verbosity of CSS/JS.
	Resize images to HTML dimensions	FEO resizes large images ahead of time to fit CSS styles and markups, thus avoiding the need to download redundant bytes.
	Minification	Minification is the process of removing comments and whitespaces in HTML, JavaScript, and CSS files, reducing the total download size.
	Responsive images	FEO's responsive images technology automatically delivers an optimal-sized image for the phone, tablet, or desktop browser that requested it, resulting in much faster page load.

	Chunked headstart	This technique utilizes chunked transfer encoding to deliver a separate <html> section in the document that contains the "static" assets and leaves "dynamic" assets in the main <html> document. The static portion of the page can start to be served while dynamic content is requested from the origin.
Faster page rendering	Asynchronous JavaScript and CSS	This technique modifies the way scripts and style sheets are embedded into the Web page, making the browser process scripts and style sheets in parallel.
	Domain sharding/connection maximization	By "sharding" hostnames (open multiple concurrent connections to the website), FEO facilitates the browser to open multiple connections in parallel allowing more requests to be satisfied simultaneously.
	User agent specific treatments	By recognizing the user agent and device type, FEO can deliver targeted optimizations for the popular devices.
	Streaming consolidation	FEO processes incoming JavaScript and CSS as it arrives, without additional HTTP requests.
	Invoke click on-touch	FEO leverages the on-touch event on mobile devices to simulate fast clicks, even on a touch screen, resulting in a significantly more responsive page.
	JavaScript pre-execution	FEO can execute much of the embedded JavaScript offline and can provide the requesting browser with a mostly static page for faster rendering.
	Deferred script loading	FEO decouples the third-party script execution from the loading of the page. By making the third-party request asynchronous, FEO ensures that a slow third-party content will not negatively impact the page rendering.
	Page prefetching/predictive browser caching	This technique loads subsequent non-homepage website content ahead of time based on users' browsing patterns.

for end-users. Mobile content acceleration can be performed through mobile site optimization or through mobile network optimization.

Mobile site optimization greatly relies on FEO for asset consolidation and payload reduction, specific to end-user devices. It reduces overall Web page size for rendering in a mobile device, reduces the number of round trips required to load objects in a mobile website, and employs techniques such as click on-touch (to simulate fast-click events). It also makes use of SPDY protocol optimization and HTML5 framework to deliver additional performance benefits for supported browsers. Further acceleration of mobile content is achieved by pushing business application logic to the edge of the network to serve appropriate advertisements, social network update, and location-specific data to mobile devices.

Mobile content acceleration is also achieved through applying optimization technique on the mobile network. Direct integration between WAN optimization hardware devices and CDN edge caches inside mobile network allows speeding up the transport of data packets at the network layer. This can achieve faster speeds from the origin over the first mile to the middle mile, and then faster transit through the CDN to edge caches that connect to the last mile and mobile gateways near the end-user [25]. In order to ensure optimized video delivery, CDNs utilize radio frequency friendly pacing technology [26]. It allows the video to be streamed in periodic bursts so that there is less signaling load and more efficient resource utilization in the radio access network (RAN). This technique also allows full utilization of high link-bitrate features of the streaming media content.

Other mobile content acceleration techniques include prioritizing certain content type/URL, increasing cell capacity and TCP throughput, video traffic differentiation, user-adaptive mobile video streaming [26], and broadcast over LTE [27].

1.4.4 Transparent Caching Convergence with CDN

CDNs handle traffic for the content providers that they have agreement with; however, they do not serve a large amount of the whole traffic, for example, unmanaged over-the-top (OTT) content over a carrier's network. One way to intelligently store popular OTT content at local storage sites and enhanced QoE is transparent caching [28–30]. Transparent caching handles unmanaged traffic and optimizes OTT content delivery. It caches popular OTT content close to the edge of the network. Thus, the content can be delivered to end-users from an operator's network, instead of retrieving it each time from a distant unmanaged content origin.

The combination of transparent caching and CDN offers a perfect solution for network operators who want to reduce the cost of serving all types of content (managed and unmanaged), improve the QoE, reduce content delivery cost, provide revenue generation, introduce new content services, and monetize the delivery. Transparent caching can be viewed as a complement to a CDN, whereby it acts as an additional content acquisition mechanism for CDN. This interpretation is aided by a component-wise view on transparent caching, comprising content caching, request interception, request redirection, and content analytics.

The drivers for CDN-transparent caching convergence are as follows:

- bring popular content closer to users in cases where the operator does not have the legal right to use its internal CDN for caching HTTP content;
- provide the CDN operator with visibility in terms of caching performance and the types of content being cached, which can be useful in both internal and external applications;
- provide reporting and analytics, that is, intelligence on user behavior that can be packaged and reused with third parties and/or used for internal service planning;
- dynamically ingest and serve content as it becomes popular, not requiring operator intervention to modify the network or the caching solution to support new services or devices; and
- provide the ability to manipulate the HTTP header directives of OTT content, without impacting or being influenced by application service logic on the origin.

The cost of adding a transparent caching capability to optimize existing capacity provides CAPEX advantages over adding more bandwidth and improves the OTT content experience more than the operator CDN services that focus on premium on-deck content. Eventually, the improved caching performance and content delivery provide content monetization and revenue opportunities to go along with the CAPEX savings.

1.4.5 Peripheral Value-Added Services

Media companies, as CDN customers, not only seek core CDN services and its VAS offering but also have requirements to deliver rich Internet applications, digital asset management, online training, video conferencing, webcasts, product demonstrations, and live events. There has been a paradigm shift among media companies from a traditional broadcast mode to a multichannel model, involving online delivery of content to a diverse and ever-changing range of devices and platforms. Nowadays, media companies expect that it can outsource all hosting (both origin and edge servers), video transcoding, DRM, multiscreen delivery, and content monetization opportunities to a CDN provider.

These new requirements are widening the definition of what a CDN is. CDN providers are offering professional services in multiple areas (e.g., application security, Analytics-as-a-Service, content awareness, network awareness, and device awareness), rather than just distributed caching and traffic management along the Internet middle mile. There are a number of new VAS offerings that CDN providers have started to offer, which in the early days of CDNs would not have been considered as CDN services. Some of these peripheral VAS offerings are the following:

- *On-the-fly transcoding* [31,32] to encapsulate content in different formats and quality for delivery to varying edge devices. It eliminates the need for media companies to pretranscode and store multiple file formats of their entire media library in an expensive storage system. With the aid of this service within a

CDN provider's infrastructure, content providers can monetize not only the most popular and widely viewed video titles but also the entire video catalog, including long tail content.

- *Ad insertion* [33–35] to embed advertisements in streaming videos, as per end-user profiles. This provides a means for pay TV operators, content programmers, and broadcasters to inject commercial and custom content into streaming video over IP. A CDN can be charged with an integrated ad insertion method to communicate with downstream ad servers and ad networks, find most suitable advertisements according to user profile, and insert them into the MPEG transport stream for ABR content formats, be it live or VoD.
- *Network PVR* [36] to record audio and video programs and later view it somewhere else via a network. It uses a DVR, with an integrated EPG, recording unit, and storage, at the provider's central location rather than at the end-users' side. In addition to device shifting capability, network PVR provides access to TV content on any connected device using the IP. A CDN can be charged with this functionality to personalize viewer content based on usage pattern, enabling personal device interactions and delivering the right content, at the right time, to the right audience.

1.5 RESEARCH ISSUES

Innovations in the CDN industry are dominated by the industry-based R&D initiatives. The focus has been on developing new technological features, along the line of industry trends described in Section 1.4. However, there are a number of research issues that would be of interest to academic researchers.

- *Cloud-based CDNs* [37] extend traditional CDNs model to integrate cloud functionalities and enhance capabilities to deliver services that are not only limited to Web applications but also include storage, raw computing, or access to any number of specialized services. Such integration creates interesting research challenges such as enhancing application scalability, system robustness, usability, access performance, data durability, and support for application security and content privacy.
- *Mobile CDNs* [24] create an exciting research area to explore content popularity in accordance with user mobility. Each end-user request is characterized by the requested content, the time of the request, and the location of the user. On this research topic, it is required to develop dynamic, scalable, and efficient content replication mechanisms that cache content on demand with respect to the locality of requests, user navigational behavior, and very high spatial and temporal demand variations.
- *Origin storage* [38,39] when integrated with a CDN's service portfolio can provide a secure multitenanted warehousing facility to allow complete origin server offload for content providers. CDNs can use this service for extended

libraries/catalog storage, catch-up TV and network-based PVR storage, and asset management. Research challenges to build a CDN origin storage service include ensuring continuous data and service availability (e.g., disaster recovery capability), guaranteeing superior storage throughput for uninterrupted content delivery over CDN, and facilitating automated storage provisioning and service workflow orchestration.

- *IP anycasting* [40] was not considered as a viable approach for request redirection for early CDNs. This is due to the lack of load awareness and unwanted side effects of Internet routing changes on the IP anycasting mechanism. Nevertheless, the emergence of route control mechanisms coupled with external intelligence to allow dynamic route selections, a load-aware anycast CDN architecture [40], and recent anycast-based measurement work [41] shed light on the possibility to realize request redirection using IP anycast (see Chapter 5 for more details). This research area can be further explored to determine the applicability of IP anycasting in real CDN deployments.

1.6 CONCLUSION

Applications of CDNs can be found in many communities, such as academic institutions, advertising media and Internet advertisement companies, data centers, ISPs, online music retailers, mobile operators, consumer electronics manufacturers, and other carrier companies. Along with the technological advancement and the changing landscape of the CDN industry, new content types and services are coming into picture. It raises new issues in the architecture, design, and implementation of CDNs. In this chapter, we have provided an overview of the CDN technologies, representative use cases, recent industry, and technology trends, and identified a few key research issues. In the next two chapters, we cover the basics of streaming media technologies. Later chapters of the book build on these three introductory chapters and elaborate some of the key CDN technologies, address specific research challenges, and present industry case studies based on real CDN deployments.

REFERENCES

1. Buyya R, Pathan M, Vakali A. *Content Delivery Networks* Vol. 9. Springer; 2008.
2. Lazar I, William T. Exploring content delivery networking. IT Prof 2001;3(4):47–49.
3. Pallis G, Vakali A. Insight and perspectives for content delivery networks. Commun ACM 2006;49(1):101–106.
4. Dilley J, Maggs B, Parikh J, Prokop H, Sitaraman R, Weihl B. Globally distributed content delivery. IEEE Internet Comput 2002;6(5):50–58.
5. Stockhammer T. Dynamic adaptive streaming over HTTP: Standards and design principles. Proceedings of the Second Annual ACM Conference on Multimedia Systems. ACM; 2011. p 133–144.

6. Balk A, Dario M, Gerla M, Sanadidi MY. *Adaptive MPEG-4 Video Streaming with Bandwidth Estimation*. Berlin, Heidelberg: Springer; 2003.
7. Puopolo S, Latouche M, Le Faucheur F, Defour J. Content delivery network (CDN) federations: How SPs can win the battle for content-hungry consumers. Cisco Internet Business Solutions Group (IBSG); 2011.
8. Latouche M, Defour J, Renger T, Verspecht T, Le Faucheur F. The CDN federation: Solutions for SPs and content providers to scale a great customer experience. Cisco Internet Business Solutions Group (IBSG); 2012.
9. Robinson D. *Buyer's Guide: Content Delivery Networks*. Streaming Media; 2012.
10. Atlantic-ACM. CDN survey, Whitepaper, USA; 2011.
11. Arlitt M, Tai J. A workload characterization study of the 1998 world cup web site. IEEE Netw 2000;14(3):30–37.
12. Informa Telecoms and Media. Content delivery networks: Market dynamics and growth perspectives, Whitepaper, UK; 2012.
13. Pathan M, Buyya R. A taxonomy of CDNs. In: Buyya R, Pathan M, Vakali A, editors. *Content Delivery Networks*. Germany: Springer-Verlag; 2008. p 33–77.
14. Canali C, Cardellini V, Colajanni M, Lancellotti R. Content delivery and management. In: Buyya R, Pathan M, Vakali A, editors. *Content Delivery Networks*. Germany: Springer-Verlag; 2008. p 105–126.
15. Rabinovich M, Zhen X, Douglis F, Kalmanek CR. Moving edge-side includes to the real edge-the clients. USENIX Symposium on Internet Technologies and Systems; 2003.
16. Sivasubramanian S, Pierre G, van Steen M. Replicating Web applications on-demand. Proceedings of IEEE International Conference on Services Computing (SCC'04); Los Alamitos, CA, USA; IEEE CS Press; 2004. p 227–236.
17. Sivasubramanian S, Pierre G, van Steen M, Alonso G. Analysis of caching and replication strategies for Web applications. IEEE Internet Comput 2007;11(1):60–66.
18. Rescorla E. *SSL and TLS: Designing and building secure systems* Vol. 1. Reading: Addison-Wesley; 2001.
19. E. Rescorla. 2000 RFC 2818: HTTP over TLS. Internet Engineering Task Force. Available at http://www.ietf. Accessed on 25 October 2013.
20. Li W-S, Hsiung W-P, Kalashnikov DV, Sion R, Po O, Agrawal D, Candan KS. Issues and evaluations of caching solutions for web application acceleration. Proceedings of the 28th International Conference on Very Large Data Bases. VLDB Endowment; 2002. p 1019–1030.
21. Grevers T Jr, Christner J. *Application Acceleration and WAN Optimization Fundamentals*. Cisco Press; 2012.
22. Akamai Technologies, Inc. Front end optimization on the Akamai intelligent platform, Whitepaper; 2012.
23. Strangeloop Networks, Inc. Strangeloop website optimization technology, Whitepaper; 2011.
24. Loulloudes N, Pallis G, Dikaiakos MD. Information dissemination in mobile CDNs. In: Buyya R, Pathan M, Vakali A, editors. *Content Delivery Networks*. Germany: Springer-Verlag; 2008. p 105–126.
25. Rayburn D. *How Mobile Acceleration Works: An Inside Look at Cotendo's Newly Announced Services*. Streaming Media Blog; 2011.
26. Reznik Y. *Video Delivery Optimization via Cross Layer Design and Video Traffic Differentiation*. InterDigital, Inc.; 2012.

27. Ericsson. LTE broadcast: A revenue enabler in the mobile media era, Whitepaper; 2013.
28. Cohen A, Sampath R., Singh N. Supporting transparent caching with standard proxy caches. Proceedings of the 4th International Web Caching Workshop; 1999.
29. Amza C, Soundararajan G, Cecchet E. Transparent caching with strong consistency in dynamic content web sites. Proceedings of the 19th Annual International Conference on Supercomputing; ACM; 2005. p 264–273.
30. PeerApp. The transparent caching primer: How network operators are taming OTT, Whitepaper; 2013.
31. Cucchiara R, Costantino G, Prati A. Semantic transcoding for live video server. Proceedings of the tenth ACM international conference on Multimedia. ACM; 2002. p 223–226.
32. RGB Networks. On the fly transcoding: Overcoming new media distribution challenges, Whitepaper; 2011.
33. Zigmond DJ, Goldman PY. Techniques for intelligent video ad insertion. US patent 6,698,020, 2004 Feb 24.
34. Bhagavath VK, O'neil JT. Method for adaptive ad insertion in streaming multimedia content. US patent 6,505,169. 2003 Jan 7.
35. Elemental Technologies, Inc. Using ad insertion to reach new media markets, Whitepaper; 2012.
36. TV1.EU – Streaming Europe. Network PVR offers viewers anytime any device viewing, Case Study; 2012.
37. Pathan M, Broberg J, Buyya R. Maximizing utility for content delivery clouds. Lecture Notes in Computer Science, Proceedings 10th International Conference on Web Information Systems Engineering (WISE'09), 5802; 2009. p 13–28.
38. Akamai Technologies, Inc. NetStorage product brief, Datasheet; 2012.
39. Level 3 Networks. Level 3 origin storage service, Datasheet; 2011.
40. Alzoubi HA, Lee S, Rabinovich M, Spatscheck O, Van der Merwe JE. Anycast CDNs revisited. Proceedings of the 17th International Conference on World Wide Web. New York, NY, USA: ACM Press; 2008. p 277–286.
41. Ballani H, Francis P, Ratnasamy S. A measurement-based deployment proposal for IP anycast. Proceedings of ACM IMC; 2006.
42. Lopez P. Mobile video optimization, Core Analysis Market Report; 2013.

2

LIVE STREAMING ECOSYSTEMS

Dom Robinson

id3as-company Ltd., Rottingdean, Brighton, Sussex, UK

2.1 INTRODUCTION

Although there are many isolated events and microsteps that have converged to evolve today's rich and versatile range of live streaming applications and technologies, there are a few milestones that demark clear step-changes of note. The live streaming systems in use today are all derived from video conferencing technologies. Largely because audio requires less bandwidth to transmit over a network than video does, it is also worth noting that voice and audio streaming predates video streaming.

This chapter contextualizes the reader by providing some background to (and clarifying a common understanding of) the term "live streaming" [1]. It takes a look at what "live" means in the context of networked telecommunications, highlighting that there may be delay between the production of the content and listener's or viewer's experience of that content, and yet that user's experience would still be considered to be of a "live" stream. This chapter also outlines and explains some of the key network considerations when producing "live streaming" over the Internet. It highlights that despite some optimization techniques being more efficient and capable of bringing the "live" experience of a receiver closer to the moment of production of the content, other less efficient network technologies have proven to be successful because of a variety of wider technical

Advanced Content Delivery, Streaming, and Cloud Services, First Edition.
Edited by Mukaddim Pathan, Ramesh K. Sitaraman, and Dom Robinson.

and economic factors. Finally, a brief overview of prevalent live streaming technologies "ecosystems" and "formats" is provided with an exploration looking at why these particular technologies are seeing ever-increasing industry-wide support.

2.2 LIVE STREAMING PRE-EVOLUTION

The birth date of "live streaming" within an Internet Protocol (IP) context is arguably the date of introduction of the "Network Voice Protocol" (NVP) [2] on Advanced Research Projects Agency Network (ARPANET). The formal specification for NVP was not published until November 22, 1977. According to that specification, it is clear that the NVP, implemented first in December 1973, had been in use for local and transnet real-time voice communication over ARPANET.

An unpublished memorandum from USC/ISI in April 1, 1981, by Danny Cohen is widely referenced as adding extensions to the Network Voice Protocol called the NVP-II or "Packet Video Protocol" [3]. This seems to mark a clear starting point for the formalization of combined real-time audio and video delivery over Internetworked systems.

In the process of compiling this history, Vint Cerf, one of the fathers of the Internet, was referenced for his recollection on the pioneers who "did the first webcasts." He pointed to the work of both Danny Cohen and Stephen Casner of ISI. Although they were clearly part of multiple teams, it is clear that Cohen and Casner had key insights into the creation of first audio streaming over what was then called the ARPANET.

On the basis of further communication with Stephen Casner, it was identified that work on transmission of packet voice over the ARPANET started in 1974. It was specific to voice rather than any audio signal because significant bandwidth compression was needed to use voice coding (vocoding) to fit in the capacity of the ARPANET. This was not voice over IP because IP did not exist yet, but it was packet voice using ARPANET protocols.

> *It was not until the early 1980's that work expanded to video when a higher capacity packet satellite network called Wideband Net was installed. The first video was, indeed, crackling black & white with variable frame rate depending upon how much of the image was changing. Later work adapted to commercial videoconferencing codecs that had been designed to work over synchronous circuits, instead of working over the packet network. These provided color and higher fidelity.*
>
> *While work on developing the packet video system occurred during the first half of the 1980s, the packet video system was not completed and operational until 1986.*
>
> *On April 1, a real-time multimedia teleconference was held between ISI and BBN using packet video and packet voice over the Wideband Net, along with the presentation of text and graphics on Sun workstations at each end. This was the first use of packet video for a working conference. Participants included BBN, ISI and SRI, plus sponsors from DARPA and NOSC.*
>
> *The teleconference was the culmination of several efforts during March. Our packet video installation at Lincoln Lab was moved to BBN for ready access by*

the multimedia conferencing researchers there. Performance of the Voice Funnel and Packet.

Video software was tuned to allow maximum throughput and to coordinate the simultaneous use of packet voice with packet video. And last but certainly not least, the Wideband Net stream service and QPSK modulation were made available to provide the high bandwidth and low delay required for good packet video.

Steve Casner (Email to author)

It is therefore safe to say that the *definitive* birthdate of live streaming over "the Internet Protocol" is April 1, 1986. However, it is clear that in the context of the ARPANET, a very similar range of streaming had been pioneered sometime before.

It is also interesting to note that these technologies took at least 10 years to evolve into the media players and production tools that have since become increasingly familiar to today's Internet browser and connected TV users.

2.3 LIVE, LINEAR, NONLINEAR

Let us begin our exploration of "live streaming" by ensuring a common understanding of the terminology and scope of the subject. Just as there are many histories of the origins of live streaming, there are many interpretations of what this actually means!

Live streaming typically has four key stages that align to form a "Workflow."

In Figure 2.1, it shows that the encoding stage converts the video feed into a suitable form for streaming (typically compressing it to "fit" within the bandwidth of the available network) and "contributes" it to the network, sending it to the publishing server. Once "acquired," the publishing server then prepares the audio or video for distribution and forwards it to a network of relays that form the content delivery network (CDN). Each client then uses a directory or schedule such as a Web page or an Electronic Program Guide ("EPG") to discover the content they want to consume, and the metadata is passed to the decoding media player. The decoding media player then connects to the distribution network, requests and receives the stream, and subsequently decodes the encoded video or audio, presenting it to the users' display.

For most people, the experience of using a simple webcam-based video conference system, such as those that Apple FaceTime or Skype can provide, has become a common experience of live streaming. The real-time facility of being able to hold a conversation in a natural way between two remote locations is a solid starting point for understanding what we mean, in common language, when we talk about video being "live."

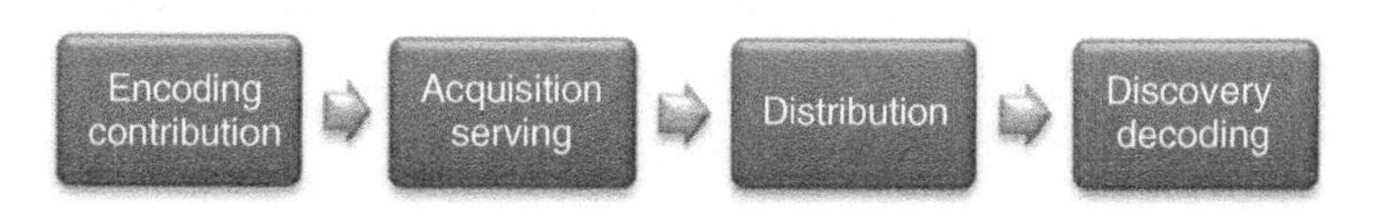

Figure 2.1 Classic live streaming workflow.

But, although it seems obvious at first, "live" is not a simple term in the context of data delivery.

In the context of data delivered electronically or digitally, the speed of light alone determines a small delay between the moment an action occurs (or a sound is made) and when it is experienced by the recipient. This delay is a combination of two effects: "Propagation delay" and "Latency." Propagation delay is a simple physical effect, specific to the length of network link that the transmission occurs over, and caused by the time the electrons or photons carrying the signal take to traverse that length, whereas latency also includes delays caused by intermediate processes within the network. We will further explore latency and these contributory intermediate processes later in this chapter; however, it is worth noting that latency is often used as a single term including propagation delay since with the exception of satellite transmission, propagation delay is usually relatively insignificant in comparison to the latency effects of processing.

Typically in telephony and real-time conversation, a maximum end-to-end latency of 150 ms is thought to be "acceptable" [4]. This is an important starting point for the understanding of what can be considered to be "live" in the context of networked audio–visual content delivery. If 150 ms latency is acceptable for two humans to hold a natural conversation, then this can most certainly be considered to be a real-time conversation. This synchronicity of communication gives the users of the system a sense that they are both with each other in "real life" despite the separation caused by the telecommunications (note that the word "Tele" means "far" in Greek).

Although this form of video is two way, obviously the key here is that events happening at either location are perceived to be seen "live" at the other remote location. Interestingly, if we now turn-off one of the two-way channels, we assume that we are watching the same live content. However, we would actually have no frame of reference, should that video from the remote location be delayed by a further few tens of milliseconds, or even minutes or hours. For a viewer at a remote location with no other way to know when the remote events occur, they are still perceived to be "live."

This can cause some confusion when talking about live video transmission. Indeed, it is possible for a viewer to, for example, make a phone call to the source and discover that the phone call has lower latency—perhaps a "real-time" experience, where the video may take considerably longer to transmit back—resulting in the caller hearing the source on the phone say hello some moments before they are seen to say hello on the video signal.

Strictly speaking, it would be better to use a slightly different term to describe what is commonly called "live video" and often the term "linear video" is used for televisual content that is transmitted to its receiver synchronously as it is created by its origin or source, while "nonlinear video" is used to describe content that is accessed "randomly" by the receiver asynchronously, or sometime after it has been created—for example, an On-Demand Movie viewing.

While the concepts are similar, the terms "linear video" and "nonlinear video" are not to be confused with linear and nonlinear editing—while they are similar and in many ways related, they are different and refer to techniques of editing video content.

2.4 MEDIA STREAMING

Despite the term having entered common vernacular, "streaming" remains a distinctly hard term to accurately and clearly define. By understanding why a "live" stream may lag behind "real life," we begin to appreciate that in the context of digital televisual communications "things must happen" to enable the communication. These "things" can be one or a series of continuous synchronous processes, all working together to bring the subject of the video to the receiver as fast as possible. Alternatively, these "things" can be a sequence of asynchronous processes that occur over a much wider timespan, perhaps occurring a significant times after the event that is the subject of the video is over, and where the audience can access the content at a later time, according to their own chosen time or the video service providers scheduled time.

In a packet network, such as the Internet, data by definition are broken up into a series of constituent packets that must be sent in a coordinated way over the network. The packets are then sequenced by the receiver, usually having any missing packets "reordered and redelivered." Then, the receiver must process those packets to reconstitute the item and restore it to being a usable communication.

Typically, when we think about these "items" of data being sent over the Internet, we think of messages, files, images, or pages. This interpretation persists for most people when they think about video, not least, because most people make extensive use of on-demand services such as YouTube and Netflix, and the impression is that some entire "film" is being downloaded in the same way an email is downloaded when you wish to read it. Our traditional perception of obtaining such data in the nonvirtual world is filled with content in the form of "letters," "books," "photographs," and so on. We have a tradition of understanding the content being preserved as discrete items in some form of medium—Intellectual Property lawyers call these "fixations." Streaming changes that.

One of the first things to be noted when trying to understand "Streaming" is what it tries to achieve. Let us take a look at mp3 audio as an example. In the early 1990s when the Internet was expanding into the domestic environment, users typically could access via dial up over standard telephone lines, and the typical access speed was 14.4 kbps. Today, domestic broadband speeds are usually faster than 1 Mbps—so over a thousand times faster than in the mid-1990s—and many are over 100 Mbps. In this new age, a single 5 MB mp3 file may only take a few seconds to download—noticeably less time than it takes to play that mp3 file—however, in the mid-1990s, when mp3 first emerged, it could take at least as long as the play duration of the file to download—so a 5 min long piece of music would take often more than 5 min between the point of choosing to listen to it and the point it could be heard. This is far from satisfactory from the user's point of view.

One of the problems was that once a download of the mp3 was started, even though much of the audio data was available on the local computer, the computer itself could not make sense of the file—it was not a discrete item.

As computer data files typically need to have a clear structure to accurately convey information, a file not only included the title and the first information about the file (the "file headers") at the front (called the Front of File data), but it also needed the last bit of data, the so-called End of File ("EOF"). Only when the EOF was received, and error checking was complete, the downloaded file was "released" to the operating system as a complete data item for use in applications such as media players. Among other things, this protected the computer from endlessly downloading data and filling up its local memory, which ultimately would have brought the computer to a standstill.

Engineers noted that *during the file transfer*, the data already received by the computer could potentially be used by the "media player application," even though there was more being delivered by the download process. The logic was if this was possible, then the listener need not wait until the download of an mp3 completed in entirety before their player could begin to process the incoming data and play the music "as it arrived."

Note that while the file was still in mid-transfer, it was being transferred by being broken into small "chunks" by the underlying packet network process. Each chunk was in its own right a small file—it just happened that the data it contained made little sense in isolation as a single "item" to the application layer technologies such as the media players. However, by accruing enough of these chunks in a buffer between the source and the point of processing, a media player application could be setup so that it really made no difference if the chunks were being retrieved from disk or from a buffer. The result was that playback of the mp3 became possible despite no EOF being present, and for as long as the buffer contained "the next" chunk that the player requested.

This continuous flow of chunks of data derived from a larger file and delivered over a packet network became known as a "Stream." It is important to note here that a continuously updated source of a stream could potentially be configured to play forever. This configuration or model is the starting point for understanding what a live stream is. In summary, it is a series of chunks of audio or video data that are being continuously generated by a source or origin (now usually called an "Encoder") and transferred (by a network of Distribution Servers) to a recipient (the "Decoder").

Live streaming is a linear process and synchronous in nature. The EOF may or may not be part of the story—some modern models have evolved so that the "chunks" may actually have many EOFs, but for the purpose of common understanding a key difference between a "live stream" and an "On-Demand" stream is that, in the case of a live stream at least, the EOF will never be transferred while the transmission is "live."

2.5 RELATED NETWORK MODELS

Earlier, we briefly looked at propagation delay and latency and the effect these have on perceptions of what "live" means. Although the propagation delay is a physical effect, as described earlier, other factors cause delays and these are more widely embraced under the description of latency. It is important to understand the Open Systems Interconnection (OSI) [8] network (Telecommunication Standardization Sector of ITU, 1994) and TCP/IP network stacks [5].

For convenience, let us consider a summary diagram of the two in Figure 2.2.

The TCP/IP model is not explicitly defined with the same details as in the OSI model, since the Internet Engineering Taskforce strives to be a consensual group rather than a standard organization, but Figure 2.2 shows a common interpretation of the

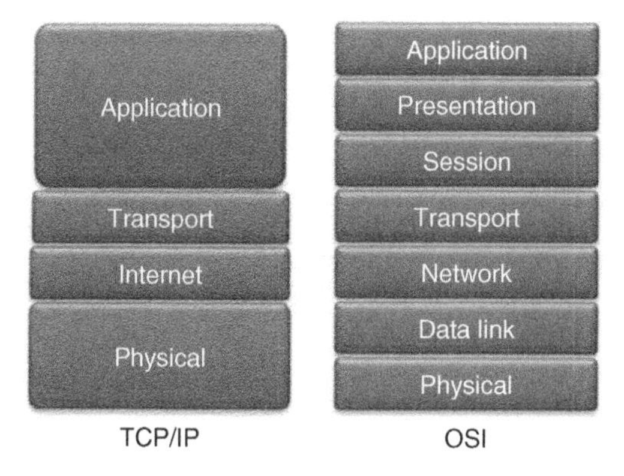

Figure 2.2 TCP/IP and OSI network layer models compared.

equivalence of the two stacks. Network engineers will talk about these layers from the bottom up, commonly referring to the physical layer (common to both) as "layer 1," but the application layer is referred to as layer 4 in the TCP/IP model, and "layer 7" in the OSI model. For the purpose of the rest of this section, the layers in the TCP/IP model will be referenced.

2.5.1 Physical Network Considerations

The propagation delay occurs in the Network Interface layer. This is where light passes through fiber and electric current flows through Ethernet (and so on). Internet networks consist of a variety of mixed forms of physical point-to-point links terminated by various telecommunications and networking systems that then interface with each other at junctions called routers. The computational processing in the terminations usually, but not always, happens as part of an electrical process or a photoelectric process with little computed logic—so, for example, fiber multiplexers operate optically based on wavelength of light, and Ethernet switches work electrically based on the sequence of bits in the header of the datagram. There may be some delay introduced, but at this stage the latency is largely affected by propagation delay more than any other form of delay.

2.5.2 Internet Layer Considerations

Once the packets are handed to the router—the "Internet Layer 2"—things start to change. The router must have a degree of "intelligence" to decide which of the multiple possible "other connected network links" that router should forward each packet. The first time a stream of data is routed, this forwarding path must be discovered and this can take a few milliseconds—possibly more. This adds latency to the time it takes these packets to reach the recipient. Once the route is discovered, the router will make a note and will store that routing in a local "directory" called a "routing table." This minimizes the processing that the router must do; however, any changes to the networks condition will cause a "reevaluation" of the route that the router can forward packets over, and again this will add some latency to the packets that are being throughput.

Routing is typically an optimized process and uses dedicated technology that can function introducing minimal latency.

It is important to understand the effects of the Network Layer in introducing propagation since when streaming live the variety of options one is presented along with the workflow can critically affect the overall process.

If the original signal is delivered over a network link that is prone to varied propagation delay or other contributing latency factors, then no matter how well constructed the onward distribution network is, it can only distribute a varied and delayed source image or sound.

2.5.3 Transport Layer Considerations

Planning the contribution feed is of utmost importance to the creation of a stable and high quality user experience in the live streaming environment. Although layer 1 and layer 2 network services are typically bought in from a network services operator, there are many choices that the Live Streaming Engineer (often called a "webcaster" in the context of streaming purely on the Internet) can autonomously make in their use of the layer 3 IP network services they buy.

Continuing our journey "up the stack," we move up from the simple IP routing on layer 2 into the "Transport" layer (layer 3) and find the Transmission Control Protocol "TCP" and its twin the User Datagram Protocol "UDP."

UDP has a variety of good uses in audio and video streaming, and for many years UDP-based transport protocols for audio and video streaming were developed assuming UDP would be the "normal" way to transmit audio and video.

UDP has no automatic retransmission process, so if a packet fails to make it over the network link it is up to the programmer to define when (or indeed *if*) this should be corrected. In the case of the large quantity of data sent in an audio or video transmission, a few lost packets in a stream are generally not missed. The end-user does not notice a few pixels of data that are not shown in a moving image—the eye and brain work together to correct this (known as "perceptual" audio and video encoding). For many years, the early streaming protocols were thus engineered with UDP transport in mind, and this led to a range of custom servers, called "media servers" appearing which specialized in packetizing encoded video into UDP datagrams, often with separate control protocols that enable client applications to communicate with these media servers to establish unique user sessions and subsequently control which video or audio stream is played, and these application control protocols offer features such as pause, stop, rewind, and play.

2.5.3.1 Applications—Transport Protocols RTSP [6], the "Real-Time Streaming Protocol," acts as a control protocol for an RTP ("Real-Time Protocol") stream and is almost invariably (although not exclusively, as we shall see later) transported on UDP. RTP essentially sequences the packets and sends them onto the network, and the recipient client, the "media player," reassembles all the packets into sequence in a buffer before playing it to the user. Missing packets are ignored, and timestamps in the packets allow for the correct timing of the playback, even if packets are missing. RTSP also allows the user to request a live stream from the server or to seek to various places in the stream if the stream is a playback of an "on-demand" file.

Although internally common on private networks and still used today for many IPTV installations, RTSP (which collectively refers to RTP too) has a few shortcomings. Natural Address Translation ("NAT"), which allows a router with a single public IP address to then provide gateway access to multiple machines, is a very common way to put enterprises or groups of computers online with a shared single Internet connection.

For a number of reasons, however, UDP is quite complex to route into a NATted local area network (LAN). The router receives the UDP from the off-site server but, without an additional application controlling the UDP packet forwarding, the router does not know where (on the LAN) to forward the packet to. For this reason, RTSP struggled for many years in scenarios where publishers wanted to deliver that content into enterprises and homes with more than one computer.

The two major vendors during this era (1996–2004), Real Networks and Microsoft, had to implement "fall-back" strategies so that when the "optimal" UDP-based RTSP streams could not be received by media players inside NATted LANs, the media players could then explicitly request the stream over RTSP using the "reliable" TCP. RTSP also requires fairly specific firewall configurations, allowing RTSP requests and responses in and out of the LAN, and the resulting stream (be it TCP or UDP) to "flow" into the LAN on ports which, by default, were often closed.

To receive the stream inside the LAN, home users would then be required to open such firewalls up—something which was usually beyond them. For enterprise admins, this has made streaming using RTSP very much an "opt in" process, which then meant building a business case for allowing streaming video and audio into the enterprise. Anecdotally this was, in the late 1990s, akin to "building a business case to bring a TV to work"—and was blocked by most network admins.

Interestingly, streaming was, in the culture of engineers of the era, seen to be something that *needed* its own transports. Because network bandwidth was a relatively scarce, commodity optimization was required and retransmission of lost video packets was one of the fundamental "waste of bandwidth" elements that contributed to that culture. UDP-based streaming protocols such as RTSP Phone Network Alliance (PNA) and even MPEG-TS (over IP) were refined to ensure that in controlled network conditions the bare minimum network utilization occurred, ensuring that one user's use of a video impacted other users on the network as little as possible.

Despite this collective work, there were some interesting external factors that eventually meant that by the end of the first decade of this century, less optimal methods for streaming have become the dominant players in the market today.

2.5.3.2 Contribution Feeds—Layers 1 and 2 There exists in the streaming and broadcast world a humorous model, the "garbage in garbage out model" ("GiGo"), which highlights that a poor source signal from the encoder to the point of origination on the distribution network means that every member of the audience will have a poor signal.

Given the stringent requirements for the initial source feed (also called a "contribution feed" and an "acquisition of a signal") from an encoder to the origination/publishing server, most contribution feeds for both live streaming and "traditional broadcast" are carried out over privately leased network links where specific optimizations and access control to preserve the high quality of the line can be effected.

Very Small Aperture Satellite (VSAT) links are sometimes tuned specifically for ad hoc contribution feeds of live streaming. Sometimes, private leased lines were optimized with UDP tunnels to ensure maximum "throughput" (and the lesser known and more important measure of "goodput"). In these environments, many aspects can be tightly controlled.

It would be unwise to use a single-shared Internet connection for the vital contribution feed (at least where the other's sharing the connection are also unknown or "uncontrolled" such as in a public Asymmetric Digital Subscriber Line ("ADSL") network or on public Wi-Fi, for example). However, some success is being shown with channel bonding and link-aggregation of multiple shared services together. It in effect vastly overprovision the required bandwidth, and "hope" that others sharing the network simply do not want to use so much of the total bandwidth at any one time of any one of the aggregated services, such that there is not enough of the remaining capacity to use for the desired contribution feed. At the time of writing, the exploding technology in this space is surrounding cellular-link aggregation—with many integrated video multiplexers and channel-bonding, cellular-link aggregators appearing—and these technologies are earning the name "CellMuxes" (contracted from "cellular multiplexing"). These devices can provide high quality video using many data services from multiple cellular service providers at the same time and employing clever application layer protocols to "trick" the transport layer 3 protocols into believing that there is a single high capacity connection available. Although they are prone to sudden losses of bandwidth, they are also prone to extended periods of overprovisioning, and with extended buffering and high latency as the compromise, they can produce relatively continuous high quality video and audio streams in ad hoc and in often mobile situations.

Generally though, the rule is that the link between the Encoder and the point of acquisition should be private and of high quality of service (QoS)—sufficient to sustain the target bandwidths—and with a clear service-level a (SLA)—particularly if the live stream is of high commercial value.

2.5.4 Distribution Networks Considerations

In the ISP subscriber networks that service the majority of the users of live streaming, no such "application-specific" layer 1 and layer 2 optimizations are possible for live streaming. In closed private networks such as Hybrid Fiber Coax (HFC) used for cable TV (which increasingly offer IPTV) or even dedicated IPTV networks, very specific optimization is possible. Also specifically *within* content distribution networks, which span multiple subscriber services (covered widely elsewhere in this book), network optimization can be fine-tuned and optimized at layer 1 and layer 2 for live streaming.

A good example of the type of optimization that may be deployed where layer 1 and layer 2 can be controlled autonomously is IP Multicast, which is a way to configure routers (which are presented with data correctly transported at layer 3) to selectively broadcast a single packet of data to many recipients at the same time as if there were only one recipient. IP Multicast is arguably the best way to scalably deliver live streams to large audiences—however, because its development occurred later than the core version 4 of the Internet Protocol (IPv4—the dominant IP used worldwide), it is also unfortunately not available in the public Internet, and while even today in IPTV and enterprises IP Multicast thrives, it has not ever enjoyed the scale or deployment it deserved in the Internet as a whole (had it done so than the migration of live TV services to over the top (OTT), public Internet services would have been dramatically accelerated).

In this way, such technologies thrive only if there is a demand. The problem has been that the demand for Internet video and audio has only grown as a response to its availability at a usable quality. Although optimizations have been forward looking and planning for great scale, the technologies and service providers that adhered to them have also been taking a long-term view at a slower pace, and not giving the emerging audiences a good quality, accessible service in the "here-and-now." This is because the cost of investing into widespread deployments of technologies, such as IP Multicast, in fact reduces network utilization while "optimizing" the network, and this has often represented a potential loss of revenue for network operators. Operators consider making revenue from selling services to publishers as inefficient, with a view that—"why deploy IP Multicast when it means one packet can be delivered in one transmission to all viewers and we can only charge once, when today's 'unicast' services means we deliver each packet to each user and charge the publisher for *every* transmission." Surprisingly, sometimes "network inefficiency" equals "revenue effective!"

2.6 STREAMING PROTOCOL SUCCESS

Returning to the issues faced by the traversal of firewalls by RTSP (see Section 2.5.3.1), for a while it caused a critical industry schism. The fallback or "failover" strategies adopted by Microsoft and Adobe were different. Both initially sought to keep publishers "confined" to their own workflow. If you published a live stream in Windows Media, it would not play in Flash Media Player and equally publishing live in Flash left users with only Windows Media Player unable to see the stream. Microsoft's Media server was widely distributed and included in all its Server licenses beyond NT4.0. The Flash server was a premium technology that had to be added to an Operating System. Considering the wide deployment of Microsoft within Enterprises, Live Streaming with the Enterprise in mind evolved around Windows Media, and Windows Media, while it had a "tendency" to veer toward proprietary implementations of standards, at least took heed of UDP, RTSP, and some of the other optimization strategies that were evolving.

In contrast, Flash was TCP only. Although this was seen to be unscalable and sub-optimal, for a variety of reasons this strategy worked extremely well in the consumer environment.

In 2005, Adobe acquired the Macromedia Flash suite of technologies and, while many engineers, myself included, saw the technology as suboptimal (for reasons outlined in a moment); the pragmatic reality was that when a user wanted to watch a stream it "just worked," and in the technology world, "just working" usually trumps anything! There was no complex configuration and it was available on most types of computer that the mass market was using at the time. Accordingly, Adobe's Flash Video format emerged extremely quickly in the early 2000s, and by 2006, Flash clearly became the dominant end-to-end suite of technologies—also known commonly as a "format."

Under the hood, Flash was far from an optimized live streaming technology. It has its own Real-Time Messaging Protocol (RTMP), which while clearly close to RTSP, supported TCP only, thus losing all the advantages of RTP's UDP protocol support in terms of network optimization, but gaining simplicity and, firewall issues aside, removing the complexity of NAT forwarding.

Initially, the proprietary RTMP locked users into the Flash ecosystem—it was the only way for audio and video stream publishers to reach the widely distributed (and free) Flash Media Player, which had gained its ubiquity as a cross platform browser plug-in that simplified presentation for graphics and animations in Web pages and included audio and video presentation capabilities. The only way to generate RTMP streams was the Flash Media Encoder/Server combination, and at first the video enCOding and DECoding (CODEC) video compression choice was limited and was very low quality. However, the simple fact that when the player opened, it would reliably open a video stream that was a "magic-bullet," which meant that the Flash ecosystem had a key economic driver that the other "more optimized" formats did not: principally it worked for *advertisers*.

Very quickly, it became interesting for publishers to add video to their sites because the Flash Player brought with it pre-roll adverts that played as soon as the Web page opened, and each advert brought the publisher money. Regardless of any other technology advantages under the hood, the owners of the publishing companies were enthused, and so Flash video gained rapid adoption. This was not without exception—the Enterprises still had to make the same "opt-in" policy decisions to allow Flash Media Player to install, and in fact even today many enterprises prefer the IP Multicast enabled Windows Media Player model to ensure low impact of traffic during live webcasts on their LANs. However, the vast majority of streaming—both live and on-demand—rapidly moved to the TCP-based RTMP.

2.7 PLATFORM DIVERGENCE AND CODEC CONVERGENCE

One would think that this would be the end of the story—however, at this stage some other factors began to come into force. The proprietary nature of RTMP increasingly locked in some publishers to the Adobe format, but it also "locked out" others: and many of these were the major broadcasters who, seeing online video "coming of age," wanted to add Internet and IP-networked video to their workflow outputs. The problem with RTMP and the Flash Media ecosystem and format is that it had a number of shortcomings. First and foremost Adobe was slow to adopt the increasingly popular h.264 video compression standard. Other technologies were faster to market with h.264, which was both undeniably better quality than Adobe's own choice of VP6 CODEC and presented a reduced risk for broadcasters looking at their long-term storage strategies for their video archives. h.264 is an international standard, while Adobe's widespread VP6 was not, and so risk-assessors within broadcasters preferred the vendor independence that h.264 offered.

In 2007, Wowza introduced their media server. It was the first independent commercially supported server that could acquire a live stream from Adobe's Flash Media Live Encoder and distribute that stream to Flash Media Player—and it came with a price ticket that was roughly a quarter of the Flash Media Server. Historically, there has been a risk of a legal case between Adobe and Wowza concerning patents and use of "proprietary" variations of the RTMP standards, but aside from that the critical step forward here was that Wowza's media server also supported other transport protocols, including RTSP, the Internet radio streaming protocols Shoutcast and Icecast, and critically MPEG transport stream (MPEG-TS).

Once the Flash ecosystem updated to include decode capabilities for h.264, the native support for MPEG-TS in Wowza was of particular significance to the Broadcast industry since it enabled their traditional TV and satellite workflows, which already used MPEG-TS for their "traditional" broadcasting systems, to "ingest" the live signals, which had been encoded in h.264, into Wowza and, by "trans-muxing" the stream from MPEG-TS to RTMP through the Wowza server, they could, in one step, simply distribute the live signals over the Internet to the ubiquitous Flash Media Players.

Although Wowza was still proprietary, its relatively open integration with many third-party encoders and its reach to Flash Media Player encouraged more and more organizations to publish to the Internet.

2.8 ADAPTIVE BITRATE (ABR) STREAMING

Despite this convergence on the h.264 CODEC, and widespread success of Flash Player underpinned by both Adobe and Wowza, there were still some significant innovations to come. Although the fundamental concept was initiated in 2002 as part of the DVD Forum, a range of technologies, collectively known as "adaptive bitrate streaming" [7] technologies, took several years to reach mainstream adoption as common Internet streaming "formats." The runaway successes of these commercial implementations that have emerged are Microsoft's Smooth streaming ("smooth") and Apple's HTTP Live streaming (HLS). Adobe has tried to keep up by introducing Adobe's HTTP Dynamic streaming, although arguably it has not seen nearly as much adoption as both Smooth (which was first to market) and HLS (which enjoyed a strong "piggy back" on the success of the iOS-based iPhone and iPad, which pretty much forced HLS into the market as the only option to live stream to those devices).

The principle aim of adaptive bitrate (ABR) streaming technology is to allow the publisher to produce several quality streams at the same time—perhaps one for mobile streaming at 300 kbps, one for domestic SD streaming over Wi-Fi at 750 kbps, and one for high definition (HD) streaming at 1.4 Mbps, for example—and to synchronize the publishing of each of these streams in such a way that if a recipient decoder wishes to switch from one bitrate to another (perhaps because the network conditions are varying), then the client player simply requests the "next" packets from the lower or higher bitrate stream, and these are sequenced seamlessly in the players buffer. This means that while the quality of the image may pixelate or increase mid-playback, there is no "transport layer" interruption of the flow of the streaming—and so the changing quality is not accompanied by a break in the continuity of the stream. This provides a much better quality of experience (QoE) for the viewer. This smooth transition from one bitrate to another was precisely why Microsoft named their technology "Smooth streaming."

Such technologies almost invariably require discrete layer 3 connections between each player and the server and so invariably use TCP as the control protocol. This comes with significant network overhead, as discussed earlier, and in fact the nature of creating a buffer to manage the bitrate switching and the decision-making processes involved all that combine to add significant latency. Although there were one or two very early approaches to try to adapt traditional media servers to support distribution of ABR streaming technologies, the role of these servers was almost universal to act as a termination point of the source contribution feed and to "chunk" or "fragment" the different bitrates into synchronized blocks of video—usually split at the video key frames along the lines of the MPEG "Groups of Pictures" (GoPs)—and then to packetize these in wrappers of HTTP transport protocol packets.

Accordingly, the "distribution technology of choice" quickly became relatively common HTTP servers, with modifications appearing for Microsoft's IIS and Apache among others. Again, from a purist's point of view, while there were many network optimization reasons not to use HTTP (which has, for example, no flow control and simply uses all available network bandwidth to transfer any given datagram), there were some critical advantages that this method of streaming introduced.

2.8.1 Adaptive Bitrate—Enterprise

The first of the concerned enterprise networks: HTTP traffic is World Wide Web (WWW) traffic—to "block" this type of streaming from a corporate network requires intelligent firewalling—and so suddenly the policy decision to block streaming in the Enterprise moved from "opt-in" to "opt-out"—and accordingly the argument behind the business case for streaming in the enterprise moved from "what is the business case for streaming" to "what is the business case for turning it off?"

The second advantage was also a critical issue for enterprises. Windows Media's own protocol (MMS), Flash's RTMP, and the "standard" RTSP were all complex to forward into the enterprise. Each essentially required a proxy server between the LAN and the wide area network (WAN), which would acquire the stream from the Internet source, and then handle all subsequent requests from within the LAN.

For video on demand (VOD), this caching required custom technology to setup, but was not expressly complex. However, when the CEO of the enterprise made a live announcement to all the staff, it became critical that the live stream was preconfigured on the proxy/gateway technology so that only a single stream would be delivered over the WAN to the thousands of recipients potentially wanting to "tune-in." And this severely limited the ability of enterprises to deploy live streaming in an ad hoc manner.

Commercial solutions were almost invariably developed from the basic Windows or Adobe Media Server SDKs, and either came as part of wider proxy server solutions (often simply as a virtual machine image running on board the proxy using a common interface) or were delivered as custom-built implementations of the same SDKs. This meant that there were no "cheap options" and again this inhibited enterprise live streaming adoption.

It is worth noting here that live traffic in enterprises is often concentrated around live "events" rather than 24/7 television-like streaming, and this causes very specific congestion issues on the networks involved. In a corporate LAN, the typical network is 100 Mbps or even 1 Gbps, so 100 users streaming at 500 kbps use a significant amount of the overall network capacity. Although 30-min on-demand file may take hours or event days or weeks to circulate an office, the number of simultaneous users will always be limited and so congestion on the network will be limited.

When the CEO makes a live announcement about forthcoming redundancies (for example) on the corporate network, however, the entire community may want to watch the stream, and this may saturate the networks in many ways—particularly in the absence of IP Multicast (which again always takes a specific and manual configuration). In the case of a modern 1 Gbps LAN, it is possible that the internal network could handle most users' requests, but it is unusual for a corporate office to have a 1 Gbps WAN connection between sites. This WAN connection, when it becomes saturated by the CEO's live stream, also prevents all "other" traffic from using the intersite WAN resources, and given these "other" reasons are usually the main reason for the network existing in the first place, this makes streaming unpopular with network administrators. (The author had to help a bank pickup the pieces after a badly configured "multicast" became an unplanned unicast and the 2000 or so viewers saturated the corporate network that also supplied the trading floor, resulting in several millions pounds of lost trades).

ABR HTTP streaming not only "found its way through the firewall" by essentially being "Web" traffic but also critically added another benefit. Even in the case of a live stream, ABR HTTP video is broken into small individual fragments of video transported on HTTP. These fragments can be cached by standard Web-proxy servers and served again to any number of users who are also requesting those streams.

Although it can mean that these proxy server caches become full of chunks of video data very quickly for the duration of a live stream, this actually significantly addresses the WAN link saturation problem—only one copy of each video fragment (for each bitrate) will be copied into the proxy server. The proxy server may still serve many separate copies of that stream out over the 1 Gbps LAN, but if that LAN is only connected to a 10 Mbps WAN, then most of that WAN link will still be available for all the other applications that network is used for—so banking/Web access/database referencing/VoIP and so on would all be relatively unaffected by the CEO's live stream since the proxy server would be doing all the serving within the LAN and requesting just a single stream over the smaller WAN link.

2.8.2 Adaptive Bitrate in the CDN and ISP

Moving up the distribution chain into the ISPs and the CDNs (a number of CDN topics are covered in depth in the other chapters in this book), HTTP presents a simplification of the distribution paradigm, and while it introduces significant latency—perhaps two or three GoP lengths (typically 8–12 s)—and while HTTP traffic itself is bursty, and cannot be managed in the same way that the superior live streaming protocols can, its very simplicity has left HTTP as the de facto way to transport streams. Incidentally, RTMP now, only a few years after ABR became commercially available, only dominates where low latency to the largest media player "reach" is the key performance indicator (KPI) for a video stream—such as for sports betting video to Web applications.

2.9 INTERNET RADIO AND HTTP

Interestingly, the success of HTTP takes us back to one of the early streaming methods: HTTP progressive download. It was not originally considered to be a live streaming protocol by many "webcasters." This is arguably no longer the case. Referring back to the mp3 streaming example at the start of this chapter and seeing HTTP-based ABR streaming come to dominate today's modern delivery strategies, it is useful to highlight that the HTTP transport is not limited to streaming ABR "chunked" or "fragmented" formats—HTTP is now widely accepted and with that the older Shoutcast and Icecast protocols used for Internet radio streaming have seen a strong renaissance in the past few years.

These are important formats—many times as much mp3 and aac encoded music and radio is streamed by HTTP progressive download to many more different technologies and devices than any form of IP streaming video—not least because the 256 kbps Internet still exhibits significantly wider reach than the 10 Mbps Internet!

Streaming radio does not make up such significant volumes of bandwidth and traffic online as streaming video; however, the audience sizes can be staggering. A single video server may break sweat when it is serving a thousand clients—and even the best server clusters may manage only 10,000 before they need to offload to a distributed CDN.

A basic Icecast server on a domestic PC can comfortably serve 20,000 streams, given suitable Network Interface Cards and connectivity. It is important not to dismiss these technologies!

2.10 CONCLUSION

This chapter has attempted to provide some critical thinking about the evolution of today's live streaming ecosystems and what a "live streaming ecosystem" actually is and delivers.

There has been some emphasis on ensuring that the contribution feed from the original video to the distribution platform is well managed—for it is in that single connection that even a little optimization has the widest benefit, and it is also there that the "webcaster" or engineer has the most autonomy to optimize.

It is clear that today's emerging standards such as MPEG-DASH and h.265 stand on the shoulders of mature technology that has already passed many tests, and that while they will rely on distribution by relatively unoptimized HTTP-based networks, there are many reasons why that has come to be. Indeed HTTP is a great reminder of the concept of "survival of the fittest" in evolutionary terms, since HTTP "fits" more use cases than any other.

Despite that, in the constraints of Autonomous Systems and Enterprise networks, it is possible to deliver live streams far more efficiently than perhaps is possible in public streaming and OTT models, and where network administrators can optimize they should consider doing so for a number of reasons.

Live streaming is a very significant part of the day-to-day use of the Internet, and while most think about the world of live video events like Obama's Inauguration and Royal Weddings, the chapter closed with a focus on Internet Radio which despite disappearing into the background ambience of many desk workers' days is a service that is used by countless millions around the world.

The next chapter gives the reader some practical opportunities to create a live stream and through this deepens technical understanding.

REFERENCES

1. Sripanidkulchai K, Maggs B, Zhang H. An analysis of live streaming workloads on the Internet. Proceedings of the 4th ACM SIGCOMM Conference on Internet Measurement. ACM; 2004.
2. Cohen D. Specifications for the network voice protocol (NVP); 1977.
3. Cohen D, Casner Stephen, Forgie JW. A network voice protocol NVP-II. USC/Information Sciences Institute 71; 1981.
4. Telecommunication Standarization Sector of ITU. 2013. International telephone connections and circuits—general recommendations on the transmission quality for an entire international telephone connection. Available at http://www.itu.int/rec/dologin_pub.asp?lang=e&id=T-REC-G.114-200305-I!!PDF-E&type=items. Accessed 2013 Jan 14.
5. Internet Engineering Task Force: Network Working Group. 1999. RFC 1122: Requirement for Internet hosts—communication layers. Available at http://tools.ietf.org/html/rfc1122. Accessed 2013 Jan 17.
6. Schulzrinne H. Real time streaming protocol (RTSP); 1998.
7. Stockhammer T. Dynamic adaptive streaming over HTTP – : standards and design principles. Proceedings of the Second Annual ACM Conference on Multimedia Systems. ACM; 2011.
8. Telecommunication Standardization Sector of ITU. 1994. Data networks and open system communications—open systems interconnection—model and notation—information technology open systems interconnections—basic reference model: The basic model. Available at http://www.itu.int/rec/dologin_pub.asp?lang=e&id=T-REC-X.200-199407-I!!PDF-E&type=items. Accessed 2013 Jan 16.

3

PRACTICAL SYSTEMS FOR LIVE STREAMING

Dom Robinson

id3as-company Ltd., Rottingdean, Brighton, Sussex, UK

3.1 INTRODUCTION

If we have a look at the key influences and drivers in the past two decades of the evolution of live streaming, it will make sense to focus on practical explorations of setting up live streams. It is only through practical experience that the reader can appreciate the diversity of skills that are required to produce a live stream. When producing a live stream, the engineer must know about everything from the commercial and technical nuances of provisioning the right telecommunications links, through the production values expected by the group for whom the content is produced, right down to the code ensuring that various technologies integrate well. While traversing through the commercial and technical details, an engineer should also understand the audio and video lighting production problems that may need to be resolved under time pressure, as well as manage crew nerves and team spirit that are all essential ingredients to a successful webcast.

Toward this objective, we can explore a few streaming technologies that can be used for free or at least for a free trial period. The aim of this chapter is to give the reader some hands-on experience in setting up real live streams. In particular, the following platforms are explored to set up a live stream:

Advanced Content Delivery, Streaming, and Cloud Services, First Edition.
Edited by Mukaddim Pathan, Ramesh K. Sitaraman, and Dom Robinson.

- A live radio stream with the free BUTT tool as the encoder, Shoutcast DNAS as the server, and almost any mp3 player you can imagine for playback. We will start with the HTML5 <audio> tags built into most contemporary browsers.
- A basic webcam video encoded using the free Flash Media Live Encoder and served using Wowza to a variety of players.
- VLC to produce an IP Multicast stream that could, in theory, scale to many users.

Obviously, this chapter can only deal with the essential engineering elements of the workflow—many aspects of live audio and video production need to be addressed separately and are beyond the scope here. However, some basic principles apply and these are included. As for the human "soft systems" management—this can take a lifetime or a moment to master! It is highly recommended that if it all seems to be going wrong and an engineer is up against the clock to "get the stream live," the workflow should be tackled from BOTH ENDS AT THE SAME TIME. If work is performed logically from one end to the other, the resolution process will almost inevitably start at the end further from the problem and take longer to discover it!

3.2 COMMON CONCEPTS IN LIVE STREAMING

Recapping from the previous chapter, a live streaming workflow is almost always the same (Figure 3.1).

However, in the practical environment, we must contextualize this with a couple of extra stages.

As you can see from Figure 3.2, before we deploy a live streaming workflow, we need to develop some understanding of the inputs into the encoding system and also the target output devices. Commonly, the technical interfaces between these stages are called the "presentations." A typical professional production crew may "present" the webcaster with the audio and video feed as a serial digital interface (SDI) signal, which must be

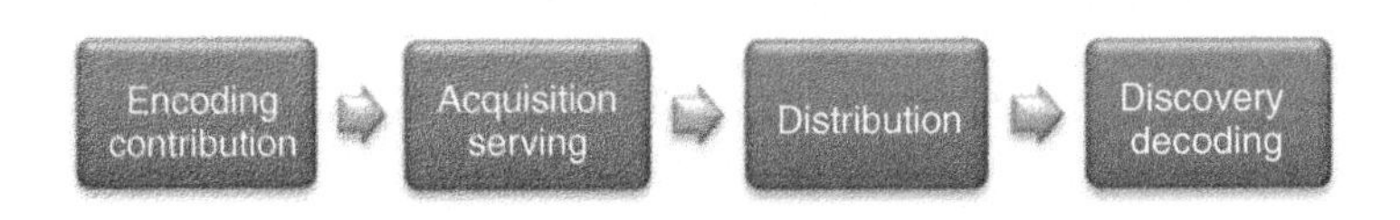

Figure 3.1 Classic live streaming workflow.

Figure 3.2 "Real-world" live streaming workflow.

captured by a "frame-grabber" video card and a sound card on the encoder device, using the device drivers to provide a raw video and audio stream to the encoding software. A simple laptop-based production may use the webcam and built-in microphone and "present" the audio and video locally to the media-encoding software. Or it may be that an external satellite integrated receiver decoder (IRD) is used to produce an IP-based MPEG-transport stream (TS), which your encoder needs to "transcode" into the specific format(s) you need for the workflow to distribute that stream to the target audience's devices.

There are a myriad of such combinations and be the event of a 20-s bulletin for the news or a 24/7 Internet TV channel, you will need to understand the nature of the source presentation at the outset.

Factors that can be effected by this include the location of your encoding equipment. A good example of this is to highlight the trouble many fledgling webcasters get into with high definition multimedia interface (HDMI) presentations. A larger and larger number of streaming media encoding devices are capable of accepting HDMI presentations, but despite all the preparatory testing of the encoding kit in the webcasters, labs, and offices, they turn up on site and find that the relatively short 20-m cable run from the production crew to their encoder seems to be incompatible. No matter what they do, the signal that the production crew send them on the HDMI cable simply does not show up on their encoder. This is simply because HDMI cable attenuates the signal significantly—it was designed to link a DVD player to a TV over a 1 m distance, and not for production cable runs of over 20 m. If you try to use it for that purpose, the signal becomes so weak that it often cannot be used by any devices.

Always defer to SDI where possible, and to be honest for long cable runs, the older format of composite video is considerably more reliable than HDMI despite not offering "HD."

3.2.1 Preparing the Contribution Feeds

The issues faced with the presentations from the production crew exemplify the range of considerations that the webcast producer must plan for. Often the presentation of the audio/video production can be deferred to the audio/video production crew. So long as the expected presentation is clearly specified, the "tradition" of each stage of the workflow "bringing the signal to" the next stage on the right (referring to Figure 3.2) means that in a production with a separate audio/visual team, the webcaster can just wait for the cable to be brought to their machine. Incidentally, this "tradition" is no "law" or formal protocol in any sense, so it is always wise to be prepared to bring your own cables and be ready to "go and get" the signal from the production stages to the left in the typical workflow.

Following on with the logic of this "tradition," however, an absolutely critical responsibility for the live stream producer/webcaster is to ensure that the connectivity between the encoding/contribution and acquisition/serving stage is accessible for the event, of sufficient capacity, and for the sole use of the webcast. As discussed in the previous chapter, any problems with this connectivity and every audience member are affected.

On the production side of the transmission, we have discussed about SDI and HDMI cable lengths in the previous section—and alluded to the use of MPEG-TS and satellites: it should be mentioned that many high profile webcasts will use the same "back-haul" (another term for contribution feed) as a live television broadcast: the TV signal will be generated at the event and sent via satellite news gathering (SNG) truck back to a television center (sometimes called "playout" or satellite hub), where it will be routed in an SDI or MPEG-TS based high quality format to a streaming media encoder and from there encoded and sent onto the Internet connection and off to the acquisition/serving stage. The advantages here are that TV SNG services are very established and very reliable, and this also means that at high capacity, high quality service Internet connection can be installed once at the television center—typically quite an expensive link—and can be reutilized for many different events with little or no adjustment to the television company's production workflow.

This is a great solution for valuable content and for situations where there is already an established TV centric workflow that requires the satellite back-haul anyway. However, live streaming is also very much about low cost event coverage for niche audiences, and for this reason there is also a common requirement to take the encoding and contribution right to the event, to compress the signal ready for distribution on site, and then take advantage of the lower prices for the contribution connectivity that this higher compression brings. So while encoding at a television center is one strategy, encoding in the field is also of key importance in this sector.

A key consideration for a live stream link at the event is usually driven by the frequency and value of the event. If the event is considered "valuable" and there is a budget for it, then a dedicated Internet connection with a committed service-level agreement (SLA) and known connectivity, in terms of IP routing, to the entry point of the distribution network should be established. "Leased lines" are optimal for this. However, such services are usually only provided for 12-month contracts and at significant expense. Obviously, this is great for an Enterprise AV studio that may have regular broadcasts from the chief executive officer (CEO) or such key player in the enterprise or from a government office, which is regularly producing bulletins. However, for a one-off event, for example, a product launch, in a place with no connectivity and which may never be used again for a live event, installing a leased line with a 12-month commitment may be an extremely expensive route to take.

In this situation of the "ad hoc webcast," two options are typically used. They are *very small aperture terminal (VSAT) IP connectivity*, where an Internet connection of the required bandwidth is delivered to the facility via satellite or *Cellular Multiplexer (CellMux)*, where a number of cellular data services are bonded together to create the effect of a single large Internet connection using the cellular networks.

VSAT has some similar advantages to the satellite SNG back-haul model outlined previously. However, it is an operation that requires line-of-sight to the satellite—often meaning that the webcaster has to arrive well in advance of the event and install the dish on a roof or other difficult location, and is a relatively skilled setup. While there are price advantages in the transmission costs when using VSAT compared to traditional TV SNG, the installation is complex and often the time spent performing such an install outweighs the cost benefit. VSAT is also relatively high latency, and some high bitrate streaming formats can struggle with that extra latency unless careful User Datagram Protocol (UDP) acceleration is deployed to ensure that Transmission Control Protocol (TCP)-based transports do not "time out" regularly.

CellMux, on the other hand, is very easy to setup. Essentially, CellMux units are just "turned on," and they find as many cellular networks as they can, bond the data channels available together, acquire the video and audio from the local SDI (or sometimes HDMI or component) ports, and transmit the stream to their counterpart "receiver," which is located "somewhere" on the Internet. The receiver in turn usually presents the resulting video in an SDI or similar production ready video format. In some ways, this is very similar in workflow terms to SNG back-haul—it fits well with TV broadcast workflows; however, the resulting SDI signal usually then needs to be encoded again to be made ready for live streaming. This "double encoding" can lead to significant quality loss. When combined with the fact that there is no SLA on a cellular network—therefore, the available bandwidth can vary unexpectedly (which CellMuxes deal with by using variable bitrate or adaptive bitrate streaming technologies themselves)—the advantage of their flexibility, speed to activate, and portability are often offset against signal variability and therefore picture quality uncertainty. They make great links for mobile use and for areas such as war zones, where the time to deploy a satellite link is impossible. They are also useful as backup links for failed higher quality planned links, but they should always be used as a second option to services that offer SLA and quality of service.

As a final rule in this section, we can identify that wired is better than wireless and private is better than shared. Beyond that, the most important thing to do is to check the service well in advance and to make sure that simple things like the power for the router and technology that provides the link is stable and has backups if at all possible. It is the band draining all the power from the unregulated mains ring in the poorly wired night club, or the journalists being wired to the Wi-Fi and turning out to use the leased line unexpectedly that can cause huge problems during live events. It is nearly impossible to always rule these things out, but it is good to know who can get the network administrator to turn everything else off if need be.

3.2.2 Constraints

Often when producing a webcast there are constraints on the connectivity—perhaps only 1 Mbps is provided, and your intention had been to provide a 1 Mbps stream and a 500 kbps stream in several formats. Or perhaps you may have to leave the site promptly after the event, but the client wants an on-demand archive visible within 30 min of the end of the event.

With careful planning, these deliverables can be catered for "within" the workflow, by creating a single origin stream to a server "in the workflow" and, for example, transcoding the stream on the server into the multiple bitrates required within the distribution process. Or, in the case the site must be cleared straight after the event, the archive could be created on the acquisition server as the live stream is acquired and made available to the distribution servers as soon as the live event finishes (allowing the team on site to pack up and leave site immediately).

The network overhead should always be accounted for during planning. A 10 Mbps pipe will typically only provide 7–8 Mbps of usable, continuous "goodput" (which is the application-usable throughput). Although "speedtest" software may indicate that level to be higher, these tests are usually HTTP based and very bursty—a long continuous stream has very different characteristics. The only true "speed test" of a contribution feed is to test with the streaming software you plan to use.

3.3 THE PRACTICALS

So far we have covered some of the preparatory considerations one must make for a generic live stream. Now let us take a few real-world examples and some actually setup live streams. We start with the most common form of live stream in use on the Internet—that of a Shoutcast live radio stream.

3.3.1 Shoutcast—Live Radio

The typical architecture for a basic Internet radio station, such as could be used for streaming online or for "in-store radio," etc., would include a simple PC that would provide the encoding of the produced audio signal. It may also be that PC runs the DJ's music software too—presenting the whole system as a single integrated computer. The encoded stream is transmitted by the laptop's encoding software over the Internet to the Shoutcast server. This server would typically be located in a data center with connectivity sufficient to sustain many individual streams. That server would, usually, have two network interface cards (NICs): one for management and the contribution feed from the encoder, and the other for distribution to the end-users.

In the supermarket or in-store model, those may be a small pre-set integrated decoder such as sold by companies such as Barix [1]. At home, for a cultural or popular music Internet radio station, users may connect using almost any media player, or Internet radios, or even the native <audio> tags in their HTML5-based Web browser—and it is this latter model that we will setup in this first practical experiment.

Here is the workflow (Figure 3.3) for the Shoutcast practical configuration.

3.3.1.1 Installing We will assume that the reader does NOT have separate machines and abundant bandwidth to attempt these experiments on, and so we will establish the entire workflow on a single PC. However, for clarity, the best way to

Figure 3.3 Shoutcast live streaming radio workflow.

run these tests would be over computers that were distributed remotely from each other—with perhaps the encoding on your laptop, the server on your desktop, and the decoding on your smartphone. We will also assume that you are using Mac OS or Windows; however, we will try to be generic about which versions. The experiments should work under most scenarios.

Let us start by downloading the three important software technologies key to the experiment:

- BUTT encoder (http://butt.sourceforge.net): Which we will use as the encoder to convert audio from the microphone or soundcard into an mp3 audio stream and send it (contribute) to the Shoutcast server. There are various versions for each operating system that can be found on the "sourceforge" link on the page we have provided. Navigate to the latest version suited to your machine and download it.
- Shoutcast (http://www.shoutcast.com): The server itself. There are various versions for a variety of operating systems. Simply download the version you need for the moment. You can run it from the desktop of your machine.
- Browser (http://chrome.com): In fact, it is possible to use any browser these days—what we are looking for is HTML5 support so that we can use <audio> tags to play the stream.

3.3.1.2 Configuring We will get the server running first. Once downloaded, unzip the folder to your desktop. Inside the folder you will find a subfolder called docs, and in that you will find a further folder called dsp_sc. This contains a very useful file called "getting_started.txt," which you should read through to augment this description. You will also see, at the top level of the folder, a file called "sc_serv_basic.conf." Open this in a text editor and change the line which reads

```
publicserver=always
```

to read

```
publicserver=never
```

and save the file. To run the Shoutcast server, open the command line and move to the folder on the desktop. Then, use the following commands from

- For Windows—`sc_serv.exe sc_serv_simple.conf`
- For non-Windows—`./sc_serv sc_serv_simple.conf`

The result should be an output looking something as in Figure 3.4.

As it can be seen, the ports listed near the end of the display give us a clue as to what is going on the application layer. The server is bound to ports 8000 and 8001. We shall explore that in more details later.

The next stage is to install the encoder. Once downloaded, simply execute the application. It will open and present a user interface that looks similar to Figure 3.5.

```
sc_serv2_mac_os_x_intel_07_31_2011 — sc_serv — 106×24
 1361896724[30] )
2013-02-26 16:39:14     I       msg:[DST 192.168.0.8:51168 sid=1] SHOUTcast 1 client connection closed (50
04 seconds) [Bytes: 79836451] Agent: 'Mozilla/5.0 (Macintosh; Intel Mac OS X 10_8_2) AppleWebKit/537.22 (K
HTML, like Gecko) Chrome/25.0.1364.99 Safari/537.22'
^C2013-02-26 18:39:30   I       msg:[MAIN] Exiting loop
2013-02-26 18:39:30     I       msg:[MAIN] Runner shutdown
2013-02-26 18:39:30     I       msg:<***>       Logger shutdown
Doms-MacBook-Air:sc_serv2_mac_os_x_intel_07_31_2011 d2$ ./sc_serv sc_serv_simple.conf
2013-02-26 18:39:32     I       msg:<***>       Logger startup
2013-02-26 18:39:32     I       msg:<***>       version 2.0.0.29
2013-02-26 18:39:32     D       msg:<***>
2013-02-26 18:39:32     I       msg:*******************************************************************
2013-02-26 18:39:32     I       msg:** SHOUTcast Distributed Network Audio Server (DNAS)
2013-02-26 18:39:32     I       msg:** Copyright (C) 1999-2011 Nullsoft, Inc.  All Rights Reserved.
2013-02-26 18:39:32     I       msg:** Use "sc_serv filename.conf" to specify a config file.
2013-02-26 18:39:32     I       msg:*******************************************************************
2013-02-26 18:39:32     I       msg:[SHOUTcast] DNAS/mac v2.0.0.29 (Jul 31 2011) starting up...
2013-02-26 18:39:32     I       msg:[MAIN] PID: 5751
2013-02-26 18:39:32     I       msg:[MAIN] Loaded config from sc_serv_simple.conf
2013-02-26 18:39:32     I       msg:[MAIN] Calculated CPU count is 4
2013-02-26 18:39:32     I       msg:[MAIN] Starting 4 network threads
2013-02-26 18:39:32     I       msg:[MICROSERVER] Listening for connection on port 8300
2013-02-26 18:39:32     I       msg:[MICROSERVER] Listening for connection on port 8301
```

Figure 3.4 A running Shoutcast DNAS server.

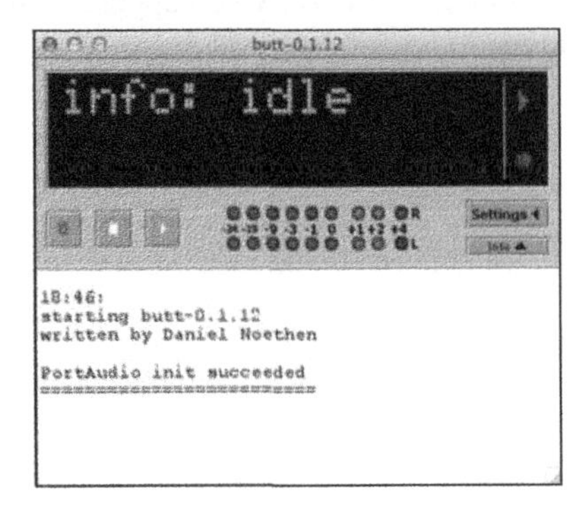

Figure 3.5 The broadcast using this tool (BUTT) user interface

BUTT is an extremely simple tool, but is also very reliable. There are many similar tools available. Clicking the "settings" tab opens the settings dialog (Figure 3.6).

This interface is very simple. However, it provides access to the key controls that will convert your analog audio feed into a live stream.

The first thing to do is to select your audio device from the drop down—and for this test, it is suggested that we simply select a microphone (assuming the laptop or headset has one). A user can select other sources such as the Line-In, and this would be the typical option if there were a sound source from an audio-mixing desk producing a feed that could be connected to the Line-In socket on the machine. For now, however, let us just use the microphone option.

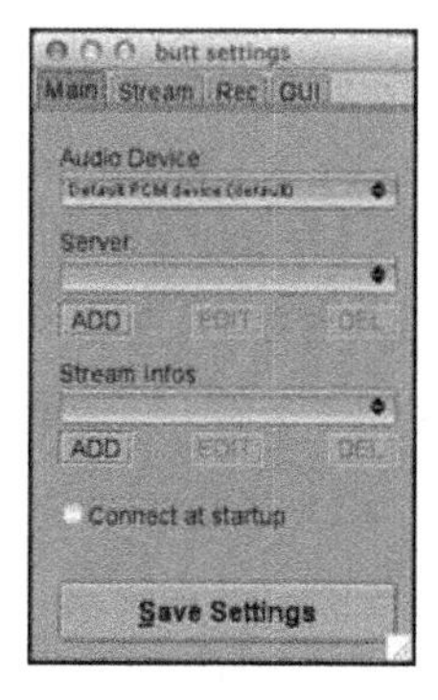

Figure 3.6 BUTT's settings interface.

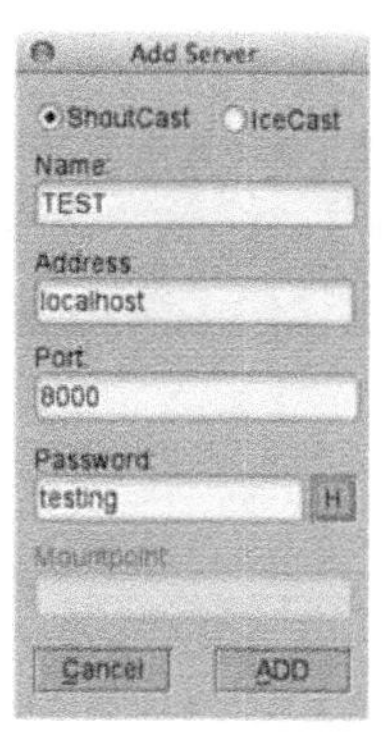

Figure 3.7 BUTT's server settings for the test.

The next thing to configure is the server. Since there is nothing listed in the drop-down list, we must click "ADD" to configure BUTT with our new Shoutcast server details as shown in Figure 3.7.

After clicking "ADD," these are submitted to BUTT's configuration record. There are a number of other configurations concerning Metadata (titles of the stream) that must be completed to provide the correct header data in the packets that get sent. These can be quickly set up as follows using "Test" or whatever details you choose (see Figure 3.8).

At this stage, after navigating back to the main window of the applications, all that remains is to press the play button. The feedback window will show some information and the outcome should look as in Figure 3.9.

So finally, we are producing a live stream. It can be seen that the VU Meter registers the sound level when you talk or make some noise.

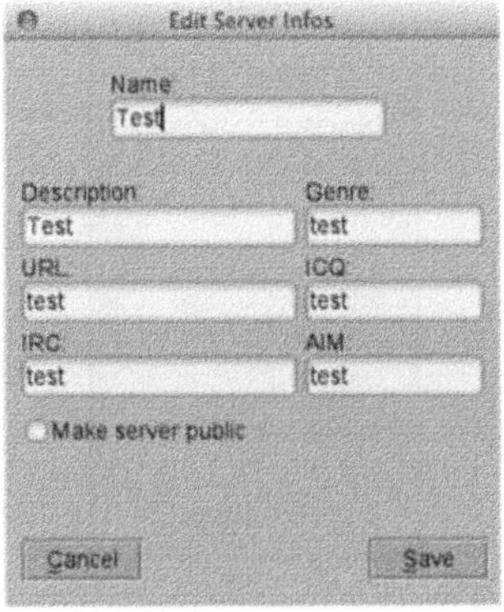

Figure 3.8 Stream metadata settings.

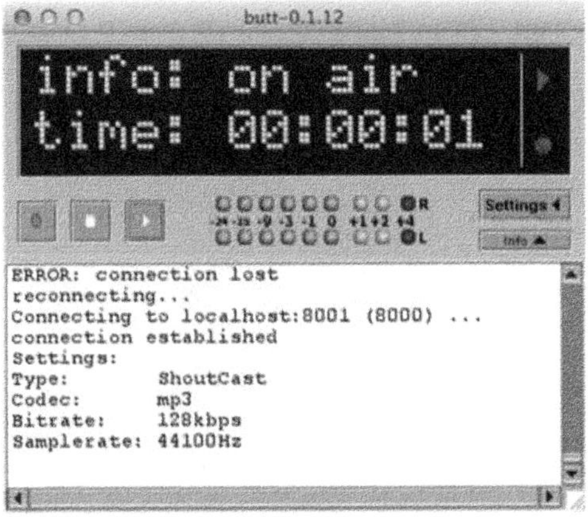

Figure 3.9 BUTT Live!

The final stage of our experiment is to hear the sound. Because HTML5 has now integrated native audio streaming into most browsers, Chrome or any HTML5 compliant browser can be used. Copy the following text into a text document and save it as a .html file (e.g., shoutcast.html):

```
<audio autoplay="autoplay" controls = "controls">
<source src="http://localhost:8000/stream/1/"/>
</audio>
```

and opening that with a browser will show the output as shown in Figure 3.10.

With some delay, voice will also be played back by the player. Congratulations, we have a working live audio stream.

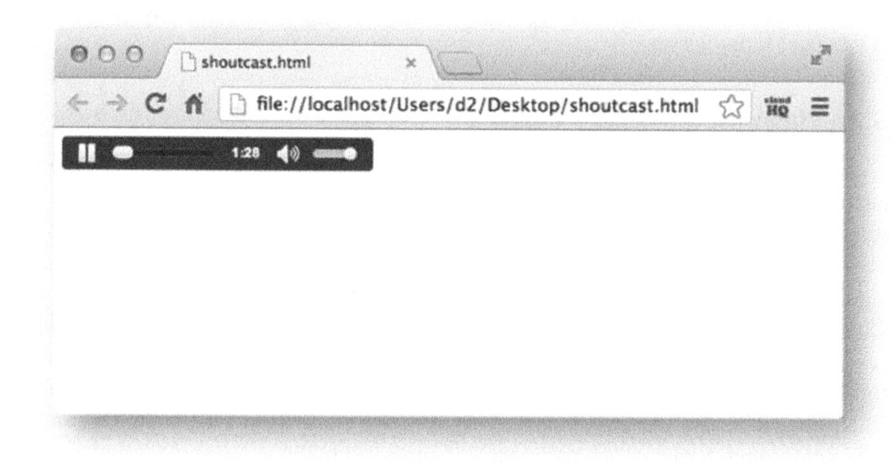

Figure 3.10 HTML5 native <audio> tag player streaming Shoutcast.

3.3.1.3 Testing Although the subject of IP routing and firewall management is vast and it is critical to live streaming, it is also beyond the scope of this book. If a user can open the firewall on his/her laptop to allow TCP traffic through port 8000, then providing the user knows his/her laptop's IP address, other devices on the user's land area network (LAN) will be able to also hear the stream if they open this Web page. Some firewalls are open within the LAN anyway so, as a quick experiment, try changing the text of the HTML file to replace "localhost" with the IP address of your laptop and then email that Web page to yourself (or pass it to your smartphone by some other means). You may just find that the stream works on your smartphone with no other modification! If it does, you are starting to build your first streaming radio station's distribution and audience reach!

Notice also that the encoder can be run on another machine and instead of "localhost" in the settings of BUTT, you could put the laptop IP address and BUTT could be located anywhere on the Internet—and so you begin to build your contribution workflow.

Also the stream URL will now work in other MP3-enabled streaming players ranging from Windows Media Player, through Winamp, to iTunes. Try them—each has an "open URL" option—enter the URL below to hear your stream, replacing "localhost" with your laptop's IP address:

```
http://localhost:8000/stream/1/
```

There are many ways to configure Shoutcast and it is a very simple, but powerful server to use. Read the configuration guides that come with it and that are widely available online. You will learn a tremendous amount about all aspects of streaming without overwhelming yourselves with too much of the additional complexity of video. The next experiment, however, will introduce you to video streaming using Flash Media Encoder and Wowza Media Server.

3.3.2 Flash Media—Live Video

The classic configuration for a live video stream always involves a live encoder, which transforms the source video and audio into a format suitable for online delivery. For this practical, we will be using the free Flash Media Encoder. The compressed stream of data is then delivered directly to the origin server (also called the acquisition server or the publishing server). The origin server's role is to ensure that it acts as a continuous point of reference for all the different distribution networks to locate the source stream that all the end-users are trying to consume. While in a very simple model the server may simply forward the streaming packets for distribution exactly as delivered by the encoder to each user, it is also possible that the origin server may "treat" the source signal to modifications and, for example, recode the compressed h.264 video and aac audio from a Real-Time Messaging Protocol (RTMP) transport into an MPEG-TS transport, and at the same time "fragment" the stream in chunks at each key frame, thereby creating options to deliver a single stream using MPEG-TS to a broadcast distribution network, while also producing an adaptive bitrate (ABR) stream that could be distributed using HTTP Live streaming (HLS) or Smooth streaming to Apple or Microsoft devices. For this practical, we will use Wowza since it is an interesting cross platform and versatile general-purpose media server, and a single source stream will be accessible on a great many devices.

We will demonstrate delivery to the HTML5 <video> tag, although highlighting the limitations of this, and also provide a pointer to the Wowza documentation outlining how to receive and decode the stream on alternative applications and devices.

3.3.2.1 Installing We will assume that the reader does NOT have separate machines and abundant bandwidth to attempt these experiments on, and so we will establish the entire workflow on a single PC. However, for clarity, the best way to run these tests would be over computers that were distributed remotely from each other—with perhaps the encoding on your laptop, the server on your desktop, and the decoding on your smartphone. We will also assume that you are using Mac OS or Windows; however, we will try to be generic about which versions. The experiments should work under most scenarios.

Let us start by downloading the key software that we will use for this experiment.

- Flash Media Live Encoder [2]: It is available for free from Adobe. It is a very common interface for webcasters to familiarize themselves with. Although it comprises a number of high end features, we will be using it with relatively few customizations. It will establish a connection with the source audio and video from the capture device available (we are assuming you have a webcam available on the test machine). Follow the installer for your OS.
- Wowza Media Server [3]: It is available with a 30-day trial. While Flash Media Live Encoder can also be used with the Flash Media server, Wowza is quick to

configure for many different outputs. Follow the installer for your OS. You will need to ask Wowza to email you a 30-day trial key as part of the download and install process.

- Safari browser [4]: As with the Shoutcast experiment, we will initially attempt delivery to HTML5 browsers. However, while Chrome offers good HTML5 <audio> support for <video> tags, this support currently extends only as far as HTTP Progressive Download (not ABRs such as HLS and Smooth), and for live streaming to Chrome or Internet Explorer it is recommended to use Flash or Silverlight players. The HLS output from the Wowza server can, however, be natively decoded from a <video> tag in Safari browser—not in the least because Safari and HLS are both Apple products. This instantly provides reach to iOS and Mac OS products since this works "out of the box," but to stream HLS live on Safari under some Windows installations, you may also need to install QuickTime (which adds some critical extra resources). For simplicity, for this practical, we will therefore use Safari to monitor our output.

3.3.2.2 Configuring Flash Media Live Encoder is extremely straightforward to configure. Figure 3.11 shows the configuration options.

Ensure the video and audio check boxes are enabled, and you have selected both a local video source (such as your webcam) and an audio source (such as your mic). Now select "h264" under the Video Format options and "mp3" from the Audio Format options. As this experiment is a local test on your PC or LAN, we do not need to adjust any of the other settings now—however, as you will be able to see that these all combine

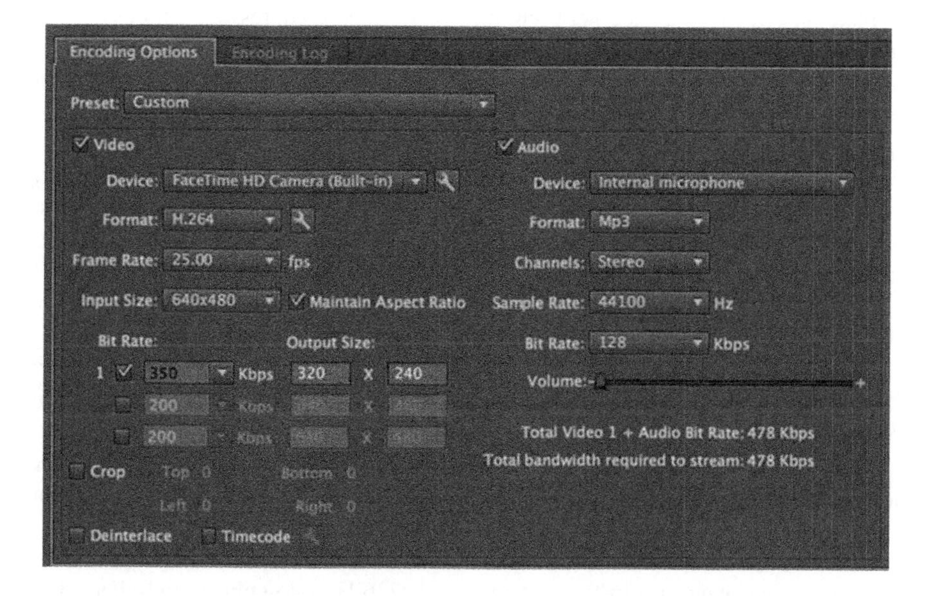

Figure 3.11 Encoding options of Flash Media Live Encoder.

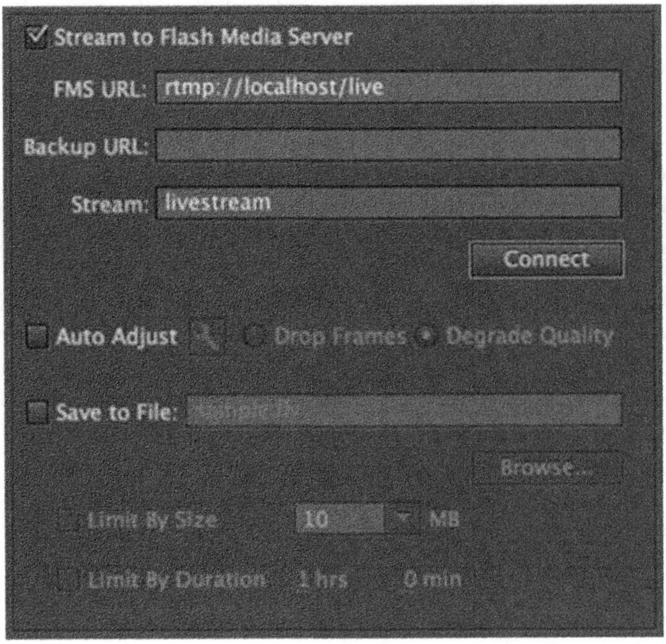

Figure 3.12 Connecting FMLE to the Wowza server.

to create an effective balance between audio and video compression (and associative quality loss) and the bandwidth utilization for the available link. When streaming from the encoder to a server over a wide area network (WAN) link, tuning these parameters becomes a fine art, and one that is attained by practice in many scenarios with many types of content.

The second part requires the local server details (see Figure 3.12). If you are working over a LAN or WAN link to connect the encoder to the server from a separate machine, then "localhost" would be replaced by the IP address of the server machine. For this test, the server is located on "localhost."

For now you will not be able to connect the encoder successfully—we must now setup the server and create an application ("live") for the encoder to connect to.

With the installation of Wowza, you will see some icons that start or stop Wowza. You will need to complete a licensing patch for your 30-day license, the very first time you run the server.

The next thing we must do is to prepare an application on the server that specifies how any incoming source stream from an encoder is to be authenticated and handled. For the purpose of this test, we will use the "live" streaming example that is provided with the Wowza installation. Full documentation of this configuration is here: http://www.wowza.com/forums/content.php?36-How-to-publish-and-play-a-live-stream-(RTMP-or-RTSP-RTP-based-encoder.

First, you must create an empty folder called "live" in each of the "applications" and the "conf" folders that are found in the Wowza Media Server's installation directory. The `applications/live` folder can be left empty, but a copy of `conf/Application.xml` file must be made and placed in the `conf/live` folder. Finally, this new copy of Application.XML must be edited (using a text editor) as follows:

- The <StreamType> property should be set to

```
<StreamType>live</StreamType>
```

- The HTTPStreamers property should be set to

```
<HTTPStreamers>cupertinostreaming,smoothstreaming,
sanjosestreaming</HTTPStreamers>
```

- The Stream/LiveStreamPacketizer properties should be set to

```
<LiveStreamPacketizers>cupertinostreamingpacketizer,
smoothstreamingpacketizer,sanjosestreamingpacketizer
</LiveStreamPacketizers>
```

- The RTP/Authentication/PlayMethod should be set to

```
<PlayMethod>none</PlayMethod>
```

With these edits, completely stop and restart the Wowza Media Server. At this point, you should now also be able to press "connect" under the server configuration options on the Flash Media Live Encoder. Providing these actions results in the indicator on the bottom left of the encoder software reporting "connected," you should also be able to "Start" the encoder, and if you do so while watching the Wowza terminal window and Encoder log, you will see the server confirming the inbound connection (Figure 3.13).

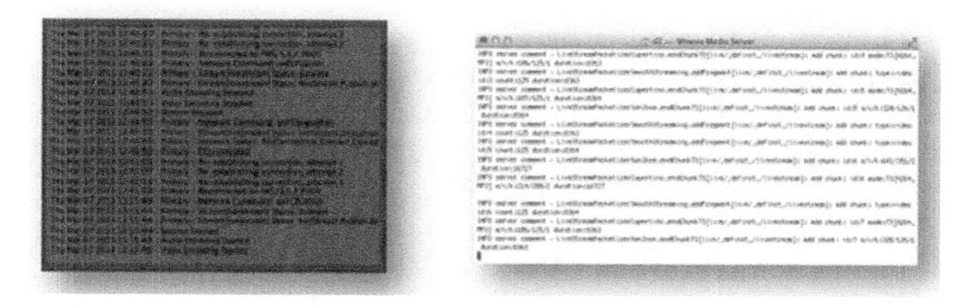

Figure 3.13 FMLE and Wowza "connected."

The final stage of the practical is to display the video. Although HTML5 provides support for <video> tags widely, this stream is a live stream and live streaming is a particular case for HTML5 <video> tags. We selected H,264 and so most of the major browsers will support h.264 decoding, and aac and mp3 are also commonly supported—however, the available transports from the Wowza server only offer HLS as a form that can be natively handled by HTML5, and only on Mac OS or iOS safari. For almost all other playback, the HTML5 <video> tag must include an embedded Flash player object or Silverlight player object as a fall back. Configuration for these can be extensive and is a chapter of a book in its own right, so for now we will just open the HLS stream in Safari.

Copying the following text into a text document, saving it as a .html file (e.g., videotag.html), and then opening it in Safari should produce a live video stream:

```
<video autoplay controls>
   <source src ="http://localhost:1935/live/livestream/playlist.
   m3u8" type="video/mp4">
</video>
```

Opening this in a Safari window should show the live stream.

3.3.2.3 Testing Because this stream is an h.264-encoded source with mp3 audio, it has significant device compatibility. Try changing "`localhost`" in `Videotag.html` to the IP address of the machine that Wowza is running on. Firewall issues aside, and assuming that your smartphone and Wowza server are on the same LAN, if you now email that VideoTag file to your smartphone (or transfer it some other way), then you will most likely see the same stream running on your smartphone. Since HTML5 is far from universally supported and mature, you will probably find exceptions to this, but both Android and iOS have means to support this type of HLS stream "out of the box."

However, you will quickly find that the HLS stream cannot be opened on many "normal" browsers. As mentioned earlier, this is a limitation of the browsers. The "fix" for this is to implement a Flash player that can be used in the place of HTML5 <video> tags.

There are many ways to do this: writing your own Flash player in Adobe's authoring tools, you can use "off-the-shelf" players such as FlowPlayer and JWPlayer to help you get the streams running in Flash players.

It is also worth noting that the Wowza Media Server installation includes a range of examples that work in varying ways on different browsers. Do test all of these: they can be found in the Wowza Media Server installation directory in the `examples/LiveVideoStreaming/` together with comprehensive `readme.html` documentation.

3.3.3 VLC IP Multicast on LAN

As discussed to some extent earlier, Multicast is an extremely efficient way to deliver a packet of data to a group of users. Essentially, you must think about what a broadcast is: the same data are sent once onto the airwaves or onto a wire, and everyone who is tuned in receives it. On a packet network, each person tunes in for a specific moment and collects data send specifically to it. It is a little like mail: each packet goes in, gets carried around, and then gets delivered. Interestingly, each packet gets broadcast on the wire, but only the specific recipient is tuned in at that moment.

IP Multicast works by asking all those who want to receive the live stream of data to ALL tune in at the moment the data are transmitted on the wire.

This means that if many people on the network want to watch the live stream, you only have to actually transmit it once and only to bits of the network where there are individuals or groups of users—making it a way to selectively broadcast to only parts of the network. Indeed, Multicast was originally called "Selective Broadcast" in Deering (grandfather of IP Multicast) and others early papers.

Let us use VLC to create and playback an IP Multicast. We want to highlight that the technical principle is probably more exciting than the result of this practical since the video quality requires considerable optimization even once the aim of this practical is completed and you have a live multicast running on your LAN.

3.3.3.1 Installing First download VLC from www.VideoLan.org—you will need a version suitable for your operating system and platform. For reference in this chapter, we are using VLC version 2.05 for Intel Mac. Ideally, install it on two machines on your home or office network. Although the experiment can be completed on one machine with two VLC instances running, given it as an experiment in network efficiency, it becomes more interesting to run this on a LAN with a number of machines.

3.3.3.2 Configuring Configuration for this practical is the easiest of the three, although perhaps conceptually the hardest to follow. First, we need to nominate one machine—with a webcam or some other video source—to be the "encoder and origin server."

Open VLC on this machine. In the File menu, click "Open Capture Device," and in the resulting window select "`Streaming/Saving`" and then click `Settings`. You should see the following dialog as shown in Figure 3.14.

Once you have clicked `Settings,` you will see the Streaming and Transcoding Options. Complete this as shown.

Ost of the configuration is self-explanatory, but the address is of particular importance. All IP addresses with a value of 224 or higher for their first number (e.g., 239.1.2.3) are reserved as Multicast addresses. You can simply choose the address at will, within the range 224.0.0.0 to 239.255.255.255. For a LAN, it is a good practice to use one starting 239 as the destination for your stream and for the group who want to join it.

The configuration is also setup for an MPEG-TS and transcoding the video to h.264 at 1.024 kbps and a stereo audio stream with 192 kbps of mp3 encoding on each channel.

Click `OK` and then click `Done`.

VLC should now simply capture the video from your webcam, compress it as shown in Figure 3.15.

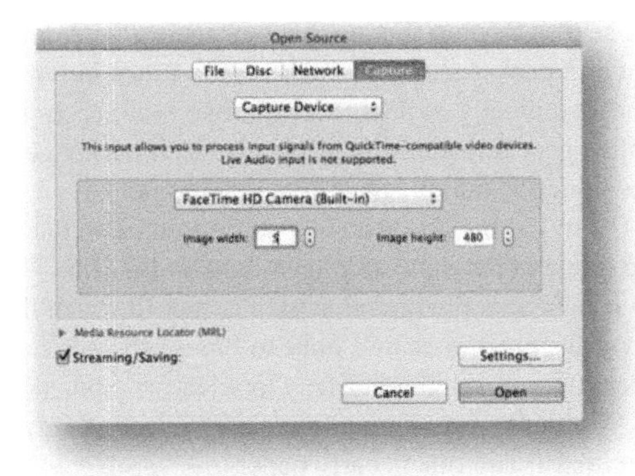

Figure 3.14 VLC capture device.

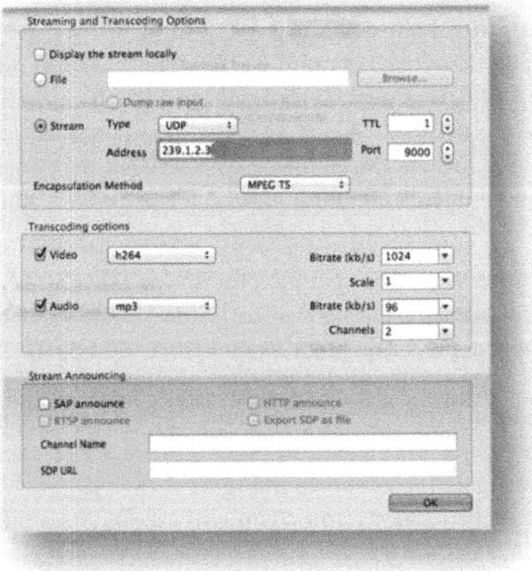

Figure 3.15 Configure the multicast.

h.264 acquires the audio from the audio device and compresses it using the mp3 CODEC, and these two compressed streams are then encapsulated in an MPEG-TS—which is tolerant to UDP transmission—and so we use the UDP stream packetizer to deliver the transport stream to the network addressed to 239.1.2.3 on an application port 9000.

Now we must add a listener—so open VLC on a second machine. Under the `File` menu, select `Open Network` and then complete the box as shown in Figure 3.16.

Figure 3.16 VLC Open network setting for client.

On clicking "Play," the stream will open and after a few seconds you will see live video.

3.3.3.3 Testing Testing the IP Multicast is interesting if you have an increasing number of machines on the LAN. Essentially you could add a relatively unlimited number of viewers of this live IP Multicast stream as you like, and the network or the encoding computer or the origin server will see any increase in load as you add more users. This system is a very basic model of how an IPTV system works.

During the setup of this, it was noted that the CODEC quality was poor and audio intermittent. The underlying implementation of the UDP and Multicast protocols is not good in VLC, which is a fun community project but generally lacks certain quality controls, and so do not be surprised if the stream is suboptimal.

VLC is a quick and easy hack to put up a multicast, and the resources it uses when it "gets going" are limited and do not get worse as you enlarge the audience. However, the fact is that VLC is not really a good tool for professional video streaming, and for Multicasting in an enterprise, Windows Media Server is probably the most successful next step for someone interested to roll out a stable and reliable IP Multicast in a large WAN or enterprise.

3.4 CONCLUSION

In this chapter, we have taken a broad practical look at the most basic steps and considerations one must make to prepare a live stream. There has been a consideration of the location of each of the major workflow components and a look at connectivity considerations.

The chapter then outlined a very basic way to setup a live Shoutcast Internet Radio stream with playback to HTML5 audio tags using the Chrome browser, and this was followed by a similarly rudimentary outline of a way to setup a live video stream using Flash Media Live Encoder, Wowza server, and specifically Safari Web browsers' HTML5 <video> tag.

As this last practical highlights, while HTML5 native live streaming is possible with some combinations of technology, even a limited amount of wider testing will reveal that multiple workflow strategies are required to reach the majority of viewers, particularly as the number of devices they want to use proliferates.

REFERENCES

1. Barix Inc. 2013. Product overview. Available at http://www.barix.com/products/exstreamer-family/barix/Product/show/exstreamer-100- 105/. Accessed 10 July 2013.
2. Adobe Systems, Inc. 2013. Flash Media Live Encoder. Available at https://www.adobe.com/cfusion/entitlement/index.cfm?e=fmle3. Accessed 15 July 2013.
3. Wowza Media Systems, Inc. 2013. Wowza Media Server. Available at http://www.wowza.com/pricing/trial. Accessed 15 July 2013.
4. Apple, Inc. 2013. Safari Browser. Available at http://www.apple.com/safari/. Accessed 18 July 2013.

4

EFFICIENCY OF CACHING AND CONTENT DELIVERY IN BROADBAND ACCESS NETWORKS

Gerhard Haßlinger

Deutsche Telekom Technik, Darmstadt, Germany

4.1 INTRODUCTION

Content delivery and especially video data transfers are the main drivers for an enormous Internet traffic growth over the past decades. The development of Internet Protocol (IP) traffic is visible in official statistics issued by administrative bodies of several countries, for example, for Australia [1], Germany [2], Hong Kong [3], and in *White Paper Series* by vendors of routing and measurement equipment, with most well-known reports by Cisco [4] and Sandvine [5]. In comparison of those studies, Figure 4.1 illustrates the common main trend, indicating traffic growth by a factor of the order 100 on a logarithmic scale for fixed IP network traffic over the past decade. Phases of steeper or slower increase rate can be distinguished in a detailed view of the curves, which often correspond to network deployment steps.

Traffic in mobile networks is included in dashed line curves, which currently accounts for only a few percent of the total IP traffic but is catching up at higher increase rates of 50–100% per year, whereas a range of 25–50% annual increase is observed in fixed networks. Measurement results on Deutsche Telekom's IP platform mainly confirm those global growth trends and also reveal major differences in the traffic mix within the daily load profile and in the upstream/downstream direction of the access [6],

Advanced Content Delivery, Streaming, and Cloud Services, First Edition.
Edited by Mukaddim Pathan, Ramesh K. Sitaraman, and Dom Robinson.

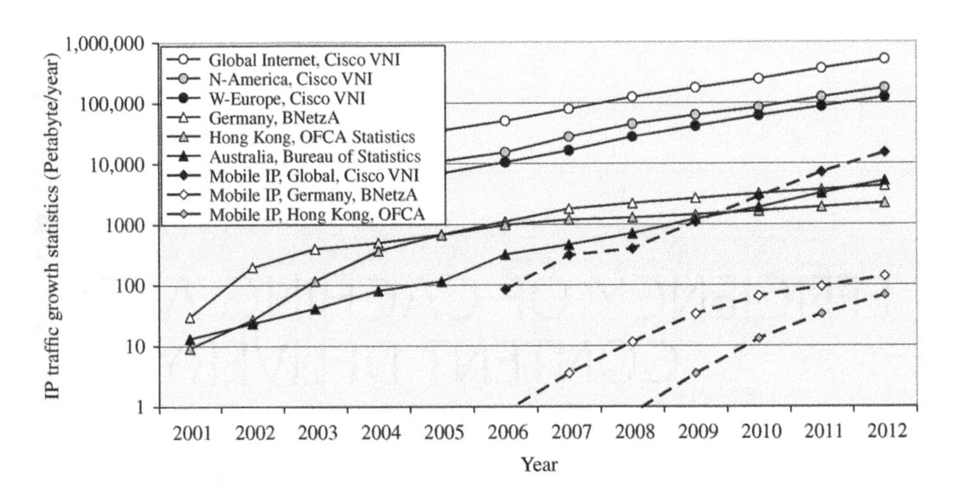

Figure 4.1 Trends in IP traffic growth reported from different sources.

where, for example, the HyperText Transfer Protocol (HTTP) traffic portion can be 10-fold higher in the busy hour in downstream direction than in the upstream at night time. Therefore, traffic statistics should clearly indicate

- if results are taken for busy hours or as daily mean values,
- which type of transport links (interconnection, core, aggregation, access, and local area network (LAN)), and
- which network domains (broadband access, corporate networks) are involved.

Such information is partly missing in reports summarized in Figure 4.1. The analysis of the application mix identifies video streaming and downloads as the currently dominant component of Internet traffic [4,7].

Traffic forecasts by Cisco also include a global cloud index [8], which estimates data center traffic and gives an impression of current and future relevance of content delivery networks. The results distinguish traffic from data centers to users, to other data centers and within data centers in a forecast until 2016. The reported volume of 438 exabyte for data center to user traffic in 2012 already covers over 80% of the complete global IP traffic, which is estimated at 520 exabyte [4]. This would mean that content delivery networks (CDNs) and data centers are dominant in content delivery over the Internet. The CDN provider Akamai announced that over 20% of today's Internet traffic is transferred on its CDN infrastructure [9].

Moreover, data center to data center traffic is estimated at about one-third of the global IP traffic and traffic within the data centers to be even fourfold higher [8]. Those ratios for traffic within and between data centers are almost constant in the forecast from 2012 to 2016, thus indicating high expenses for data center traffic management. Consequently, the benefits of content delivery from servers and caches close to the user do not come for free, being partly compensated by internal processing and a lot of cross traffic.

Nevertheless, there are main benefits of caching from the perspective of the content and network providers as well as the users that make a steadily extending CDN and data center infrastructure profitable to be seen in

- better scalability and adaptation to user demand for content delivery from multiple distributed servers avoiding bottlenecks of a single server with limited bandwidth,
- reduction of traffic in core networks and especially on expensive interconnection and peering links due to shorter transport paths,
- lower delay, higher availability, and throughput of content delivery for the users.

In this chapter, we investigate efficiency aspects of caching systems. After discussing applications and performance aspects of caching, in general, we start with a closer look on user request pattern for Internet content that can be characterized by Zipf laws with positive effect on cache hit rates. Our focus in the main part is on strategies to put the most relevant content into the cache. The influence of caching strategies on the hit rate is evaluated revealing large differences between simple approaches and more elaborate methods, which combine statistics on requests in the past with adaption capability to changing popularity of objects. Finally, we extend the scope to a broader discussion of traffic management and optimization via content delivery (CDN) and peer-to-peer (P2P) overlays on the network infrastructure.

4.2 OPTIONS AND PROPERTIES FOR WEB CACHING

Traditionally, there are two main application areas for caching [10]:

- Caching by operation systems of computers in fast storage in order to shorten delays in data access and
- Web caching in order to shorten transport paths.

In comparison of both areas, the management and performance evaluation of Web caching seem much more complex because of variable-sized, Internet-based content versus fixed size pages, variable benefit of Web caching depending on the transport paths from original sources and changing congestion conditions on those paths over time. Moreover, a number of preconditions have to be taken into account for Web caching as listed in Reference 11, including avoidance of stale content in caches after updates on original websites.

On the other hand, different parts of a network may be chosen for the positioning of caches. Distributed caching and CDN systems have been built up and further integrated into virtualized clouds or data centers as a common caching solution over many locations. There are several options for locating caching and CDN nodes in the broadband access network infrastructure of the Internet, in particular.

- Global content delivery networks
  Popular websites and service platforms are usually supported by a CDN with global footprint. The content platform and the CDN are partly operated under the same administration as for Google/YouTube or the content provider is supported as a customer of a global CDN provider. Large CDN providers have installed servers at many Internet exchange points worldwide with peering connections to network providers.
- ISP-operated CDNs
  Large ISPs and network providers often include caching systems and CDNs within their own infrastructure to make content available closer to the users [12]. Even if ISPs cannot reach global coverage, service for locally relevant content can also be important, especially in small countries and communities separated, for example, by language [13]. The efficiency of caches close to the user can benefit from local communities with their own preferences. In general, caching options within ISP networks have to be integrated into the steadily ongoing process of network planning and bandwidth upgrading in order to cope with increasing Internet traffic [14,15]. The overall goal is to maximize resource utilization in the operational network and thus to minimize capital expense (CAPEX) and operational expense (OPEX) including energy consumption while guaranteeing the required quality of service level and failure resilience.
  Caching was successfully introduced already before 2000 for HTTP-based Web transfers, even if the cache hit rate and thus the saving potential in transport capacity are limited by a nonnegligible amount of requests to one-timers, that is, objects on Internet servers that are addressed only once over long time spans [16] and further obstacles. After 2000, caching efficiency declined during a phase of dominant P2P traffic, which did not make use of Web caches, except for a caching option for eDonkey P2P file sharing via HTTP [13].
- Caches for load reduction on expensive links
  A special use case of Web caches is for reducing load on expensive links. When a network provider has to pay for the traffic volume over peering connections or has transcontinental links with limited bandwidth and high costs, then it is often profitable to install a caching system especially for those links.
- Client-side caches on the user equipment
  The role of caches on end systems should not be underestimated. A study on client-side caching [17] revealed that caches on terminals can save over 20% of user-generated traffic because of repeated requests of a user to the same Web pages over time, which then can be reloaded without noticeable delay. Therefore, it is important to provide information about the update frequency and expiry of Web content [11,18].

The different caching options are partly utilized in parallel under different administration and are independent of each other by global CDN providers, network providers, and end devices, although their effects are interrelated. Client-side caching may be considered almost independent of network caching, as the access profile of a single user determines its efficiency [17], whereas the aggregated access pattern of many users is relevant for network caches. On the server side, the content providers surely have an influence on the usage of client caches and can help to optimize caching of their content on end systems.

Caching by CDN and network providers is closely related, but administration boundaries are an obstacle for coordination of available caches in different domains. From the perspective of network providers, a lot of incoming traffic is delivered by a few over-the-top (OTT) providers, for example, via global CDNs by Google and Akamai. In addition, traffic from many other websites with small contributions also amounts to a considerable volume. Most of the latter traffic can be supported by transparent caching, whereas additional caching of traffic from external CDNs by network providers is not efficient without close coordination. Global content and CDN providers would lose control of the delivery and usage of their content when a transparent cache of a network provider would serve part of the users without even giving feedback, for example, for access statistics to the content provider.

Alternatively, large global CDN providers offer to extend their CDN into the network providers' broadband access infrastructure, for example, with Google global caches <https://peering.google.com/about/ggc.html>. However, network providers must involve a number of different caching systems to cover traffic from all relevant OTTs while losing control over traffic management and optimization in their network domain. Therefore, global CDN traffic is partly supported by network providers based on bilateral cooperation agreements, but often traverses broadband access networks via best effort peering. Currently, standardized support for CDN interconnection is considered by a CDNI working group at the Internet Engineering Task Force (IETF), but without focus on use cases relevant to network providers [19].

Finally, Web caches largely differ from caches in operating systems regarding access pattern. Operating systems often have highly correlated and periodical access pattern when data are periodically addressed in fixed or slowly varying sequences within programming loops. On the other hand, requests by a large user community to content on the Internet do not show periodical subsequences, but are better characterized by randomness in terms of independent and identically distributed requests to the available objects. Zipf laws have been observed for request distribution expressing the popularity scale of objects. Zipf laws have very favorable effect on the caching efficiency.

4.3 ZIPF LAWS FOR REQUESTS TO POPULAR CONTENT

The user behavior when requesting data and services on the Internet is following Zipf-like distributions, which concentrate a major portion of activity on a small subset of the most popular items. Such behavior makes it efficient to distribute popular data over caches and servers even of limited size and to deliver data from many sources that are located closer to the user. In extreme cases, the release of a single item of new software, for example, Windows or iOS updates, a video, or a news announcement of considerable volume can cause spontaneous traffic spikes when a large population is interested, as often observed in flash crowd events [20].

More generally, Zipf-like request distributions have been measured manifold on Internet platforms and in other scenarios, whenever a large population has access to a large set of objects, such as videos, books, or other goods. Social networks, links on Internet websites, and the topology of routing exchange points also indicate a strong relevance of a few statistically highest ranked nodes, such that a fraction below 10% of items attracts most of the access activity. According to a Zipf law, the number of requests $A(R)$ to an item in rank R in the order of highest popularity is given by

$$A(R) = aR^{-\beta}(a > 0; b > 0),$$

where the parameter $\alpha = A(1)$ is the maximum number of requests observed for an item in the statistics. Alternatively, the same term holds for the probability distribution or the fraction of accesses to the item in rank R if α is set to match a normalization constraint $\Sigma_R A(R) = 1 \Rightarrow \alpha = 1/\Sigma_R R^{-\beta}$. The exponent β determines the decay in access frequencies and thus the variance of the distribution, which increases with β. The fraction of accesses to top elements is growing with the exponent β. Accesses to the top 1% are varying in a range of 10–40%. A century ago, Vilfredo Pareto introduced a related form of distributions for property and income over the population, expressing major influence of a few extreme outliers on the entire mean value and the variance [5].

The relevance of Zipf laws has been confirmed in numerous case studies, for example,

- for page requests on popular websites by Breslau et al. [21] in measurement yielding a parameter range $0.64 < \beta < 0.85$,
- for access to popular content delivery platforms, including YouTube [22,23] with $\beta \rightarrow 1$, America Free TV [24], Amazon and P2P networks [25,26], and
- for cross-references on the Internet and in literature databases, for relationships in large social networks, or for the frequency of words in a long text [24].

On the other hand, studies on popularity also show deviations from Zipf distributions over the entire range. A good adaptation for the top ranked items goes on account of deviations for the mass of seldom accessed items [21,23,24,27]. The discrepancy between matching the top items and the mass can also be derived from the fact that the sum $\Sigma A(R)$ for $R = 1, \ldots, N$ does not converge for growing N. Therefore, including a very large number N of items would completely change the relevance of the top items such that their influence becomes negligible compared to the mass [6]. Therefore, alternative distributions have been proposed, for example, by Reference 28. In a study on accesses to IPTV channels [29], utility functions of different shapes are suggested with more parameters, yielding a better fit to include the popularity trend of the items over time.

4.4 EFFICIENCY AND PERFORMANCE MODELING FOR CACHES

The previously discussed Zipf law request behavior of users is a major factor in favor of caching efficiency. The most important performance measure for the performance of caching is the hit rate, that is, the fraction of requests that can be served from the cache [10,30]. Caches in data centers, CDNs, and other network domains can store only a very limited fraction of the content available on the Internet even if their storage capacity is in the order of terabytes.

We investigate selection methods for data to be stored in the cache. In order to obtain high hit rates, it is essential to put the most popular data into the cache. In principle, we can distinguish an initial filling phase followed by a replacement phase as soon as the cache storage is fully exploited. Network caches are located at nodes that forward aggregated data for a user population in broadband access networks. In the filling phase, any cacheable data requested by those users are passing by the cache node and may be copied into the cache if the copy process does not lead to relevant extra cost and load in the network. Afterwards, replacement strategies determine the remaining cache content, which are a classical research topic with applications also in operation systems and databases of computer and server systems [30]. We introduce a cache model for evaluating the hit rate based on the following assumptions (Independent Reference Model [10]):

- There is a set of N objects $O = (o_1, o_2, \ldots, o_N)$, which includes all cacheable data that are requested over a considered time period by the user population.
- The cache has a fixed size for storing M objects as a limited subset of all considered objects $M << N$.
- We assume that requests address the objects according to a random, independent and identically distributed (i.i.d.) sequence, such that the next request refers to an object o_k with probability p_k. Zipf distributions $p_k = \alpha k^{-\beta}$ are considered as the most relevant case, where the objects are ordered according to their popularity $p_1 \geq p_2 \ldots \geq p_N$. These assumptions characterize the case of static popularity.
- In general, popularity of objects may change over time in a dynamic popularity scenario. Then, different probability distribution $p_k(j)$ can be introduced for the jth access.

In the static popularity case, it is optimal to hold the M most popular items in the cache, since the hit rate h is then given by the sum of request probabilities for all items in the cache, which is maximal for caching the top M objects: $h_{\max} = \Sigma_{k=1}{}^{M} p_k$.

We use the achievable hit rate $h_{\max}$ for static popularity as performance criterion in comparison of cache replacement strategies. The request probabilities of objects are not known a priori but have to be estimated based on information on the request statistics of the past. When we count the requests of each object and put the objects with the highest count into the cache, then this least frequently used (LFU) replacement strategy converges to the optimum hit rate $h_{\max}$ over time.

On the other hand, a statistics over all requests in the past is not adequate for changing popularity of objects, since popularity of objects is partly decreasing over time, whereas new items are appearing spontaneously in top ranges from time to time. Therefore, refined caching strategies with limited memory over the past are preferable, which are proposed and evaluated in the next section. This also includes a closer look on the effect of dynamic popularity with regard to the cache hit rate as well as the influence of objects of different sizes. Since the impact of dynamic popularity and object size on the hit rate is finally estimated to be small for user requests to content on the Internet [22,25,26,31], we evaluate caching strategies based on the previous caching model with static popularity and unique size objects starting with least recently used (LRU).

4.5 EFFECT OF REPLACEMENT STRATEGIES ON CACHE HIT RATES

A broad variety of investigated cache replacement methods is classified and discussed in a survey [30], including the simple and most usual LRU principle. LRU replaces the item with the longest time span from a previous request out of the cache whenever a request addresses a new item not found in the cache. LRU caching corresponds to a stack with the requested object always being put on top.

A recent technology block [32] of the Dropbox platform for uploading/downloading and sharing content confirms that LRU is applied for caching in Android and iOS mobile Dropbox clients. Moreover, LRU is compared to alternatives of most frequently used (MFU) and LFU with evaluation of worst-case scenarios. Those scenarios refer to deterministic request sequences, where LRU degrades to zero hit rate for periodic access sequences over more than M different objects. However, such cases may be relevant to paging and other caching applications on a single computing system, whereas caches for a large population in broadband access networks with random Zipf-like requests distributions show completely different performance characteristics and worst-case scenarios as derived in the sequel.

4.5.1 Worst-Case LRU Deficits for Web Caches with Static Popularity

We illustrate deficits of LRU starting from the smallest cache size $M = 1$. Then LRU is most inefficient, if all items have negligible request probabilities except for the top item:

$$p_1 >> \varepsilon \geq p_2 \geq p_3 \geq \dots \geq p_N$$

For static popularity, the optimum hit rate $h_{\max} = p_1$ is achieved by keeping the most popular item constantly in the cache. With LRU strategy, the most recently requested item is in the cache of size $M = 1$ and the top item is found in the cache only with probability p_1 or otherwise, with probability $1 - p_1$, another seldom requested item. We obtain the LRU hit rate

$$h_{\text{LRU}} \leq p_1 \cdot p_1 + (1 - p_1) \cdot \varepsilon \approx {p_1}^2$$

Thus, the absolute deficit of the LRU caching hit rate can be as large as $h_{\max} - h_{\text{LRU}} \approx 25\%$ for $p_1 = 0.5$. The relative difference $h_{\text{LRU}}/h_{\max} \approx p_1$ can be even arbitrarily large for small p_1, for example, 1%/10% = 0.1 for $p_1 = 0.1$.

Next, we show that such a poor LRU performance can be encountered not only for $M = 1$ but also for arbitrary cache size. For larger caches of size M, we consider the worst-case LRU scenario for static popularity of objects under following assumptions:

- There is a set $I_{Pop} = \{o_1, o_2, \ldots, o_M\}$ of M popular objects with request probability p/M.
- On the other hand, negligible request probabilities ε ($\varepsilon << p$) are assumed for the remaining $N - M$ objects. Nevertheless, there are many of them and their request probabilities sum up to $1 - p$:

$$p_1 = p_2 = \ldots = p_M = p/M >> \varepsilon = p_{M+1} = p_{M+2} = \ldots = p_N$$

Then, again the optimum hit rate for static popularity is $h_{max} = p$ when all M popular items are kept constantly in the cache.

The LRU hit rate for network caches of size M has been investigated in several studies. Although an analytical formula is given, for example, in Reference 33, its exact evaluation is infeasible for caches of realistic size. Useful approximations have been derived instead [34,35]. However, in the previously defined worst-case scenario with only two object classes, an exact analysis of the LRU hit rate is still feasible by the simple iterative approach used in Reference 33. Our results in Figure 4.2 reveal that worst-case LRU hit rate deficits are even slightly increasing with cache size up to a maximum of 28.9% for $p \approx 58\%$. The range where LRU is 25% below the optimum is extending from $0.38 \leq p \leq 0.78$ when $\varepsilon < 0.005$ and $M \geq 5$.

4.5.2 LRU Hit Rates for Measured Traces of YouTube Video Requests

In addition to the worst-case analysis, we have evaluated LRU hit rates for requests to YouTube videos based on measurement traces made available and analyzed in many details by Cha et al. [22,33], see also Reference 36 for another YouTube caching study leading to similar results. Figure 4.3 compares LRU with optimum hit rates based on the

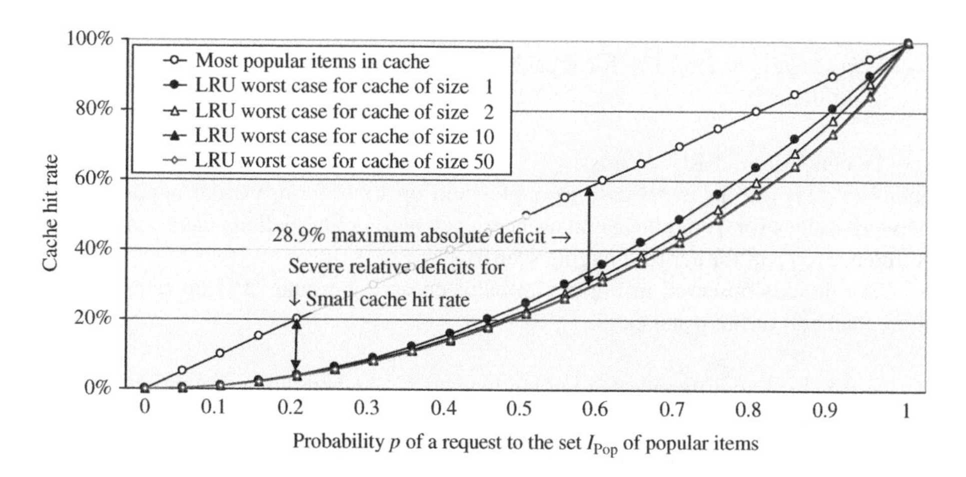

Figure 4.2 Worst-case LRU cache hit rate analysis for static popularity of objects.

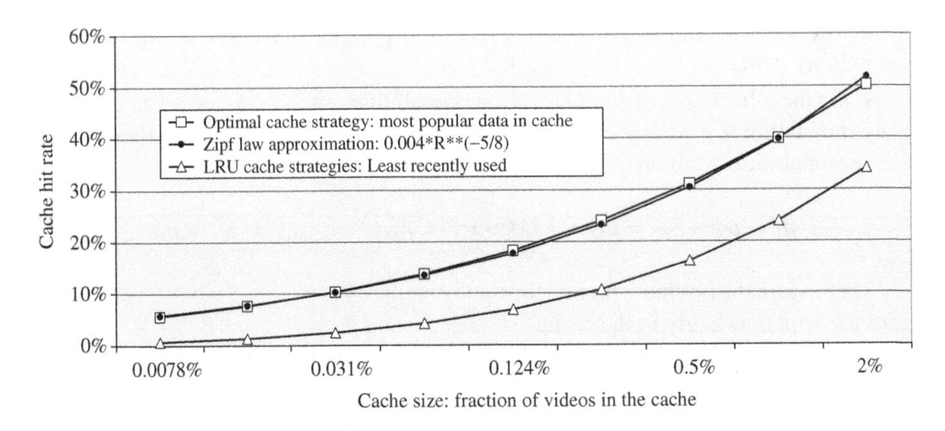

Figure 4.3 Hit rates depending on the cache size evaluated from a YouTube trace [22].

provided measurement trace for 3.7 billion requests to 1.65 million videos. In Figure 4.3 the curve for the optimum static strategy of holding the most popular videos in the cache includes hit rates for different cache sizes from 0.0078% to 2% of the 1.65 million videos, where 50% cache hit rate can already be achieved with only 2% of the videos being cached.

The measurement curve for the optimum static strategy is adapted by a Zipf law distribution $A(R) = \alpha R^{-\beta} = 0.004R^{-5/8}$ whose relative deviation does not exceed 3%.

The deficits observed in Figure 4.3 between optimum and LRU hit rates are still more than half of the worst case:

- The absolute gap for LRU is beyond 15% when the optimum hit rate is in the range of 30–50% and
- the relative differences become arbitrarily large for small caches, for example, $h_{\text{LRU}}/h_{\text{max}} \approx 2\%/10\%$ for a cache size of about 500, that is, 0.031% of 1.65 million.

In conclusion, LRU is missing a considerable portion of almost 30% of the achievable hit rate in worst-case examples and up to 15% in a usual application of network caches for YouTube. Similar experience of 10–20% deficits has been made in References 37, 38 for demand paging scenarios.

The deficits observed in Figure 4.3 between optimum and LRU hit rates are still more than half of the worst case:

- the absolute differences are going beyond 15% when the optimum hit rate is in the range of 30–50% and
- the relative differences become arbitrary large for small caches, for example, $h_{\text{LRU}}/h_{\text{max}} \approx 2\%/10\%$ for a cache size of about 500, that is, 0.031% of 1.65 million.

In conclusion, LRU is missing a considerable portion of almost 30% of the achievable hit rate in worst-case examples and still >10% in a usual application of network caches for YouTube. Similar experience of 10–20% deficits has been made in [37,38] for demand paging scenarios.

4.5.3 Limited Influence of the Object Size on Web Cache Performance

The influence of the size of objects on caching performance is limited and often secondary. In principle, caching of a large object may be treated as caching of j objects of unit size of, for example, 1 MB, where all j unit objects have the same popularity rank and score as the original object. A large object requires j-fold storage but the delivery from the cache also saves j-fold amount of traffic. In this regard, the benefit from caching does not depend on object size but mainly on the hit rate, with the exception of the following aspects:

- In principle, bin-packing problems are relevant when several objects of different sizes can be put into the cache with lowest score or priority rank, and when only a subset of combinations of those objects fits into the cache. But usual cache sizes are in the terabyte range and thus large enough to store more than 1000 objects. Then, bin-packing problems are almost negligible for the total hit rate, since only a few more objects can be included by fitting optimizations.
- Data delivery from the cache involves some overhead depending on the performance and load of the cache server. The overhead is relatively higher for smaller items. Therefore, caches may introduce a minimum object size and refuse storing smaller objects, since not much traffic can be saved by supporting small data flows. On the other hand, the original website may have the same overhead and longer delay to deliver a small object. Therefore, signaling, Domain Name Servers (DNS) and other control messages usually get special quality of service (QoS) support including caching in order to speed up the start phase of IP transport services even if they are of small size.

In general, more aspects may be considered in the decision on the objects to be put into and served from a cache, including information about the cost saving as compared to delivery from another available source, as well as other content- or even user-specific properties.

4.6 REPLACEMENT METHODS BASED ON REQUEST STATISTICS

The LRU hit rate deficits make it worthwhile to consider cache replacement strategies including more detailed statistics on past accesses. An adaptive step in this direction is proposed in Reference 38 by subdividing the cache space into two parts: one being managed by LRU and the other based on more precise information from statistics on past requests. A survey on cache replacement strategies briefly presents variants in this category, which are denoted as frequency/recency-based methods in Reference 30. We study two basic alternatives of cache replacement strategies using statistics with limited memory of the past [33]:

- Sliding window:
  The cache holds those items that have the highest request frequency in the sliding window of the last K requests.
- Geometrical aging:
  The cache holds items with the highest sum of weights for previous requests, where the kth request in the past is assigned a geometrically decreasing weight $\rho^k (0 < \rho < 1)$.

Both strategies provide one parameter, that is, the size K of the sliding window and the aging factor ρ, which determine the relevant backlog of the memory into the past. For $K \le M$, the window statistics is restricted to those objects that are also in an LRU cache, whereas all other objects have zero frequency count. This leaves most of the replacement strategy open and requires an additional tie-breaking decision to determine the item to be evicted out of many with the same zero score, where LRU is a useful alternative. Geometrical aging with $\rho \le 1/2$ is also equivalent to LRU, since then the last weight is dominant and thus puts the recently requested object on top.

On the opposite extreme, both sliding window with $K \to \infty$ and geometrical aging with $\rho \to 1$ approach an infinite count statistics over the past. In case of a static access distribution, they converge to the optimum strategy of placing the most popular items into the cache. When the cache is filled at the start without much backlog information being available, the short-term statistics may deviate from long-term behavior.

In practice, the popularity of items is slowly changing [22,25]. Then, an unlimited statistics has the disadvantage to keep those objects in the cache, which have been very popular and stay within the first M ranks in the long-term statistics, even when they recently degraded in popularity and should be removed. Therefore, a finite backlog is mandatory to make a cache capable of responding to changing popularity. A window of size K stands for a fixed backlog, whereas geometrical aging implicates a continuously declining influence of the past, where the sum $\Sigma_{j \ge 1} \rho^j = \rho/(1-\rho)$ determines the mean backlog. Thus, we have to set $\rho = K/(K+1)$ for a corresponding mean backlog of K in geometrical aging. The parameter K for the number of request can be converted into a time span when the number of requests per hour, day, or month is known. For network caches serving a large population, thousands or even millions of requests have to be included in a one-day window.

4.6.1 Implementation Effort of Replacement Methods

4.6.1.1 Least Recently Used A main advantage for the widely applied LRU strategy is the simple and efficient implementation as a double-chained stack of M objects. An update per request is done at a constant effort for putting the requested object on top of the LRU stack.

4.6.1.2 Sliding Window The sliding window statistics requires some more, but still constant updating effort per new request. The count of the requested object is incremented and the count of another object is decremented, corresponding to the last request that is dropped from the sliding window. The list of cached objects eventually has to be reordered according to both modified request counts. As a main additional implementation effort, the sliding window has to be stored as a cyclic list of size K, in order to determine the oldest request in the sliding window and to overwrite it by the recent request.

4.6.1.3 Geometrical Aging On first glance, an update per request for geometrical aging has to increment the weight of the new requested item and to multiply the aging factor ρ to the weights of all items, which would mean an enormous $O(N)$ effort. But instead of decreasing the weight of all other items by a factor ρ, we can equivalently increase the weight only for the new requested item by a reciprocal factor $1/\rho$ each time, that is, the weight to be added for a new request is initiated with 1 and then is growing to $(1/\rho)^k$ after k requests. This procedure has the same aging effect due to an inflation of weights for new requests. The only problem is that the weight will approach the maximum number in floating point representation over time. Therefore, the weights have to be readjusted by dividing them by $(1/\rho)^k$ when a threshold is reached. In this way, the updating effort for geometrical aging is reducing to the order $O(\log(M))$ for reinserting the requested object at an upward position in the cache.

The survey [30] refers to another ρ-aging method, where the scores of all items are reduced by a factor ρ not for each request but in longer time intervals, for example, once an hour, thus essentially reducing the effort per request.

4.6.2 Influence of Dynamic Popularity on Cache Hit Rates

Except for the YouTube measurements in Figure 4.3, a static constant popularity based on request probabilities of objects is assumed, although the popularity of cached objects is changing over time. In principle, such changes make future requests less predictable and reduce the cache efficiency with exceptions of periodical and anticipated changes that may be efficiently handled by prefetching.

The popularity of content on the Internet is usually classified as a two-phase up-and-down going development. New objects are rising in popularity until a maximum is reached and popularity is fading away afterwards. The initial increase is often steep up to flash crowd behavior [20], which ends at a peak level that is often held almost constant for some time, followed by a smooth decrease [22,25,31].

However, those phases of changing popularity are observed on the timescale of days, weeks, or months [25,26]. The measurement study [26] on file popularity in the Gnutella network shows only 1–3% of change in the top 100, top 1000, and top 10,000 files within a daily drift. If a cache keeps the request statistics over a day in a sliding window for the next day, then 1–3% change regarding the top M objects would probably lead to only a small degradation in the hit rate. The popularity profile is then partly predictable from the age of an object based on past request statistics development. Such prediction can be exploited to improve the proposed frequency/recency replacement strategy.

4.6.3 Prefetching

Prefetching of data is another option to improve cache efficiency due to foreseeable changes in popularity [39]. Prefetching is especially useful when new content is announced or advertised to be published at a certain date, and then can be distributed earlier on CDNs and data centers. For IPTV applications and media centers, most of the preproduced program for the next day can be put on caches in advance.

Prefetching can have different effects on the traffic load: in the worst case, bandwidth and storage may be wasted when prefetched content is not requested as expected, whereas systematic prefetching outside of peak traffic periods has smoothing effect on traffic profiles and improves resource utilization.

4.7 GLOBAL CDN AND P2P OVERLAYS FOR CONTENT DELIVERY

We complement the study on efficiency aspects of single caches with a broader perspective on content delivery on the Internet across network domains and protocol layers.

During the last decade, P2P networks and global CDNs were carrying major portions of Internet traffic as overlays of different types [4,13]. P2P systems interconnect the terminals of their users involving only a minimum of own network and server infrastructure, whereas CDNs are based on globally distributed servers whose connectivity can be supported in a virtual network. Both approaches introduce their own network and traffic management functions within the overlay, which can be organized independently of the network layer as a main advantage for development of new overlay services in P2P systems or clouds.

Nonetheless, the efficiency of CDN and P2P overlays depend on the appropriate embedding and cross-layer awareness of functions from the application down to the infrastructure level. P2P and CDN overlays have essentially improved content delivery over a single server hosting of content. Both approaches distribute the content and make it available from many sources in order to avoid bottlenecks when content is spontaneously becoming popular with requests from a large user population. The applicability of CDN and P2P overlays is not restricted to Web browsing and file transfer, but includes most of the broadening spectrum of Internet services especially when large data volumes are transferred for video streaming and IPTV [40].

P2P networks are known for their highly scalable adaption of source capacity on peers to the user demand and their ability to follow flash crowd popularity in short time with exponentially growing data exchange rate between peers. In this way, P2P networks also generate a huge traffic volume that was dominant for several years and still is relevant in Internet traffic statistics. Measurement of the ports on transport layer shows an increasing and currently dominant portion of traffic via the HTTP protocol [19]. On the other hand, P2P sources for downloading are chosen more or less randomly and cause unnecessary long transport paths to the destinations.

CDNs are more efficient in shortening transport paths on a global scale, but are not optimizing over several network domains separated by administrative boundaries. Investigations of transfer paths in Akamai's CDN [9,41] show how users are redirected from the original website through a hierarchical server farm consisting of thousands of servers to a delivery node in the proximity of the client. In this way, the reliability and user experience are improving. The CDN connection to a user can even be handed over to another server through prepared backup paths if performance measurement indicates problems on the current path or for load balancing in the CDN server cluster. The infrastructure of global CDNs is currently further developing toward virtualized cloud systems offering services to content providers and users.

The study by Su et al. [39] confirms that CDNs are efficient in abbreviating transport paths and delays while also stabilizing throughput. In measurement on aggregation links on Deutsche Telekom's broadband access network, we studied round trip times in the handshake at the start of Transmission Control Protocol (TCP) connections [6]. In a comparison of transmissions for large P2P networks and for downloads supported by global CDNs, we experienced that the mean round trip times are 2.5-fold longer in P2P networks. Moreover, 10% of P2P transmission round trip times is longer than 1 s and therefore inappropriate for real-time applications.

Challenges of cross-layer and cross-domain inefficiencies for P2P as well as CDN overlays are addressed in two working groups at the IETF on application layer traffic engineering (ALTO) and CDN interconnection (CDNI) [19]. The ALTO working group is focused on avoiding unnecessarily long transport paths for P2P and other applications by providing information on content sources in the near of a requesting user via cooperative servers. The CDNI working group deals with standardized data and control information exchange between CDNs and caching infrastructures, which are involved in an end-to-end transport path crossing different administration domains.

4.7.1 Combined CDN, Caching, and P2P Overlay Approaches

While CDN and P2P distribution schemes can be seen as competitive alternatives in current and future Internet architectures, combined CDN–P2P content delivery is a promising way to exploit the advantages of both principles [24,40,42,43]. CDNs can provide content locally at servers in different regions, while an additional P2P overlay including CDN servers as primary P2P sources can provide more throughput, which is scalable with the user demands and helps to save bandwidth for the CDN connectivity.

Some streaming services have implemented and evaluated hybrid CDN–P2P content distribution networks, which are flexible and profitable especially for regions with high bandwidth costs for CDN server connections [43]. Several CDN solution and service providers have integrated hybrid CDN–P2P approaches. The industry forum on digital video broadcasting also addresses hybrid CDN–P2P delivery technology in current reports with references to related projects and providers [40].

4.7.2 Caching in Fixed and Mobile Networks

Two main differences are obvious in comparing caches in fixed and mobile networks [44]:

- Traffic volumes in mobile networks are still essentially smaller than in fixed networks.

- The access bandwidth of mobile users via air interfaces is more expensive than in fixed access networks.

Curves for the IP traffic development in fixed and mobile networks in Figure 4.1 indicate that mobile traffic currently amounts for less than 5% of the total IP traffic, although with increasing tendency because of essentially higher growth rates, currently driven by fourth-generation LTE deployment. As a conclusion, traffic reduction due to caching is mainly relevant for fixed networks and can be handled separately for mobile networks.

Moreover, applications and popular content on mobile networks partly differ from fixed networks, although high volume downloads for mobile devices, for example, software upgrades are often off-loaded via fixed and wireless local area network (WLAN) access points at higher speed. Therefore, typical high volume mobile network traffic is partly transferred and relevant for fixed networks. Fixed/mobile convergence also shows opposite trends of LTE deployment with hybrid access in rural areas, which are not adequately supplied with fixed network capacity. In those areas, fixed network traffic profiles and user behaviors are expected to be transferred to wireless, although not to mobile networks.

The cost-saving effect of CDN and caching in mobile networks is reduced by the fact that air interfaces are an expensive and bandwidth-limiting resource. Caches on the user equipment remain as the only way to reduce the required transmission capacity even on the air interface. Unfortunately, storage and computing power are often limited on mobile devices, but should be fully exploited, since investigations indicate a 20% saving potential for client-side caching with enough storage on PCs and laptops [17].

4.8 SUMMARY AND CONCLUSION

Distributed server and caching systems for delivery of the steadily increasing volume of content on the Internet have opened new potential for traffic engineering and QoS support in terms of scalability, avoidance of bottlenecks, shorter transport paths, and delays. The efficiency of caching is promoted by a Zipf law concentration of requests from large user populations on a small fraction of highly popular content and services, which make deployment of small caches profitable.

In the main focus of this study, we evaluate the impact of cache replacement methods on the cache hit rate. For Web caches, we identify the maximum hit rate for static popularity in the independent reference model as most relevant, although the caching strategies must adapt to dynamic changes in the set of relevant objects at a low or moderate pace in the timescale of days or months.

The evaluations show substantial deficits of the LRU caching principle of up to 28.9% hit rate reduction below optimum in the worst case and still 10–15% hit rate deficit for caching of YouTube videos in measurement traces of user requests. Therefore, it seems worthwhile to take some more expenditure in caching strategies for including count statistics on user requests to objects in the past, which approach the optimum hit rate for static popularity on account of higher implementation and updating effort.

We propose the sliding window and geometrical aging strategy to obtain statistics over a limited time horizon into the past to make them adaptive to dynamically changing popularity. Their implementation effort seems affordable and still has optimization potential for further study confirming results for similar variants in the literature.

From the perspective of network providers, caching is becoming more attractive since the P2P traffic portion is decreasing while the HTTP traffic portion again has become dominant over recent years. Caches can be efficient in different network domains, for example, for expensive peering links or in the aggregation. Moreover, client-side caching on user equipment and in nanodata centers should be fully exploited as an ideal option to completely avoid part of the data transfers.

ACKNOWLEDGMENTS

This work has been performed partially in the framework of the European EU ICT STREP project SmartenIT (FP7-ICT-2011-317846) <www.smartenit.eu> as well as the MEVICO project founded by the German Federal Ministry of Education and Research (BMBF).

REFERENCES

1. Australian Bureau of Statistics. 2013. Pages on Internet activity.
2. Bundesnetzagentur (BNetzA). 2012. Annual reports on statistics and regulatory activities (in German). Available at www.bundesnetzagentur.de.
3. Office of the Communications Authority (OFCA) of the Government of Hong Kong. 2013. Statistics of Internet traffic volume. Available at www.ofca.gov.hk/en/media_focus/data_statistics/internet/.
4. Cisco Systems. 2013. Cisco visual networking index, white paper series. Available at www.cisco.com.
5. Reed WJ. The Pareto, Zipf and other power laws. Econ Lett 2001;74(1):15–19.
6. Haßlinger G, Hartleb F. Content delivery and caching from a network provider's perspective. Special Issue on Internet Based Content Delivery. Comput Netw 2011;55:3991–4006.
7. Sandvine Inc. 2012. Global Internet phenomena report, white paper series. Available at www.sandvine.com.
8. Cisco Systems. 2012. Cisco global cloud index, forecast and methodology 2011–2016, white paper. Available at www.cisco.com.
9. Akamai. 2013. State of the Internet. Quarterly Report Series. Available at www.akamai.com.
10. Bahat O, Makowski AM. Optimal replacement policies for non-uniform cache objects with optional eviction. Proceedings of IEEE Infocom; San Francisco, CA, USA; 2003. http://www.abs.gov.au/ausstats/abs@.nsf/Lookup/8153.0/. Accessed 25 June 2013.
11. Fielding R, Nottingham M, and Reschke J. 2013. Hypertext transfer protocol HTTP/1.1: Caching, Internet-draft, work in progress. Available at tools.ietf.org/html/draft-ietf-httpbis-p6-cache-22. Accessed June 25, 2013.

12. Kamiyama N, Mori T, Kawahara R, Harada S, and Hasegawa H. ISP-operated CDN. 14th NETWORKS Telecom. Network Strategy & Planning Symposium; Warszawa, Poland; 2010.
13. Haßlinger G. ISP platforms under a heavy peer-to-peer workload. In: Steinmetz R, Wehrle K. Proceedings of Peer-to-Peer Systems and Applications; Springer LNCS 3485; 2005. p 369–382.
14. Haßlinger G, Nunzi G, Meirosu C, Fan C, Andersen F-U. Traffic engineering supported by inherent network management: Analysis of resource efficiency and cost saving potential. Int J Network Manage 2011;21:45–64.
15. Haßlinger G, Schnitter S, Franzke M. The efficiency of traffic engineering with regard to failure resilience. Telecommun Syst 2005;29(2):109–130, Springer.
16. Williams A, Arlitt M, Williamson C, Barker K. Web workload characterization: Ten years later. In: Tang X, Xu J, Chanson ST, editors. *Web Content Delivery*. Springer; 2005. p 3–22.
17. Charzinski J. Traffic properties, client side cachability and CDN usage of popular web sites. Proceedings of 15th MMB Conference; Essen, Germany; Springer LNCS 5987; 2010. p 182–194.
18. Fielding R, Gettys J, Frystyk H, Masinter L, Leach P, and Berners-Lee T. 2000. Hypertext transfer protocol - HTTP/1.1, Request for Comments 2616. Available at www.rfc-editor.org/rfc/rfc2616.txt. Accessed 26 June 2013.
19. Internet Engineering Task Force (IETF). 2013. Working group on application layer traffic optimization (ALTO)/Working group on CDN interconnection (CDNI)/Working group on HTTPbis. Available at tools.ietf.org/wg/alto/charters/ tools.ietf.org/wg/cdni/charters/tools.ietf.org/wg/httpbis/charters. Accessed 26 June 2013.
20. Jung J, Krishnamurthy B, Rabinovich M. Flash crowds and denial of service attacks: Characterization and Implications for CDNs and websites. Proceedings of 11th World Wide Web Conference, Honolulu, Hawaii, USA; 2002.
21. Breslau L, Pei C, Li F, Phillips G, and Shenker S. Web caching and Zipf-like distributions: Evidence and implications. Proceedings of IEEE Infocom; 1999.
22. Meeyoung Cha, Haewoon Kwak, Pablo Rodriguez, Yong-Yeol Ahn, and Sue Moon. I tube, you tube, everybody tubes: Analyzing the world's largest user generated content video system, Internet measurement conference IMC'07; San Diego, USA; 2007.
23. Gill P, Arlitt M , Li Z, Mahanti A. YouTube traffic characterization: A view from the edge, Internet measurement conference IMC'07; San Diego, USA; 2007.
24. Eubanks M. The video tsunami: Internet television, IPTV and the coming wave of video on the Internet, Plenary talk, 71. IETF Meeting; 2008. Available at www.ietf.org/proceedings/08mar/slides/plenaryt-3.pdf. Accessed June 27 June 2013.
25. Raffaele Bolla, Mirko Eickhoff, Krys Pawlikowski, and Michele Sciuto. Modeling file popularity in peer-to-peer file sharing systems. Proceedings of 14th ASMTA Conference; Prague, Czech Republic; 2007. p 149–155.
26. Zhao S, Stutzbach D, Rejaie R. Characterizing files in the modern Gnutella network: A measurement study. SPIE/ACM Proceedings of Multimedia Computing and Networking; 2006.
27. Haßlinger G, Hartleb F, Beckhaus T. User access to popular data on the Internet and approaches for IP traffic flow optimization. Proceedings of ASMTA; Madrid, Spain; Springer LNCS 5513; 2009. p 42–55.
28. Lei Guo, Enhua Tan, Songqing Chen, Zhen Xiao, and Xiaodong Zhang. Does Internet media traffic really follow Zipf-like distributions? ACM Sigmetrics; 2007.

29. Kandavanam G, Botvich D, Balasubranmaniam S. PaCRA: A path-aware content replication approach to support QoS guaranteed video on demand service in metropolitan IPTV networks, IEEE/IFIP Network Operations & Mgmt Symposium NOMS; 2010. p 591–598.
30. Podlipnik S, Böszörmenyi L. A survey of web cache replacement strategies. ACM Comput Surv 2003:374–398.
31. Cohen B. 2003. Incentives build robustness in BitTorrent. Available at bitconjurer.org/BitTorrent/bittorrentecon.pdf. Accessed July 7, 2013.
32. Panchekha P. Caching in theory and practice, Dropbox Tech Block. 2012. Available at https://tech.dropbox.com/2012/10/caching-in-theory-and-practice. Accessed 07 July 2013.
33. Haßlinger G, Hohlfeld O. Efficiency of caches for content distribution on the Internet. Proc. 22. Internat. Teletraffic Congress; Amsterdam, The Netherlands; 2010.
34. Dan A, Towsely D. An approximate analysis of the LRU and FIFO buffer replacement schemes. SIGMETRICS Perform Eval Rev 1990;18:143–152.
35. Fricker C, Robert P, Roberts J. A versatile and accurate approximation for LRU cache performance. IEEE Proceedings of 24th International Teletraffic Congress; Kraków, Poland; 2012.
36. Braun L, Klein A, Carle G, Reiser H, and Eisl J. Analyzing caching benefits for YouTube traffic in edge networks: A measurement-based evaluation. IEEE Network Operations & Management Symposium; Maui, Hawaii, USA; 2012.
37. Janapsatya A, Ignjatovic A, Peddersen J, Parameswaran S. Dueling CLOCK: Adaptive cache replacement policy based on the CLOCK algorithm. IEEE Proceedings of the Conference on Design, Automation and Test, Dresden, Germany; 2010. p 920–925.
38. Megiddo N, Modha DS. ARC: A self-tuning, low overhead replacement cache. Proceedings of the 2nd USENIX conference on file and storage technologies; San Francisco, CA, USA; 2003. p 115–130.
39. Antonis Sidiropoulos, George Pallis, Dimitrios Katsaros, Konstantinos Stamos, Athena Vakali, and Yannis Manolopoulos. Prefetching in content distribution networks via web communities identification and outsourcing. World Wide Web 11; Springer; 2008. p 39–70.
40. Digital Video Broadcasting Project (DVB). 2009. Internet TV content delivery study mission report, DVB Document A 145. Available at www.dvb.org/technology/standards/A145_Internet_TV_Content_Delivery_Study.pdf. Accessed July 7, 2013.
41. Su A-J, Choffnes DR, Kuzmanovic A, Bustamante FE. Drafting behind Akamai. IEEE/ACM Trans Netw 2009;17:1752–1765.
42. Huang C, Wang A, Li J, Ross K. Understanding hybrid CDN-P2P. Proceedings of NOSSDAV; Braunschweig, Germany; 2008. p 75–80.
43. IRTF-75 Presentation. 2009. P2P live streaming for the masses – deployment and results. Stockholm: P2P Research Group. Available at www.ietf.org/proceedings/75. Accessed on 8 July 2013.
44. Eisl J Haßlinger G. The effect of caching for mobile broadband Internet access (in German). Proceedings of Mobile Communication Conference; Osnabrück, Germany; 2012.

5

ANYCAST REQUEST ROUTING FOR CONTENT DELIVERY NETWORKS

Hussein A. Alzoubi[1], Michael Rabinovich[1], Seungjoon Lee[2], Kobus Van Der Merwe[3], and Oliver Spatscheck[2]

[1]*Case Western Reserve University, Cleveland, OH, USA*
[2]*AT&T Labs—Research, Florham Park, NJ, USA*
[3]*University of Utah, Salt Lake City, UT, USA*

5.1 INTRODUCTION

As the Internet continues to grow, the need to distribute digital content to end-users in a timely and efficient manner became a priority for most Web content providers. Accomplishing this goal is challenging because of the often bursty nature of demand for such content [1] and also because content owners require their content to be highly available and be delivered in timely manner without impacting presentation quality [2]. Content delivery networks (CDNs) (e.g., Akamai, Limelight) have emerged over the last decade as an answer to this challenge and have become an essential part of the current Internet apparatus. In fact, Akamai alone claims to deliver between 15% and 30% of all Web traffic [3].

The basic architecture of most CDNs consists of a set of CDN nodes distributed across the Internet [4]. These CDN nodes serve as content servers' surrogates from which clients retrieve content from the CDN nodes using a number of standard protocols.

Advanced Content Delivery, Streaming, and Cloud Services, First Edition.
Edited by Mukaddim Pathan, Ramesh K. Sitaraman, and Dom Robinson.

The key to the effective operation of any CDN is to direct users to the "best" CDN node, a process normally referred to as "redirection" or "request routing" [5]. Redirection is challenging because not all content is available from all nodes, not all nodes are operational at all times, nodes can become overloaded or the network path from the user to the nodes can become congested, and perhaps most importantly, a user should be directed to a node that is in close proximity to the user to ensure satisfactory user experience.

A keystone component of not only CDNs but also the entire Internet infrastructure is the Domain Name System (DNS). Its primary goal is to resolve human-readable host names (such as cnn.com) to hosts Internet Protocol (IP) addresses. Virtually all Internet interactions start with a DNS query to resolve a hostname into IP addresses. Suppose a user wants to retrieve some Web contents from a website on the Internet. The first step for this is for the Web browser to send a DNS query to the user's *local DNS* (*LDNS*) server to obtain the hostname of the requested Uniform Resource Locator (URL). Unless the hostname is stored in its local cache, the LDNS in turn sends the request (by navigating through the DNS infrastructure) to the *authoritative DNS* (*ADNS*) that is responsible for the requested name. The ADNS server maintains the mapping information of host names to IP addresses and returns the corresponding IP addresses back to the LDNS. The LDNS saves the resolution in its local cache and forwards that resolution to the client. The Web browser then stores the resolution in its own cache and proceeds with the HyperText Transfer Protocol (HTTP) interactions by establishing a session using the provided IP address.

CDNs fundamentally rely on DNS to *reroute* user communication from the origin servers to the CDN infrastructure. A typical technique to achieve this goal leverages DNS protocol's name aliasing, which is done through a special response type, a *CNAME* record. As part of service provisioning, the origin site's ADNS is configured to respond to DNS queries for host names that are *outsourced* to the CDN not with the IP address of the origin server but with a CNAME record specifying a hostname from the CDN's domain. For instance, consider a website foo.com that wants to outsource the delivery of an object with URL http://images.foo.com/pic.jpg to a CDN cdn-x.com, as shown in Figure 5.1. When a client (say, Client 1 in the figure) accesses the above object, it sends a DNS query for images.foo.com to its LDNS server (step 1 in the figure), which, after traversing the DNS infrastructure, ultimately forwards it to foo.com's ADNS (step 2). Upon receiving this query, the ADNS responds with a CNAME record listing hostname images.foo.com.cdn-x.com (step 3). The requester (the LDNS that had sent the original query) will now attempt to resolve this new name using another query, which will now arrive at the ADNS for CDN x.com, as this is the domain to which the new name belongs (step 4). The CDN's ADNS now can resolve this query to any IP address within its platform (135.207.24.11 in the figure, step 5). The LDNS returns this response to the client (step 6), which then proceeds with the actual HTTP download from the prescribed IP address (step 7). The end result is that the HTTP download has been redirected from foo.com's content server to the CDN platform, which will provide the desired content either from its cache or, if the object is not locally available, first obtaining it from the origin server (and in this case storing it in its cache for future requests).

Now that the user request for content has been effectively diverted to the CDN platform, a key question for the CDN is to which node within its infrastructure to route

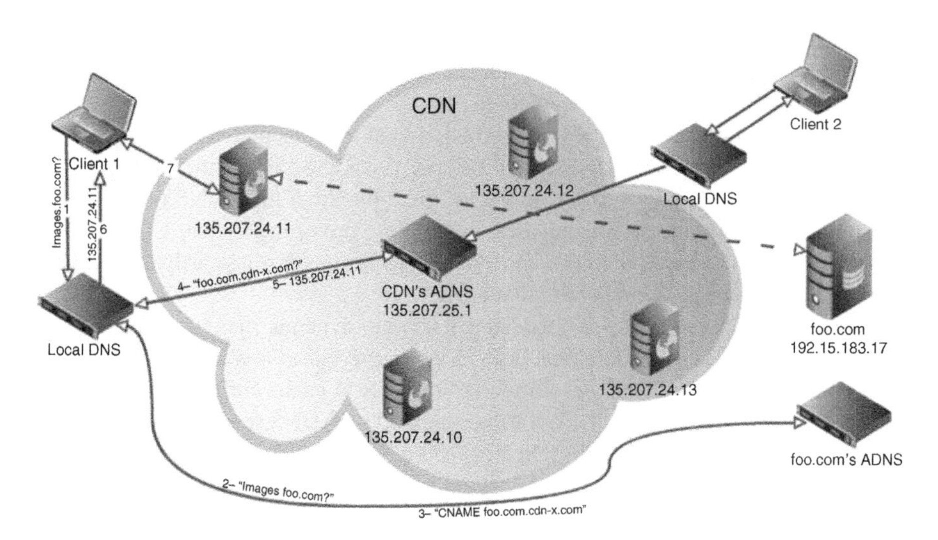

Figure 5.1 DNS-based request routing.

the user's request. This chapter is devoted to mechanisms and algorithms to answer this question. The chapter is based, in part, on the material that appeared in our ACM TWEB article [6].

5.2 CDN REQUEST ROUTING: AN OVERVIEW

This section outlines two basic mechanisms for request routing in CDNs: DNS based, as the most commonly employed in today's CDNs and anycast on which we focus later in this chapter. For other mechanisms, such as HTTP-based redirection, refer to Reference 5.

5.2.1 DNS-Based Request Routing

As illustrated in the previous example, DNS is used to divert user communication from the origin site to the CDN platform. Once a DNS query for an outsourced URL arrives at the CDN's ADNS, the CDN provider can direct different HTTP requests to different content servers by returning different IP addresses to different queries/users. For instance, in Figure 5.1, the CDN's ADNS returns IP address 135.207.24.11 to the LDNS serving Client 1 and 135.207.24.12 to the LDNS serving Client 2.

This commonly forms the basis for transparent client request routing in CDNs as well as in other replicated services (e.g., cloud computing). In this context, the goal of the CDN's ADNS is to respond with the IP address of a CDN node that is not overloaded and satisfies other, often contradictory, objectives such as proximity to the client, the quality of the network path, business and cost considerations, and so on. As a result, the algorithms and policies involved in DNS-based request routing and load balancing in CDNs are quite complex and still a subject of active research, which often form the "secret sauce" of various CDN providers.

DNS-based redirection, however, exhibits several fundamental limitations. First, in DNS-based redirection, an ADNS of a CDN only sees the LDNS of an end-user and makes a redirection decision based on the information about the LDNS (e.g., latency) [5]. In practice, however, not all end-users are in close proximity to their LDNS servers [7–9]. Therefore, a good CDN node for a LDNS server is not necessarily a good CDN server for all of the end-users behind that LDNS.

Second, the DNS system was not designed for very dynamic changes in the mapping between host names and IP addresses. Both LDNS servers and browsers (as well as other client applications) store DNS resolutions in their local cache for a certain period of time called the time-to-live (TTL)—a value supplied by ADNS servers along with their responses. This aspect in DNS complicates load distribution decisions for the CDN by limiting the granularity of its control over load balancing. Note that CDNs cannot lower TTL excessively because this could hurt user's performance by increasing the frequency of DNS resolutions by the clients (as an example, Akamai returns TTL of 20 s).

Third, the number of end-users behind an LDNS server is unknown to the CDN, hence the amount of load on a CDN node resulting from a single redirection decision can be difficult to predict. One possible workaround is for the ADNS to use very small TTL values to shorten the extent of the DNS response reuse. But as mentioned earlier, CDNs cannot lower TTL excessively; moreover, it is not unusual for LDNS servers and browsers to cache and reuse DNS responses beyond the TTL specified by the ADNS [10], which further complicates load distribution decisions for the CDN.

Finally, among other dimensions for selecting the best CDN node for a given DNS request, DNS-based redirection requires the ADNS to know which CDN nodes are topologically close to the originator of a given DNS query (i.e., LDNS). However, collecting such knowledge on the Internet requires a complex measurement infrastructure and raises the barrier of entry for new CDN providers.

5.2.2 Request Routing Using IP Anycast

IP anycast refers to the ability of the IP routing and forwarding architecture to deliver packets to the closest among a set of possible endpoints. In IP anycast, the same IP address is assigned to multiple endpoints, and packets to this IP address are routed to the endpoint that is the nearest to the sender based on the shortest (or otherwise most preferred) path from the network-routing perspective [11].

In an anycast setup, endpoints are typically configured to provide the same service. For example, IP anycast is commonly used to provide redundancy in the DNS root–server deployment [11]. Similarly, in the case of a CDN, all endpoints with the same IP anycast address can be configured to be able to serve the same set of contents. In this case, DNS would still be used to divert clients' requests to the CDN platform, but in step 5 in Figure 5.1, the CDN's ADNS would return the anycast IP address, the same one to all clients, and let IP anycast distribute these requests among the replicated endpoints.

Although attractive to content distribution because it "naturally" performs proximity request routing, anycast was deemed infeasible in the CDN context in the past due to the following reasons [5]. First, underlying routing changes can cause anycast traffic to be rerouted to an alternative anycast endpoint in the middle of a communication session, which will disrupt any session-based communication such as Transmission Control Protocol (TCP). Second, anycast selects the endpoint for a given sender based only on the routes as reflected in the routers' routing tables and therefore is unable to react to network conditions. Third, IP anycast is not aware of server load, and therefore cannot react to node overloading conditions.

5.2.3 Request-Routing Mechanisms in Practice

One issue in DNS-based request routing as discussed so far is that ADNS operated by the CDN, which performs request routing, can become a central bottleneck in the system. This bottleneck can manifest itself as a performance choke point and can add overhead for long-haul interaction onto the critical path of user request processing. Indeed, even if a client downloads content from a nearby CDN node, its preceding DNS query may potentially incur interaction with a faraway central ADNS server. Two leading CDN providers, Akamai and Limelight, effectively address this issue but use very different mechanisms to do so.

Akamai implements its ADNS infrastructure as a two-level distributed system, comprising the "root" centralized server and a large number of widely distributed "leaf" servers. The root server is registered as the ADNS for the Akamai domain and therefore receives initial queries from every LDNS server. However, instead of resolving these queries to an IP address, the root redirects each requesting LDNS to a nearby leaf server by replying with an appropriate NS-type response. An NS record in the DNS protocol allows an authoritative DNS server to delegate name resolution to another DNS server, and this record carries its own TTL. In Akamai, the root server assigns long TTLs to these redirecting messages, effectively assigning LDNSs to their leaf DNS servers for a prolonged period of time. It is these leaf DNS servers that perform the ultimate hostname resolution of queries from LDNSs into IP addresses of CDN nodes. Since different LDNSs are assigned to different leaf servers, the load for name resolution is shared. Because leaf DNS servers are widely distributed and LDNSs are redirected to nearby leaf servers, the resolutions do not incur long-haul communication. Finally, because the NS records have long TTL, most DNS queries involve interactions with leaf servers only, without long-haul communication with the root.

In contrast, Limelight utilizes hybrid IP anycast/DNS request routing. In contrast to Akamai, which maintains CDN nodes at thousands of locations around the world, Limelight concentrates its capacity in a relatively small number (around 20) in extremely large data centers. Each data center deploys an ADNS, and all ADNS servers share the same IP address, which is registered as the ADNS for Limelight's domain. When an LDNS issues a DNS query for a Limelight-accelerated hostname, IP anycast delivers this query to the ADNS at the nearest data center from the network perspective. Then, this ADNS resolves the query to an IP address of a CDN node within the same data center. In this way, the load for DNS query resolution is shared among ADNSs in different data centers; Limelight avoids the need to know the global network topology because IP anycast organically delivers DNS queries to the nearest data center, and subsequent communication occurs with a CDN node within the same data center. And there is no possibility of session disruption due to route changes because anycast is only employed at the DNS resolution stage, which is not session-oriented.

Still, the fundamental limitations of DNS-based request routing remain in both Akamai and Limelight mechanisms: the proximity-based server selection is done relative to LDNS and not the client that performs actual download, the server selection occurs at coarse granularity not fully under control of the CDN provider, and the effect of server selection decisions is not always easy to predict due to unknown number of clients behind a given LDNS server. The rest of this chapter discusses anycast approaches to request routing that address many of the anycast limitations mentioned earlier.

5.3 A PRACTICAL LOAD-AWARE IP ANYCAST CDN

There are two basic approaches to CDN deployment. Standalone CDNs, such as Akamai, connect their servers to various ISPs; thus maintaining presence in multiple autonomous systems. A CDN operated by a global ISP, such as AT&T's ICDS content delivery service, can have sufficient footprint while deploying its nodes at various points within the ISP's own autonomous system. This section presents an architecture for anycast-based CDNs that target the latter deployment alternative.

Our architecture leverages this deployment to address many of the anycast limitations mentioned earlier. In particular, it allows full consideration of server load within its anycast-routing mechanism.

Figure 5.2 shows a simplified view of a load-aware anycast CDN. In this example, an IP anycast address is assigned to two CDN nodes, A and B. Provider edge (PE) routers are located at the edge of the AS and act as the ingress and egress points between the AS and neighboring networks. In other words, traffic directed to any of these CDN nodes has to ingress the network through one of the PEs. Once traffic destined to the anycast address enters the network, it is delivered to one of the CDN nodes according to the routes installed by the *route controller* that is central to our approach [12]. The route controller makes a decision on the "best" CDN node for each ingress point and installs appropriate routes by taking into account CDN server load, internal network conditions, and overall cost.

More specifically, the load-aware anycast CDN functions as follows. CDN nodes A and B advertise the same IP anycast address into the network via BGP (the standard Internet routing protocol) through PE_0 and PE_5, respectively. PE_0 and PE_5 in turn advertise the anycast address to the route controller, which is responsible to install the "appropriate" route to every other PE in the network (PE_1 to PE_4 in the figure), which determines both the CDN node to which a given PE will deliver traffic destined to the anycast IP address and the internal path from this PE to this CDN node. These PEs further advertise the fact that they can deliver traffic to the anycast address via eBGP (the component of BGP for route advertisement across autonomous systems) sessions with peering routers

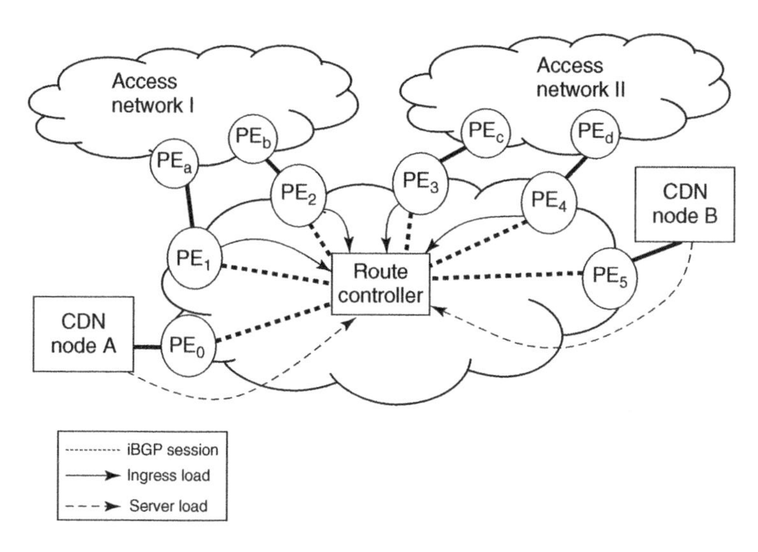

Figure 5.2 Load-aware anycast CDN architecture.

PE_a to PE_d in the neighboring networks so that the anycast address becomes reachable throughout the Internet. Requests for content on a CDN node will follow the reverse path. Thus, a request will come from an access network and enter the CDN provider network via one of the ingress routers PE_1 to PE_4. Such request traffic will then be forwarded to either PE_0 or PE_5 and then to one of the CDN nodes. On the basis of the load feeds (from both ingress PEs and CDN nodes) provided to the route controller, it can decide which ingress PE (PE_1 to PE_4) to direct to which egress PE (PE_0 or PE_5) and therefore to which CDN node. By assigning different ingress PEs to appropriate CDN nodes, the route controller can minimize the internal network costs of servicing the requests and distribute the load among the CDN nodes.

The route controller uses the mechanism from Reference 12 to enact this PE assignment. This involves preinstalled Multiprotocol Label Switching (MPLS) tunnels [13] between every PE pair in the network. The route controller then uses measurements and decision algorithms (described in Section 5.4) to activate routes from PE_i to PE_j as needed; the controller only needs to signal PE_i to start using an appropriate MPLS label. In this approach, route change is an atomic operation and does not involve any other routers along the path. For example, in Figure 5.1, to direct packets entering through PE_1 to the CDN node B, the route controller would signal PE_1 to activate the MPLS tunnel from PE_1 to PE_5; to send these packets to node A instead, the route controller would similarly activate the tunnel from PE_1 to PE_0.

Note that AT&T recently partnered with Akamai on content delivery, while also retaining its ICDS platform. In this partnership, Akamai provides the CDN's authoritative DNS servers and edge servers, which are deployed within AT&T's data centers. AT&T provides the data centers and the network connectivity between the data centers and AT&T's PE nodes. Our approach fully applies to this cooperative environment as long as the load of Akamai edge servers is reported in a timely manner to AT&T so that AT&T can activate appropriate tunnels within its network.

In summary, our approach utilizes the BGP-based proximity property of IP anycast to deliver client's packets to nearest ingress PEs. These external portions of the paths of anycast packets are determined purely by inter-AS BGP routes. Once packets enter the provider network, it is the route controller that decides where these packets will be delivered through mapping ingress PEs to content servers. The route controller makes these decisions taking into account both network proximity of the internal routers and server loads.

5.3.1 Benefits and Limitations

A major advantage of this design over DNS-based CDNs is that this architecture redirects actual client demand rather than LDNS servers, and thus it is not affected by the distance between end-users and their local CDN servers. Further, this design is not impacted by DNS caching and can make redirection changes quickly as soon as PEs learn about new routing decisions. This design also eliminates the need for determining proximity between CDN nodes and external destinations.

Revisiting anycast limitations discussed in Section 5.2.2, this architecture effectively solves the problem of load-awareness of the CDN nodes. It also partially addresses the session disruption and obliviousness to network path quality problems. Indeed, because the internal routes are under full control of the CDN provider, it can take internal path quality into account when activating internal routes, and the internal route changes can be enacted so as to minimize session disruption (see Section 5.4). Furthermore, a previous study of IP anycast [14] has found that most anycast IP prefixes have very good affinity, that is, anycast-addressed packets would be routed along the same path toward the anycast-enabled network. Thus, session disruption due to external route changes is unlikely. The remaining limitation is obliviousness to the external path quality. Even this aspect could potentially be controlled by manipulating eBGP advertisements from the PEs to their peers, although this aspect remains unexplored at the time of this writing.

The granularity of load distribution offered by our route control approach is at the PE level, unless a more complex mechanism is used [15]. For large tier-1 ISPs, the number of PEs is typically in the high hundreds to low thousands. A possible concern for our approach is whether PE granularity will be sufficiently fine grained to adjust load in cases of congestion. Our results [6] indicate that even with PE-level granularity, we can achieve significant performance benefits in practice.

5.4 MAPPING ALGORITHMS

In our architecture, the route controller collects load feeds from all the PEs and CDN nodes and, based on these measurements, periodically (every *remapping interval*) reexamines the PE-to-server assignments and computes new assignments if necessary using a remapping algorithm. The algorithm for assigning PEs to CDN nodes has several—often contradicting—objectives. One objective is to minimize the network cost of serving requests. Another is to flexibly remap PEs to CDN nodes to distribute the user demand among the CDN nodes so that every CDN node stays within its capacity constraints. Finally, an important goal is to minimize the service disruption due to load balancing. In this section, after formally formulating the problem, we present two algorithms that prioritize these objectives differently.

5.4.1 Problem Formulation

We formulate our mapping problem as an integer linear problem. Our system has m servers, where each server i can serve up to S_i requests per time unit. A request enters the system through one of n ingress PEs, and each ingress PE j contributes r_j amount of requests per time unit. We consider a cost matrix c_{ij} for serving PE j at server i. Since c_{ij} is typically proportional to the distance between server i and PE j as well as the traffic volume r_j, the cost of serving PE j typically varies with different servers.

The objective we consider is to minimize the overall cost without violating the capacity constraint at each server. The problem is called Generalized Assignment Problem (GAP) and can be formulated as follows:

$$\text{Minimize} \sum_{i=1}^{m}\sum_{j=1}^{n} C_{ij}X_{ij}$$

$$\text{Subject to} \sum_{i=1}^{m} X_{ij} = 1, \forall j$$

$$\sum_{j=1}^{n} r_j X_{ij} \leq S_i, \forall i$$

$$X_{ij} \in \{0,1\}, \forall i,j$$

where indicator variable $x_{ij} = 1$ iff server i serves PE j, and $x_{ij} = 0$ otherwise.

Since we want x_{ij} to be an integer, finding an optimal solution to GAP is NP-hard, and even when S_i is the same for all servers, no polynomial algorithm can achieve an approximation ratio better than 1.5 unless $P = \text{NP}$ as shown by Shmoys and Tados [16]. They also present an approximation algorithm (referred to as the ST-algorithm in this chapter) for GAP, where the objective value of the solution matches that of an optimal feasible solution, but the load on an individual server may exceed the capacity constraint by up to max r_j.

5.4.2 Minimum Cost Algorithm

Algorithm 5.1 Minimum Cost Algorithm

```
INPUT: CurrentLoad[i], OfferedLoad[j], Cost[i][j] for each
  server i and PE j
```

```
Run expanded ST-algorithm
{Post Processing}
repeat
  Find the most overloaded server i;
  Let Pi be the set of PEs served by i. Map PEs from Pi to i starting from largest Offered-Load,
until i reaches its capacity
  Remap(i, {the set of still-unmapped PEs from Pi})
until None of the servers is overloaded OR No further remapping would help return
Subroutine Remap(Server: i, PE Set: F):
for all j in F, in the descending order of OfferedLoad: do
  Find server q = arg mink Cost[k][j] with enough residual
  capacity for OfferedLoad[j]
  Find server t with the highest residual capacity
  if q exists and q != i then
     Remap j to q
  else
     Map j to t {t is less overloaded than i}
  end if
end for
```

Algorithm 5.1 begins by running what we refer to as *expanded ST-algorithm* as follows: we first run ST-algorithm with given server capacity constraints, and if it could not find a solution (which means the load is too high to be satisfied within the capacity constraints at any cost), we increase the capacity of each server by 10% and try to find a solution again. In our experiments, we set the maximum number of tries at 15, after which we give up on computing a new remapping and retain the existing mapping scheme for the next remapping interval. However, in our experiments, the algorithm found a solution for all cases and never skipped a remapping cycle.

Note that ST-algorithm can lead to server overload by up to max r_j. In practice, the overload volume can be significant since a single PE can contribute a large request load (e.g., 20% of server capacity). Thus, we use the following postprocessing.

We first identify the most overloaded server i, and then, among all the PEs served by i, find the set of PEs F (starting from the least load PE) such that server i's load becomes below the capacity S_i after offloading F. Then, starting with the highest load PE in F, we offload each PE j to a server with enough residual capacity, as long as the load on server i is above S_i. (If there are multiple such servers for j, we choose the one with minimum cost to j, although other strategies such as best-fit are possible.) If there is no server with enough residual capacity, we find server t with the highest residual capacity and see if the load on t after acquiring j is lower than the current load on i. If so, we offload PE j to server t even when the load on t goes beyond S_t, which will be fixed in a later iteration.

Once the overload of server i is resolved, we repeat the whole process with the most overloaded server. Note that the overload comparison between i and t ensures the monotonic decrease in the maximum overload of the system and therefore termination of the algorithm either because there are no more overloaded servers in the system or "repeat" postprocessing loop could not further offload any of the overloaded servers.

5.4.3 Minimum Disruption Algorithm

Algorithm 5.2 Minimum Disruption Algorithm

INPUT: CurrentLoad[i], OfferedLoad[j], Cost[i][j] for each
  server i and PE j
Let FP be the set of PEs mapped to non-overloaded servers.
  {These will be excluded from re-mapping}
For every non-overloaded server i,
  set server capacity Si ← Si −CurrentLoad[i]
For every overloaded server i and all PEs j
  currently mapped to i, set Cost[i][j] ← 0
Run expanded ST-algorithm to find a server for PEs ∈ FP
  {This will try to remap only PEs currently mapped to
  overloaded servers while they prefer their current server.}
{Post Processing}
repeat
  Find the most overloaded server i Map PEs ∈ Pi − FP to i,
  starting from the largest OfferedLoad, until i reaches
  it capacity
  Remap(i, {the set of still-unmapped PEs from Pi})
until None of the servers is overloaded **OR** No further remapping
  would help

While Algorithm 5.1 attempts to minimize the cost, it does not take the current mapping into account and can potentially lead to a large number of connection disruptions. To address this issue, we present another algorithm, which gives connection disruption a certain priority over cost.

Algorithm 5.2 divides all the servers into two groups based on their load: overloaded servers and non-overloaded servers. The algorithm keeps the current mapping of the non-overloaded servers and only attempts to remap the PEs assigned to the overloaded servers. Furthermore, even for the overloaded servers, we try to retain the current mappings as much as possible.

Specifically, for the PEs that need to be remapped due to server overloads, we use ST-algorithm to minimize the costs, but by manipulating input to the algorithm in two ways to minimize the disruption. First, for each non-overloaded server i, we consider only its residual capacity as the capacity S_i in ST-algorithm. This allows us to retain current PEs mapped to the servers while optimizing costs for newly assigned PEs. Second, for each overloaded server j, we set the cost of servicing its currently assigned PEs to zero. Thus, current PEs will be reassigned only to the extent necessary to remove the overload.

As described, this algorithm reassigns PEs to different servers only in overloaded scenarios. It can lead to suboptimal operation even when the request volume has gone down significantly and a simple proximity-based routing would yield a feasible solution with lower cost. One way to address this is to exploit the typical diurnal pattern and perform full remapping once a day at a time of low activity (e.g., 4 AM every day).

Another possibility is to compare the current mapping and the potential lowest cost mapping at that point, and initiate the reassignment if the cost difference is beyond a certain threshold (e.g., 70%).

5.5 EVALUATION

We evaluate the efficiency and feasibility of our approach using data from a production single-AS CDN. Specifically, by combining information from several logging points in the network, we produced a data set where each record has detailed information about an HTTP request and response such as ingress PE where the request entered the network, the IP address of the cache server that processed the request, request size, response size, and the time of arrival. Depending on the logging software, some servers provide service response time for each request, while others do not. In our experiments, we first obtain sample distributions for different response size groups based on the actual data. For log entries without response time, we choose an appropriate sample distribution (based on the response size) and use a randomly generated value following the distribution. The CDN service under study contained eight CDN nodes.

We used the concurrent requests being processed by a CDN node as the load metric that we need to control in our experiments. We used the request arrival time and the corresponding service response time to determine whether a request is currently being served; we further combined that information with the information on which ingress PE request has entered the platform to determine the number of concurrent requests for each ingress PE.

IP anycast delivers the request from the client to the nearest entry point into the CDN's autonomous system, resulting in the shortest external path from the client to the CDN's network from the network perspective. Therefore, our immediate goal is to minimize network delays inside the CDN's autonomous system. As the internal response path is always degenerate independently of our remapping (it uses hot-potato routing to leave the AS as quickly as possible), the network proximity between the client's ingress PE and server is determined by the request path. Thus, we used the request path as our cost metric to reflect the proximity of request processing.

We obtained the distance matrix between every server and every ingress PE from the CDN in terms of air miles and used it as the cost of processing a request. While we did not have access to the full topological routing distances, the latter are known to be highly correlated with air miles within an autonomous system since routing anomalies within an AS are avoided. Topological routing distances, if available, could be equally used in our design. Given the air miles d_{ij} between server i and ingress PE j, we use the product $r_j d_{ij}$ as the cost c_{ij} of serving requests from PE j at server i.

Another input required by ST-algorithm is the capacity S_i of each server i. To assign server capacity, we first analyze the log to determine the maximum aggregate number of concurrent requests across all servers during the entire time period in the log. Then, we assign each server the capacity equal to the maximum aggregate concurrent requests divided by the number of servers. This leads to a high load scenario for peak time, where we have sufficient aggregate server capacity to handle all the requests, assuming ideal load distribution. Note that server capacity is simply a parameter of the load balancing algorithm (i.e., load high watermark that the algorithm is trying to avoid exceeding), and in practice would be specified to be below the servers' actual processing limit. We refer to the latter as the server's physical capacity. Specifically, the server capacity computed as described earlier is just under 1600 concurrent requests and, in the experiments reported here, the physical capacity is set to 1.6 times the server capacity or 2500 concurrent request.

5.5.1 Simulation Environment

We used CSIM (http://www.mesquite.com) to perform our trace-driven simulation. CSIM creates process-oriented, discrete-event simulation models. We implemented our CDN servers as a set of facilities that provide services to requests from ingress PEs, which are implemented as CSIM processes. For each request that arrives, we determine the ingress PE j, the response time t, and the response size l. We assume that the server responds to a client at a constant rate calculated as the response size divided by the response time for that request. In other words, each request causes a server to serve data at the constant rate of l/t for t seconds. Multiple requests from the same PE j can be active simultaneously on server i. Furthermore, multiple PEs can be served by the same facility at the same time.

To allow flexibility in processing arbitrary load scenarios, we configured the CSIM facilities that model servers to have infinite capacity and very large bandwidth. We then impose capacity limits at the application level in each scenario. Excessive load is handled differently in different systems. Some systems impose access control so that servers simply return an error response to excess requests to prevent them from affecting the remaining workload. In other systems, the excessive requests are admitted and may cause overall performance degradation. Our simulation can handle both these setups, although we only present results for the setup without access control, where we admit all the requests and simply count the number of *overcapacity requests*. An overcapacity request is a request that at the time of its arrival finds the number of existing concurrent requests on the server to already equal or exceed the server's physical capacity limit.

As the result of request redirection, a request may use a server different from the one used in the trace, and its response time may change, for example, depending on the server load or capacity. In our experiments, we assume that the response time of each request is the same as the one in the trace no matter which server processes it as a result of our algorithms.

The scale of our experiments required us to perform simulation at the time granularity of one second. To ensure that each request has a nonzero duration, we round the beginning time of a request down and the ending time up to whole seconds. We assume that the route controller runs the remapping algorithm every 120 s.

5.5.2 Schemes and Metrics for Comparison

We experiment with the following schemes and compare the performance:

- *Trace Playback (PB).* We replayed all requests in the trace without any modification of server mappings. In other words, PB reflects the current CDN routing configuration.
- *Simple Anycast (SAC).* This is the "native" anycast, which represents an idealized proximity routing scheme. Each request is served at the geographically closest server.
- *Simple Load Balancing (SLB).* The goal is to minimize the difference in load among all servers without considering the cost.
- *Advanced Load Balancing, Always (ALB-A).* Always attempts to find a minimum cost mapping as described in Algorithm 5.1.
- *ALB, On-Overload (ALB-O).* Minimize connection disruptions as described in Algorithm 5.2. Specifically, it only reassigns PEs currently mapped to overloaded servers and performs full remapping only if the cost reduction from full remapping would exceed 70%.

In SAC, each PE is statically mapped to the closest server, and there is no change in the mappings across the entire experiment run. SLB and ALB-A recalculate the mappings every 120 s (the remapping interval).

Owing to space limitation in this chapter, we show a brief set of our simulation results. For a detailed list of the results, please refer to our previous paper [6].

Figure 5.3 shows how the five schemes distribute requests across the eight CDN nodes. Figure 5.3(a) presents the current behavior of the CDN (PB). It is clear that some servers process a disproportionate share of load (e.g., server 4 handles up four to five times the load of other servers). This indicates current overprovisioning of the system and an opportunity for significant optimization. SAC (Figure 5.3(b)) does not take the server load into account, and thus ends up with a few nodes serving virtually no requests while the heaviest loaded node is handling more than 6000 concurrent connections. SLB (Figure 5.3(c)) distributes the load well, although that behavior comes at a high cost in terms of connection disruption and long-distance communication (detailed in later figures). Both ALB-A and ALB-O algorithms (Figure 5.3(d) and Figure 5.3(e), respectively) effectively prevent overloading condition throughout the experiments.

Figure 5.4 shows the average air miles for each scheme normalized by that of PB (i.e., the figure plots the ratio of the air miles under a redirection algorithm in question over the air miles exhibited in PB).

In SAC, a PE is always mapped to the closest server, and the average air miles for a request is always the smallest. This can be viewed as the optimal cost one could possibly achieve. In the other extreme, since SLB balances the load among servers without taking cost (i.e., air miles) into account, it leads to the highest cost (although as we could see in Figure 5.3(c), it achieves nearly perfect load balance among all the nodes). ALB-A is nearly optimal in cost when the load is low (e.g., at 6 AM) because in this case each PE can be assigned to the closest server. As the traffic load increases, however, not all PEs

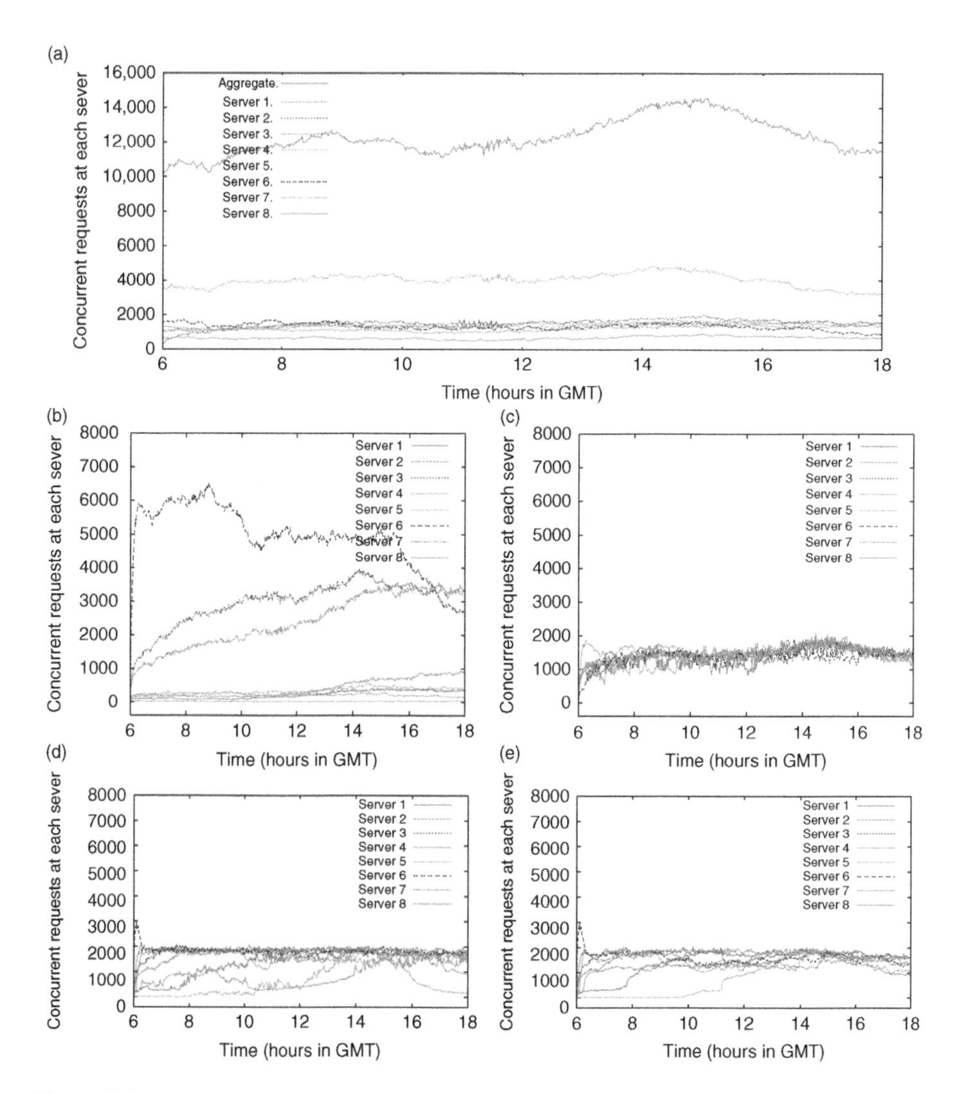

Figure 5.3 Number of concurrent requests for each scheme. (a) PB, (b) SAC, (c) SLB, (d) ALB-A, and (e) ALB-O.

can be served at their closest servers without violating the capacity constraint. Then, the cost goes higher as some PEs are remapped to more distant servers. ALB-O also finds virtually optimal-cost mapping in the beginning when the load is low. As the load increases, ALB-O behaves differently from ALB-A because it attempts to maintain the current PE server assignment as much as possible, leading to increased costs of ALB-O mappings.

Finally, we consider an impact of the potentially serious drawback of anycast request routing, which is that it may disrupt connections that are active at the time of route

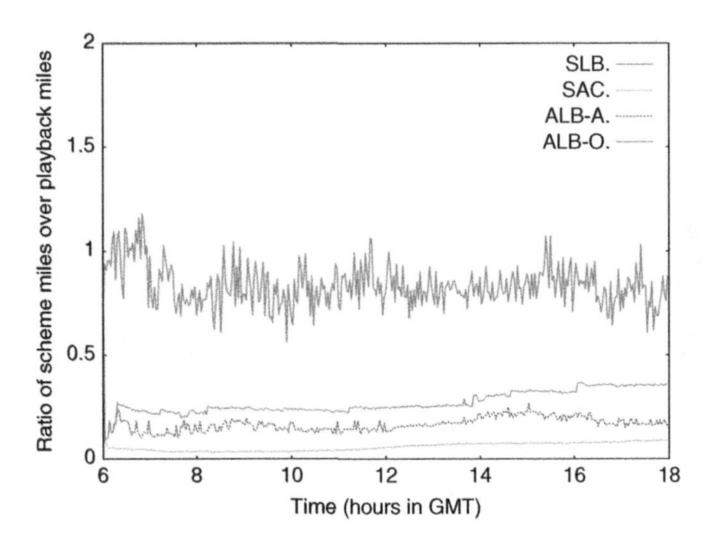

Figure 5.4 Average miles for requests calculated every 120 s.

changes. In the context of our architecture, remapping PEs to new CDN nodes can lead to disruption of active connections on the remapped PEs, which means that the corresponding client requests will fail. The basic trade-off here is that if we never remap PEs, there is no possibility of connection disruption but neither there is any load balancing, which can lead, again, to failed requests due to overloaded CDN nodes. The end result in both cases is therefore similar, and ultimately we are interested in minimizing the total number of affected requests from both causes. We therefore consider both connection disruptions due to PE-to-server remappings and the incidents of overcapacity requests, that is, the requests that bring the total number of concurrent requests at the node over its physical capacity (2500 concurrent requests).

The results are shown in Figure 5.5. Since SAC only considers PE-to-server proximity in its mappings, SAC mappings never change and thus connection disruption does not occur. However, by not considering load, this scheme exhibits many overcapacity requests (over 18% of all requests). In contrast, SLB always remaps to achieve balanced load distribution. As a result, it has no overcapacity requests but a noticeable number of connection disruptions.

The overall number of negatively affected requests is much smaller in SLB than in SAC. Both ALB-A and ALB-O show significant further improvement in the number of affected connections. Furthermore, by remapping PEs judiciously, ALB-O reduces the disruptions by an order of magnitude over ALB-A without affecting the number of overcapacity requests. Specifically, the disruption we observed in our experiments is negligible at 0.04% for the ALB-O algorithm (which we ultimately advocate).

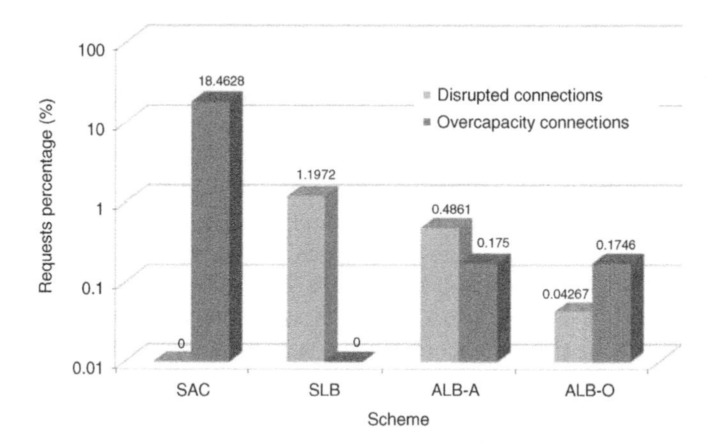

Figure 5.5 Disrupted and overcapacity requests for each scheme (*Y*-axis in log scale).

5.6 IPv6 ANYCAST CDNs

The transition to IPv6 has been accelerating as the supply of large unallocated IPv4 address blocks has been exhausted and the allocated IP address space is rapidly running out. Once IPv6 gains wider adoption, the Internet will need CDNs that operate in the IPv6 environment. Although IPv6 increases the size of the IP address, it retains the spirit of the DNS protocol and IP routing; therefore, the general mechanisms behind CDN request routing considered above apply equally to both IPv4 and IPv6. However, IPv6 offers additional functionality that can be leveraged to implement request routing in a more flexible manner. In this section, we present a general lightweight IPv6 anycast protocol for communication utilizing connection-oriented transport, and then use this protocol to design an architecture for an IPv6 CDN based on anycast request routing. This design relies heavily on IPv6 mobility support, and we begin with outlying this mechanism.

5.6.1 IPv6 Mobility Overview

IPv6 Mobility (MIPv6) allows a client (the "correspondent node," or CN, in mobile IP parlance) to communicate with a mobile node (MN) using the mobile's node home address. The mechanism from the high level is as follows. The MN is represented by two addresses: the permanent home address belonging to the MN's home network and a care-of address that belongs to the currently visited ("foreign") network, which can change as a result of node mobility. The home network maintains a special router endowed with mobility support, called home agent. As the MN moves from one foreign network to another, it keeps updates to its home agent, keeping it informed of its current care-of address.

The correspondent node starts communication by sending the first packet to MN's home address. As this packet enters MN's home network, it is intercepted by the home agent, which tunnels this packet to the mobile agent using MN's current care-of address. To take the home agent out of the loop for subsequent communication, MIPv6 allows the mobile agent to execute *route optimization* with the correspondent node. For security reasons discussed shortly, the MN initiates route optimization by sending two messages to the correspondent node: HoTI ("home test initiation") and CoTI ("care-of test initiation"). HoTI is sent through the home agent and CoTI directly. The CN responds to each message with a corresponding test message: HoT ("home test") through the home agent and CoT ("care-of test") directly. Each message contains pieces of crypto material, both of which are needed to construct a puzzle ("binding management key") that the CN would require to complete the protocol. Once the MN receives both messages, it constructs the puzzle and includes it in a special *binding update* (BU) field in a mobility header in the next data packet to the CN. This packet also includes the MN's care-of address as its source address and home address are in its destination option (DST OPT) header.

When the CN receives this packet and verifies the puzzle, it stores the binding in a special cache. Any subsequent packets that an application running on the CN addresses to the home address will be rewritten to list the corresponding care-off address as the destination address moves the home address into a special packet header. A reverse transformation happens within the IP layer of the MN, so that the application code at both ends of the communication is oblivious to any mobility-related issues including the fact that the effective address of the MN (i.e., care-of address) changes.

Observe that the correspondent node cannot simply update its binding once it receives a packet with new home to care-of address mapping. Indeed, this could enable any malicious node to hijack the communication by sending a packet to the CN that maps the MN's home address to the attacker's own address as a new care-of address. By sending HoT and CoT messages and routing one of them through the home agent, the CN verifies that the host possessing the new care-of address is properly associated with the home agent.

Moreover, HoTI and CoTI messages are also important because without them, an attacker could mount a reflected denial of service attack on the MN's home agent. The attacker could simply send packets to the CN with the MN's home address and some fake care-of addresses, causing the CN to bombard the home agent with HoT messages. In MIPv6, the CN only sends these messages in response to HoTI messages received from the home agent, that is, when asked by the recipient.

5.6.2 Using MIPv6 for CDN Request Routing

Previous work proposed using IPv6 mobility support to implement CDN request routing [17] as well as for general anycast [18]. For request routing, Reference 17 uses IPv6 Mobility *BUs* to direct client to one of the CDN nodes. Specifically, in the approach sketched in Reference 17, the client starts its Web download from the CDN platform by opening a TCP connection to the IPv6 address of a *request router*, which acts like a home agent. The latter tunnels this request (i.e., TCP SYN segment) to a

CDN node selected for this client, which responds to the client with the SYN-ACK TCP segment using its own IP address as the source address (which serves as care-of address of the mobility node) but including the original address of the request router as the home agent address and also supplying a BU in the IPv6 mobility header. The client's IP layer then remembers the binding between the two addresses and uses this new IP address of the CDN node as the destination address for subsequent communication while also providing the original request router address in the DST OPT header.

The request routing mechanism sketched in Reference 17 does not address security issues mentioned in Section 5.6.1, especially the crucial vulnerability that a malicious server can hijack client's Web interaction. This issue has been addressed in Reference 18, which again leveraged IPv6 mobility mechanisms, namely, the HoTI/CoTI/HoT/CoT protocol, in the context of implementing general anycast. This scheme involves a complex protocol with a two-level, hand-off communication from the home agent to the so-called contact node and then to the final anycast endpoint. At each level, the full HoTI/CoTI/HoT/CoT protocol is executed for security purposes. This scheme aims at avoiding any modifications to both the correspondent node and home agent, and also at making the anycast endpoint selection oblivious to upper layer protocols: packet delivery can switch to another anycast endpoint at any time in communication.

Both schemes above are free of drawbacks of both DNS-based and IPv4 anycast-based request routing. In contrast to DNS-based request routing, CDN node selection occurs for the actual client and not its LDNS, can be done individually for each request (thus there is no issue with unpredictable amount of load being redirected), and also is done at the time of the request (removing the issue of coarse granularity of control). In contrast to anycast-based request routing, request routing can fully reflect both CDN node load and network path conditions, and there is no possibility for session disruption. However, the first approach does not address security issues while the second, as discussed later, is unnecessarily heavy weight for the CDN context.

5.6.3 Lightweight IPv6 Anycast for Connection-Oriented Communication

We now describe a lightweight IPv6 anycast that derives its efficiency from leveraging connection-oriented communication, which happens to be the predominant mode of Internet communication in general and of CDN-accelerated communication in particular.

To deploy IPv6 anycast, the platform must setup a set of *anycast servers* that share the same anycast address and a set of *unicast servers* each with its own unicast address. These sets need not be distinct. In fact, it would not be uncommon to have a single set of servers where each server has both the shared anycast address and its own unicast address. Neither would it be uncommon to have a single or a small number of anycast servers that act as request routers: they would receive the first packet from a client, select a unicast server for subsequent communication, and hand off the connection to the selected unicast server. Each anycast server has a preinstalled secure channel to each unicast server.

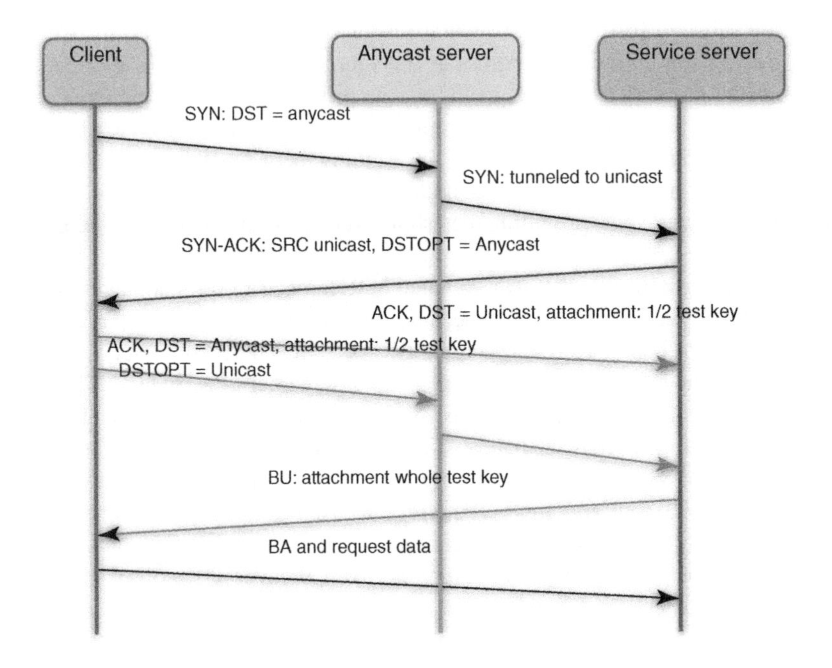

Figure 5.6 TCP interaction for an IPv6 anycast server.

Figure 5.6 shows the message exchange in setting up anycast communication using TCP as a connection-oriented transport protocol although our scheme can be adapted to any session-oriented communication that includes a session establishment phase.

Similar to any TCP connection, the client starts by sending a SYN packet to the announced IP address of the service, in our case the anycast address. Once the anycast server receives the SYN packet, it selects the unicast server to handle the incoming connection and passes the packet to this server via a secure tunnel. The unicast server responds to the client with a SYN-ACK packet using its own unicast address as the source address and piggybacks the anycast IP address in the DST OPT header.

A SYN-ACK packet with a DST OPT informs the client that it has reached an anycast service. In order to establish the connection, the client needs to verify that the source of the SYN-ACK packet (unicast server) is an authentic representative of the originally intended service (as represented by the anycast address, to which the client addressed its initial SYN packet and which is also reported in the DST OPT field). To this end, the client issues tests very similar to HoT and CoT messages. It sends the CoT message to the unicast server directly and the HoT message to the anycast address, to be tunneled to the unicast server reported in the DST OPT.

Once the unicast server receives both HoT and CoT messages, the server combines the tests and creates a BU message in the same way as an MN would in MIPv6. Once the client receives the BU message, it activates the binding entry at the IP layer's binding cache and replies back with BA (binding acknowledgment) message along with the piggybacked first chunk of application data. The binding occurs fully at the IP layer, which means that it is transparent to transport to higher layers. The connection is now established between the client and the unicast server. The upper layers at the client continue directing communication to the anycast address, which is then mapped securely to the unicast address at the client's IP layer.

The above protocol does away with HoTI and CoTI messages. The DST OPT on the SYN-ACK packet triggers the HoT/CoT test instead. However, the denial of service attack against the home agent (the anycast server in our context) does not apply in our case. Indeed, to cause a reflected HoT message, the attacker would have to send a well-timed SYN-ACK packet with correct sequence number acknowledging the correspondent's SYN. Otherwise, this message would be discarded. Further, any unexpected SYN-ACKs would be discarded by the client, so the attacker can at most induce one spurious HoT message per a TCP connection being opened by the client. Even if the attacker can time its message and guess a sequence number, only a small number of such messages would be generated, which could not cause a DoS attack.

The security of our approach is further enhanced by the IPv6 requirement that all addresses be provider-specific. The corollary is that anycast and all unicast addresses in our approach must share a common prefix. This further prevents a malicious outside node to pretend to be a member of the anycast group. By checking for the common prefix restriction, the client can often detect or discard malicious SYN-ACKs from nonmembers right away without issuing HoT/CoT messages. In all other aspects, our protocol provides the same protection as MIPv6.

Our approach is more humble and thus lightweight than the versatile anycast from Reference 18. We do not support in-flight TCP hand-off to another server: we believe this is an overkill for the CDN case (the primary intended application of our mechanism). Indeed, one can always issue a subsequent range request using a new connection. Thus, the server can simply reset the connection to affect a hand-off as proposed in Reference 19. We also give up another goal of versatile anycast, which is to keep the client unmodified. The reason is that, given CDN clout, requesting clients to download a patch is not unreasonable: this is already often done through invitation to install a download manager to speed up performance. Finally, we do not need to authenticate clients to servers during route optimization because CDNs do not authenticate clients at the IP layer either. Thus, at least in the context of CDNs, our approach to anycast does not weaken current security properties.

Finally, while we overload the MIPv6 mechanism with support for lightweight anycast, it does not prevent the regular mobility support of MIPv6. Indeed, the client can always move to another network and perform regular MIPv6 route optimization by a full HoTI/CoTI/HoT/CoT protocol. The only exception is a short period while the TCP connection is being established. If the client migrates to another network within this period, it simply needs to reopen its TCP connection anew or delay the execution of route optimization until the original TCP connection is fully established. Similarly, although we do not expect the server to be mobile, it too can, in principle, migrate to another network and execute the full HoTI/CoTI/HoT/CoT protocol with the client—again, as long as this happens outside TCP connection establishment.

5.6.4 IPv6 Anycast CDN Architecture

The anycast mechanism from the previous section can serve as the basis of a CDN with various architectural flavors. We now present one such CDN architecture as an example.

This architecture assumes multiple data centers distributed across the Internet, the deployment approach exemplified by Limelight, and AT&T (but not Akamai, which pursues a much more dispersed approach). A datacenter consists of an *anycast gateway*, which subsumes the actions of the anycast server in the anycast protocol as described in Section 5.6.3, and a number of edge servers each of which is responsible for serving content on behalf of the anycast service, and which acts as unicast servers in our anycast protocol.

Each anycast gateway maintains information of the current load on all servers in its datacenter. The anycast gateway is also aware of and maintains secure unicast tunnels to gateways from all other data centers as well as to every edge server in its own data center. This internal network of anycast gateways is used, as we will see shortly, for global load distribution and management.[1] Finally, each gateway knows the address blocks used in each data center for edge server unicast addresses, so it can immediately tell which valid unicast address belongs to which data center.

Figure 5.7 shows the basic architecture and interactions in such a platform, depicting for simplicity only two data centers, DC1 and DC2. The gateways in each data center

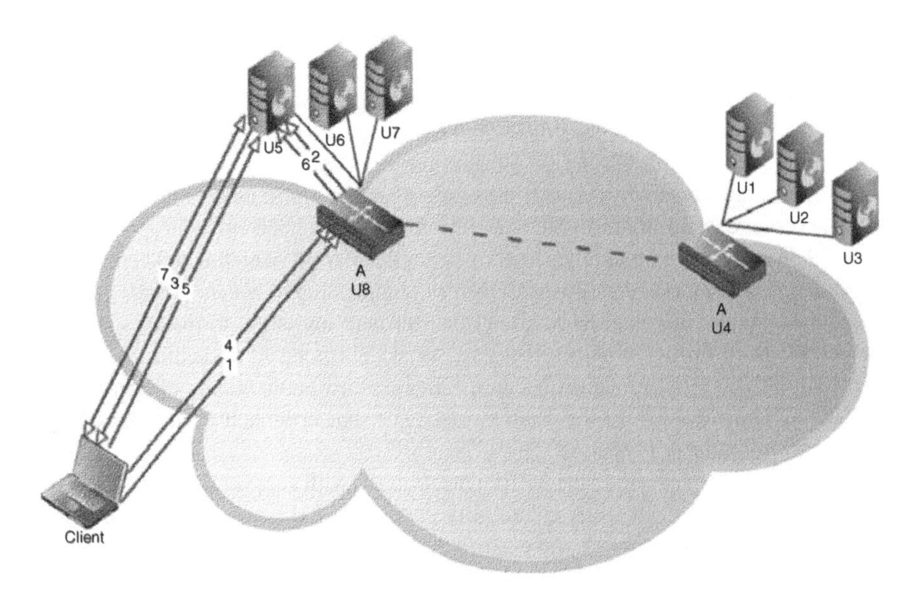

Figure 5.7 IPv6 anycast CDN.

[1]Although in principle more elaborate routing among gateways, for example, using distributed hash tables, can reduce the number of inter-gateway tunnels, in practice the number of data centers is usually small enough to just assume a full mesh of pairwise tunnels.

share an anycast address *A*, and every host in the platform including gateways also maintains a unicast address *U1–U8*.

When a client tries to retrieve content delivered by this CDN, the client starts with a DNS request that eventually get resolved to the anycast address A. Next, the client sends the TCP SYN packet to this address, which is delivered by virtue of anycast to the "nearest" datacenter, in our case DC2, and thus to the gateway *U8*. Assuming there are non-overloaded local edge servers, the gateway chooses such an edge server, say, *U5*, to serve the actual content to the client. The gateway then passes the SYN packet to this edge server, which responds to the client with a SYN-ACK packet with DST OPT header as described in Section 5.6.3, thus triggering the rest of the anycast hand-off.

If all local edge servers are overloaded, the anycast gateway will utilize the internal network of anycast gateways to tunnel the SYN to a neighbor gateway as shown in Figure 5.8. In principle, the SYN packet can be handed off from one gateway to the next until it finds a data center with enough capacity to hang the connection, although in practice we do not expect many such hand-off hops.

Once the SYN packet reaches the closest datacenter with space capacity (DC1 in the figure), its gateway (*U4* in our case) forwards the SYN packet to a non-overloaded local edge server, say, *U3*. *U3* responds to the client with a SYN-ACK with the anycast address *A* in the DST OPT header. The client will now generate HoT and CoT messages, but by virtue of anycast, the HoT message will likely be delivered to the original gateway *U8* and not to the one local to edge server *U3*. In fact, it is possible that due to a route change, the HoT message will be delivered to an entirely different gateway. But since the unicast address of the edge server is included in the DST OPT field of the HoT message,

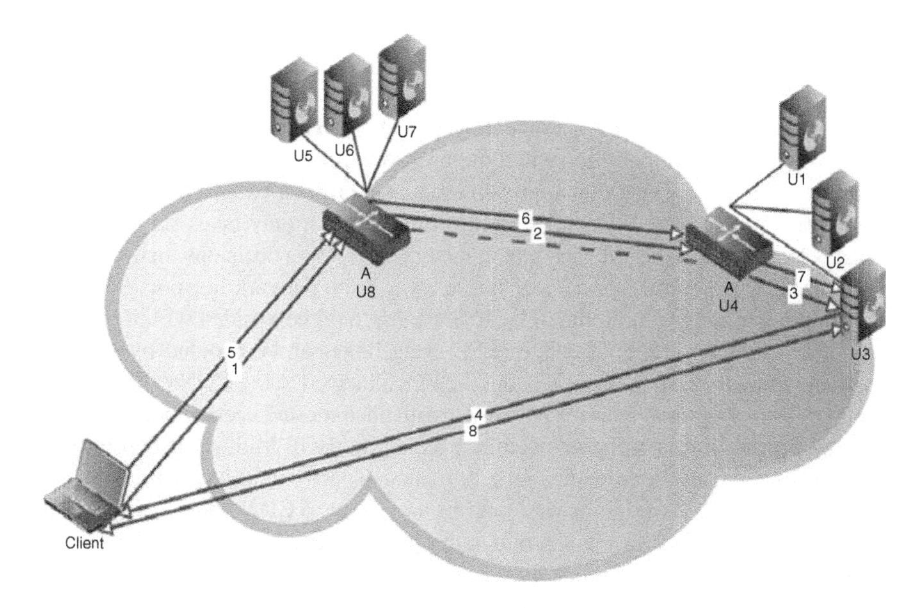

Figure 5.8 Redirection in IPv6 anycast CDN.

the receiving gateway can map this unicast address to the data center containing it and then route the HoT message to the proper anycast gateway, in this case *U4*. The rest of connection establishment with *U3* is already explained in previous sections.

5.7 DISCUSSION AND OPEN QUESTIONS

Despite increased distribution of rich media content via the Internet, the average size of "traditional" Web objects remains relatively small. This means that download sessions for such Web objects will be relatively short-lived with little chance of being impacted by any anycast remappings in our IPv4 anycast CDN architecture. The same is, however, not true for long-lived sessions, for example, streaming or large file download.

In our IPv4 architecture, we can deal with this by making use of an additional application-level redirection mechanism after a particular CDN node has been selected via our load-aware IP anycast redirection. A client will perform a DNS lookup, which will be resolved to an IP anycast address. The client will attempt to request the content using this address and, at this point, the CDN node will respond with an application-level redirect message (e.g., HTTP 302 response) containing the unicast IP address associated with this CDN node, which the client will use to retrieve the actual content. The redirection interaction is very short and is amenable to our anycast scheme. The actual long-lived content delivery session occurs over the unicast address associated only with this CDN node and will not be impacted by any routing changes along the way. Although the additional overhead associated with application-level redirection is clearly unacceptable when downloading small Web objects, it is less of a concern for long-lived sessions where the startup overhead is amortized.

In parallel work, we presented an alternative approach to handle extremely large downloads using anycast, without relying on HTTP redirection [19]. Instead, the approach is to recover from a disruption by reissuing the HTTP request for the remainder of the object using a range HTTP request. The CDN could then trigger these disruptions intentionally to switch the user to a different server midstream if the conditions change. However, that approach requires a browser extension.

An open issue in our IPv4 CDN architecture is that it only takes server load and internal network conditions into account in its server selection decisions. In other words, our approach does not attempt to steer traffic away from network hotspots outside the CDN's autonomous system. Some of these hotspots could be avoided within our architecture. For example, outgoing congested peering links can be avoided by setting an appropriate route for response traffic from the PE connected to the CDN node (e.g., PE_0 in Figure 5.2) to an egress PE on the AS border with uncongested peering links. Incoming congested peering links can be avoided by exchanging BGP Multi-Exit Discriminator (MED) attributes with appropriate peers. A related aspect is that our redirection decision is based primarily on the forward path from a client to CDN nodes. However, due to route asymmetry, this may not produce the optimal reverse path used by response packets. These mechanisms remain unexplored at the time of this writing.

The focus of work is on making the case that anycast CDNs are a practical alternative to a DNS-based CDN redirection system even within IPv4. One obvious question, which we could not answer with the data available, is the quantitative side-by-side comparison between the two mechanisms. To answer this question, one needs to be able to evaluate the difference between the "external" proximity from the client to the end of the CDN network given in the CDN and anycast CDN, and how it affects the performance from the client perspective.

The remapping algorithm in our IPv4 CDN architecture can route requests among edge servers only at the granularity of the entire PEs where requests enter the platform. In principle, request routing could be done at the individual ingress interface level of the PEs; this however requires more complexity in the PEs, as well as significant increase in the search space for the optimization algorithm. A detailed study would be required to understand the potential benefits and trade-offs involved.

None of these issues arise in an IPv6 anycast CDN. In this case, an explicit hand-off to a selected endpoint occurs using route optimization. The subsequent packets are routed using the unicast address of the endpoint with no possibility of anycast-induced disruption. In selecting the endpoint for a given request, the CDN can take any factors, including the network path quality, into account, and perform such selection at the granularity as low as individual connections if so desired.

However, transitioning to an IPv6 CDN presents its own interesting questions. To support IPv4-to-IPv6 transition, the DNS protocol allows DNS messages to be exchanged over IPv4 while defining separate message types to let the LDNS request, and ADNS respond with the IPv6 IP address for a queried domain name. The intent is to allow the client request both IPv4 and IPv6 resolutions, and let the ADNS choose between the two protocols. Now consider Limelight-style proximity routing where the CDN deploys anycasted ADNS servers at each data center and relies on IP routing to deliver client's requests to the nearest ADNS, and then delivers content from the same data center. Unfortunately, IPv4 and IPv6 routes may exhibit very different proximity characteristics: the nearest data center according to IPv4 may be far from optimal in the IPv6 realm. Therefore, this request routing mechanism may result in suboptimal server selection.

Another pitfall relates to the stage in request processing at which the decision between IPv4 and IPv6 realms is made. As an example, consider a content provider, say, firm-x.com that wants to have its content delivered via IPv4 using one CDN (cdn1.com) and via IPv6 using another CDN (cdn2.com). It would seem natural for firm-x.com's ADNS to return different CNAME records to different DNS queries. For a query requesting an IPv4 address ("type-A" queries), it would return CNAME containing, for example, firm-x.com.cdn1.com as the canonical name and for a query requesting an IPv6 address ("type-AAAA" query), firm-x.com.cdn2.com. Unfortunately, DNS resolvers typically do not key cached CNAME records to the query type, and these two responses could be conflated for subsequent queries when these queries utilize a shared resolver. Thus, more complex means of separating the two IP versions are needed: firm-x.com may need to have two distinct sets of URLs for the same content or return the same CNAME regardless of the IP version, and let the CDN select between the versions. Of course, the latter alternative precludes the possibility of using different CDNs for different IP versions.

5.8 CONCLUSION

This chapter discusses request-routing mechanisms in modern CDNs, and, in particular, argues for viability of IP anycast as the basis for request routing.

In IPv4, anycast was considered infeasible for this purpose because it can disrupt HTTP connections and is oblivious to edge server load and network path conditions. We employ recently developed route control mechanisms within an autonomous system to make anycast request routing aware of edge server load, and we also show that load-aware request routing can be done with negligible connection disruption.

In IPv6, several mechanisms for session-stable anycast have been previously described. We discuss these mechanisms and also outline a new approach, which leverages connection-oriented nature of the Web traffic to make the anycast hand-off more secure than some of these mechanisms and simpler than others.

Anycast removes a number of limitations of DNS-based request routing, and we hope techniques discussed here will prompt reassessment of its suitability in modern CDNs.

REFERENCES

1. Jung J, Krishnamurthy B, Rabinovich M. Flash crowds and denial of service attacks: Characterization and implications for CDNs and web sites. Proceedings of 11th WWW Conference; 2002.
2. Reibman A, Sen S, Van der Merwe J. Network monitoring for video quality over IP. Picture Coding Symposium; 2004.
3. Akamai. 2013 http://www.akamai.com/html/technology/index.html. Accessed 15 Oct 2013.
4. Rabinovich M, Spatscheck O. *Web Caching and Replication*. Addison-Wesley; 2001.
5. Barbir A , Cain B, Douglis F, Green M, Hofmann M, Nair R, Potter D, Spatscheck O. Known content network (CN) request-routing mechanisms. RFC 3568; July 2003.
6. Alzoubi H, Lee S, Rabinovich M, Spatscheck O, Van Der Merwe J. A practical architecture for an anycast CDN. ACM Trans Web 2011;5(4):17.
7. Mao Z, Cranor C, Douglis F, Rabinovich M, Spatscheck O, Wang J. A precise and efficient evaluation of the proximity between web clients and their local DNS servers. USENIX Annual Technical Conference; 2002.
8. Shaikh A, Tewari R, Agrawal M. On the effectiveness of DNS-based server selection. INFOCOM; 2001. p 1801–1810.
9. Alzoubi H, Rabinovich M, Spatscheck O. The anatomy of LDNS clusters: Findings and implications for web content delivery. Proceedings of the 22nd International Conference on World Wide Web; 2013.
10. Pang J, Akella A, Shaikh A, Krishnamurthy B, Seshan S. On the responsiveness of DNS-based network control. Proceedings of Internet Measurement Conference (IMC); October 2004.
11. Hardie T. Distributing authoritative name servers via shared unicast addresses. IETF RFC 3258; 2002.
12. Verkaik P, Pei D, Scholl T, Shaikh A, Snoeren A, Van der Merwe J. Wresting control from BGP: Scalable fine-grained route control. 2007 USENIX Annual Technical Conference; June 2007.

13. Ross K, Kurose J. *Computer Networking: A Top-Down Approach*. Addison-Wesley; 2003.
14. Ballani H, Francis P, Ratnasamy S. A measurement-based deployment proposal for IP anycast. Proceedings of the ACM IMC; October 2006.
15. Lee S, Van Der Merwe J. Multi-path load balancing using route controller. US patent 8175006, 2012 May.
16. Shmoys D, Tardos E. An approximation algorithm for the generalized assignment problem. Math Prog 1993;62:461–474.
17. Acharya A, Shaikh A. Using mobility support for request routing in IPv6 CDNs. 7th International Web Content Caching and Distribution Workshop (WCW); 2002.
18. Szymaniak M, Pierre G, Simons-Nikolova M, van Steen M. Enabling service adaptability with versatile anycast. Concurr Comput Pract Exp 2007;19(13):1837–1863.
19. Al-Qudah Z, Lee S, Rabinovich M, Spatscheck O, Van der Merwe J. Anycast-aware transport for content delivery networks. Proceedings of the 18th International Conference on World Wide Web. ACM; 2009.

6

CLOUD-BASED CONTENT DELIVERY TO HOME ECOSYSTEMS

Tiago Cruz, Paulo Simões, and Edmundo Monteiro

Faculdade de Ciências e Tecnologia da, Universidade de Coimbra, Coimbra, Portugal

6.1 INTRODUCTION

Cloud computing is the result of several technological advances, such as virtualization and the pervasive availability of broadband network access, enabling a multitude of new services and support infrastructures. Specifically considering telecommunication operator infrastructures, private cloud services are already widespread, supporting key services such as content provisioning, authentication, authorization, accounting, and management.

Cloud paradigms encompass not only content delivery but also the whole home ecosystem, where many users consume content and the associated services. This is a peculiar ecosystem, where different devices coexist (e.g., set-top boxes (STBs), Internet Protocol (IP) telephones, storage devices, media players, classic desktops, smart meters, home automation devices, mobile devices) behind the Residential Gateway (RGW) and the broadband access network.

For this specific ecosystem, over-the-top (OTT) content delivery solutions may strongly depend on operator support, especially in countries where operators themselves also compete against OTT service providers with their own content and services.

Advanced Content Delivery, Streaming, and Cloud Services, First Edition.
Edited by Mukaddim Pathan, Ramesh K. Sitaraman, and Dom Robinson.

This happens because the delivery of services to the home ecosystem depends on aspects as diverse as management, quality of service (QoS), protocol translation, and routing, some of which are directly controlled by operators and out of reach for OTT providers.

Also important is the emergence of virtualization technologies, which affect not only the content backend but also the whole end-to-end support infrastructure, up to the home consumer edge. This deeply impacts the cloud-sourcing model, in terms of service delivery and the entire support infrastructure used for transport and consumption.

This chapter addresses the delivery of cloud-based content to home ecosystems, providing a broad perspective on key challenges, industry practices, and recent trends. It is organized as follows: Section 6.1 discusses the challenges of delivering cloud services to home networks. Section 6.2 describes possible approaches for virtualizing parts of the access network infrastructure and the RGW, reducing context segmentation between the home local area network (LAN) and the operator domains. Section 6.3 discusses how to use device and service virtualization to improve the delivery of services to the home environment. Section 6.4 discusses future trends and Section 6.5 concludes the chapter.

6.2 BRINGING CLOUD SERVICES TO HOME: STATE OF THE ART

This section presents a state of the art on the delivery of cloud services to the home LAN environment, its device and service ecosystem and how it relates with telecommunication operators and third-party service providers. It also presents an overview of access network, mediation, and home LAN technologies, explaining how both mutual influence and new technology trends shape their evolution.

6.2.1 The Home Device and Service Ecosystem

The original home LAN directly descends from the workgroup network paradigm that became popular in the 1990s. Not more than a mashup of loosely networked devices, they were created to ease sharing of resources such as storage space or printers. Later, RGWs were added to provide Internet access, service mediation, routing, and firewall capabilities, paving the way to the subsequent introduction of converged IP-based services such as VoIP (voice over IP) and IPTV (Internet pay TV).

Eventually, the residential LAN ecosystem evolved into a dynamic environment supporting services such as resource sharing, voice communications, media distribution, and home automation, where devices as diverse as PCs, STBs, IP telephones, smartphones, media players, smart TVs, or storage appliances coexist, providing access to a wide array of services. Still, all this diversity comes at the price of complexity. Delivering content and services to the home ecosystem requires adequate management (and provisioning) of involved devices and services, QoS traffic differentiation, proper routing and Network Address Translation (NAT) configuration, and possible mediation and/or conversion between Internet and LAN protocols, among others. In order to support the whole array of currently available content services, the home LAN became too complex to be autonomously managed by the average consumer. The growing dependence on converged IP-based services (such as triple-play) further aggravates this scenario, since the reliability and performance of these IP-based services depend on the proper configuration and operation of all equipment in the service delivery path, including endpoints (such as STBs and IP telephones) and RGWs.

As most customers are unable or unwilling to manage devices and associated services, operators need to remotely manage them to ensure adequate performance. In fact, even for the minority of users able to perform such duties, operator involvement is still necessary for services requiring differentiated handling (QoS and/or flow isolation) on the path between the providers and the home LAN. The old Internet service provider (ISP) model where operator responsibility ended at the edge of the access network—leaving everything else beyond to the customer's responsibility—is no longer an option.

Although legitimate customer privacy concerns, most triple-play customers already have ISP-managed devices on their LAN with customized configurations and/or firmware that they cannot control. The industry has developed a number of protocols and frameworks for this purpose, among which Broadband Forum's CPE WAN Management Protocol (CWMP) [1] stands out as the *de facto* standard for operator-based remote management of equipment and services in the customer premises.

With two fundamental business models competing for the home service space ecosystem—ISP-sourced and OTT—the present situation constitutes a natural advantage to ISPs. This is not happening by chance: as broadband service paradigms transitioned from the data-based, single-service model to the consolidated *n*-play service model, telecommunication operators were always wary of protecting their prevalence as the single-service provider, using their ownership of the access network infrastructure to gain a competitive lead. Also, due to years of regulatory limitations and lack of adequate policy management mechanisms, telecommunication operators had to endure the burden of constantly updating their infrastructure without getting a fair share of the revenue of the services using it, a situation that also contributed to their unwillingness.

As owners of the access network, telecommunication operators play a decisive role in delivering third-party services to the home environment. In this sense, operator indifference toward services might be as bad as countermeasures taken against them, such as shaping. QoS-sensitive services such as VoIP or IPTV, for instance, generally require operator involvement in order to ensure adequate end-to-end performance.

Multitenant management mechanisms could relieve these issues by fostering shared management interfaces for both operators and third-party service providers. This has been the subject of research [2,3] and also, to a certain extent, suggested in recent Broadband Forum's specifications for northbound interfaces on CWMP management servers [4] and policy control mechanisms for partner control functions and applications [5]. Up to this date, however, such mechanisms are not in use.

6.2.2 The Access Network, RGWs, and the Home LAN

The expansion of high speed broadband access networks enabled a new breed of services, such as converged *n-play* offers or cloud services that displace traditional split-medium communication and service delivery models in favor of converged

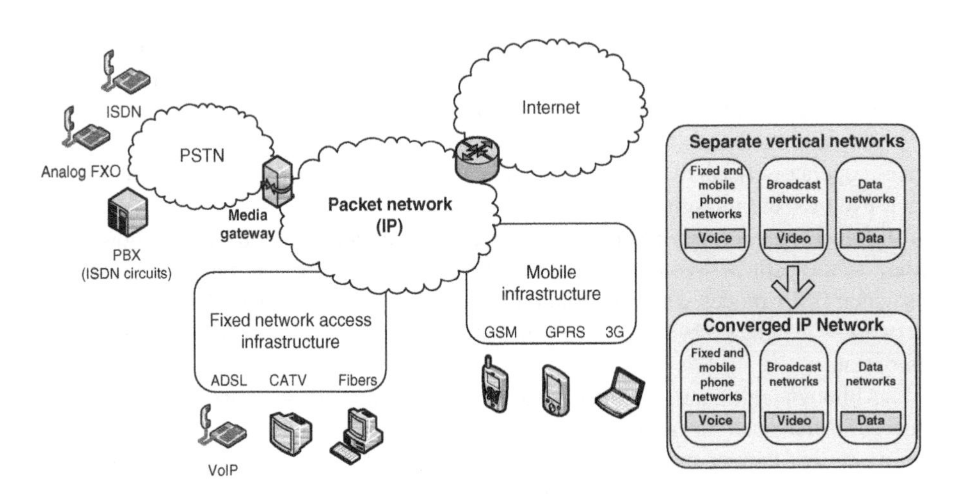

Figure 6.1 Converged IP network (adapted from [8]).

everything-over-IP approaches. Together with the increased pressure to compensate for decreasing revenue from traditional services (such as traditional telephony), this motivated an integrated development in the architecture of operator networks toward optimization of diversified service delivery—shifting from the legacy vertical multiple infrastructure model to a unified network supporting converged IP services (see Figure 6.1).

RGWs play an important role in the scope of home LAN environments, mediating traffic, providing security, and becoming bridgeheads for a multitude of purposes, from management to services. This happens because RGWs are placed in a strategic location in the frontier between the access network and the home LAN domains. Nonetheless, it is more than a simple question of physical location, as the RGW is in the intersection of three fundamental realms: service, connectivity/management, and home LAN [6].

As the RGW becomes an increasingly capable device with considerable computing resources in terms of processing power and memory, its functions are also evolving from providing simple connectivity services (in the data-only service model) to a more sophisticated device fulfilling different roles and delivering multiple services such as converged *n-play*, content distribution, or femtocell-based indoor mobile telephony. To support this multiservice model, RGWs became modular devices whose abilities can be customized using software components that can be added, removed, or upgraded using management mechanisms such as CWMP component management functionalities [7].

Virtualization is also gaining increased relevance in the access network. In addition to traditional virtualization applications, a new trend is emerging toward the entire virtualization of the supporting infrastructure. In order to reduce costs and take advantage of recent fiber to the premises (FTTx, fiber to the x) deployments, operators are considering reshaping the boundaries between the home LAN and the access network, by virtualizing and moving the RGW to their data centers [9]. Figure 6.2 illustrates this approach.

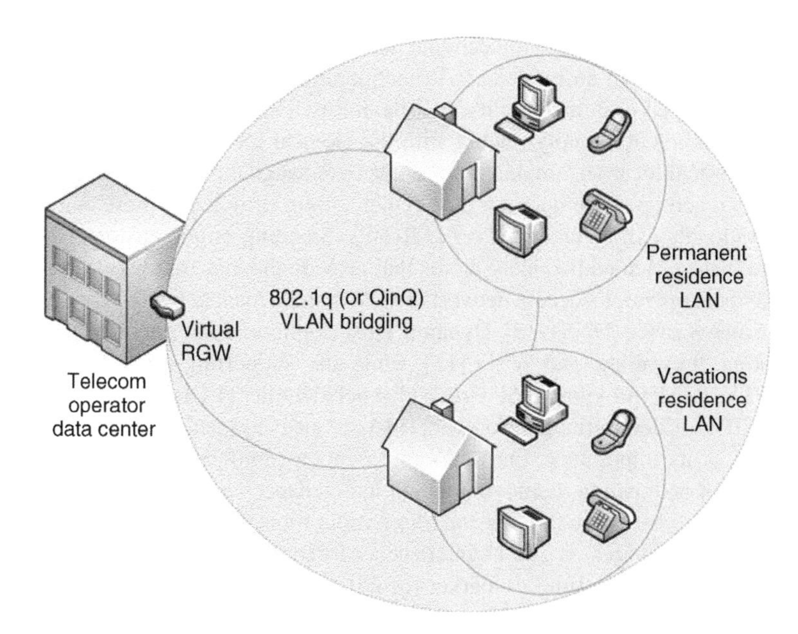

Figure 6.2 Single subscriber layer 2 domain in virtualized RGW scenario.

Once the operator infrastructure becomes bridged at layer 2 with the customer premises on Ethernet Passive Optical Networks (EPONs), it is possible to use 802.11q VLANs (virtual LANs) [10] to create per-subscriber network domains spawning several households. Telecommunication operators are still discussing the implications of this proposal, in order to define a reference architecture for its implementation [11].

Following the same rationale, several devices may also be virtualized, including STBs—moving all interface and service logic to the cloud—and local storage devices—merging them with cloud-based storage services [12].

6.3 VIRTUALIZING THE ACCESS NETWORK INFRASTRUCTURE

Technical advances in the field of virtualization, together with the increased penetration of FTTx, are encouraging operators to rethink the end-to-end infrastructure (core, edge, and access network), in order to reduce costs and improve flexibility and manageability. In this section, we describe an RGW virtualization approach that moves most of its functions to the operator infrastructure (using virtualization and private clouds).

6.3.1 Improving the Access Network: A Rationale for Virtualization

So far, the physical access network infrastructure has remained relatively excluded from the cloud paradigm, since many of its components strongly depend on physical location.

Nevertheless, it is still possible to decouple hardware-dependent functionalities (which cannot be moved) from software-based functionalities, moving the latter to the data center, for enhanced cost, availability, and flexibility. Furthermore, in some cases, the location constraints are simply related with the "logical location" of the components, not their "physical location," making it possible to redesign the logical network in order to extend its reach up to the operator data center—thus virtualizing those components.

This rationale can be extended up to RGWs. Standing on the customer premises, RGWs are feature-rich embedded systems that provide the interface between the home network and the operator's access network. RGWs handle local network services such as Domain Name Service (DNS) [13], Dynamic Host Configuration Protocol (DHCP) [14], NAT, routing, firewalling, and Wi-Fi [15], while also supporting value-added services such as IPTV (IGMP proxying [16], virtual channel identifier (VCI)/VLAN management [17]) and SIP (Session Initiation Protocol [18] VoIP gateways and/or terminal adapters).

In spite of its importance, the RGW represents a significant burden for the operator because of acquisition, deployment, and maintenance costs. The modern RGW is complex, prone to hardware failures and misconfiguration, constituting a single point of failure that often requires on-site maintenance. Moreover, the RGW is a critical limiting factor in terms of the time to market for introduction of new services—operators depend on RGW manufacturers to introduce support for new services, a penalty aggravated by the subsequent need to remotely upgrade thousands or millions of devices. It is also difficult for the operator to keep a homogeneous set of RGWs, which affects manageability even if adopting a single model from a single vendor; minor firmware and hardware revisions gradually compromise uniformity. In extreme cases, operators have been even forced to mass replacement of RGWs, in order to support new services.

In this scenario, it is attractive to consider virtual RGW (νRGW) approaches. Complex services and demanding network functionalities can be moved to the operator's infrastructure while remaining functionalities, such as Wi-Fi and bridging between the access network and the home LAN, can be provided by a much simpler device.

Implementing a νRGW involves decoupling device hardware and software functionalities, while extending the logical reach of the customer's home LAN to the data center (see Figure 6.3). The νRGW concept transcends a mere transfer of a handful of services from the (physical) RGW to the operator infrastructure. To start with, the remaining bridging device must be drastically simplified, in order to effectively reduce the associated capital expenditure (CAPEX) and operational expenditure (OPEX). Second, the access network infrastructure and the data center need to accommodate thousands or millions of logical networks, in order to link the home network of each customer with its νRGW. Finally, the virtualization technologies at the data center need to efficiently support a large number of νRGWs, taking into account that the νRGW is different from the typical virtualized server. It demands less computing resources but more network performance (e.g., to support routing/bridging, NAT, and/or firewalling). In this perspective, the potential benefits of virtualizing the RGW are manifold:

- *For the operator* it reduces CAPEX, since the cost of the remaining device is lower (especially if merged with the Optical Network Terminal (ONT)). OPEX is also improved by reduced on-site maintenance and simplified service creation and

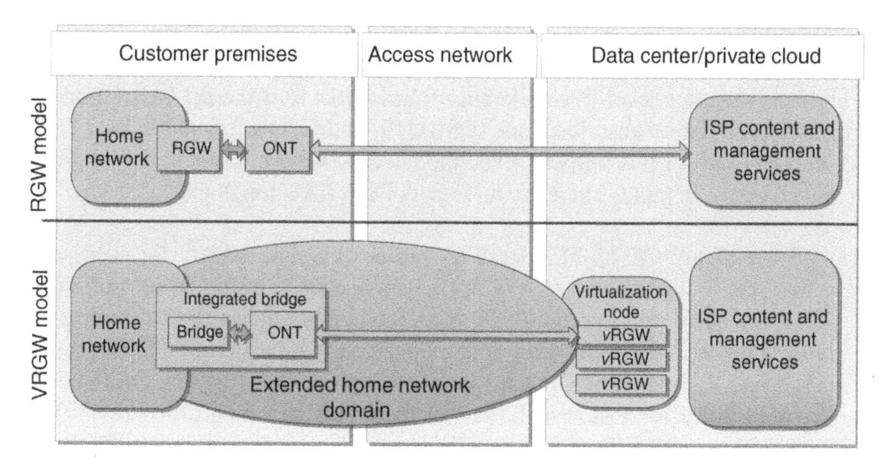

Figure 6.3 Classic RGW versus virtualized RGW.

support. The operator also becomes less dependent from hardware manufacturers (the virtual machine (VM) is, by nature, hardware independent), eliminating the heterogeneity of RGW models and makers. The operator can keep a unified image for all VMs, therefore, easing and accelerating the introduction of new services.

- End-users benefit with reduced costs, increased reliability for services and infrastructure alike, and potential access to a new array of services and features (such as roaming LAN access across different premises, as described in Section 6.1).
- *For cloud service providers*, this concept brings the possibility to improve or rethink existing services, while paving the way for new proposals. For instance, a service provider can establish a service-level agreement (SLA) with an operator to bring its content delivery network (CDN) servers up to the operator datacenters, from which it can directly provide services and content right inside the subscriber LAN domain.

The *v*RGW concept does raise security concerns. Extending the home network up to the data center may lead to illegitimate probing of domestic network traffic, performed by the operator at the data center. Yet, this risk already happens with current setups where the operator remotely controls the physical RGW, which can be remotely used for illegitimate probing. Either with virtual or physical RGWs, if the user does not trust the operator its only option is to hide the domestic network behind an additional firewall.

6.3.2 Virtualization Support at the Access Network

This section describes the technical aspects of logically extending the home network from the customer premises up to the operator data center, across the access network, in order to enable the implementation of the *v*RGW concept.

Logically extending each customer's home network to the data center is a considerable challenge, for scalability and manageability reasons, involving resource virtualization at the aggregation level. Two reference frameworks for Ethernet-based broadband aggregation on Digital Subscriber Line (DSL) [19] and GPON (gigabit PON) [17] scenarios can be used for this purpose, even though they were not specifically developed with this objective in mind. There are three possible VLAN topologies:

- *Service VLAN (N : 1).* This topology defines a single VLAN for each service (mapped in the provider S-VLAN), which carries traffic to all subscribers. This fits quite well into the IPTV model, where multicast IPTV is delivered in the same VLAN for all subscribers. Specifically considering GPON scenarios (Figure 6.4), where the Optical Line Termination (OLT) and the ONT share the responsibility for access node VLAN requirements (as specified by [17]), the ONT adds an S-VID (S-VLAN ID) or translates an incoming S-VID tag for upstream traffic. There is always an S-VID between the aggregation node and the access nodes. The aggregation node allows any upstream frames with an S-VID to pass-through.
- *Customer VLAN (1 : 1).* According to this topology, there is one VLAN per subscriber (customer VLAN, C-VLAN), which carries all its traffic, trusting on the edge router to manage thousands of VLANs. This topology is not efficient for IPTV, as identical streams for multiple subscribers are carried across the network.

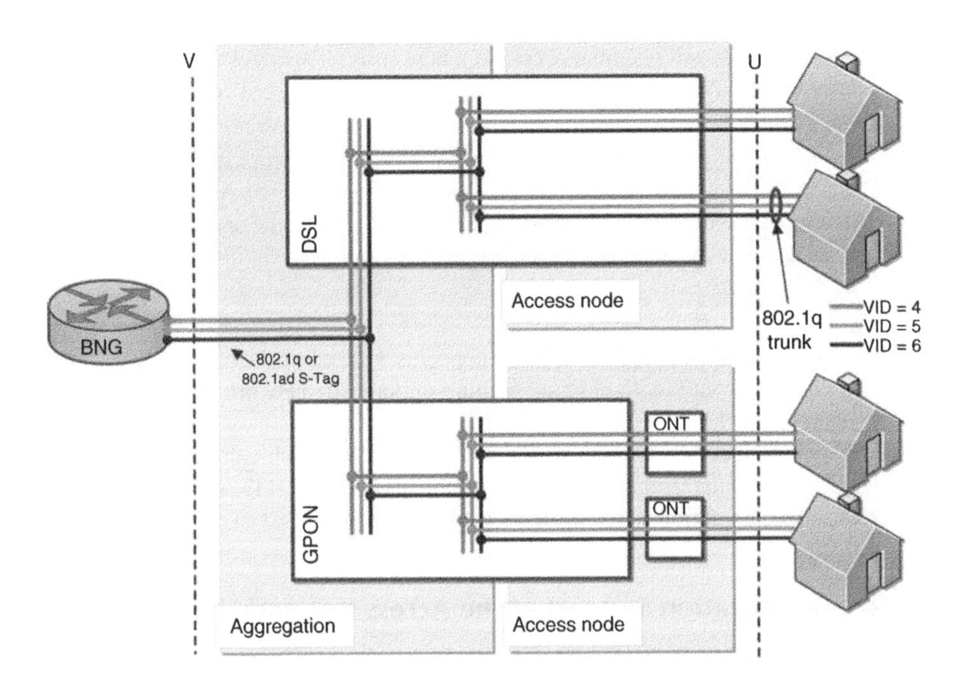

Figure 6.4 VLAN per service model (adapted from [20]).

Specifically considering GPON scenarios, the ONT maps each 1 : 1 VLAN into a unique U interface. Each U interface can map into one or more 1 : 1 VLANs. The ONT always adds a tag to untagged frames or translates an incoming Q-Tag in the upstream direction. Tag assignment at the V interface may follow two variations. The first variation corresponds to *single tagging*. For single-tagged VLANs at V, the ONT is provisioned to add an S-VID or translate an incoming tag into an S-VID, and the aggregation node passes through the tag. However, the 12-bit VLAN identifier only supports up to 4095 subscribers, affecting overall scalability.

Double tagging (depicted in Figure 6.5), using *Q-in-Q* VLAN stacking [10] (initially standardized by 802.1ad [21] and later incorporated into the 802.1Q standard [10]) makes it possible to increase VLAN capacity at the aggregation level. For double-tagged VLANs at V, the ONT is provisioned to assign a C-VID (C-VLAN ID) or translate an incoming tag into a C-VID, and the aggregation node adds the S-VID. The operator core network will carry traffic with double-tagged, stacked VLAN headers, retaining the VLAN and layer 2 protocol configurations of each subscriber.

- *Customer VLAN with Multicast VLAN*. This topology is a mixture, retaining the benefits from the two previous approaches, using a shared multicast VLAN to carry specific services such as IPTV traffic [22] while the rest of the traffic is delivered using the 1 : 1 topology of customer VLAN.

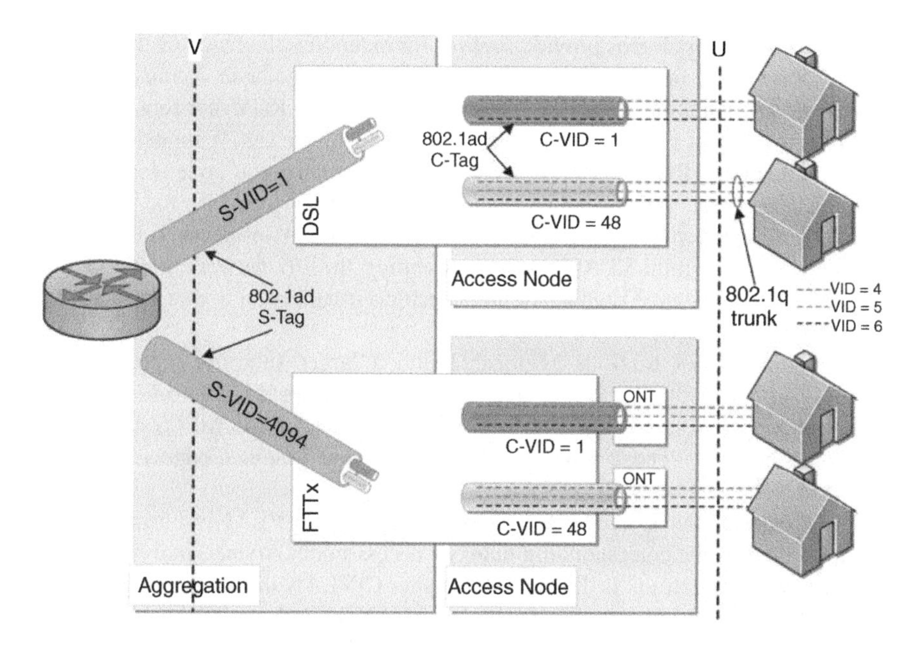

Figure 6.5 VLAN per subscriber model (adapted from [20]).

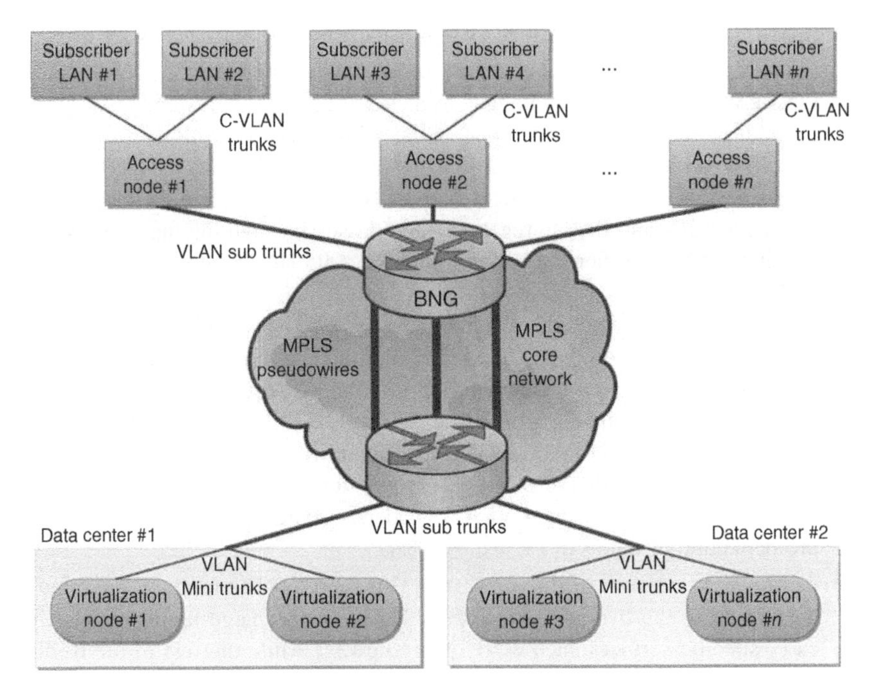

Figure 6.6 *v*RGW-enabling network architecture.

Overall, these topologies provide support for extending the reach of the customer home LAN up to the operator data center, providing both the layer 2 connectivity and the network segmentation necessary for implementing the *v*RGW concept. Figure 6.6 presents an overall network architecture that complies with the *v*RGW model. According to this architecture, *v*RGW instances are hosted by virtualization nodes located at the operator's data centers (DC#1 and DC#2, in the case of Figure 6.6).

Each virtualization node is responsible for a set of *v*RGW instances. It is connected to the network by a mini-VLAN trunk representing the *I/O* network traffic of every individual *v*RGW instance running in that specific virtualization node (i.e., the corresponding set of C-VLANs).

Each mini-VLAN trunk is aggregated into a larger subtrunk representing all C-VLANs on a specific data center. This subtrunk is encapsulated into Multiprotocol Label Switching (MPLS) pseudowires (PWs [23]) at one of the MPLS edge routers of the core network. Those PWs are then transported to other edge routers—the Broadband Network Gateways (BNGs).

At the exiting BNGs, each PW is converted again into VLAN subtrunks, each of them connected to the corresponding network access node. At each individual access node, each VLAN subtrunk is divided into smaller C-VLAN trunks carrying the traffic of each individual subscriber (separated by one VLAN for each service being delivered).

At the customer premises, on the ONT, each VLAN is untagged and mapped into a port in one unmanaged bridge with a wireless access point (the ONT and this bridge are not represented in Figure 6.6, for sake of simplicity). At this point, all the equipment inside the subscriber LAN may connect to the stripped-down bridge and obtain a valid IP address from the DHCP daemon running on the *v*RGW.

This architecture extends the logical reach of the subscriber LAN to the operator's data center using the technologies already in place. Moreover, the joint use of VLAN stacking, VLAN trunks, and PW makes it possible to handle the high number of VLANs in a scalable manner required by thousands or millions of subscribers.

6.3.3 Virtual Residential Gateways

From a virtualization environment point of view, the RGW has humble computing requirements (modest processors and a few hundred megabytes of RAM), with the possible exception of network interfaces. Overall, this means that mass virtualization of RGWs can be performed using off-the-shelf virtualization platforms, further reducing their computing requirements while opening a wealth of possibilities.

On a first approach, the RGW may be virtualized and migrated to the virtualization nodes without extensive rearrangements, with each *v*RGW remaining isolated from the other ones sharing the same private cloud. Later, this solution can be improved by consolidation, moving some services previously handled by the RGW to a context of colocation or functional distribution, as depicted in Figure 6.7. Eventually, the *v*RGW can be trimmed down to a bare instance with minimal networking and management capabilities, an execution environment and a basic service set, retained for customer privacy or functional reasons. Application-level gateways such as SIP gateways, DNS cache servers, and content caching mechanisms may benefit from this approach to enhance manageability, functionality, and footprint. Heavy-duty network functionality such as NAT and firewalling may be moved outside the *v*RGW and relocated to carrier-grade mechanisms. DHCP may also be replaced by a relay agent [24].

CWMP management support for *v*RGWs departs from the vertical segmentation model used for physical device management instead of going for a proxied approach.

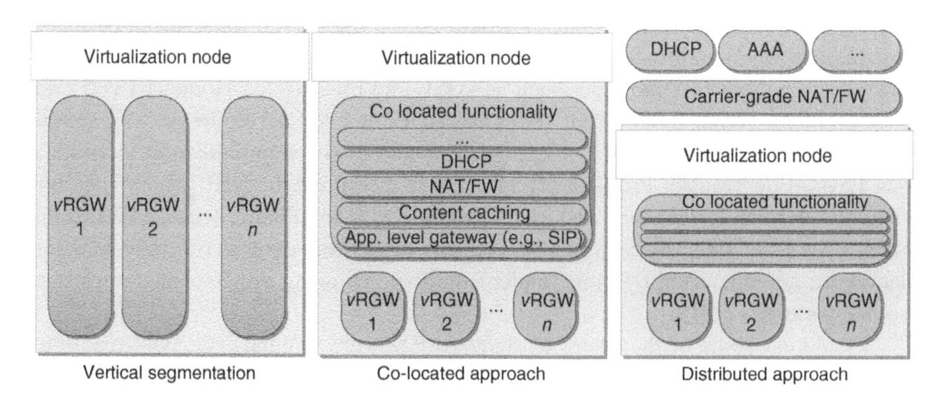

Figure 6.7 Architectural optimizations for *v*RGWs.

Each *v*RGW runs a low overhead management component that interfaces with a full-fledged CWMP agent hosted at the hypervisor level, decoupling the CWMP logic from specific *v*RGW management mechanisms. The configuration of *v*RGWs can be shared between the customer and the operator, allowing the customer to access an operator-provided Web interface to define user-accessible parameters (e.g., DHCP pools). The configuration profile for each customer is activated on instantiation of the corresponding *v*RGW.

6.4 VIRTUALIZATION FOR CLOUD SERVICE DELIVERY TO HOME

This section discusses how device and service management can be used to improve the delivery of cloud-sourced services to the home environment, with innovative service proposals and device concepts being brought together to introduce new approaches to content and service delivery to households. Altogether they hint for a *post-OTT* business model, based on a collaborative paradigm involving operators and third-party service and content providers to ensure end-to-end service delivery performance.

6.4.1 Thin-Client Devices for Broadband Service Appliances

As an alternative to the conventional standalone appliances, it is possible to bring the benefits of managed thin-client technology to all kinds of service appliances by providing remote OS boot capabilities on broadband access network environments [25], using the Preboot eXecution Environment (PXE) protocol [26].

PXE is the *de facto* standard for network boot firmware, allowing a PC to download and execute an agent—the Network Bootstrap Program (NBP)—over a LAN at boot time for deployment, diagnostic, or bare metal recovery. PXE can also be used to support completely stateless thin-clients [27] whose operating environment is downloaded from the network when powered up, instead of using local firmware.

When originally conceived, the use of PXE outside LAN environments was not envisaged, since the download latency would be too high, even for small boot agents with limited functionality—a premise that has changed because of the rise of broadband access network technologies such as xDSL and Fiber-to-the-Home (FTTH). Yet, other problems remain: first, PXE relies on a DHCP LAN server to get its configuration parameters (through DHCP option tags) and also for initialization; second, it uses Trivial File Transfer Protocol (TFTP) [28] to download the NBP, which is an low resilience, unreliable, insecure, and inefficient protocol; and finally, because PXE is not secure—its Boot Integrity Services [29] specifications for server verification and validation are not supported by most implementations, allowing impersonated servers to provide tampered NBPs.

To address some of the issues of standard PXE, the original PXE boot read-only memory (ROMs) may be replaced with Reference 30. Management is provided using CWMP, enabling a management server (Auto-configuration Server or ACS) to configure all PXE-related parameters, mapped on the RGW device data model. The CWMP agent

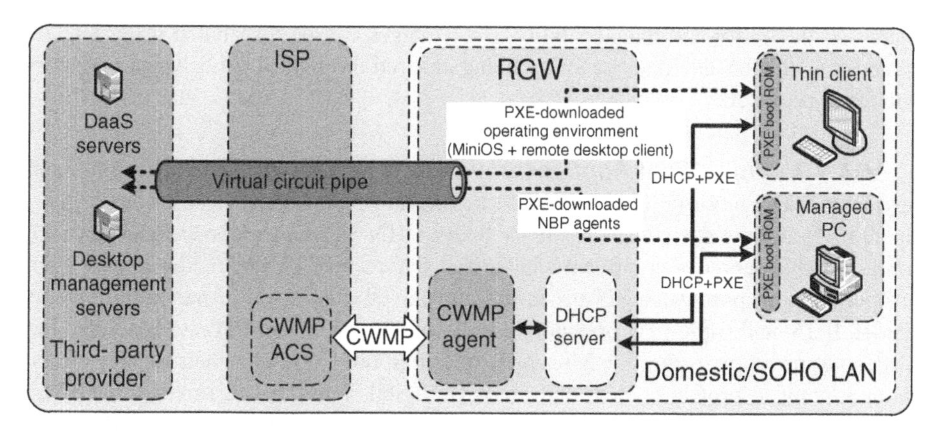

Figure 6.8 PXE-based broadband desktop management.

of the RGW uses this data to configure the embedded DHCP server for PXE operation (Figure 6.8).

To enhance PXE operation, an ISP may also use CWMP to configure a private virtual circuit pipe in order to offer QoS guarantees to PXE and the related management traffic and services. This enables the establishment of SLA agreements between operators and third-party providers to allow end-to-end differentiation of service traffic. Application scenarios for this solution include several use cases, which is discussed later.

6.4.1.1 Outsourced PC Management This solution can support the implementation of complementary recovery and diagnostics mechanisms for conventional PCs, in many different contexts, enabling domestic or business users to subcontract PC management to third parties, working in cooperation with operators (through SLAs) and using of a combination of management servers and CDN infrastructures to support PC management services such as bare metal recovery, automated OS upgrades or remote diagnostics, cutting operations costs by reducing both user downtime and on-site interventions. This could also pave the way for ISPs to start bundling managed PCs to their commercial offers for home and SOHO customers, reduce after sales costs, and increase customer satisfaction.

6.4.1.2 Stateless Thin-Clients for Desktop-as-a-Service Applications The Desktop-as-a-Service (DaaS) concept applies to the delivery of a desktop environment as a subscribed service. In this scope, this solution enables replacing standalone PCs with stateless thin-clients for DaaS, turning desktop computing into a secure and managed service. Instead of using conventional thin-clients with support for specific remote desktop protocols (RDPs) (such as RDP [31] or independent computing architecture (ICA) [32]) in firmware, these devices download their entire operating environment from the DaaS provider (which can be the ISP itself or a third party) at boot, not requiring firmware updates to support new protocols or functionality, with increased flexibility and manageability. Boot images can be distributed using a CDN backend, with QoS and manageability being ensured through SLAs between operators and DaaS providers

6.4.1.3 Thin-Client Appliances: Virtual IPTV Set-Top Box The thin-client appliance is a variation of the previously described concept, which can be used to support the virtualization of devices such as IPTV STBs. IPTV STBs are embedded devices with a considerable degree of complexity that are used to receive TV and media content from an operator, transported using IP streaming, and played in a TV set. As part of its business model, IPTV and triple-play operators usually bundle their own STBs (often locked to a specific media platform, like Microsoft's Mediaroom [33]) with their service offers, in an attempt to protect content and preserve digital rights. It is estimated that STBs constitute the largest share of IPTV operator's CAPEX and OPEX, in some cases up to 70% of CAPEX [34].

As such, it becomes evident that potential cost reductions could be obtained by simplifying the STB. In this line, a solution has been proposed (see Figure 6.9) for virtualizing existing STBs, using the remote boot capabilities previously introduced to transform them into stateless and robust thin-client devices with minimal hardware and devoid of any firmware. Once the STB is turned into a thin-client appliance, the service logic, functionalities, and interface can be moved to the service provider infrastructure.

This solution leverages the capabilities of HTML5 and JavaScript to provide the STB service interface (rendered in a full-screen Web browser which is part of the NBP payload) in a transparent way to the end-user, which interacts with the virtualized STB as it already does with existing STBs. To ensure DRM (Digital Rights Management) and scalability, multicast video reproduction is provided by a standalone media player, together with an IGMP [16] proxy component on the RGW, for multicast media streams. HTML5 video support will be considered once DRM is standardized.

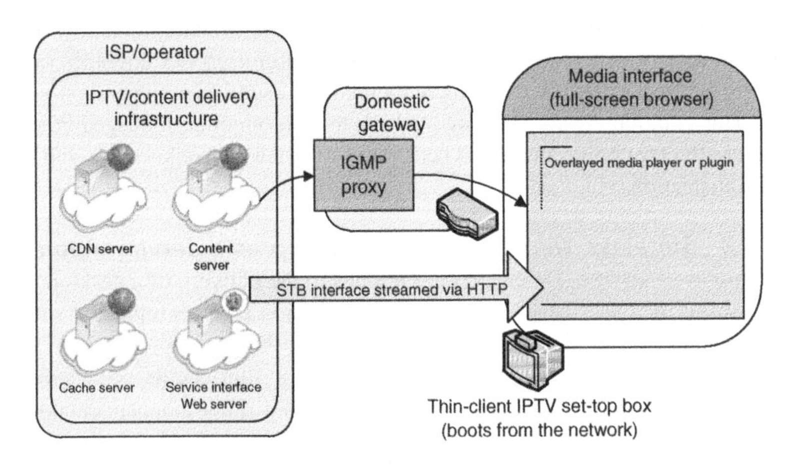

Figure 6.9 Thin-client IPTV STB.

This virtual STB concept can be used by both operators and third-party service providers—yet, for the latter, operator involvement is required to provide infrastructure and management support for both QoS, routing/multicast and service traffic isolation.

6.4.2 Content Provider Media Streaming for DLNA Devices

This managed service [35] consists of a solution for seamlessly delivering operator or third-party media content for media devices inside the residential LAN, supporting existing protocols and media delivery frameworks, like Universal Plug and Play Audio Video (UPnP AV) [36] and DLNA (Digital Living Network Alliance) [37]. It turns the RGW into a managed mediator for both operator-provided and Internet media content, provided as UPnP AV/DLNA resources accessible on the domestic LAN (Figure 6.10).

The benefits of this approach are manifold: first, it helps overcome the limitations of the DLNA specifications and their core UPnP AV functionalities, which were designed for use on LAN environments, being unable to properly operate across the LAN boundaries, over broadband access networks; second, it helps reducing the clutter on the residential LAN generated by the increased number of devices deployed to provide access to

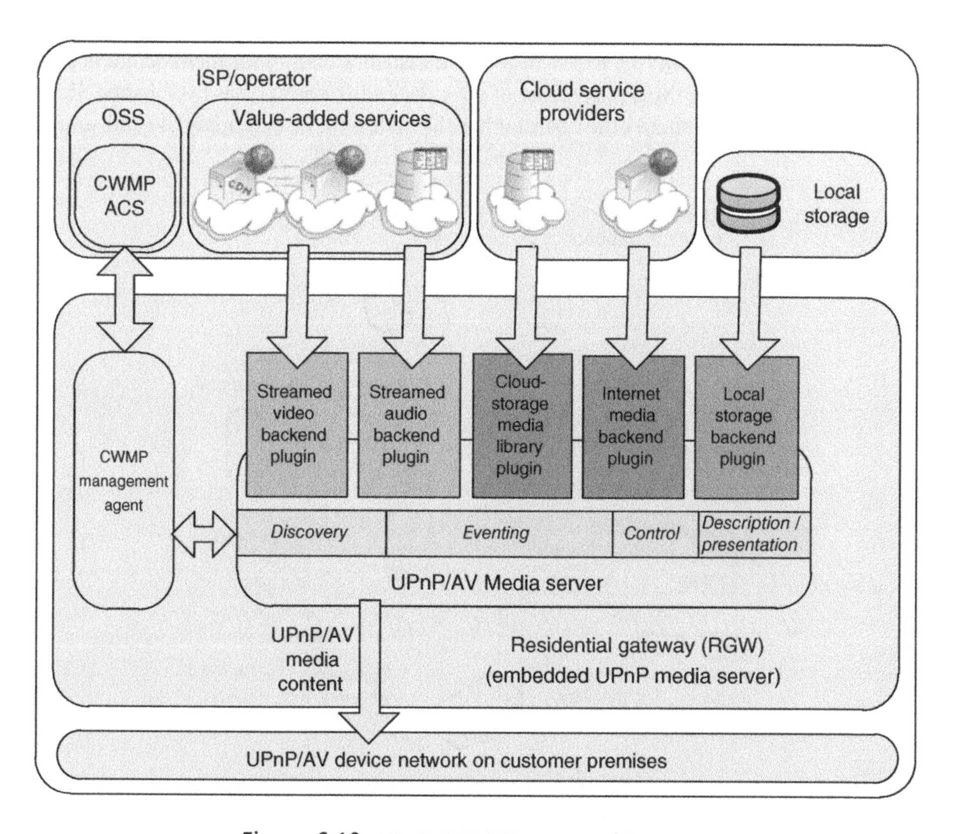

Figure 6.10 UPnP AV RGW server architecture.

specific media services from Internet or content providers; finally, by adopting a standard framework for media delivery, it also becomes possible to avoid the problems arising from protocol or Application Program Interface changes for specific media services—a situation that can render devices useless until a firmware update is available, something which may never happen, depending on their support status.

This solution allows for seamless and secure media distribution to a wide range of devices that already support those protocols, without the need for pushing another protocol stack or specific device into the domestic LAN for such purpose, extending the reach of the DLNA device ecosystem beyond the LAN scope.

The media server embedded on the RGW is based on a modular architecture that makes use of plug-ins (which can be dynamically deployed and activated) to deliver a wide range of contents and services available outside the LAN environment—abstracted as media items residing on a regular UPnP AV media server advertised on the domestic LAN. CWMP is used to manage the UPnP AV media server embedded on the gateway, allowing the operator to customize which specific plug-ins are available and the configurations of each active instance. For this purpose, a CWMP bridging subagent was developed with the purpose of mapping the configuration of the UPnP AV media server and its plug-ins into the CWMP data model of the RGW, so that the operator ACS can remotely manage all media server properties.

Using this framework, the RGW becomes a distribution hub for media content provided by the operator or third-party content providers with whom it has established SLAs (Figure 6.11). In this scenario, the operator has the means to provide adequate end-to-end

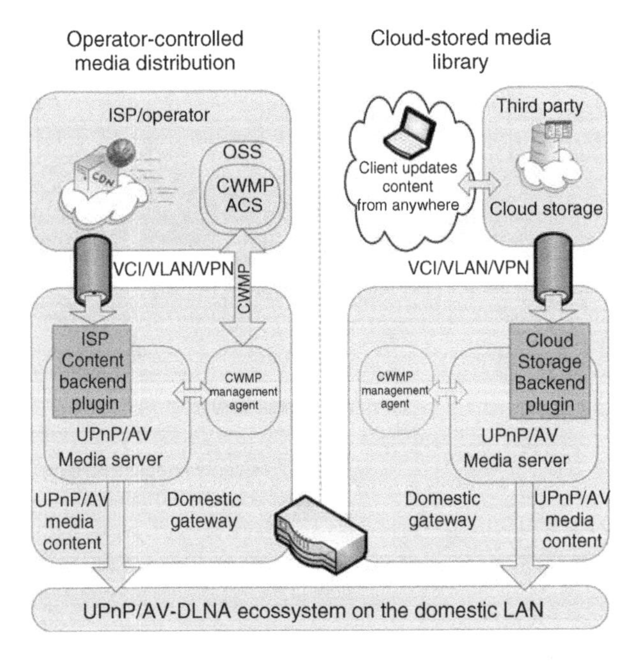

Figure 6.11 Application scenarios for DLNA-integrated content delivery.

security and QoS transport features, enabling it to deliver media content to all kinds of UPnP AV or DLNA-enabled media playing devices inside the subscriber premises—in a controlled way, without directly exposing the backend infrastructure.

This solution can also be used to provide access to third-party Internet media service content or cloud service media libraries, enabling users to retrieve and play such content on DLNA-compliant devices. As an example of the latter case, a user traveling abroad may sync its photo collection to the cloud storage or media library service to share with his family back at home, which will be able to watch the pictures on a Smart TV or other UPnP AV/DLNA renderer.

This feature can be marketed a value-added service from the operator, which can establish SLAs with third-party content providers to expand its offer. When compared with unmanaged solutions for Internet-to-DLNA streaming, such as Skifta [38], it offers several advantages such as device neutrality, not requiring any special applications nor dedicated mediation devices, and also going beyond the OTT paradigm by involving operators to ensure QoS and management support.

6.4.3 Improving Cloud Storage for Media and Content Distribution

This solution [12] consists of an operator-managed hybrid storage solution that combines the benefits of standalone storage appliances with cloud storage. For this purpose, each RGW is transformed in a storage hub and given its own local storage resources (solid state or hard disk) whose contents are made accessible to the customer premises LAN using standard protocols such as SMB/CIFS [39], FTP [40], or HTTP(s) [41] and kept synchronized with a cloud storage container on the operator infrastructure.

This approach as several advantages: first, it relieves users from the task of configuring and managing their own storage devices; second, it takes advantage of the fact that the RGW is permanently powered on (especially in triple-play environments) in order to provide connectivity services for the LAN, therefore eliminating the need for a separate appliance (and its cost of ownership); and third, it provides redundancy and reliability by replicating data to a virtual container located outside the customer premises, on the storage provider infrastructure, instead of relying on self-contained data replication methods, such as RAID (which can be used for extra redundancy). In addition, this solution has the ability to stream content to the customer premises network at LAN speeds (once it is synchronized), while supporting disconnected operation. Versioning is also supported, allowing the user to recover previous versions of a specific file.

Every RGW associated with the same storage service subscription is allowed to synchronize with the central repository, making this solution adequate for both multibranch SOHO infrastructures and home users with a service subscription for a single household. Roaming users might also use native client applications to access their storage container—Figure 6.12 illustrates the proposed architecture.

In this context, storage service management interfaces and eventing mechanisms are supported by CWMP. Each RGW incorporates a specific integration component that acts as module of the RGW CWMP agent, interacting with the ISP ACS (the CWMP management server) and with the local storage service components. On the operator

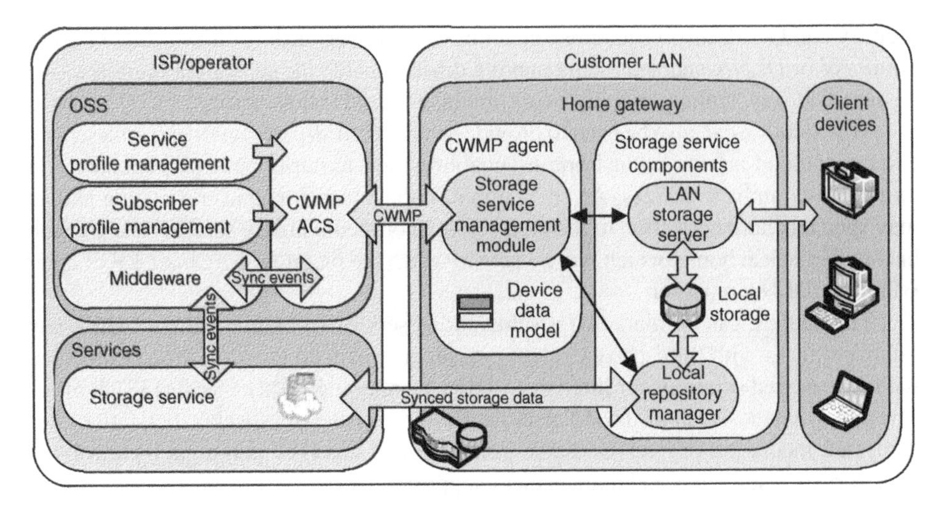

Figure 6.12 Integration architecture for the managed cloud storage service.

side, the ACS interfaces with the storage service front-end by means of a middleware layer (a message-oriented queuing system) for passing events back and forth between them. This solution may also enable new value-added services.

6.4.3.1 Storage Synchronization for Domestic Users or Enterprise Branches A domestic user with several residences could use this storage service to keep a synchronized (and safe) document and media repository between all households. This service could also be integrated with a CDN backend to distribute software updates, being marketable by operators in communication service bundles for small and medium enterprises, providing a managed solution with end-to-end QoS enforcement.

6.4.3.2 Hybrid NAS for Home Users Most low cost dedicated NAS appliances are not redundant by nature, being little more than network-attached disks with dedicated firmware—as such, this storage model could be used to add redundancy, replicating data to an external storage container. In this perspective, an operator could market this service in a converged service bundle, supported by RGWs or dedicated NAS devices.

6.4.3.3 Content Distribution for Home Users This storage service could be used by an operator as a content delivery mechanism for media, integrated with a CDN backend. Users could, for instance, buy movies from an online store and have it placed on their storage container, propagated to all devices associated to the same account. Also, this service could be used to provide software and firmware updates to devices that the user has acquired, right to the customer premises.

6.5 FUTURE TRENDS

Although IPv4 still dominates the majority of broadband access network environments, most operators are planning the move to IPv6. IPv6 brings new challenges because of the need of both infrastructure/protocol compliance and interoperability between IPv6 and IPv4. For this purpose, guidelines have been published by both the Internet Engineering Task Force (IETF) [42] and the Broadband Forum [43]—with development still undergoing in the latter case [44,45]. While the transition to IPv6 might have a mid-to-short term architectural impact (IPv4 NAT will likely disappear with time, with several customer LANs remaining IPv4-based for long), its influence does not significantly affect the proposals presented on this chapter, which remain pertinent regardless of the IP protocol version.

Also, with the emergence in some countries of the multiprovider, multiservice paradigm, telecommunication operators still resist the idea of allowing third parties to provide services using their infrastructure. Yet, this service model shift is proving to be disruptive, causing a profound change in services and its perception. We are presently in a transition stage that is evolving at different paces: while in Europe the most common scenario is still based on single-sourced, operator-provided services (mainly data, voice and television), in some other countries (like United States), third-party services are beginning to emerge. It is expected for the broadband service paradigm to evolve into an ecosystem where multiple providers coexist together with the telecommunications operator that provides connectivity, infrastructure management, and its own value-added services.

6.6 SUMMARY AND CONCLUSION

This chapter provided an overview on the subject of cloud service delivery in broadband access networks, unveiling its relationship with topics as diverse as technologies, standards, devices, and services. It has covered some of its most relevant issues, as well as current trends, also emphasizing the importance of the access network and the RGW as instrumental mediators between communication and content/service providers in a multiservice scenario. Virtualization techniques for the access network and application scenarios for device and service virtualization were also discussed as well, illustrating the potential of virtualization technologies for cloud-sourced content delivery.

ACKNOWLEDGMENTS

This chapter is partially derived from the authors' past work, especially Reference 46. We would like to acknowledge support from PT Inovação and Project QREN ICIS (Intelligent Computing in the Internet of Services—CENTRO-07-0224-FEDER-002003).

REFERENCES

1. Broadband Forum. TR-069—CPE WAN Management Protocol spec. v1.1, Amendment 4; July 2011.
2. Royon Y. Environments d'exécution pour paserelles domestiques [PhD Thesis]. Institut National des Sciences Apliquées (INSA/INRIA); December 2007.
3. Balemans H, Hartog F, Önnegren J, Smedt A. Concurrent remote management of CPE by multiple service providers. Proceedings of the Broadband Europe Conference (BBEurope 2006); 2006 Dec 11–14; Geneva, Switzerland; 2006.
4. Broadband Forum. TR-131: ACS northbound interface requirements, issue 1; November 2009.
5. Broadband Forum. TR-134: Broadband policy control framework (BPCF), issue 1; July 2012.
6. Yvan R, Pierre P, Stéphane F, Serafeim P, Humberto A, and Dirk Van de Poel T. Multi-service, multi-protocol management for residential gateways. Proceedings of the Broadband Europe Conference (BBEurope 2007); Antwerp, Belgium; 2007.
7. Broadband Forum. Component objects for CWMP, TR-157 Amendment 5; November 2011.
8. Fontes F. Operator's network evolution and NGN. Proceedings of AICT 2005 (Advanced Industrial Conference on Telecommunications); Lisbon, Portugal; July 2005.
9. Bastos F, Cruz T, Laranjeira A, Monteiro E, Reis N, Simões P. An architecture for virtualized home gateways. Proceedings of IFIP/IEEE International Symposium on Integrated Network Management (IM 2013); Ghent, Belgium; May 2013.
10. IEEE 802.1 Working Group. IEEE Std. 802.1Q-2011, media access control bridges and virtual bridged local area networks; 2011.
11. Abgrall D. D1 - virtual home gateway, EURESCOM P2055 Study Report; September 2011.
12. Bastos F, Cruz T, Laranjeira A, Monteiro E, Rodrigues J, Simões P. Managed hybrid storage for home and SOHO environments. Proceedings of IFIP/IEEE International Symposium on Integrated Network Management (IM 2013); Ghent, Belgium; May 2013.
13. Mockapetris P. Domain names—implementation and specification, STD 13, RFC 1035. USC/Information Sciences Institute; November 1987.
14. Droms R. Dynamic host configuration protocol, IETF RFC 2131; March 1997.
15. IEEE 802.11 Working Group. IEEE 802.11: Wireless LAN medium access control and physical layer (PHY) specifications; 2007.
16. Cain B, Deering S, Kouvelas I, Fenner B, and Thyagarajan A. Internet group management protocol, Version 3, IETF RFC 3376; October 2002.
17. Broadband Forum. TR-156: Using GPON access in the context of TR-101; September 2010.
18. Rosenberg J, Schulzrinne H, Camarillo G, Johnston A, Peterson J, Sparks R, Handley M, and Schooler E. SIP: Session initiation protocol, IETF RFC 3261; June 2002.
19. Broadband Forum. TR-101: Migration to Ethernet-based broadband aggregation, issue 2; July 2011.
20. Young G. Broadband Forum overview with focus on next generation access. 14th UK Network Operators' Forum Meeting; September 2009.
21. IEEE 802.1 Working Group. IEEE Std. 802.1ad-2005, virtual bridged local area networks Amendment 4: Provider bridges; March 2006.

22. Joseph V, Mulugu M. *Deploying Next Generation Multicast-Enabled Applications: Label Switched Multicast for MPLS VPNs, VPLS, and Wholesale Ethernet*. Morgan-Kaufmann; 2011. ISBN: 0123849233.

23. Bryant S, Pate P. Pseudo wire emulation edge-to-edge (PWE3) architecture, IETF RFC 3985; March 2005.

24. Patrick M. DHCP relay agent information option, IETF RFC 3046; January 2001.

25. Cruz T, Monteiro E, Simões P. Integration of PXE-based desktop solutions into broadband access networks. Proceedings of the 6th International Conference on Network and Services Management (CNSM 2010); Niagara Falls, Ontario, Canada; IEEE Press; 2010. p 182–189.

26. Intel Corporation. Preboot execution environment (PXE) specification version 2.1; September 1999.

27. Cruz T, Simões P. Enabling PreOS desktop management. Proceedings of the IM'2003 (IFIP/IEEE Int. Symposium on Integrated Network Management); Colorado Springs; May 2003.

28. Sollins K. The TFTP protocol (Revision 2), IETF RFC 1350; July 1992.

29. Intel Corporation. Boot integrity services API version 1.0; December 1998.

30. Anvin HP, Connor M. x86 network booting: Integrating gPXE and PXELINUX. 2008 Ottawa Linux Symposium; 2008.

31. Microsoft Corporation. Remote desktop protocol: Basic connectivity and graphics remoting specification, Revision 19; June 2010.

32. Harder J, and Maynard J. Technical deep dive: ICA protocol and acceleration; July 2009.

33. Ketcham M, Mirlacher T, Monpetit M. IPTV: An end to end perspective. Acad Publish J Commun 2010;5(5):358–373. DOI: 10.4304/jcm.5.5.358-373.

34. Husain S, Siebert P. Business case and technical requirements for IPTV STBs. Proceedings of the IEEE International Symposium on Broadband Multimedia Systems and Broadcasting; 2009 May 13–15; Bilbao, Spain; 2009.

35. Bastos F, Cruz T, Laranjeira A, Monteiro E, Simões P. A framework for Internet media services delivery to the home environment. Springer's J Netw Syst Manage March 2013;21(1):99–127. DOI: 10.1007/s10922-012-9228-2.

36. UPnP Forum. UPnP AV architecture: 1 for UPnP Version 1.0; 2008.

37. DLNA Consortium. DLNA networked device interoperability guidelines; 2009.

38. Qualcomm, Inc. 2001. Skifta media-sharing service. Available at http://www.skifta.com. Accessed 2013 Apr 01.

39. Microsoft Corp. Microsoft SMB protocol and CIFS protocol overview; March 2011.

40. Postel J, and Reynolds J. File transfer protocol, IETF RFC 959; October 1985.

41. Fielding R, Gettys J, Frystyk H, Masinter L, Leach P, and Berners-Lee T. Hypertext transfer protocol—HTTP/1.1, IETF RFC 2616; June 1999.

42. Asadullah S, Ahmed A, Popoviciu C, Sakvola P, and Palet J. ISP IPv6 deployment scenarios in broadband access networks, IETF RFC 4779; January 2007.

43. Broadband Forum. BroadbandSuite 4.1. Available at http://www.broadband-forum.org. Accessed 2013 Apr 01.

44. Broadband Forum. TR-242: IPv6 transition mechanisms for broadband networks; August 2012.

45. Broadband Forum. WT-296: IPv6 transition mechanisms test plan, Working Text; 2013.

46. Cruz T. A management framework for residential broadband environments [PhD dissertation]. Department of Informatics Engineering, University of Coimbra, Portugal; 2011.

7

MOBILE VIDEO STREAMING

Ram Lakshmi Narayanan, Yinghua Ye, Anuj Kaul, and Mili Shah

Nokia Siemens Networks, Mountain View, CA, USA

7.1 INTRODUCTION

It is widely observed that video content contributes to major portion of the Internet traffic. Both user-generated content (UGC) traffic from YouTube [1] and premium traffic such as Netflix [2] are major contributors in the United States. It is anticipated that the video traffic is expected to increase by many folds due to the increase in penetration of smartphone and tablets [3]. Such a sudden explosion of video traffic makes network unmanageable. Lessons learnt from deploying fourth-generation mobile network such as long-term evolution (LTE) and 3G mobile network such as wideband code division multiple access (WCDMA) technologies have shown that video streaming does not perform well over mobile broadband networks. Major reasons for such poor performance are too many parallel developments in services, applications that are not network friendly and demand more data, and smartphones consuming energy than they required. To manage network traffic, wireless operators move from flat rate billing to the tiered billing. Though such measure is considered temporary, but it clearly puts the requirement to improve the network infrastructure and service utilization.

Advanced Content Delivery, Streaming, and Cloud Services, First Edition.
Edited by Mukaddim Pathan, Ramesh K. Sitaraman, and Dom Robinson.

Video streaming service has strict requirements for latency and bandwidth throughout the video play. To satisfy these requirements, mobile operators have increased their server and network capacity. However, wireless signal strength varies with respect to location, time, and environment; delivering guaranteed bandwidth to video streaming application in such nonuniform wireless network condition becomes a challenge.

Traditionally, to ensure guaranteed service to users, operators measure packet loss, jitter, delay, available bandwidth, and other network-related parameters as part of quality of service (QoS). However, standard techniques and protocols to improve QoS are not widely deployed in the Internet, video streaming is delivered as best effort traffic over the Internet. Measuring network-level information alone is not sufficient to assess quality of video service; therefore, the perceived user quality as quality of experience (QoE) needs to be included when network and services are designed [4].

Mobile networks have constraints such as energy consumption, limited radio spectrum, radio propagation characteristics, and interference. It is not possible to deliver video streaming data without adaptation at client or server or network. The objectives of this chapter are describing the challenges in delivering video in mobile networks, presenting various adaptation technologies for mobile video streaming, and identifying a few potential research problems to improve mobile video steaming service. This chapter is organized as follows: Section 7.2 gives an overview of mobile broadband architecture; basic video streaming protocols are described in Section 7.3; Section 7.4 presents how video optimization services have been done in mobile networks; Section 7.5 demonstrates several content delivery network (CDN) deployment scenarios in mobile operator networks; cloudification has become mainstream in providing network services, and in Section 7.6, we present cloud-based video streaming architectures. Finally, Section 7.7 pinpoints future research problems to improve mobile video steaming service.

7.2 MOBILE BROADBAND ARCHITECTURE

3G and 4G mobile broadband networks are shown in Figure 7.1 and are composed of IP core network, radio network, and transport network. For simplicity, many of the network functions and interfaces are omitted. Transport network is also called as wireless backhaul network; it connects radio network and Internet Protocol (IP) core network. The transport network can be fixed or wireless network depending on the deployment. In most cases, the transport backhaul has limited capacity.

3G radio access network (RAN) shown in Figure 7.1(a) is composed of NodeBs and radio network controller (RNC), whereas LTE radio network shown in Figure 7.1(b) consists of eNodeBs. The functions of RAN include radio resource management, radio transmission and reception, channel coding and decoding, and multiplexing and demultiplexing. 3G packet core consists of gateway GPRS support node (GGSN) and serving GPRS support node (SGSN) while mobility management entity (MME) and serving and packet gateway (S/P-GW) are part of evolved packet core (EPC) in LTE. GGSN/SGW/PGW function is to manage IP sessions and perform various supplementary services such as legal interception gateway functions, policy-based routing, and charging functions. The mobility sessions are controlled and managed by MME in LTE and SGSN in 3G network, respectively.

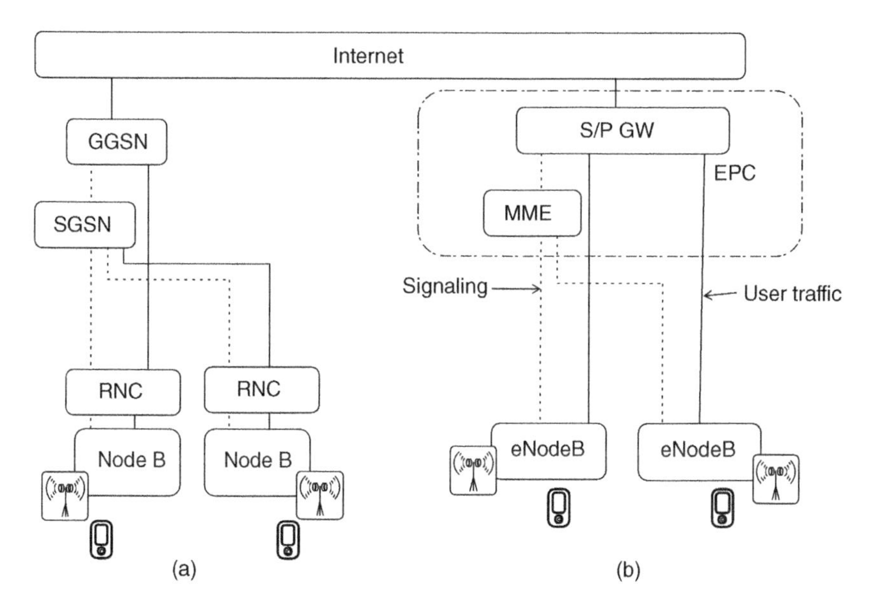

Figure 7.1 Mobile broadband networks (a) 3G network and (b) LTE network.

7.3 VIDEO STREAMING PROTOCOLS

HyperText Transfer Protocol (HTTP) and Real-Time Streaming Protocol (RTSP) are two popular protocols used for video streaming. RTSP uses Real-Time Transport Protocol (RTP) and it is often blocked by firewalls; thus, it is not widely used. HTTP uses Transmission Control Protocol (TCP) to ensure that reliable data transfer and CDNs are enabled to deliver HTTP services over the Internet. Thus, HTTP, a stateless protocol, which is well established and widely accepted for Web browsers, firewall traversal, Network Address Translation (NATs), and CDNs, becomes suitable choice for video streaming technology.

7.3.1 Progressive Streaming Over HTTP

Progressive download is a technique wherein video starts playing on the client devices without waiting for entire video download. The expectation is that video must start immediately after user has selected to watch a video content and should not stall during the play. Figure 7.2 shows an example of HTTP message flow for progressive streaming.

In Figure 7.2, when user requests a specific video, HTTP GET message 1 for the requested video is sent to video server located geographically close to the client [5]. This HTTP GET message contains a unique video identifier. After receiving the GET message, the video server replies with an HTTP redirect message 2. This message redirects the client to the video server from which the video will be streamed. Client sends HTTP GET message 3 to request the content. Upon receiving HTTP GET message 4, the server starts streaming video content. Client does not wait to download entire video

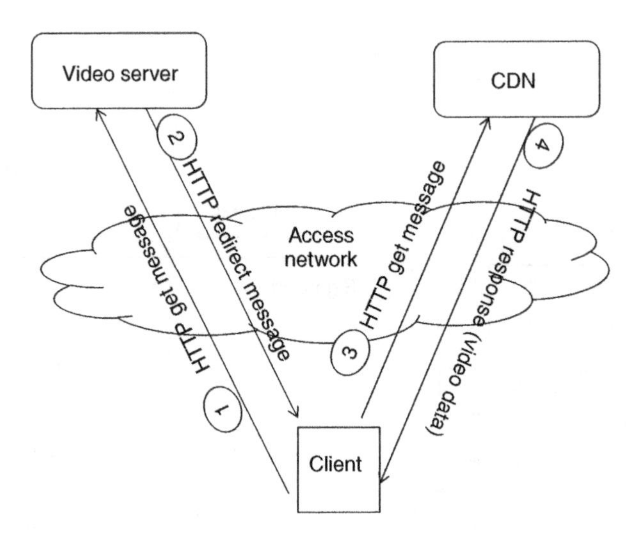

Figure 7.2 Video streaming message flow.

data to start the video. As soon as the required amount of video data is available to the client, video playback starts while downloading of the video data continues.

7.3.1.1 Buffer Management For on-demand progressive streaming service, several improvements have been made to get better QoE in both client and server. Here, key QoE metrics are initial video start time, the number of stalls, and the duration of each stall. For better QoE, several buffer management techniques on client have been adopted. The principle behind the buffer management is to supply client application with sufficient data to avoid video rebuffering during play. Amount of data filled in client buffer is divided into three levels, namely minimum, lower threshold, and upper threshold. The key is to choose correct values for these defined levels. If minimum level in client buffer is too small, video playback would start instantly but cannot tolerate network inconsistencies. And if it is too large, though it can restrain network inconsistencies, initial playback time would be more which would have bad user experience. Based on selection of these values, different patterns of progressive streaming download are observed [5].

GREEDY BUFFERING. In this scheme, video starts playing when minimum data are available in the buffer. There is no maximum limit, and the buffer depth depends on application implementation. During the video playback, server continues to send data till the client buffer is full. Here, server sends the data to client buffer as quickly as possible. The advantage of this approach is its simple implementation, but the disadvantage is data wastage if user terminates the session prematurely.

CONTROLLED BUFFERING. This scheme uses all three thresholds, namely, minimum, lower, and upper threshold. Initially, server sends data to client buffer as fast as possible to reach minimum threshold for video start. When requested video bitrate is lesser than available bandwidth, downloading of data will be faster. Then server conducts throttling until the upper threshold value is reached, client requests server to stop sending data. Meanwhile, buffer depletes as the playback continues. As soon as the buffer level reaches lower threshold, client requests server to start sending data again. This pattern repeats till entire video is streamed. Reducing data wastage and providing high resilience to network inconsistencies are advantages of this implementation.

7.3.2 Adaptive Bitrate Streaming

In progressive streaming, due to lack of adaptation to changing network condition, rebuffering often occurs when network bandwidth reduces. Adaptive bitrate (ABR) streaming overcomes the shortcoming of progressive streaming, supports adaptive bitrate switching based on network conditions for both live and on-demand sessions, and minimizes the wastage if user prematurely terminates the video session. Adoption of ABR-based streaming has increased since the introduction on both fixed and mobile networks. Since Chapter 2 has described ABR streaming ecosystems in detail, in this section, we only provide a quick overview of ABR streaming protocol. ABR streaming process is divided into three phases, namely, media creation and processing; segment selection and streaming; and client adaptation. Figure 7.3 illustrates ABR media creation and processing. The raw video file generated via digital video camera or camcorders is encoded at different bitrates. The encoded files are further fragmented

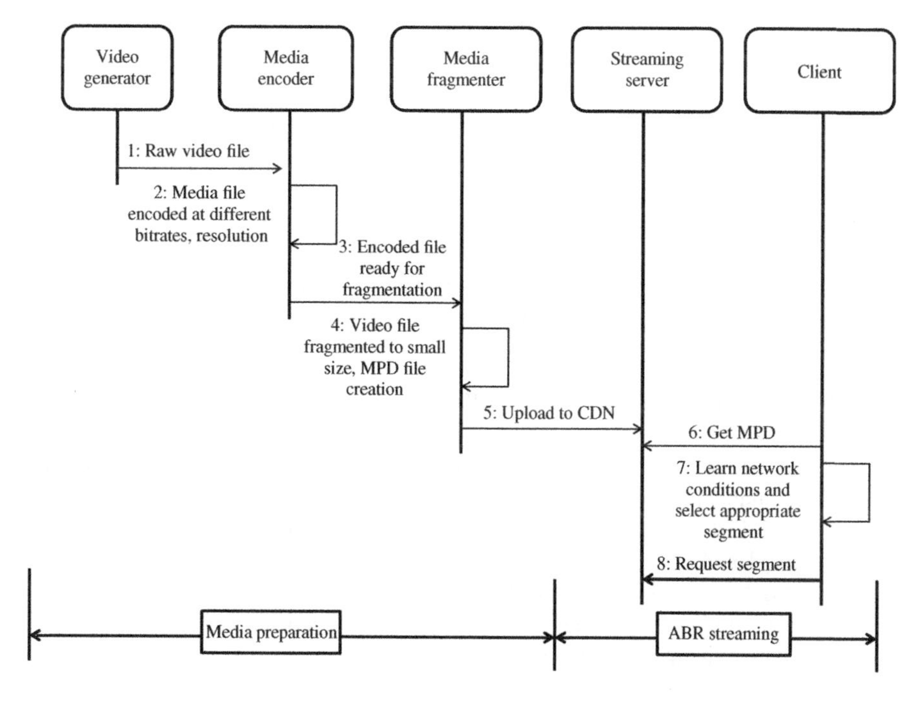

Figure 7.3 Media creation and processing.

into smaller files called segments or chunks or fragments. Typically, the size of each segment contains video duration from 3 to 10 s. After fragmentation, server generates Media Presentation Description (MPD) file, it contains information related to the segments, such as video bitrate, CODEC used, duration of the segments in seconds, video resolution, and location of the segments. Then video is pushed to the CDN for delivering the files.

After the video files are made available to CDN, client application can make request for these contents through HTTP get. The server responds with MPD file, client parses MPD file, and selects appropriate bitrate segment based on its adaptive algorithms. To achieve smooth starting, usually client starts with the lowest bitrate segment, then it gradually requests for higher bitrate segment till the bitrate of segment matches the available bandwidth. The time taken by the application to adjust from the lowest quality video to desired quality of video for a given network bandwidth is called initial reaction time. Initial reaction time is an important parameter to measure the quality of adaptation algorithm. After the first successful adaptation, client continuously learns the network conditions and computes available bandwidth for future request.

When network get congested, client adapts and requests lower bitrate chunks, and vice versa. When to switch UP and DOWN based on network condition is a continuous learning process. As network changes are time variant and nonpredictable, how quickly the application learns about such network changes and adapts to the appropriate bit stream is critical for client algorithm design. When higher fluctuations in bandwidth are observed, too frequent switching may result in downloading multiple segments for same content. This further creates data wastage and leads more network congestion. Client algorithm must be optimized to avoid frequent switching and also late switching.

Smooth Streaming Server from Microsoft, Dynamic Streaming by Adobe, and HTTP Live Streaming by Apple are variants of ABR technology [6]. Several studies were conducted and evaluated for video streaming performance of these streaming technologies in terms of observed rebuffering events, the number of switching events, average bandwidth utilization, and time-varying network conditions [7,8]. It is evident even though these variants have similar functionalities, but they are not compatible in operations. 3GPP has proposed dynamic adaptive streaming over HTTP (DASH) as part of standards to ensure interoperability among these implementations [9–11].

7.4 VIDEO OPTIMIZATION SERVICES

Original video that have been coded in high frame rate, high resolutions, and formats may not be suitable for mobile networks. It is difficult to stream smoothly due to heterogeneity in mobile devices types, supported formats, and significant bandwidth variations on radio link. To address these issues from mobile network, video optimizer (VO) and CDN-based delivery techniques have been adopted. This section illustrates how these techniques are applied in mobile networks to deliver video streaming in an effective manner while maintaining good QoE.

VO is a technique that is commonly adopted by the operator. Principle behind video optimization is to transform one or more video properties to adapt to network conditions and device capabilities without noticeable degradation to QoE. Optimizer is a middle-box residing transparently between user terminal and video server. VO provides two functions, namely, media aware optimization such as transcoding, and network adaption-based optimization, such as for instance traffic pacing, smart caching, and rate control [12,13].

7.4.1 Video Transcoding

Video transcoding is an operation that modifies source video streams to target video stream based on transcoding inputs. In order to convert each source video stream, network adaptation function inside VO performs deep packet inspection (DPI) of client request, collects video request type, device preference, learns various network conditions, and combines all these information to come up with appropriate input transcoding parameters. Transcoders are broadly classified as homogeneous and heterogeneous transcoders [14]. In homogeneous transcoders, transcoding happens without changing to video format, whereas heterogeneous transcoders performs conversion of source to different target video formats. Video transcoding converts from one video stream to another video streams and includes functions such as format conversion, bitrate conversion, spatial resolution reduction, temporal resolution reduction, and error resilience. To support transcoding functions, three types of transcoders architectures exist, namely, cascaded-pixel transcoder, open-loop transcoder, and closed-loop transcoder.

- *Cascaded-Pixel Transoder*. In cascaded-pixel transcoder, source video stream is fully decoded first, then encoded again to target video format. The quality of the target video stream that is achieved by this transcoding operation is high, and user is not able to notice degradation. The operation of decoder–encoder is computationally intensive and not suitable for real-time streaming application.
- *Open-Loop Transcoder*. To overcome the computation complexity of cascaded-pixel transcoder, open-loop transcoder is used. In open-loop transcoder, on the fly, information from the source video such as variable length is decoded to extract the variable length code words referring quantized Dropped-cell Rate (DCR) coefficients, along with macroblocks. Then, it is translated to target video bit stream by remapping video motion vectors, requantizing to target bitrate, and applying variable length encoding. Remapping process involves predictive coding technique wherein the coded frame is predicted from previous frames and only the prediction error is coded, and the prediction error accumulates during the entire Group of Pictures (GOP) period. When decoding is done at the client, it results in drift errors. This is the drawback of this approach.
- *Closed-Loop Transcoder*. It addresses drift error problem by taking approximation of cascaded-pixel transcoder and feeding it as feedback to compensate for initial prediction errors. Closed-loop transcoder provides balance between quality and computational load compared to cascaded-pixel domain and open-loop transcoders.

7.4.2 Network Adaptation

Network adaptation performs following activities:

1. *Transcoder Inputs*. In order to dynamically adapt network changes and support device diversity, transcoder needs the following information as inputs:
   - *Available Bandwidth to Users*. It depends on the degree of network congestion and user's subscription.
   - *Device Type*. This decides the screen size of the target play.
   - Video format information contained as part of metadata in first few packets of source video stream.
   - *Content Policy*. This depends on the agreement between mobile operator and content provider, this policy could include whether the content is allowed to be modified/cached.
2. *Media Streaming Functions*. For on-demand video sessions, even though the transcoder is adapting the content, it still requires network-level smoothing functions to minimize wastage of data and ensure that the client does not experience rebuffering. Some of the popular techniques are (i) pacing of packets used to avoid excessive buffering of data in the client buffer and wastage of data. (ii) Throttling the available bandwidth or delivering the data in shorter burst based on network conditions.

VuClip has confirmed the effectiveness of VO in mobile network [14]. In VuClip, 30 days of traces from video servers were collected, the traces reveal that there are more than 2000 device model type, tens of different video format resolutions, and large variation in number of supported operating system and native CODEC. It is observed that mean video duration of 162 s, and video traffic distribution follows Zipf-like distribution pattern. VuClip performs transcoding depending on user devices and their capabilities, and it has been observed that around 73% of requested video content is HD quality that requires optimization.

Studies have shown that transcoding combined with caching saves mobile bandwidth [16]. When a client requests for video stream, the data are fetched from the video server and transcoded to target bit stream on the fly. There is an option to store either the source or target video bit stream locally to serve future requests. The advantage of caching the transcoded bit stream is to avoid repeated computation-intensive transcoding operation. But this approach shifts the problem of compute to storage of video content, and we need to have algorithms to intelligently cache only popular content.

Caching content at network intermediary does not involve content owner. It may break video content provider business model. They run the following supplementary service for their revenue generation and improving their services:

- Advertisement.
- Analytics to improve the content quality and online user rating of video service.
- Reporting and removal of inappropriate user-generated content from the network.

7.4.3 Video Optimizer Deployment

Effectiveness of VO depends on where it is positioned. VO can be inside the core network, radio network, or outside of the operator network. A few deployment architectures are presented later.

7.4.3.1 VO as Part of Packet Core Network 3G network is used as an example to illustrate this deployment. VO can be either integrated to GGSN or collocated with GGSN. In the integrated case, GGSN performs all VO functions and sends the optimized video traffic to user. When VO is collocated with GGSN, media aware functions are handled by VO, and network adaption functions are handled by GGSN. Such functionality split increases the scalability of the service. The shortcoming of this approach is when mobile bandwidth changes over time and location, VO does not adapt to such changes [17,12].

Even though VO has been proved to be effective to deliver video streaming for various devices under dynamic network conditions, It still has the drawbacks given altering the content is required, (i) VO functionality breaks net neutrality, (ii) only partially solves video streaming delivery problem, (iii) does not address latency issue and does not work for encrypted contents. Hence, operator-hosted, CDN-based approaches are explored to address these drawbacks.

7.5 OPERATOR-HOSTED CDN

Mobile traffic increase prompts to have CDN inside the mobile network. It is observed that latency in the mobile network is at least 70 ms [15]. Studies have shown that geographical location dominates latency and throughput [18]. Therefore, it is desired to move the content close to mobile users. Content for CDN is prepared by content owner. The content preparation is also called as ingestion, this process includes the following steps: (i) take raw video streams and convert them to different encoding formats; (ii) fragment the content in case of ABR, (iii) creates the video files, and (iv) push video files to CDN.

Like VO, effectiveness of CDN can be determined by where it is placed in the network and how it improves network and service performance.

7.5.1 CDN as Part of Mobile Packet Core Network

In this architecture, CDN network is collocated with GGSN. The advantage of this solution is that data can be prefetched to CDN prior to the use. How the data are prefetched involves many combinational factors including user preference, network load, location of user, and video recommendation system. The effectiveness of algorithms is still a research topic.

CDN at mobile core does not give substantial gain because internal transport backhaul network has main bottleneck for bandwidth and delay. When concurrent accesses are made to popular content, multiple copies of content have to be delivered from the core network, this results in transport network congestion.

7.5.2 CDN as Part of Radio Access Network

Making CDN as part of RAN improves the QoE of users and reduces the transport backhaul usage. This solution requires RAN perform DPI and do intelligent prefetching of content. Cache hit ratio versus cache size needs to be dimensioned. There are several approaches to learn the data on the fly and populate the CDN [19], wherein base station acts as a rendezvous point for video streaming traffic, and the user content is prefetched from the CDN, and for subsequent access, the content is delivered from the local CDN. Coded distributed caching [20] proposes an architecture for small cell-based networks. Small cells are deployed by the operator to share the load of macrobase stations. In this architecture, storage capacity inside each small cell is used as a CDN cache. There is always a challenge to use storage in small cells for CDN because the number of users behind the cells is limited, and the probability that users request same video is low.

7.5.2.1 Hierarchical Mobile CDN This is a hybrid approach to overcome the transport network problem and increase the cache hit inside the mobile network, CDN are deployed at both RAN and packet core. Netflix encourages operators to deploy the CDNs based on their suitable architecture, so as to minimize the traffic crossing the operator network [21]. By considering hierarchical CDN distribution has shown that there is 24% increase in cache hit ratio and 45% savings in transport network capacity [22].

Irrespective of where we place CDN, the following are the key findings that need to be considered to improve effectiveness of CDN:

- Understanding video usage and its traffic pattern is key for deploying CDN.
- Studies have shown that almost 80% of mobile data traffics are generated by fewer users and come from fewer cells. Effectiveness of CDN depends on the combination of traffic and user concentration [23]. Netflix content provider prepares same movie content into 120 different downloadable formats, and such different types of files are worth storing at centralized CDN as opposed to small cell nodes [24].

7.5.3 Mobile CDN Use Case Deployment

Having described various deployments of CDN for mobile networks, studies in Reference [25] show that there is a significant benefit in bringing the content inside the mobile networks. The improvement is achieved by combining information including user preference, device preference, location, and flow information as part of mobile CDN delivery. Until now, operator-hosted content has not become so popular even though we see operators around the world are experimenting converged services to deliver TV, voice, and video over mobile networks. On the basis of the following public References 26–28, there are the few CDN solutions proposed by vendors for mobile operators: (i) Telekomsel Indonesian operator has chosen an integrated packet core with CDN from Ericsson and Akamai [26]; (ii) CDN solutions are being deployed as part of Radio Access Cloud Service (RACS), vendors such as NSN Liquid Applications and CDN are providing solutions to manage flows at radio access level [27]; (iii) There are several video delivery providers such as Jetstream is offering solution that can be integrated at different points inside mobile networks.

7.6 CLOUD-BASED VIDEO STREAMING

Cloud computing has proven its success in delivering enterprise services over the network due to its capabilities to support elastic computing, ubiquitous access, low cost, easy maintenance, and pay-as-you-go. Here, elastic computing refers to the abilities that services can dynamically utilize computing resources in the cloud based on real-time demands.

Recently, mobile vendors and operators are working toward moving mobile video optimization services into cloud extensive studies have been done in cloud-based mobile video optimization space [29–32]. In this section, we describe two important topics in this area: mobile cloud video optimization service offered by operators and scalable video coding (SVC)-based mobile cloud streaming techniques.

7.6.1 Cloud-Based Mobile Video Streaming Service

The basic principle of such service is to leverage the elastic computing feature of the cloud to overcome the scalability issues faced by hardware box-based mobile video optimizer. For instance, in the case of traffic spikes, cloud-based mobile video optimization service allows operators to expand video optimization capacity instantly. The key mobile network element to connect mobile networks and cloud needs to support DPI function and cloud access. GGSN/P-GW, RNC, and eNB can be candidates for such key network element. DPI function recognizes the video session via inspecting the first HTTP request URL. Once video session is identified, HTTP request is forwarded to mobile cloud video optimization service. Upon receiving the request, cloud-based optimizer, further requests the content from content server inside CDN, performs video optimization, and sends the optimized content to UE through mobile networks.

Figure 7.4 illustrates a typical architecture of cloud-based mobile video optimization service, where mobile video traffic comes to mobile network through GGSN/PGW from operator cloud. Mobile video cloud optimization service resides inside operator cloud and consists of mobile video cloud optimizers (MVCO) and mobile video cloud controller (MVCC). MVCO are responsible for performing video optimization. MVCC performs the following functions:

- *Monitoring and Dynamic Provisioning MVCOs*. When it detects MVCO is overloaded, it launches new MVCO instance. On the other hand, when MVCOs have light load, MVCC consolidates video streaming sessions to fewer MVCOs and shuts down unnecessary MVCOs. This elastic feature can address the scalability issue of hardware-based solution, and is expected to drastically reduce the cost.
- *Load Balancing*. The first HTTP request for each video session always goes through MVCC. Depending on the load of each running MVCO instance, MVCC distributes sessions across multiple MVCO instances.

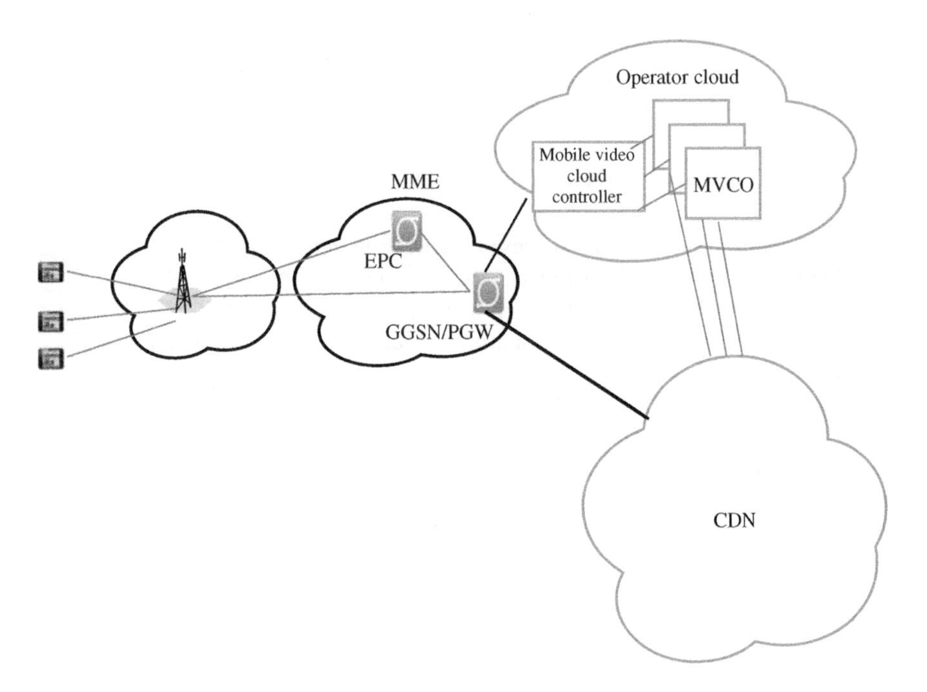

Figure 7.4 High level architecture of cloud mobile video optimization service.

When VOs are moved to cloud, how to achieve effective dynamic MVCO provisioning becomes critical. Since it takes time to launch MVCO instance, MVCC needs to accurately predicate the needs of MVCO based on real-time load condition and video optimization demand ahead of time.

In order to achieve good QoE, mobile video streaming needs to address the challenges from mobile networks, devices and cloud-based optimizers. Two important parameters for QoE are examined, that is, video reaction time and rebuffering.

Video reaction time is directly related to end-to-end network latency and the processing delay that MVCO takes to optimize video. End-to-end network latency consists of the delay in mobile network and the delay from video optimizer to CDN. Given latency in mobile network can be at least 70 ms in LTE network, how to reduce latency between operator cloud and CDN and manage the processing delay of MVCO becomes critical for achieving desirable video response time. To address latency problem between MVCO and content source, MVCO should fetch content from the closest CDN. Managing the processing delay of MVCO is more challenging. There are at least three potential approaches to improve MVCO processing delay:

- Parallel transcoding/transrating technique can be adapted to process large files, where MVCC splits the large files into multiple small chunks, and sends these small chunks to different MVCOs to enable parallel video optimization processes. This kind of parallel transcoding technique demands high computation power from MVCC and brings scalability issue to MVCC.

- Local cache is utilized to store optimized contents to avoid repetitive video processing in the case that the same optimized contents are requested. Whether local cache is beneficial depends on cache hit ratio. Given the limited cache size of MVCO, smart cache algorithms are demanded.
- Content servers have started to support various content formats that are optimized for diversified mobile devices, optimization service should be able to identify first HTTP request for such contents and allow UE to access contents directly from CDN. Hence, processing delay at MVCO can be eliminated for already accessing optimized contents.

The leading cause for rebuffering is network bandwidth variations. In mobile network, UE perceived bandwidth keeps changing due to the fluctuation of wireless link and UE mobility. Unfortunately, in current MVCO architecture, once the targeted video stream format is decided, it does not change during video optimization process for a given video session. Since current MVCO cannot provide adaptive optimization based on dynamic bandwidth changes at UE, rebuffering is still a challenging issue

7.6.2 SVC-Based Mobile Cloud Streaming

SVC is Annex G extension of the H.264/MPEG-4 AVC video compression standard and a key enabler to provide adaptive video optimization based on network dynamics [31]. SVC defines base layer (BL) and enhanced layers (ELs) by exploiting the scalability from temporal, special, and quality requirements. If only BL is delivered, the receiving end can decode with the lowest quality; the more ELs are delivered, the higher quality can be achieved. [31–34] have investigated SVC-based mobile cloud streaming where SVC proxy/agent proxy converts video content from original format into SVC format, then utilizes multilayer encoding feature of SVC to dynamically adjust stream content according to the bandwidth of the receiving end.

Figure 7.5 presents a high level architecture of SVC-based mobile cloud streaming. The streaming request from UE is directed to user agent/proxy. The latter requests streaming content from content server inside CDN and stores the content in Media Data module. SVC transcoding controller divides the content into small chunks, and then instructs multiple cloud transcoding services to transcode these chunks into SVC files. These SVC files can be cached for other UEs. Finally, user agent/proxy dynamically adjusts EL selections based on time-varying UE bandwidth and streams them with BL to UE.

SVC-based architecture can well support both HTTP PD and ABR streaming techniques. To make architecture more effective, the following issues need to be addressed:

First, the accurate bandwidth predication schemes need to be developed. In Reference [34], UE periodically reports signal to interference plus noise ratio (SINR), delay, jitter, packet loss, and so on to user agent/proxy, and the latter derives UE bandwidth information based on periodic UE reports. Another bandwidth predication approach is done at UE, and UE reports its anticipated bandwidth to the cloud.

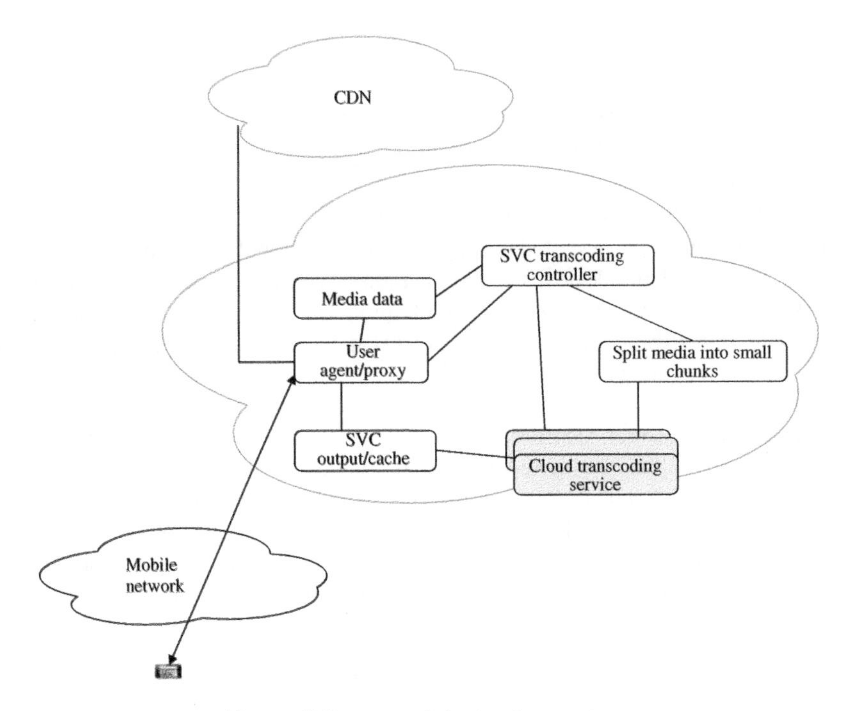

Figure 7.5 SVC mobile cloud streaming.

The comparison of these two approaches can help come up with scalable and accurate bandwidth predication algorithms. Furthermore, how to efficiently report UE information with minimal overhead also requires some attentions.

Second, how to effectively select the appropriate EL based on dynamic changing bandwidth is an open research problem. Each EL can have different resolutions. Low resolution ELs cannot match bandwidth fluctuation well and lead to bandwidth waste or packet loss, while high resolution ELs can fit the fluctuation of bandwidth well according to [34]. However, there is a tradeoff to introduce high resolution since it demands more transcoding power. How to balance these constraints becomes an interesting research topic.

In order to achieve cost-effective, fast, and efficient SVC cloud-based video streaming, a systematic approach has to be investigated with the consideration of anticipated bandwidth, EL resolution adaptation, transcoding speed/jitter, resource constraints, and overhead.

7.7 FUTURE RESEARCH DIRECTIONS

There are many open problems yet to be addressed. In this section, a few problems are highlighted as future research studies.

Contextual Analysis and Streaming. Factors that influence streaming include user preference, user devices configuration, available energy, user location, service subscription, recommendation from video recommender system, and so on. How to effectively use these factors to deliver mobile video streaming service has been explored as a research topic. To improve streaming service by using these factors involve machine learning approach and do predictive analysis. General principle behind machine learning is to pick up a known model and make predication, and let the model adapt as it learns from the data set. After initial learning phase, machine learning algorithm can give a good estimate to guide VO and CDN.

Cross-Layer Optimization. Idea of cross-layer optimization is old, existing layered protocol triggers are not optimal for rapid changing network conditions. For example, when radio signal becomes weak and the available bandwidth goes suddenly to a lower value, upper layer and application take time to react. Without doing cross-layer optimization makes the video user experience unacceptable. Several studies [35–37] show how the UE stack and application must interact with each other in order to provide smooth QoE to the users. Most of the techniques rely on underlying triggers such as distortion in signal, frame errors in radio link layer, and TCP/IP packet spacing, and use these triggers as input directly to video client application.

Multimedia Broadcast Multicast Service (MBMS). MBMS is a 3GPP standard aiming to provide multicast service inside RAN. Though standard existed for long time, there is no compelling application to deploy it. Now, real-time streaming videos become potential use cases to explore such services. There is a clear need to have joint architecture, protocols to stream real-time video delivery over wireless that utilize multicast transport and trans-slicing techniques.

Power-Aware Video Streaming. Power is a fundamental problem as per Koomey's law energy efficiency of a system gets doubled in 18 months but the required processing capacity doubles once in a decade [38,39]. Several protocol improvements have been proposed to increase battery life in 3G and 4G networks [40]. However, increase in usage of background applications and adoption of high resolution display make battery optimization techniques ineffective. From various studies, we learnt that battery drain characteristics is not linear mobile phones, and mobile devices have different processing capabilities, display sizes, audio, and radio network capabilities, and it is hard problem to isolate and evaluate when studying application [41–43]. It is an open research problem to build devices and video streaming applications that are energy efficient [44].

Streaming over Heterogeneous Networks. Cellular networks are congested, and there are proposals to look into handover to small cells, such as Wi-Fi hotspots and Femto cells [45]. When operators deploy VO or CDN or caching inside their RAN networks, frequent handovers between macro- and small cells demand new strategy for buffering, and anchor point for transcoding needs to be managed. At this moment, the standards are being developed in parallel for Wi-Fi and LTE small cells, and the goal is to ensure seamless mobility for applications.

ACKNOWLEDGMENTS

This work is supported by Nokia Siemens Networks. Opinions, findings, or recommendations expressed in this paper are from the authors and do not reflect the views of Nokia Siemens Networks.

REFERENCES

1. YouTube. 2012. Available at www.youtube.com. Accessed 2013 Jul 12.
2. Netflix. 2012. Available at www.netflix.com. Accessed 2013 Jul 12.
3. Cisco Visual Index. 2012. Available at http://blogs.cisco.com/sp/cisco_visual_-networking_index_forecast_annual_update/. Accessed 2013 Jul 12.
4. ITU. ITU-T Recommendation P.910. Subjective video quality assessment methods for multimedia applications. Geneva, Switzerland: International Telecommunication Union; 1999.
5. Haddad M, Altman E, El-Azouzi R, Jiménez T, Elayoubi S, Jamaa S, Legout A, Rao A. A survey on YouTube streaming service. VALUETOOLS; May 2011.
6. Akhshabi S, Begen A, Dovrolis C. An experimental evaluation of rate-adaptation algorithms in adaptive streaming over HTTP. ACM Multimedia Syst 2011:157–168.
7. Riiser H, Bergsaker HS, Vigmostad P, Halvorsen P, Griwodz C. A comparison of quality scheduling in commercial adaptive HTTP streaming solutions on a 3G network. Proceedings of the 4th Workshop on Mobile Video; 2012. p 25–30.
8. Muller C, Lederer S, Timmerer C. An evaluation of dynamic adaptive streaming over http in vehicular environments. Proceedings of the 4th Workshop on Mobile Video; 2012. p 37–42.
9. Stockhammer T. Dynamic adaptive streaming over HTTP: Standards and design principles. Proceedings of ACM MMSys; 2011. p 133–144.
10. 3GPP. 3GPP TS 26.244. Transparent end-to-end packet switched streaming service (PSS); 3GPP file format (3GP); 2013.
11. 3GPP. 3GPP TS 26.234. Transparent end-to-end packet switched streaming service (PSS); Protocols and codecs, release 12; 2013.
12. Miller RB, Hearn JP, Purzynski C, Cuervo F, Scheutzow M. Mobile video delivery using network aware transcoding in an LTE network. Bell Labs Tech J 2012;16(4):43–61.
13. Ma KJ, Bartos R, Bhatia S, Nair R. Mobile video delivery with HTTP. *IEEE Communication Magazine*; April 2011.
14. Ishfaq A, Wei X, Sun Y, Zhang Y. Video transcoding: an overview of various techniques and research issues. IEEE Trans Multimedia 2005;7(5):793–804.
15. Liu Y, Li F, Guo L, Shen B, Chen S. A server's perspective of Internet streaming delivery to mobile devices. Proceedings of IEEE Conference on Computer Communications (INFOCOM); March 2012.
16. Maheshwari A, Sharma A, Ramamritham K, Shenoy P. TranSquid: Transcoding and caching proxy for heterogenous e-commerce environments. Proceedings. Twelfth International Workshop on Research Issues in Data Engineering: Engineering E-Commerce/E-Business Systems; 2002. p 50–59.
17. Liu Y, Ci S, Tang H, Ye Y, Liu J. QoE-oriented 3D video transcoding for mobile streaming. ACM Trans Multimedia Comput Commun Appl 2012;8(3).

18. Leighton T. Improving performance on the internet. Commun ACM 2009;52(2):44–51.
19. Nossenson R. Base station application optimizer. International Conference on Data Communication Networking; July 2010. p 1–6.
20. Golrezaei N, Shanmugam K, Dimakis AG, Molisch AF, Caire G. Wireless video content delivery through coded distributed caching. ICC 2012:2467–2472.
21. Temkin D. Netflix open connect. 2012. NANOG 55, CDN Panel. Available at http://www.nanog.org/meetings/nanog55/presentations/Tuesday/Temkin.pdf. Accessed March 2013.
22. Ahlehagh H, Dey S. Video aware scheduling and caching in the radio access network. International Conference on Communication; June 2012. p 7082–7087.
23. 2012. Informa Report. Understanding today's Smartphone user - Part I and Part II. Available at http://www.informatandm.com/wp-content/uploads/2012/08/Mobidia_final.pdf. Accessed 2013 Jul 12.
24. 2013. Netflix encoding types. Available at http://vimeo.com/52637219. Accessed 2013 Jul 12.
25. Yousaf F, Liebsch M, Maeder A, Schmid S. Mobile CDN enhancements for QoE-improved content delivery in mobile operator networks. IEEE Netw 2013;27(2):14–21.
26. 2011. Ericsson and Akamai integrated CDN. Available at http://www.telecomsemea.net. Accessed 2013 Jul 15.
27. 2013. NSN Liquid application. Available at http://www.nokiasiemensnetworks.com/portfolio/liquid-net/intelligent-broadband-management/liquid-applications. Accessed 2013 Jul 15.
28. 2013. Jetstream. Available at http://www.jet-stream.com/mobile-cdn/. Accessed 2013 Jul 15.
29. Hoffman J, Kaul A. *Centralized vs. Distributed EPC and Role of SDN and Cloud*. ABI Research; 2012.
30. Dihal S, Bouwman H, Reuver M, Warnier M, Carlsson C. Mobile cloud computing: State of the art and outlook. Emerald Insight 2013;15(1):4–16.
31. Huang Z, Mei C, Li E, Woo T. CloudStream: Delivering high-quality streaming videos through a cloud-based SVC proxy. Proceedings of IEEE INFOCOM 2011; 2011 Apr 10–15; Shanghai, China; 2011.
32. Wang F, Liu J, Chen M. CALMS: Cloud-assisted live media streaming for globalized demands with time/region diversities. Proceedings of IEEE Infocom 2012; 2012 Mar 25–30; Orlando, FL, USA; 2012.
33. Lai C, Wang H, Chao H, Nan G. A network and device aware QoS approach for cloud-based mobile streaming. IEEE Trans Multimedia 2013;15(4):747–757.
34. Wang X, Chen M, Yang T, Leung V. AMES-cloud: A framework of adaptive mobile video and efficient social video sharing in the clouds. IEEE Trans Multimedia 2013;15(4):811–820.
35. Vangelis G, Nancy A. Cross-layer design proposals for wireless mobile networks: a survey and taxonomy. IEEE Commun Surveys Tuts 2008;10(1):70–85.
36. Miao G, Himayat N, Li Y, Swami A. Cross-layer optimization for energy-efficient wireless communications: A survey. Wireless Commun Mobile Comput 2009;9(4):529–542.
37. Ness BS, Rayadurgam S. A tutorial on cross-layer optimization in wireless networks. IEEE J Selected Areas Commun 2006;24(8):1542–1463.
38. Koomey JG. Implications of historical trends in the electrical efficiency of computing. IEEE Ann Hist Comput 2011;33(3):46–54.

39. Shearer F. *Power Management in Mobile Devices*. Newes Publication; 2007. ISBN: 0750679581.
40. 3GPP. 2013. 3GPP specification. Continuous connectivity for packet data users. 3GPP TF 25.903. Available at http://www.3gpp.org/ftp/Specs/html-info/25903.htm. Accessed 2013 Mar 1.
41. Zhang J, Wu D, Ci S. Power-aware mobile multimedia: A survey (invited paper). J Commun 2009;4(9).
42. Vallina-Rodriguez N, Crowcroft J. Energy management techniques in modern mobile handsets. IEEE Commun Surveys Tuts 2013;15(1).
43. Carroll A, Heiser G, An analysis of power consumption in a smartphone, USENIX; 2010.
44. Siekkinen M, Ashraful M, Hoque JK, Alto M. 2013. Streaming over 3G and LTE: How to save smartphone. Energy in radio access network-friendly way. Available at http://users.tkk.fi/~siekkine/pub/siekkinen13movid.pdf. Accessed 2013 Jul 15.
45. 3GPP. 2013. 3GPP specification. Local IP access and selected IP traffic offload. 3GPP TR 23.829. Available at http://www.3gpp.org/ftp/Specs/html-info/23829.htm. Accessed 2013 Jul 15.

PART II

CDN PERFORMANCE MANAGEMENT AND OPTIMIZATION

8

CDN ANALYTICS: A PRIMER

Timothy Siglin

Braintrust Digital, Inc., Harriman, TN, USA

8.1 INTRODUCTION

Let us consider a scenario where a content delivery network (CDN) provider has successfully delivered, as part of a larger CDN federation, the "next big thing" in live online video entertainment to a global audience. As the primary CDN, the provider's job was to deliver both to other CDNs and directly to key access networks. From the provider's perspective, the event was a success, and the team congratulates each other for a job well done.

Long before the janitors sweep the floor and put away the chairs at the entertainment venue, however, it is time to get to arguably work on the most important phase of the provider's services: billing. To do so, a CDN must be able to measure both its metered capabilities and the overall quality of experience (QoE) of those who viewed the content. Some of these measurements occur in real time during the event itself, while others are compiled after the completion of the event and compared against metrics provided by other service providers as well as by industry benchmarks.

This scenario leads directly into the world of analytics, where measurement is key to successful revenue realization. Analytics are the measurements against which a CDN

Advanced Content Delivery, Streaming, and Cloud Services, First Edition.
Edited by Mukaddim Pathan, Ramesh K. Sitaraman, and Dom Robinson.

service is judged both in the moment and in the weeks and months to follow. CDN providers may very well "measure twice, bill once" to bastardize a carpentry truism.

In this chapter, we cover CDN analytics tools and explore a variety of analytic practices and their practical implications, including new methods designed to analyze emerging technologies such as adaptive bitrate (ABR) video delivery via HyperText Transfer Protocol (HTTP).

8.2 WHY MEASURE?

CDNs measure for a variety of reasons, both internal and external. Internal measurements focus on the overall CDN delivery cycle from available transit capacity to storage efficiency and to cache hit ratios. As for example, EdgeCast CDN has six key criteria [1], it uses to pitch customers on the idea of analytics:

1. *Know It All.* Access constantly updated information about server performance, user demographics, and bandwidth utilization.
2. *Know It Now.* With live events, often there is no time to wait for server log processing. Use real-time statistics and take action—in the moment.
3. *Improve Efficiencies.* Leverage in-depth information to optimize a website from both cost and performance perspectives.
4. *Find and Fix Problems.* Detect errors and other delivery issues before they become major showstoppers.
5. *Understand the Audience.* See how users are interacting with the website content, where they are coming from, and how long they stick around.
6. *Drive Customer Loyalty.* Ensure an excellent user experience by leveraging information about successes and failures.

These key criteria are important for customer satisfaction, but what are the reasons for internal analytics and reports? Three key points stand out and are covered later in the chapter. They are quality of service (QoS), troubleshooting, and delivery optimization.

8.3 WHAT DO WE MEASURE?

What to measure, when to measure it, when to report it, and what to do with the report are key decision points when it comes to proper CDN analytics. This is true for both internal assessment and external reporting.

8.3.1 Internal Assessment

For internal assessment, the CDN operator looks at a number of key factors surrounding delivery.

8.3.1.1 *Quality of Service* The move toward over-the-top (OTT) content puts pure-play CDNs and accesses network operators directly against one another. If a pure-play CDN hosts the origin server for live or highly popular on-demand content, its goal is to disseminate that content to as many points of presence (PoPs) as it can. The pure-play CDN will distribute the content as close to the highest number of potential viewers as it can and then push that content OTT onto the local service provider's access network. The pure-play CDN is at a disadvantage as the "best effort" delivery across an access network may result in a lower QoE and no guaranteed QoS between the PoP and the end-user's video player.

The local provider receives no benefit from this highly popular content traversing its network, as most of the PoPs reside—and interconnections occur—at major peering points. So recent moves by operators to build out their own CDNs (often called O-CDNs or Telco CDNs) are understandable from both cost abatement and QoS standpoint. Using techniques such as transparent caching, which is covered in depth in the earlier chapter on CDN basics, the O-CDN leverages temporary content storage deep within the operator's access network to serve up this content to multiple network subscribers, without requiring the content to traverse the peering point multiple times.

The operator, in contrast to the pure-play CDN, can exploit last-mile ownership as part of the O-CDN. Yet even the O-CDN faces challenges for QoS, when compared to an operator's own Internet pay TV (IPTV) offerings: the multicast-enabled delivery of IPTV within an operator's own network is optimized for the highest quality delivery. OTT content may be delivered at any number of quality levels given current network conditions. This is especially true if the OTT content is set up for HTTP-based ABR delivery, as multicasting and ABR delivery are mutually exclusive. Yet if OTT content is to successfully compete with established pay TV offerings, it must be delivered in as consistent a manner as IPTV live linear content. This is a challenge for both access network operators and pure-play CDNs.

The ideal O-CDN will take into account both the deep (potentially one-hop) and transparent caching, as well as initial placement of highly popular content. Once the content is launched, there's an overall risk of flash points and hot spots, both of which can be revealed via deeper troubleshooting analysis.

8.3.1.2 *Troubleshooting* Whether they are called "service irregularities" or hiccups or some other common name, the issues surrounding flash points and hot spots have three key elements: asset issues, geography-region issues, and specific PoP failures.

ASSET ISSUES. A number of asset-related issues can cause delays in delivery or even outright failures. On the delay front, content that is partially cached in a CDN's cache can cause issues if the end-user chooses to evoke trick-play commands such as fast forward or rewind. In this case, the CDN needs to bypass the cache and request content from the origin server, which will often be off-network incurring additional latency. One could argue that prefetching any content n seconds before or after the point at which a user has chosen to view is good policy, but it is both impractical and inefficient in many instances.

In addition, asset issues for ABR-based content pose a particularly unique challenge, as the access network's transparent cache may hold parts of multiple bitrates/resolutions, but will not contain the entirety of any bitrate or resolution. Velocix, an Alcatel-Lucent O-CDN service offering for operators (often called licensed CDN), takes an interesting approach to this particular issue. Velocix creates logs for events related to all objects in the CDN, regardless of the origin server, format, or even streaming protocol. Reports for HTTP-based ABR content anticipate the issue of ABR content spanning several files, where all may have originated from "an origin sever that may not be under the control of the CDN operator."

The way that Velocix handles this issue is by allowing the O-CDN operator to bundle assets (files) together, via the paths to the origin server, and then treat them as a single object. This means, regardless of whether the ABR content is segmented into 3 separate bitrates/resolutions or 20, the logging system notates access to all the assets as a single entity.

Another area where asset issues rear their head is in the corruption of content at the cache. In this case, the CDN must choose between comparing content from the origin server to the cache via a rerequest to the origin server, comparing a single cache's asset with that of another cache, or simply flushing the cache and then rerequesting the same content from the origin server to deliver to the end-user and refill the cache.

A final asset issue to consider is whether content itself cannot be served to a wider swath of end-users. This is likely to occur when a content format is new and less ubiquitous or when the content itself does not match a particular specification of the format specification. An area where this issue continues to arise is in the delivery of OTT content to "smart" televisions as each smart TV manufacturer implements its underlying OTT architecture differently, hoping that developers will dive in and create apps for the particular CE manufacturer's ecosystem.

Geographic-Regional (Geo-regional) Issues. Not all CDNs have equal coverage across the globe. In fact, many CDNs are dominant in only one or two countries, with reasons ranging from licensing (incumbent telecom providers in Europe, for example) to cost (access network leases in Asian countries are often prohibitively expensive) to limited viewership (a Portuguese-centric CDN may see little benefit of beefing up its network in German-speaking countries, for instance). For those CDNs who have not ventured into the world of unilateral agreements in emerging regions—or even CDN federation, which is covered in the chapter on next-generation CDNs—the issue of maintaining QoS in a noncore geography is very important. The reasons range from a concerted effort to win business in a particular geography to servicing the needs of a multinational corporate client.

Geography issues often stem from two factors: connectivity and overall infrastructure. While Australia, Europe, the United States, and parts of Asia enjoy very solid infrastructure, many emerging markets face challenges within even urban topologies and existing data centers. Add to this the fact that end-user connectivity varies widely even within particular geographies, and it is easy to understand why analytics and reporting are so crucial to maintaining a consistent QoE for a CDN's premium content customers.

TABLE 8.1 Summary of Key Facts on Akamai and Limelight

Description	Akamai	Limelight
Year founded	1998	2001
Revenues (percentage of overall pure-play CDN revenues)	66	14
Core- or edge-centric	Edge	Core
Sustained throughput (Gbps average)	2	85–90
10 GbE connectivity	Limited	Extensive
Caching locations (estimate)	1200	250
Servers	25,000	6000
Peer interexchanges	900	900
Cities with PoPs	760	60
Operator CDN (O-CDN) brand	Aura	Deploy
Licensed CDN or SaaS	Both	Both

PoP Issues. Many CDNs live or die by the strength of their PoP footprint. While important for incumbent carriers and O-CDNs, the distribution of POPs is critical for pure-play CDNs.

Let us look at two examples, both from US-based pure-play CDNs—Akamai and Limelight Networks. Akamai and Limelight have directly competed with one another for well over a decade. Table 8.1 shows the comparison of a few key facts about each company.

Notice that Akamai, the dominant player in the pure-play CDN space, has approximately 750 cities in which it houses servers, and its total number of servers is at the 25,000 mark. Limelight, by comparison, offers 6000 servers in 60 cities. This means that Akamai deals with about 33 servers per PoP, while Limelight deals with about 100 servers per PoP. These numbers are approximate, of course, since neither company makes their per-PoP server count known.

Limelight might also argue that it does not need as many servers, since it has an equal number (900) of peering arrangements compared to Akamai, and that its network can sustain an average throughput, some 20 times faster than its bigger competitor. These arguments may be sound for the type of pure-play CDN that Limelight maintains, but it still means the company will face either geo-regional or PoP issues. Likewise, Akamai will have to maintain a higher number of PoPs compared to Limelight, making PoP cluster and stand-alone server analytics all the more important.

The primary question both examples drive home is that a CDN must not just monitor connection between its core and its PoPs, or between its PoPs and the end-user, but ultimately must be able to determine which PoPs produce the highest rate of failed connections. And then they must determine why.

In isolation, each of the issues—asset, geo-regional, and PoP—are important, but together they form a detailed viewpoint for troubled spots within the CDN's footprint, whether dormant, emerging, or current.

Skytide, a CDN analytics platform, refers to this as a "compounding" [2] of the discrete issues, noting that a greater level of specificity in the three areas "will allow the network operator to quickly and precisely identify the sources of quality problems." By this, they mean filtering known issues down into a compound question that asks whether a specific PoP or multiple PoPs are having trouble delivering specific content to specific geographical areas.

8.3.1.3 Delivery Optimization With the number of PoPs growing, and the commensurate rise in connectivity between each PoP and the CDN's core network, the question of load balancing becomes more critical. Three factors are important when analyzing load balancing for optimal delivery. They are described in the following.

LOAD BALANCING. Transparent load balancing works best when content needs to appear to be coming from a single entry domain name emanating from what appears to be a single IP address. One of the strong points of a CDN is the ability to mask the fact that a single origin server may be replicated to multiple servers and multiple IP addresses. This is often done to insure high availability, but has the added benefit of providing fail-over support.

Load balancers work on the concept of a listener, which use a back-end instance to confirm that a front-end action should take place, based around the combination of a port number and a protocol. A load balancer is, therefore, represented by a domain name system (DNS) name and a set of ports, but some solutions allow a custom domain name to be specified that masks the DNS name.

One recent trend in load-balancing solutions that is pertinent to CDN analytics is the movement to use cloud-based load balancing. As legacy load-balancing tools give way to cloud-based solutions such as Amazon's elastic load balancer (ELB), the benefit is that "each load balancer can distribute requests to multiple EC2 instances" within a specific EC2 region. However, given Amazon's self-imposed geographical constraints for elastic computing, a single ELB cannot span multiple EC2 regions reminding CDN administrators and architects that cloud-based load balancing is still in its infancy.

Elastic load-balancing instances can be accessed, according to Amazon [3], simply by "pasting the DNS name generated by ELB into the address field of an Internet-connected web browser." The benefit for CDN routing comes in the intelligence that can be built into load balancing in the form of intelligent routing requests (IRR) that are masked from an end-user via custom domain names that resolve to a load balancer's DNS name.

IRR routing differs from traditional load balancing, which often could be described as passive request-routing scales. Passive request-routing is essentially using DNS's inherent ability to list many servers in response to a lookup, meaning all edge servers can be listed and the end-user is connected, somewhat at random, to the content at any one of the listed nodes.

From a live stream standpoint, though, operators have to provision all server locations, even if the entire audience is local to one particular server. This inefficiency coupled with the fact that a server failure removes the DNS list, requiring considerable extra time to propagate, means that passive request-routing trouble spots occur at two major points: the start of a live event and—in the case of a large sporting event—just after some major event occurs on the court or field. The two points share a common problem:

at the start of a game or at a major event within the game, the rush of viewers joining can overwhelm the system.

Using proper analytics and predefined business and routing rules, intelligent request routing evaluates each user via logic in the supervisor layer, routing a user to an optimal server from which to serve their session. Criteria may include geography, rights, availability, resilience, latency, or even localization for alternate or local languages.

At this point, one may wonder about the concept of just increasing bandwidth capacity to a particular robust server. The two main problems with that approach are a potential for underutilization as well as a single point-of-failure weakness. A third issue, particular to the concept of intelligent request routing (IRR), is a lack of predefined granularity to accommodate a surge in requests. In other words, even with IRR, if the logic that steers those requests is not granular enough to route many thousands of request to less-utilized servers over a short time span, all users will suffer connection problems.

An initial uniform resource locator (URL) is often simply a query into a database infrastructure, yet many structured query language (SQL) servers or server-side scripts will behave oddly if presented with a sudden high volume burst of user requests, manifesting itself with as little as a few hundred users a second. So while the number of simultaneous requests may seem trivial in terms of CDN capacity, it is often big enough for an ill-designed database query request to overwhelm the IRR system, resulting in an overflow that pushes viewers on to any available server in a passive cascading approach.

Without a way to simulate such loads, using key analytic data as a baseline, even multiple CDNs really do not help the problem as the key to intelligent routing is letting the IRR decide where to best point the user.

Dynamic Bandwidth Allocation and Pricing. The question of routing content over a fixed bandwidth pipe, versus allocating additional bandwidth to move priority content in a timely manner—even at peak times—is of key interest to CDN operators and its customers. After all, one benefit of a CDN is the ability to handle significant end-user loads, such as when a new mobile operating system is placed into service and millions of smartphone operators choose to download the update at the same time.

Here, we briefly cover how analytics allow CDN customers to see the financial impact of moving content via dynamic bandwidth allocation. In some instances, content delivery between internal servers or even PoPs is constrained at peak bandwidth cost times, resulting in a much higher premium for moving priority content at those times. An analytical model for dynamic pricing has even spawned a patent [4], in which a dynamic pricing engine is coupled to a single Web location, which itself is associated with the content provider.

The patent suggests that, during the course of the CDN "delivering content objects from a content provider to a recipient," the operator may wish to add a second Web location to the delivery mix to balance delivery loads. If the first Web location is unable to deliver a requested content object one or more times, an automatic request is made of the second Web location to assist in delivering the content object to a recipient one or more times, and the dynamic pricing engine then "automatically produces the delivery quote, which is affected by at least one of the content object or a Web location of the recipient system."

In other words, content delivered from more than one location can be tracked for internal or external billing purposes, allowing dynamic pricing—whether incremental or premium—to be used for critical content object deliveries.

8.3.2 External Reporting

For external reporting, there are two types of reports: historical and projected, generated by CDN analytics tools. Usage reports focus on the content delivered. The metrics can be aggregated across all delivery nodes in the CDN or broken down by object, customer, delivery appliance, or geographic location going down to the granularity of the metropolitan area. Example reports are described in the following.

8.3.2.1 Asset- or Object-Based Reports These reports focus on content itself (often called assets or objects) and how well these assets are delivered. Unlike the internal analytics, which cover assets from a PoP or intra-CDN standpoint, these external-facing reports deal with information the content owner cares most about which resolutions, bitrates, and formats were consumed—in aggregate—across a variety of end-user devices and geographies. Further detail on geography is part of geography-based reports (later), but these asset-based reports cover a high level trending across regions. In addition, content owners need to know what modifications to make to transcoding workflows, so these reports help answer the question of which resolutions, bitrates, and formats continue to trend upwards or downwards. Trend analysis requires access to historical data spanning several months.

Ultimately, the content owner wants to know if any assets are not being properly monetized, and if those assets should be purged all together from caches due to lack of overall interest (e.g., last week's news show or last season's episodes), then they best utilize available storage and transport. In this way, asset-based reports essentially double as quota management reports, allowing content owners to best use their budgeted CDN resources.

8.3.2.2 Customer-Based Reports Assets produced for a wide array of devices—mobile, desktop, set-top box, and even big-screen viewing—are key candidates for cross-promotion. Whether it be in featured playlists or on social media platforms, these widely dispersed assets offer the next level of granularity: customer reports. In some instances, whether due to privacy laws or limited measurement capabilities, customer reports are merely the reverse of an asset-based report, meaning that the report shows a customer demography and the assets consumed within that demographic region, all overlaid onto an interactive global map.

Other customer reports, though, include session information that covers geography, types and resolution/bitrate of consumed content, and devices, overlaid against known customer details and social media ties. One company, TeliaSonera, refers to this type of report as its advanced content analytics, providing interaudience information for specific tracked assets that include referrer and host data, location, and loyalty—the latter a combination of repeat visits and "stickiness" on that particular visit.

8.3.2.3 Geography-Based Reports While CDN providers look to determine which geographies and PoPs require additional optimization, the content owner is concerned with the metaquestion around geographies: does content popular in one part of the globe resonate with viewers in another part of the globe.

Some content have universal appeal—think of the British royal wedding, Cricket or Football World Cup, Olympics, or even the stratospheric, sound-barrier-breaking free fall of Felix Baumgartner [5]—but for every cricket or football match there are thousands of videos that fail to catch on across geographies. Content owners need near-real-time updates on large-scale live events or premium on-demand content launches, to best judge overall monetization and edge performance on a regional and city-by-city basis.

8.3.2.4 Delivery Appliance Reports Content owners do not need to understand the intricacies of the CDN provider's network, but they grow increasingly interested in the inner workings when content is not being properly delivered. To address the need for transparency, a number of CDNs now provide both core- and edge-performance analytics.

At the core, analytics monitor transfer rates, cache hit ratios, and bandwidth for simultaneous unicast connections [6]. Often these reports are provided on an hourly basis, so that a content owner can track simultaneous viewers, progress of large-file delivery, and status of network cache utilization. At the edge, content owners want to know how both their Web portal and the content delivered via the portal are faring, so these reports offer insight into server or cluster performance per data center or PoP as well as granular details on cache configuration and primary protocol performance (e.g., HTTP, Real-Time Messaging Protocol (RTMP)) and any security concerns with rights management or intrusion.

The end goal of every CDN provider is to offer these external statistics in real time. Many CDNs have chosen a dashboard model to present current-state graphs and charts covering object delivery, simultaneous connections, bandwidth usage, and overall network health. Figure 8.1 shows an example dashboard for CDN reporting and analytics, using Splunk.

8.4 WHAT ABOUT BUSINESS INTELLIGENCE?

The rise of business intelligence (BI) platforms and services, some of which are owned by media server product manufacturers, led to increasing uncertainty around the choice of focused analytics for a particular type of content delivery (e.g., video) or more generic BI-based analytics.

BI is intended to give insight into a number of key elements of a business. It works well for product-based businesses and for a limited number of repetitive service businesses. Advocates claim that BI is better than standard analytics as "the primary objective of BI" is to "support effective, timely, and actionable business decision-making." For the BI proponent, BI encompasses measurement, analytics, and reporting, similar in some ways to CDN analytics, but inclusive of collaboration tools and knowledge management.

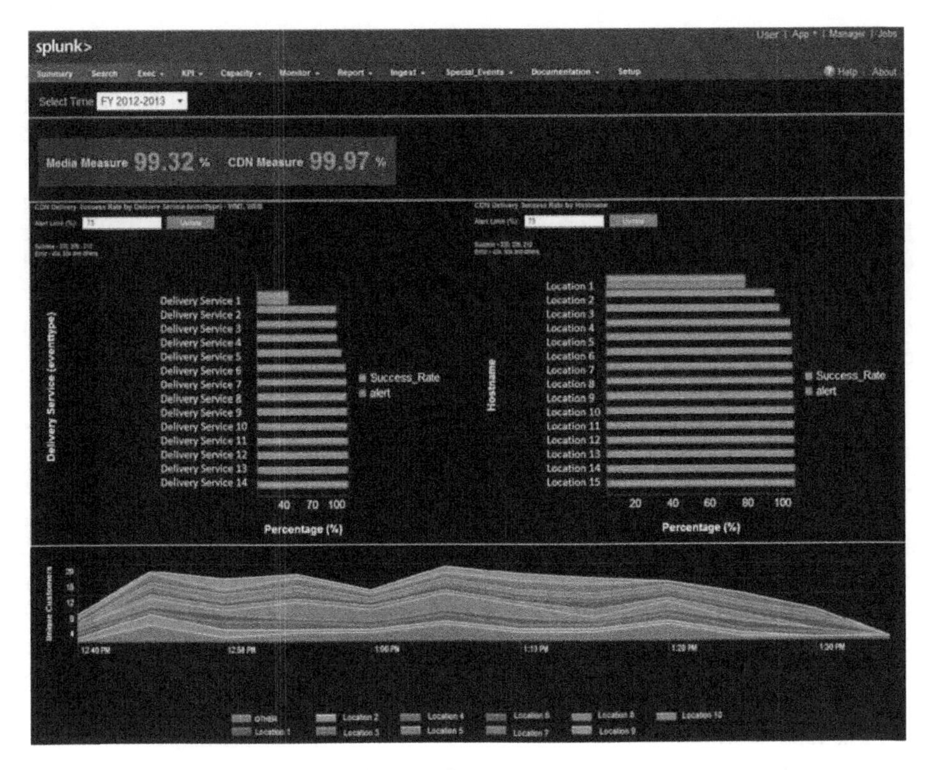

Figure 8.1 Example dashboard for CDN reporting and analytics.

BI advocates acknowledge that the best decisions are made once the raw data are "turned into information that is accurate, timely, and actionable." The timely aspect is where—from a firehose of information that encapsulates video streaming—the BI bottleneck occurs. BI is designed around dashboards geared toward tracking widgets and products, not the massive flow of bits needed to deliver timely media content.

Skytide, a company with a long history in analysis of CDN-based online media delivery, notes five key objections to BI versus more traditional CDN analytics.

The first objection revolves around the sheer amount of data that must be processed. Thinking back to the "stateless environment" discussion around ABR, most BI systems are not geared to deal with the terabytes of data flowing through a CDN every hour.

The second objection to BI is the timeliness of reports, or the time it takes to turn data into actionable options. The major bottleneck here is known as extract–transform–load (ETL) which requires extraction from the current logs, transformation from disparate log eXtensible markup languages (XMLs) to a more rigid database structure, and then loading of the database queries into a dashboard. Skytide argues that it is not necessary to perform the ETL process, and that a better approach is to tune the analytics system to a diverse group of data types (e.g., manifests from a variety of media server types) rather than extract each data type to a central, proprietary data structure.

The third objection, parallel to the issues surrounding ETL, is one of data integrity. Skytide argues that the data warehousing portion of a BI solution "can introduce data loss and inaccuracies" not present in a solution that does not perform ETL.

The fourth and fifth objections complement each other: rapid implementation and total cost of ownership. Most BI solutions are bespoke, even between similar companies in similar industries, where CDN analytics have a baseline set of measurements that are rapidly implemented and then enhanced to add additional value. The total cost of ownership, Skytide argues, comes from a combination of baseline measurements and analytics coupled to custom reports.

The sixth and final objection is that CDN customers want to provide reports to their business partners and detailed billing to their customers, but that these reports must be filtered. Most BI solutions, according to Skytide, do not natively possess this information filtering capability on the reporting side, even though filters and analytics are a key part of BI's appeal to businesses. CDN solutions, on the other hand, provide easy customization of portals for customers and business partners alike.

8.5 MEASURING STATELESS DELIVERY

Video delivery has always had an advantage when it comes to analytics, as streaming protocols required specialized servers to deliver unicast content to a client player. Whether it was standard-based Real-Time Transport Protocol/Real-Time Streaming Protocol (RTP/RTSP) or proprietary protocols such as RTMP, each unicast required session initiation between the server and client device, allowing discrete analytics on a session-by-session basis.

This stateful approach allowed many media server companies to infuse their server products with a myriad of analytic tools. With the advent of ABR technologies (discussed in detail Chapters 2 and 3), however, the need for specialized media servers to deliver video content has given way to the use of standard and more generic HTTP servers serving dynamic resolution video content in hundreds or thousands of small segments.

HTTP-based delivery scales well, but the downside to this scalability means that these small video segments traverse port 80 along with other Web traffic (e.g., HTML files, images, CGI/scripts). In other words, the concept of a stateful environment was lost when HTTP-based segmented video delivery emerged. A typical problem, then, for a CDN or last-mile provider would be tracking discrete video consumption across a home gateway for two users each watching ABR content. Given the network fluctuations and a standard configuration of three resolutions/bitrates per video program, the CDN would need to track six different segments. Some ABR manifest files, too, only represent correlative segments via hash information, making rationalization all the more difficult.

The new reality of a stateless environment for video delivery is a daunting analytics challenge, but a variety of new tools have emerged to offer initial virtual session tracking for the brave new stateless world.

Some analytic tools monitor log details around each stream delivered via the network. Records can be acted upon as delivery occurs or fed back to larger data collection systems to view aggregate traffic results. These session management tools can be used for both live and on-demand video delivery. More information is presented in Section 8.7.

In most ABR streaming solutions, the same manifest file is delivered to each viewer, and client-side controls are relied upon to provide personalized viewing experiences. In order to provide per-session analytics, however, these newer ABR tools create a unique manifest file per content initiation request and a stateful session is created. Monitoring tools then manage and monitor the extent of a single viewing session by creating an internal per-session table and mapping all requests (both client and network sides) into a log.

Before addressing the direct analytic benefit of creating a per-session manifest/table of requests, let us briefly consider how authentication benefits from this approach. Not only can the tools be used to apply subscriber authentication to approve a viewer for initial access to content housed within the CDN, but it can also be used for resolution-specific authentication in which the viewer is approved to view specific resolutions or streams.

In addition, some ABR analytics tools also offer time- and data-related delivery restrictions that limit per-session views to a specific number of minutes or megabytes or even device type.

Besides conforming sessions to content owner business rules, authentication also plays a role in mapping of the sessions to a log. These logs, known by a variety of names, detail all the records of a single session for auditing, diagnostics, and reporting purposes. The elements in a session log—such as the session detail record SDR used in a product like Seawell's Spectrum [7]—can be as high as 30 discrete elements, as noted in Table 8.2.

Of particular interest to an operator are the fields BytesSent, MinPlaybackRateAchieved, MaxPlaybackRateAchieved, and AveragePlaybackRate. The fields mentioned on the previous page reveal not just how much data are transmitted to the viewer but also how QoE changed on a chunk-to-chunk (request-to-request) basis. A

TABLE 8.2 Representative Elements in Seawell's Spectrum Session Logs

Field	Description
UUID	Unique client identifier across multiple instances
URL	Manifest URL as received from client
Client ID	Client IP address noted by system
ConnectionTime	Time at which client connected and session entry was created
ThrottleRate	Bitrate client session is rate limited based on policy or load
CacheDuration	Cache maximum age set for client cache
BytesSent	Total bytes of video + audio data sent to the client in this session
MinPlaybackRateAchieved	Minimum bitrate delivered to the client in this session in bps
MaxPlaybackRateAchieved	Maximum bitrate delivered to the client in this session in bps
AveragePlaybackRate	Average bitrate achieved in this session in bps

simple comparison, also on a per-chunk basis, against the resolution requested reveals gaps in Transmission Control Protocol (TCP) delivery ability for requested content, as well as an overall QoE mapping for a viewer's OTT experience.

Spectrum's ability to apply real-time rules to ABR sessions helps improve the quality of video delivered to viewers. SDRs are logged separately from the rest of the spectrum logs in easy-to-read CSV format and an application programming interface (API) allows access to the logs, to generate SQL database records for later retrieval to add value for auditing and historical reports.

8.5.1 Other ABR QoE Tools

SeaWell's Spectrum is not the only QoE or ABR testing tool on the market. Another one of interest is the Spirent Avalanche. The Avalanche provides standard HTTP statistics as well as ABR statistics that measure video manifest/playlist and fragments transport delays. By keeping count of bitrate requests and receives, Avalanche generates a patent-pending adaptive streaming score, or a QoE index, to provide an indicator of the end-user's video-viewing experience. In addition to an adaptive streaming score, Avalanche uses a proprietary algorithm that allows Avalanche ABR streaming clients to "automatically shift to different video bit streams based on network conditions, including bandwidth, delays, and various network impairments."

Besides use in establishing criteria for shifting from one bitrate to another, the tool can also be used to test at a constant bitrate, a key benefit for networks that maintain a sustained throughput higher than the highest ABR resolution. Another test methodology Spirent's Avalanche provides is ABR video client emulation. Its streaming clients, according to Spirent, "emulate real-world video clients that test server functionality, from sending manifest/playlist files to streaming video over TCP to the clients." In addition, emulation also supports digital video recorder (DVR) and "trick-play" commands such as play, seek, and pause.

A final note on ABR analytics: load balancing and stress testing are new areas for ABR technologies that differ from classic load testing of Web page content such as HTML, common gateway interface (CGI), scripts, images, and SQL queries. As such, the pace at which new tools emerge for analyzing and reporting stateless sessions will quicken over the next few years.

8.6 BILLING ANALYTICS

Two approaches to billing analytics dominate the CDN industry. The first is an approximation, known popularly as 95/5, while the second is a strict metering of bandwidth.

8.6.1 95/5 or the 95th Percentile

One approach to billing, used by large CDNs, is 95/5 or the 95th percentile approach. Akamai [8] describes it this way: 95/5 is the billing and measurement methodology, shorthand describing a process of determining the 95th percentile of usage or the uncompressed equivalent as measured by Akamai over 5-min intervals. The 95/5 methodology is used to measure usage of storage, concurrent users, and services billed in Mbps, Gbps, or any other bit per second methodology. CDNs use 95th percentile to factor peak usage into their pricing, approximating peak bandwidth utilization, but some analytic firms feel its core concept is flawed since it offers no compelling reason for a CDN's customer to shift usage to network's off-peak time.

8.6.2 Metered Bandwidth

The objection to 95/5 argues that metered bandwidth, including dynamic pricing for peak versus off-peak times, inherently causes CDN customers to shift their delivery to off-peak time periods. It is not hard to understand why dynamic pricing is instantly recognizable to anyone who pays for Internet service by the meter, whether it be by time or by megabit. This approach, though, raises the specter of net neutrality [9], as the ability to track utilization in many ABR scenarios would require a form of deep-packet inspection more akin to metering by traffic type rather than just by overall streams or megabytes delivered.

Projected reports deal with usage quotas and possible threshold impact. Quota management reports allow content owners and service providers to know how much of their budgeted CDN resources they are using. Network appliance reports are hardware usage reports that show the benefit provided by the CDN.

8.6.3 Metered Processor Time

An additional emerging approach—metered processing time—is worth a mention, although it is still in its infancy, given an intentional move to platforms as a service (PaaS) by companies such as Amazon (CloudFront, EC2) and Microsoft (Windows Azure, Windows Azure Media Services). The overall PaaS model moves the entire operating system to the cloud, offering elastic and dynamic computing for a variety of tasks. Within the streaming video industry, these tasks include a single-stream ingest, which is then transcoded to multiple bitrates, resolutions, and formats, creating a "virtual head end" in the cloud. Within the website hosting industry, elastic computing offers both site acceleration and the ability to repurpose content on-the-fly for a variety of devices and device types. It is expected that this PaaS approach encroach on a number of CDN providers' core services in the coming future.

8.7 CDN ANALYTICS TOOLS

A previous section in this chapter detailed external reporting, but what about inter-CDN performance? There are several third-party reporting platforms—Cedexis, Gomez, and Keynote are three major players—which provide CDN performance analytics from a third-party perspective. An example of these constantly updated reports can be found, on a country-by-country basis, at the Cedexis website [10]. Third-party reporting can then be compared to a CDN's own reporting to its customers.

Table 8.3 provides a sampling of companies that offer CDN analytics tools.

TABLE 8.3 Summary of CDN Analytic Tools

Company	Description
Cedexis	Application and performance analytics insight into CDN and cloud-hosted applications
Cisco VNI	Video analytics (specific to Cisco videoscape suite)
Gomez	CDN analytics
Guavus	Predictive analytics for communications service providers (CSPs)
IBM Analytics	Business analytics (BI)
Kaltura	Video analytics
Keynote	CDN analytics
Pivotal	Predictive analytics for CSP and cloud-hosted applications
Ooyala	Video analytics
Skytide	CDN analytics
Splunk	Video and nonvideo analytics
VidYard	Video analytics

8.8 RECENT TRENDS IN CDN ANALYTICS

We now look at a few recent trends in CDN analytics. One major trend—which is affecting everything from acceleration of content delivery to basic video transcoding—is the move toward cloud-based services. Analytics services are no different, with new analytics-as-a-service (AaaS) options being launched each month.

One such example is HP's new analytics service, built atop its cloud-based HAVEn "big data" analytics platform. Customers are using the services to model Big Data concepts before purchasing very expensive data warehousing solutions, but the HAVEn solution is also used for more mundane tasks such as image, audio, and text recognition.

Another key trend is application analytics. Pivotal, mentioned in Table 8.3, offers a service on its PaaS that not only accelerates delivery of data to popular apps but also analyzes the ways to further speed up applications using CDNs to position app-based content where it is most likely to be consumed. Other well-known PaaS solutions are Amazon's EC2 and Microsoft's Windows Azure.

A third trend is the combination of customer satisfaction indices with network performance analytics. These solutions judge the efficacy of dynamic site acceleration against perceived customer satisfaction, using index ratings such as net promoter score (NPS). As one textbook on this topic notes [11] "However [NPS] is implemented, the hope is that movements in NPS are positively correlated with revenue growth for the company." Such solution combining NPS and network performance metrics is vital as it not only provides an accurate measure of customer expectation for existing products but also identifies gaps and indicates what future products/services can be brought out to the market.

A final trend in CDN analytics is actually a throwback to a much earlier form of analytics: click-through or response analytics in terms of advertising "impressions." As more and more video content is delivered online, the revenue debates between advertising-driven content and subscription-based delivery continue unabated. A recent research study [12] by comScore and Pretarget—two companies tasked with measuring ad impressions—noted that when "a report provides ad *impressions*, it is really referring to the number of times an ad was requested from that specific server," not the number of times it was viewed. As such, companies are now beginning to track placement of a video ad (e.g., does it appear above-the-fold on a website?) and whether customers actually click on a video ad or merely just hover over the ad. The latter is important, according to both companies, and should count as active engagement with the advertisement [13].

8.9 CONCLUSION

In this chapter, we covered CDN analytic tools and looked at key questions regarding why analytics are important for both internal CDN operations and external customer and marketing reasons. We also explored what measurements offer insight into potential trouble spots. Finally, we assessed a variety of analytics practices and their practical implications, including new methods designed to analyze emerging technologies such as ABR video delivery via HTTP.

Analytics may not sound like the most compelling part of a CDN provider's job, but it is both necessary—as online audiences attempt to grow to the size of television audiences—and pragmatic as CDNs become more commonplace, with only their performance analytics setting one apart from another.

REFERENCES

1. Edgecast. 2013. Edgecast analytics: Benefits. Available at http://www.edgecast.com/services/analytics. Accessed 10 Nov 2013.
2. The four keys to telco CDN success, Whitepaper; Skytide; 2011.
3. 2012. Elastic load balancing: Developer guide (API 2012-06-01. Available at http://docs.aws.amazon.com/ElasticLoadBalancing/latest/DeveloperGuide/using-domain-names-with-elb.html. Accessed 10 Nov 2013.
4. Gordon M, Raciborski N. WO 2007016707 A3: Dynamic bandwidth allocation. Patent Application PCT/US2006/031144. 2009 Apr 30.
5. Tierney J. 24 miles, 4 minutes and 834 M.P.H., all in one jump. *The New York Times*, p. A15, October 15, 2012.
6. Walkowiak K, Rak J. Simultaneous optimization of unicast and anycast flows and replica location in survivable optical networks. Telecommun Syst 2013;52:1043–1055, 2013.
7. 2013. Spectrum overview. Available at http://info.seawellnetworks.com/resources/spectrum-overview. Accessed 8 Nov 2013.
8. 2013. Akamai services. Available at http://www.akamai.com/service. Accessed 2013 Sep 12.
9. Crowcroft J. Net neutrality: The technical side of the debate: A white paper. ACM SIGCOMM Comput Commun Rev 2007;37(1):49–56.
10. Cedexis. 2013. Cedexis report. Available at www.cedexis.com/country-reports/. Accessed 12 September 2013.

11. Jeske D, Callanan T, Guo L. Identification of key drivers of net promoter score using a statistical classification model. In: Jao C, editor. *Efficient Decision Support Systems—Practice and Challenges From Current to Future*. InTech; 2011.

12. comScore.com. 2012. For display ads, being seen matters more than being clicked. Press release. Available at http://www.comscore.com/Insights/Press_Releases/2012/4/For_Display_Ads_Being_Seen_Matters_More_than_Being_Clicked. Accessed 2012 Apr 24.

13. Pretarget.com. 2011. Position paper. When is an ad impression not making an impression? Available at http://aws.pretarget.com/whitepapers/Opportunity-to-View-Position-Paper.pdf. Accessed 12 September 2013.

9

CDN MODELING

Tolga Bektaş[1] and Ozgur Ercetin[2]

[1]*University of Southampton, Highfield, Southampton, UK*
[2]*Sabancı University, İstanbul, Turkey*

9.1 INTRODUCTION

Effective and efficient use of resources available to a content delivery network (CDN) has never been so relevant, particularly in the light of the ever-increasing amount of content that has to be distributed on a network with limited resources, including bandwidth and storage space. Indeed, Cisco systems predict that global Internet Protocol (IP) traffic will reach 1.3 zettabytes ($= 1.3 \times 10^{21}$ bytes) annually, implying a fourfold growth from 2011 to 2016 [1]. Moreover, stringent requirements on quality-of-service (QoS) mechanisms and the necessities for CDN operators to compete in the market require optimal decisions to be made around guaranteed service levels and pricing.

The aim of this chapter is to describe and detail some of the fundamental problems arising in CDNs relevant to the optimization of resource management, allocation, and pricing. This chapter also includes related problems arising in video-on-demand (VoD) content delivery and looks at such problems from an optimization perspective. The chapter adopts a tutorial style, as with the rest of the book and covers basics of optimization modeling and techniques, with illustrations of how they can aid in solving some of the challenging problems arising in CDNs or CDN-like environments, such as VoD.

Advanced Content Delivery, Streaming, and Cloud Services, First Edition.
Edited by Mukaddim Pathan, Ramesh K. Sitaraman, and Dom Robinson.

The rest of this chapter is structured as follows. The following section covers basics on mathematical modeling and optimization techniques, including branch-and-bound, decomposition, and heuristics. Section 9.4 describes some of the more fundamental optimization problems arising in content delivery and ways of effectively solving these problems. Visionary thoughts for practitioners are offered in Section 9.5, and this is followed by future research prospects in Section 9.6. Conclusions are stated in Section 9.7.

9.2 BASICS ON MATHEMATICAL MODELING AND OPTIMIZATION

Many of the problems arising in CDNs can be modeled in the following form:

$$(P) \quad \text{Minimize } \mathbf{f}(\mathbf{x}, \mathbf{y}) \text{ subject to } \mathbf{Ax} + \mathbf{By} = \mathbf{d}, \mathbf{x} \in \mathbf{X}, \mathbf{y} \in \mathbf{Y}$$

where $\mathbf{x}$ and $\mathbf{y}$ are column vectors of variables and $\mathbf{f}(\mathbf{x}, \mathbf{y})$ is a function to be minimized. In the remainder of the exposition, we assume that $\mathbf{f}(\mathbf{x}, \mathbf{y})$ is a linear function in the form $\mathbf{cx} + \mathbf{dy}$ but mention some cases where it need not be linear. $\mathbf{A}$ and $\mathbf{B}$ are matrices of constraint coefficients and $\mathbf{d}$ is the column vector of right-hand side values, all with appropriate dimensions. $\mathbf{X}$ and $\mathbf{Y}$ are nonempty sets in which variables $\mathbf{x}$ and $\mathbf{y}$ are, respectively, defined. If $\mathbf{f}$ is linear in $\mathbf{x}$ and $\mathbf{y}$, and $\mathbf{X}$ and $\mathbf{Y}$ are both continuous, then P is a linear program (LP). In addition, if $\mathbf{X}$ or $\mathbf{Y}$ is discrete, then P is called an integer linear program (ILP). A special case of a linear program is where a subset of the variables is required to take binary values, in which case P is called a binary (or a 0-1) program.

LPs of even very large scale can nowadays be routinely solved to optimality using the Simplex method [2], implemented in some of the off-the-shelf solvers. ILPs are more challenging in their nature because of their combinatorial nature and particularly if they are of large scale. A number of optimal solution techniques have been devised for their resolution, including branch-and-bound, column generation [2], branch-and-cut, and branch-and-price [3], as well as other specialized techniques, two of which relevant to our discussion will be further explained. These two techniques are based on decomposition and are particularly well suited for models resulting from CDN applications, which tend to be large scale due to the sheer size of the network or content that often needs to be taken into account in the resource allocation planning.

Before the details on these techniques are presented, we also mention an alternative class of solution methods for problems for which an optimal solution is either not sought or difficult to find. Heuristics, or metaheuristics, are often designed to quickly produce good quality solutions for difficult optimization problems and employ various (guided) mechanisms to search the solution space to discover near-optimal solutions. For an overview of heuristic algorithms, the reader is referred to Reference 4.

9.2.1 Benders Decomposition

Benders decomposition is a method that reformulates an optimization problem such that it splits into two problems, namely, a master problem (MP) and a subproblem, ideally in such a way that the latter will further decompose into smaller problems that are easier to solve than the original problem. To illustrate, consider problem P where a variable fixing $\mathbf{y} = \hat{\mathbf{y}}$ is applied to be able to rewrite the problem in the following form:

$$\text{Minimize } \mathbf{d}\hat{\mathbf{y}} + \Theta \text{ subject to } \mathbf{y} \in \mathbf{Y}$$

where $\Theta(\hat{\mathbf{y}})$ corresponds to the optimization problem or, more precisely, the *subproblem*,

$$\text{Minimize } \mathbf{cx} \text{ subject to } \mathbf{Ax} = \mathbf{d} - \mathbf{B}\hat{\mathbf{y}},\ \mathbf{x} \in \mathbf{X}$$

Given that $\Theta(\hat{\mathbf{y}})$ is linear and continuous, one can replace it with its dual using a vector $\mathbf{w}$ of dual variables, corresponding to each of its constraints:

$$\text{Maximize } \mathbf{w}(\mathbf{d} - \mathbf{B}\hat{\mathbf{y}}) \text{ subject to } \mathbf{wA} \leq \mathbf{c}$$

We assume that the feasible region of $\Theta(\hat{\mathbf{y}})$ is nonempty and define Λ and Φ as the set of extreme points and rays, respectively. Then, P can be reformulated in the following way, namely the MP:

$$\text{Minimize } z + \mathbf{dy}$$

$$\text{subject to } z \geq \mathbf{p}(\mathbf{d} - \mathbf{By}) \quad \mathbf{p} \in \Lambda \tag{9.1}$$

$$0 \geq \mathbf{r}(\mathbf{d} - \mathbf{By}) \quad \mathbf{r} \in \Phi \tag{9.2}$$

$$\mathbf{y} \in \mathbf{Y}$$

Constraint (9.1) is called an optimality cut and corresponds to an optimal solution of $\Theta(\hat{\mathbf{y}})$ when feasible, which is also feasible for P providing an upper bound. Constraint (9.2) is named as the feasibility cut for every infeasible solution $\mathbf{r}$. In practice, sets Λ and Φ are likely to be rather large due to the number of optimal or infeasible solutions of $\Theta(\hat{\mathbf{y}})$, resulting in a large number of cuts making the MP difficult to be solved. A good strategy to solve the MP lies in starting out with a restricted version where only small subsets of constraints (9.1) and (9.2) are present. An optimal solution $\mathbf{y}^*$ of the restricted MP, the value of which is a lower bound for P, can then be used to generate either an optimality or a feasibility cut through solving $\Theta(\mathbf{y}^*)$. Through such an iterated process, cuts can be added "on-the-fly" until convergence is ensured. The reader is referred to Reference 5 for additional details on the procedure.

9.2.2 Lagrangean Relaxation

Models such as P often include set(s) of constraints such that if "complicating" or "linking" constraints were not present, then P would be "easy" to solve. Lagrangean relaxation is a popular approach for solving models of this kind, where complicating constraints are relaxed (albeit with a penalty) such that the resulting problem can be solved efficiently. Furthermore, relaxed problems often naturally decompose into smaller problems. To illustrate this, consider a relaxation of P where constraints $\mathbf{Ax} + \mathbf{By} = \mathbf{d}$ are relaxed using dual variables (also known as Lagrangean multipliers) ξ resulting in

$$(R_\xi) \quad \text{Minimize } \mathbf{cx} + \mathbf{dy} + \xi(\mathbf{Ax} + \mathbf{By} - \mathbf{d}) \text{ subject to } \mathbf{x} \in \mathbf{X},\ \mathbf{y} \in \mathbf{Y}$$

or equivalently,

$$(R_\xi) \quad \text{Minimize } (\mathbf{c} + \xi\mathbf{A})\mathbf{x} + (\mathbf{d} + \xi\mathbf{B})\mathbf{y} \text{ subject to } \mathbf{x} \in \mathbf{X},\ \mathbf{y} \in \mathbf{Y}$$

which decomposes into two problems, one defined only in $\mathbf{x}$ variables and the other defined only in $\mathbf{y}$ variables. We note that the optimal solution value of R_ξ is a lower bound on the optimal solution value of P, and the problem of finding the best possible lower bound, namely to maximize R_ξ, is a piecewise linear optimization problem for which one often resorts to nondifferentiable optimization techniques.

Although the Lagrangean relaxation scheme described earlier is able to generally find lower bounds in short computational times, the solution of R_ξ does not yield a feasible solution to problem P. Lagrangean heuristics are often devised for this purpose, which make use of the information provided by the relaxed problem and "repair" the relaxed (therefore often infeasible) solution to arrive at a feasible solution for the original problem. Further details on Lagrangean relaxation can be found in References 6, 7.

9.3 VIDEO-ON-DEMAND APPLICATIONS

VoD is one of the more prominent applications of CDNs used for on-demand home entertainment, remote learning and training, video conferencing, and news on demand [8]. These applications primarily differ from other CDN services by their requirement of significant amounts of bandwidth. A VoD service is a special type of electronic content delivery service in that it deals specifically with the distribution of videos (e.g., movies) to a number of geographically distributed users. In other words, a VoD service can be described as a virtual video rental store in which a user has the option to choose and watch any program on request, at the convenience of their time. Interactive VoD services over the user a fine-grained control, enabling them to pause, resume, fast rewind, and fast-forward the video. Applications of such services are not limited to home entertainment and can be extended to banking applications, education, and home shopping. A VoD can be regarded as a special CDN where significantly large amounts of data (multimedia files) are to be distributed; hence, bandwidth and server capacities pose tight constraints. These services require special networks that are capable of supporting such high bandwidth applications (such as cable networks).

A complete VoD system consists of three fundamental components that may be stated as the storage servers, network on which the system is built, and the user interface. The servers hold programs or families of programs, and the underlying network can be modeled either as a tree or as a fully meshed graph.

Figure 9.1 shows two possible architectures for VoD applications. In the centralized setting, there exists a central server acting as the main storage unit holding all programs.

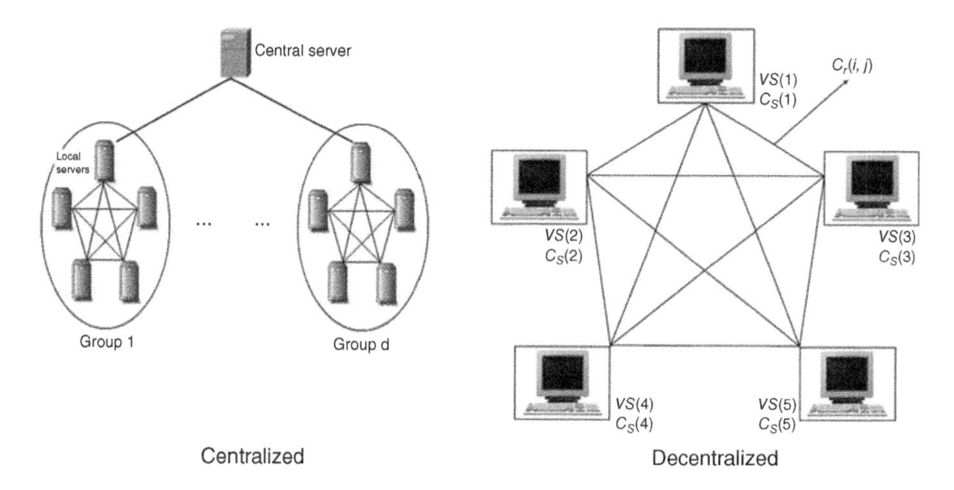

Figure 9.1 Two possible architectures for video-on-demand applications.

Connected to the central server, there are groups consisting of local video servers. Each group is a fully meshed network, that is, units in the group are all connected to each other. Each user is connected to a local server, although users can watch programs transparently from other local servers in the group. However, this incurs an additional cost. In contrast to the decentralized setting, all servers collectively store all programs and each request is directed to and served by a suitable server holding a copy of the requested program. This is made possible by what is termed as *cooperative caching*.

A VoD provider may either choose to offer its services as *data-centered* or *user-centered* [9]. The former is in general called broadcasting, where the provider broadcasts the programs through a single channel in specific time periods and the user has to tune into the channel to receive the program. In this case, the user is a passive participant to the system and has no control over the program. In contrast, the user-centered approach specifically dedicates a channel and bandwidth to the user through which the program is immediately transmitted on request. Moreover, the user has complete control over the session. Although the former approach requires less system resources and is less expensive, the latter has a higher quality. There are also hybrid approaches, such as batching, where the provider collects similar user requests in specific time intervals, which are then served using a single video stream. In this case, the user has to wait after issuing the request and does not have a control over the program. For more details on this topic, we refer the reader to the survey in Reference 9.

9.3.1 Dynamic Streaming Policies

Live streaming is becoming increasingly popular as more and more enterprises want to stream on the Internet to reach a worldwide audience. Common examples include radio and television broadcasts, live events, and multimedia conferencing. Transporting the streaming bits across the Internet from the encoder to the end-user without a significant loss in stream quality is a critical problem that is hard to resolve, since server and network bottlenecks may impede with high quality delivery of streams. For example, in January 2009, Akamai hosted President Obama's inauguration event, which drew seven million simultaneous viewers worldwide with a peak aggregate traffic of 2 terabits per second. The Internet is designed as a best-effort network with no quality guarantees for communication between two endpoints, and packets can be lost or delayed as they pass through congested routers or links. This can cause the stream to degrade, producing "glitches," "slide-shows," and "freeze ups" as the user watches the stream.

The content delivery infrastructure has the potential to alleviate these problems by moving content from the origin server toward the edge servers and intermediates (replica/surrogate servers). For live content streaming, no caching is involved, but content may still need to be directed to certain edge servers. A distributed infrastructure for live streaming in a CDN is shown in References 10, 11. As depicted in Figure 9.2, this infrastructure is different from object distribution for regular HyperText Transfer Protocol (HTTP) traffic. Streams typically originate in a customer's encoder. A collection of machines, termed entry points, act as origin proxies of customer's encoders. Stream data are then sent out to a larger collection of machines, termed set reflectors, which in turn propagate this data to edge regions. Streaming servers are organized in groups (called edge regions), deployed in individual data centers and tied together by a private local area network (LAN). Regions are deployed widely so as to provide good geographical and network coverage. The use of set reflectors is mainly for scalability reasons. It is not necessary to deliver many unpopular streams to an edge region if they are not needed there. The set reflectors implement a subscription mechanism. Once the set reflector receives a subscription request for a certain stream, it propagates this request to the appropriate entry point where the stream enters the CDN.

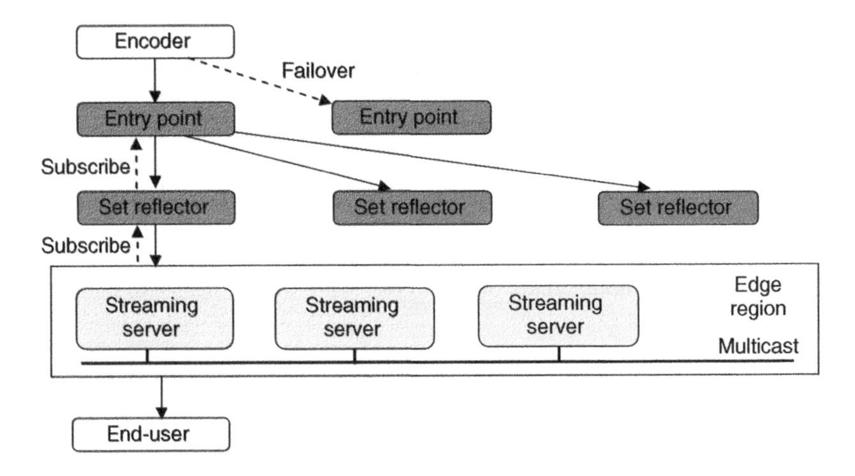

Figure 9.2 Architecture for streaming applications.

9.4 OPTIMIZATION PROBLEMS IN CONTENT DELIVERY AND VoD SERVICES

9.4.1 Resource Management and Allocation in CDNs

There are three main problems that arise in managing and allocating resources in a CDN, which are summarized later. For an excellent overview of the technologies used within CDN, the reader is referred to Reference 12.

9.4.1.1 Proxy Server Location Problem Being one of the fundamental and earliest studied problems in designing CDNs, the *proxy server location problem* consists of finding the number and location of a given number of proxy servers to be deployed, such that predefined measure (e.g., flow of traffic, average delay, total cost) is minimized. The problem is also referred to as the *replica server placement*, *mirror placement*, or *cache location* problem. Mathematical models for this problem are based on, or variations of, the uncapacitated p-median or facility location problems [13,14]. The problem of placing transparent caches is described in Reference 15. The objective function considered therein is one of cost-minimizing nature and considers the case where the requested content is not found in a specific caching server.

9.4.1.2 Request Routing Routing in a computer network refers to sending data from one or more sources to one or more destinations so as to minimize the total traffic flowing on the network. Requests can be served either through unicast [16] or multicast routing. For a detailed review of the latter as well as a survey of combinatorial optimization applications, we refer the reader to Reference 17. Request routing in a CDN is the process of guiding a client's request to a suitable proxy server that is able to serve the corresponding request. The problem is formally defined as selecting a proxy server to address a request for an object such that a cost function is minimized. For a simplified mathematical formulation of the problem, the reader may refer to Reference 18.

9.4.1.3 Object Placement Previously mentioned studies assume that the content held in the origin server is entirely replicated onto the caching servers (in case of which the caching servers are usually referred to as replicas or mirrors). Unfortunately, this may not always be possible in situations where the objects are significantly large in size (i.e., multimedia files) and only a partial replication can be performed due to the limited storage capacity of the caching servers. In this case, any caching server can only hold a subset of the content. Determining which objects should be placed at each caching server under storage capacity restrictions is known as the object placement problem. The reader is refereed to Reference 19 for a mathematical model of this problem, and the more recent work by Khan and Ahmad 20 for an extensive comparison of various methods, including those based on branch and bound, Benders decomposition, heuristics, and a number of game theoretical approaches to solve the problem.

Since the early 2000s, the literature on resource allocation CDNs has substantially grown in that mathematical models and techniques have started looking at more complex problems, including simultaneous consideration of one or more of the three main problems above, or variations thereof to include more complex constraints such as those concerning delay. These are considered in detail in the remainder of this chapter.

9.4.2 Request routing

In general, users are located at mutually exclusive networks, which may correspond to the networks of possibly different Internet service providers (ISPs). When a user requests a Web page, the request is first directed to the request-routing subsystem of a CDN to which the owner of the Web page has subscribed to during DNS lookup. The request-routing subsystem (RRS) matches every user request with the most appropriate surrogate server in CDN. This decision depends among other factors on the surrogates' cache content, their current network loads, proximity to the user networks, and their service prices. The surrogates' distance to a user network and their current network loads determine the delay experienced by the users of that particular network. This delay can be estimated by the RRS by such mechanisms as periodic polling of the servers and by some *a priori* topology information. Furthermore, the RRS keeps information about the cache content of the surrogates. This information can be effectively gathered by employing mechanisms such as bloom filters or delta updates [21]. Once the RRS determines a suitable surrogate to service the user request, it informs the user of the IP address of this surrogate. Finally, upon receiving the IP address of the surrogate, user opens an HTTP connection with that particular surrogate for the delivery of the requested object.

A content provider and a CDN make a service-level agreement (SLA), which specifies the maximum average delay observed by the users accessing the publisher's content. Owing to recurring costs of leased lines subscribed to by the surrogates and fixed costs of procuring servers, each surrogate charges a fee for each user request served. The objective of the RRS is to minimize the total service cost (and thus, maximize the profit) while satisfying all the content provider SLAs.

Let D_j denote the maximum average delay the users requesting objects from content provider j should experience as specified in the SLA. The content of the content provider consists of I objects, where object i is distributed among the set of $S_{ij} = \{s_0, s_1, \dots\}$ surrogate servers. Let s_0 denote the origin server for the object. The request arrival rate for object i in publisher j from user network n is given by λ_{ij}^n. The delay between surrogate s and network n is d_s^n. Each surrogate charges a fee for every user request served. The unit price of service by surrogate s is p_s. Let $x_{ij}^n(s) = 0$ or 1 be the decision variable denoting the surrogate serving the user requests from network n for object i in content provider j.

The objective of the CDN is to minimize the total service cost, while satisfying the individual average delay bounds of the publishers. The following optimization problem (P) describes this objective:

$$\underset{x}{\text{Minimize}} \sum_i \sum_j \sum_{s \in S_{ij}} \sum_n p_s \lambda_{ij}^n x_{ij}^n(s)$$

subject to

$$\sum_i \sum_{s \in S_{ij}} \sum_n \lambda_{ij}^n \; d_s^n \; x_{ij}^n(s) \le D_j \qquad \forall_j$$

$$\sum_{s \in S_{ij}} x_{ij}^n(s) = 1 \qquad \forall n, i, j$$

We now show how Lagrangean relaxation can be used to obtain a tight lower bound to the aforementioned request-routing problem. The dual problem of (*P*) with respect to the delay constraint is determined as follows. Let the dual problem (*D*) be defined as follows:

$$Z_D(\alpha_j) = \underset{x_{ij}^n}{\text{Minimize}} \left[\sum_i \sum_j \sum_{s\in S_{ij}} \sum_n p_s \lambda_{ij}^n x_{ij}^n(s) + \sum_j \alpha_j \right] \left(\sum_i \sum_{s\in S_{ij}} \sum_n \lambda_{ij}^n \, d_s^n \, x_{ij}^n(s) - D_j \right)$$

subject to

$$\sum_{s\in S_{ij}} x_{ij}^n(s) = 1 \qquad \forall n, i, j$$

where $\alpha_j \geq 0$ is the Lagrangean multiplier. It is easy to see that the routing decision given by the solution to $Z_D(\alpha_j)$ is

$$x_{ij}^n(s) = \begin{cases} 1 & \text{if } s = \arg\min_{s\in S_{ij}} \{p_s + \alpha_j d_s^n\} \\ 0 & \text{otherwise} \end{cases}$$

By weak duality, a lower bound for (*P*) as the solution of max $\{\alpha_j : Z_D(\alpha_j)\}$ can be determined. Since $Z_D(\alpha_j)$ is nondifferentiable, a subgradient projection method is needed, where the subgradient of $Z_D(\alpha_j)$ is $g_j^k = \Sigma_i \Sigma_n \Sigma_{s\in Sij} \lambda_{ij}^n d_s^n x_{ij}^{n,k}(s) - D_j$. The iterations are $\alpha_j^{k+1} = \alpha_j^k + s^k g_j^k$, where $\alpha_j^k \geq 0$ and s^k is the step size. The new iterate may not improve the dual cost for all values of the step size; however, if the step size is small enough, the distance of the current iterate to the optimal solution set is reduced.

We can also determine a simple feasible (upper bound) solution to (*P*) as follows. Let $d_{\min}^j = Dj/\Sigma_i \Sigma_n \lambda_{ij}^n$. If $x_{ij}^{n,k}(s') = 1$ for s' such that $d_{s'}^n \leq d_{\min}^j$, then the delay constraint is satisfied. Let $S'(i,j,n) = \{s \in S_{ij} : d_s^n \leq d_{\min}^j$ denote the set of surrogates satisfying this property for the corresponding object for the users in network n. Furthermore, if we choose s'' as $s'' = \text{argmin}\{p_s : s \in S'(i,j,n)\}$, that is, for each user network and object such that $p_{s''}$ corresponds to the minimum price offered by the surrogates satisfying the delay constraint, then we get an upper bound to the problem (*P*).

Given the lower and upper bounds, one can devise an iterative suboptimal algorithm. In this algorithm, initially starting from the routing scheme resulting in the upper bound, the total cost is reduced iteratively by reducing the slack in the delay constraint. While increasing the total average delay to the delay bound, new surrogates that reduce the cost the most with a minimum increase in the associated delay are chosen. This is performed by calculating for each user network a "benefit coefficient," $w^n(k, l)$, which corresponds to the reduction in total cost per unit increase in delay by switching from the current surrogate assignment to a new surrogate. A greedy reassignment of the user networks maximizing this coefficient is performed while maintaining the delay bound.

On a similar note, Pathan, Broberg, and Buyya [22] investigated the request-routing problem in *content delivery clouds*. Content delivery clouds give content delivery services over the existing clouds; hence, they can provide significant cost savings for the content providers. However, unlike a fully featured CDN, they do not provide capabilities for automatic replication, geographical load redirection, and load balancing. MetaCDN realizes a content delivery cloud, providing the required features for high performance content delivery [23]. When a user requests content, MetaCDN chooses an optimal replica for content delivery, thereby ensuring satisfactory user-perceived experience. The ultimate goal is to improve the utility of MetaCDN's content delivery services, where the utility is the quantification of the traffic activities in MetaCDN and represents the usefulness of its replicas in terms of data circulation in its worldwide distributed network. Pathan, Broberg, and Buyya [22] formulate the utility maximization problem with quantitative expressions and devise a utility-based, request redirection policy.

9.4.3 Proxy server location and object placement

Earlier work on handling the proxy server location and object placement problem include Reference 24 where the optimal number and location of proxies along with the placement of replicas of a single object on the installed proxies are considered. In this problem, a maximum number of potential proxies are given and the network has a tree-like topology. In Reference 25, the authors discuss the joint problem of proxy server placement and object placement in a CDN, subject to a budget constraint.

More recent work on this problem is done by Luss 26, who studies the problem of determining optimal location of servers and the optimal assignment of program families to servers within a VoD network, where multiple assignments are permitted. In other words, programs may be assigned to multiple servers. The problem is defined on a tree network and the objective is to minimize the total cost of servers, program storage, and bandwidth, such that all demands for programs are satisfied. In Reference 26, the author describes a mathematical model for this problem, but argues that, due to the size of the model even for small networks, using general-purpose integer programming software will be impractical. Instead, by exploiting the tree-like structure of the network, Luss [26] describes a dynamic programming formulation and algorithm as an efficient way to solve the problem.

The growth for significant amount of data flowing over CDNs, particularly in applications such as VoD, an effective management of bandwidth resources, becomes critical so as to provide a good quality service to users. In this setting, a CDN sits on multiple trees and a server at the root of each tree broadcasts several programs through the tree. Each link a on the network has a limited bandwidth capacity shown by w_a and carries a set of programs shown by P_a. For each program p, at least B_p units of bandwidth must be allocated on all links. The delivery quality of digital content is often measured by a performance function $F_p(r_{ap})$ that is continuous and strictly increasing, with the bandwidth r_{ap} allocated to the link a over which program p flows into the destination node. Some examples to $F_p(r_{ap})$ for content delivery applications are given in Reference 26, such as

i. $\pi_p + \delta_p r_{ap}$
ii. $\ln(\pi_p + \delta_p r_{ap})$
iii. $\pi - \delta_p/(r_{ap} + B_p)$

where π and δ are parameters that could be program dependent. Function (ii) shows a diminishing return of satisfaction for larger bandwidth, whereas function (iii) is for near-VoD applications [27].

Within such a setting, Luss [27] studies the problem of allocating bandwidth in an equitable manner, that is, an allocation ensuring that no performance function for an object value can be feasibly increased without degrading an equally poor or worse function for another object. This is referred to as a *lexicographic maximin optimization problem*. Luss [27] also describes an algorithm for this problem where the performance function may be program or node-dependent, but the algorithm itself is not necessarily polynomial. Later work by Luss [8] presents an algorithm that is truly polynomial for a class of performance functions, which repeatedly solves, for each link in the network, a single-link bandwidth allocation model and presents two algorithms to solve this model.

9.4.4 Request routing and object placement in VoD networks

Proxy location problems often appear at a strategic level of planning and are periodically revised, that is, twice a year [26], whereas the joint problem of request routing and object placement lies at a more operational level as the changes in allocations are more frequent. This is particularly the case with request routing, however, due to the interdependency between the two problems, solving them jointly has benefits to reap in terms of a more efficient use of resources. The joint problem is of a complex nature particularly when a complete network is considered. Such a problem within a VoD network has been described in Reference 28 and later on studied in Reference 29. In this problem, the aim is to place each program on a number of servers, such that the total cost of storage and transmission of the programs in the network is minimized, and the demand of each node for each program is satisfied. As in Reference 26, multiple assignments are also allowed in this setting. For efficient solution methods for this problem, see Reference 28 for a heuristic algorithm, Reference 29 for a Lagrangean relaxation scheme to generate lower bounds coupled with a Lagrangean heuristic to generate upper bounds, and Reference 30 for enhancements to the algorithm and a more scalable version of the latter.

The streaming policies adopt a different architecture as shown in Figure 9.1. Andreev et al. [31] investigated the overlay network design and routing problem for this architecture. The architecture of the overlay network allows for distributing a stream from its entry point to a large number of edge servers with the help of reflectors, and thus alleviating the server bottleneck. An overlay network can be represented by a tripartite digraph $N = (V, E)$ and a set of paths Π in N that are used to transport the streams. The node set V consists of a set of sources S representing entry points, a set R representing reflectors, and a set of sinks D representing edge servers. Physically, each node is a cluster of machines deployed within a data center of an ISP on the Internet. Given a digraph $N = (V, E)$ that represents a deployment of sources, reflectors, and sinks, and given a set of live streams, the construction of an overlay network involves computing a set of paths Π that specify how each stream is routed from its source to the subset of the sinks that are designated to serve that stream to end-users. Overlay network construction can be viewed as an optimization problem to minimize cost, subject to capacity, quality, and reliability requirements. The cost of the overlay network comprises transmission cost of sending traffic over the Internet and fixed costs such as the amortized cost of buying servers. The capacity constraints reflect resource limitations of the nodes and links, such as bandwidth, central processing unit (CPU), memory, or the maximum number of district streams a node can transmit simultaneously. Meanwhile, the quality constraint often depicts how much packet loss a stream can tolerate before there is significant loss in video quality. Finally, the reliability constraint requires that multiple copies of the stream are available in the Internet against failures in the network.

Given this general model, Andreev et al. [31] developed an integer programming formulation for the overlay network design problem for streaming applications given below.

$$\min_{s.t.} \sum_{i\in R} r_i z_i + \sum_{i\in R}\sum_{k\in S} c_{ki}^k y_i^k + \sum_{i\in R}\sum_{k\in S}\sum_{j\in D} c_{ij}^k x_{ij}^k$$

(1) $y_i^k \le z_i \ \forall i \in R,\ \forall k \in S$

(2) $x_{ij}^k \le y_i^k \ \forall i \in R,\ \forall i \in D,\ \forall k \in S$

(3) $\sum_{k\in S}\sum_{j\in D} x_{ij}^k \le F_i z_i \ \forall i \in R$

(4) $\sum_{j\in D} x_{ij}^k \le F_i y_i^k \ \forall i \in R,\ \forall k \in S$

(5) $\sum_{i\in R} x_{ij}^k w_{ij}^k \ge W_j^k \ \forall i \in D,\ \forall k \in S$

(6) $x_{ij}^k \in \{0,1\}, y_{ij}^k \in \{0,1\}, z_i \in \{0,1\}$

In this model, y_{ik} is the indicator variable for the delivery of the kth stream to the ith reflector, z_i is the indicator variable for utilizing reflector i, and x_{ij}^k is the indicator variable for delivering the kth stream to the jth sink through the ith reflector. Constraints (1) and (2) force a payment for the reflectors being used and to transmit packets only through reflectors that are in use. Constraint (3) gives the maximum number of streamed that can be served by each reflector. Constraint (4) provides a useful cutting plane in the linear program rounding algorithm used to solve this optimization problem as discussed in Reference 31. Constraint (5) is the "weight constraints" that capture the end-to-end loss requirements for QoS. Constraint (6) is the integrality constraint for the variables.

The set of paths Π that is the output of overlay network construction can be extracted from the solution to the aforementioned IP by routing a stream from its source k through reflector $i \in R$ to sink $j \in D$, if and only if x_{ij}^k equals 1 in the IP solution. The cost objective function that is minimized represents the total cost of operating the overlay network and is the sum of three parts: the fixed cost of using the reflectors, the cost of sending streams from the sources to the reflectors, and the cost of sending streams from the reflectors to the sinks.

9.4.5 Request routing and object placement in CDNs

Within a CDN, the joint problem of request routing and object placement takes a slightly different form because of the existence of an "origin server" and a more hierarchical representation of clients and proxy servers as compared to the VoD setting described in the earlier section. The main assumption is the existence of a central origin server that is able to hold all objects. Then, when a client makes a request for an object, it is served from either the associated server if the object is already stored there or the origin server via the path from the corresponding proxy server to the origin server at the expense of an additional transfer cost.

To give a more formal description, we follow the modeling framework described in Reference 32. More specifically, there exists a complete network $G = (V, E)$, where V is the set of nodes and $E = (\{i,j\} : i,j \in V, i \neq j)$ is the set of links. The node set V is further partitioned into three nonempty subsets I, J, and $S = \{0\}$, where I is the set of clients, J is the set of nodes where proxy servers are installed, and S is a singleton containing the origin server. The set of clients may be composed of ISPs, corporate firms, universities, and so on. In such settings, it is usually assumed, without loss of generality, that no client can directly access the origin server (e.g., for security reasons). As with VoD, with each link $(i,j) \in E$ is associated with a nonnegative unit transfer cost denoted by c_{ij}. The cost may be an indicator of, say, unit bandwidth cost, number of hops, and so on. We denote d_{ij} by the delay representing the amount of time required to retrieve data between nodes i and j.

The capacity of a server located at site j is s_j. If large objects are to be distributed, then the capacity can be defined in terms of physical storage, which will be the bottleneck in such a situation. If not, it can be defined as the total bandwidth a server may support. We define K as the set of objects located in the origin server and assume that the size of each object $k \in K$ is b_k. Also we consider that the probability of client $i \in I$ requesting object $k \in K$ over a given time interval is denoted by p_{ik}.

Finally, we assume that there are QoS mechanisms in place such that, for any object, the total time for a requested object to reach the client should not exceed a given threshold. This threshold, say D, can be defined separately for each client, for each object, or both, depending on the agreed SLA between the content publisher and the commercial CDN. Let $T(b_k, d_{ij})$ denote the delay caused by transferring object k over link (i,j) and let $T'(b_k, d_{ij})$ denote the additional delay in requesting the object from the origin server should it not be available at the local proxy server.

There are two decisions to be made: one is request routing and the other is object placement. For the former, a binary variable x_{ij} is used that is equal to 1 if client $i \in I$ is assigned to proxy server $j \in J$, and 0 otherwise. The other binary variable, denoted z_{jk}, is 1 if proxy server $j \in J$ holds object $k \in K$, and 0 otherwise. Then, the following formulation can be used to place objects and assign each client to a single proxy server such that the total cost of providing a delivery service over the network is minimized, which we refer to as model F.

$$\text{Minimize} \sum_{i \in I} \sum_{j \in J} \sum_{k \in K} \left(b_k \lambda_{ik} c_{ij} x_{ij} z_{jk} + b_k \lambda_{ik} \left(c_{ij} + c_{j0}\right) x_{ij} \left(1 - z_{jk}\right)\right)$$

subject to

$$\sum_{j \in J} x_{ij} = 1 \qquad \forall i \in I$$

$$\sum_{k \in K} b_k z_{jk} \le s_j \qquad \forall i \in J$$

$$\sum_{j \in J} T\left(b_k, d_{ij}\right) x_{ij} z_{jk} + \sum_{j \in J} T'\left(b_k, d_{ij}\right) x_{ij}\left(1 - z_{jk}\right) \le D \quad \forall i \in I,\ j \in J,\ k \in K$$

$$x_{ij} \in \{0, 1\} \qquad \forall i \in I, j \in J$$

$$z_{jk} \in \{0, 1\} \qquad \forall i \in J, k \in K$$

The objective function of the formulation above minimizes the cost by looking at two mutually exclusive cases for every triple ($i \in I, j \in J, k \in K$). The first case arises when proxy j holds object k (i.e., $z_{jk} = 1$) in which client i receives their request. In the other case, that is, $z_{jk} = 0$, the second part of the objective function becomes active and takes into account the additional cost c_{j0} of retrieving the object from the origin server. The first set of constraints ensure that a client can only be assigned to a single proxy server, whereas the second set of constraints guarantee that the capacity limits on the proxy servers are not exceeded. The third set of constraints are written in a similar vein to the objective function, and model the situation that, regardless of where a requested object is sourced from (i.e., either locally or centrally), the total time spent by a client does not exceed D. The last two restrictions model the integrality of the decision variables.

One challenging aspect of this formulation is the quadratic nature of the objective and the constraints. Bektaş, Oguz, and Ouveysi [32] show that the QoS-related constraints can be handled *a priori* through excluding those variables from the formulation that would result in a violation of these constraints. Let A_{ij} denote a set of objects for a client–proxy pair that can be served without violating the delay limit. Then, the QoS constraints simply reduce to $x_{ij} \le z_{jk}$ for all $i \in I, j \in J$, and $k \in A_{ij}$.

We now show how Benders decomposition can be applied to solve this problem. First, a new variable φ_{ijk} is defined for every triple ($i \in I, j \in J, k \in K$) to denote the multiplication $x_{ij} z_{jk}$ appearing in the formulation and hence is also binary *ipso facto*. This variable is actually an indicator of whether client i is routed to proxy server j holding object k or not. Then, the following model is an exact linearization of model F (see Reference 32 for a proof).

$$\text{Minimize} \sum_{i \in I} \sum_{j \in J} \sum_{k \in K} \left(b_k \lambda_{ik}\left(c_{ij} + c_{j0}\right) x_{ij} - b_k \lambda_{ik} c_{j0} \varphi_{ijk}\right)$$

subject to

$$\sum_{j \in J} x_{ij} = 1, \qquad \forall i \in I$$

$$\sum_{k \in K} b_k z_{jk} = \le s_j \qquad \forall j \in J$$

$$\varphi_{ijk} - x_{ij} \leq 0 \qquad \forall i \in I, j \in J, k \in K$$

$$\varphi_{ijk} - z_{jk} \leq 0 \qquad \forall i \in I, j \in J, k \in K$$

$$x_{ij} - z_{jk} \leq 0 \qquad \forall i \in I, j \in J, k \in A_{ij}$$

$$x_{ij} \geq 0 \qquad \forall i \in I, j \in J$$

$$z_{jk} \in \{0, 1\} \qquad \forall i \in J, k \in K$$

$$\varphi_{ijk} \in [0, 1] \qquad \forall i \in I, j \in J, k \in K$$

To apply Benders decomposition, we consider fixing all z_{jk}, for all $j \in J, k \in K$, to either 0 or 1 as z^*_{jk} such that for each proxy server j, the capacity constraints are satisfied. The resulting problem, termed the Benders *subproblem*, after the variable fixing is shown below:

$$\text{Minimize} \sum_{i \in I} \sum_{j \in J} \sum_{k \in K} \left(b_k \lambda_{ik} \left(c_{ij} + c_{j0} \right) x_{ij} - b_k \lambda_{ik} c_{j0} \varphi_{ijk} \right)$$

subject to

$$\sum_{j \in J} x_{ij} = 1 \qquad \forall i \in I \qquad (\alpha_i)$$

$$\varphi_{ijk} - x_{ij} \leq 0 \qquad \forall i \in I, j \in J, k \in K \qquad (\theta_{ijk} \geq 0)$$

$$\varphi_{ijk} \leq z^*_{jk} \qquad \forall i \in I, j \in J, k \in K \qquad (w_{ijk} \geq 0)$$

$$x_{ij} \leq z^*_{jk} \qquad \forall i \in I, j \in J, k \in A_{ij} \qquad (\varsigma_{ijk} \geq 0)$$

$$x_{ij} \geq 0 \qquad \forall i \in I, j \in J$$

In the Benders subproblem shown above, to the right of the constraints are the dual variables. For each client $i \in I$, let F_i denote the set of proxy servers for which it is possible to request any object without violating the QoS constraints. Similarly, let H_i denote the set of proxy servers j holding objects k (i.e., $z^*_{jk} = 1$) that are only possible to request from the origin server while respecting the QoS constraints. It turns out that the Benders subproblem decomposes into smaller subproblems, one for each client $i \in I$, each of which can be solved by easily, as shown later.

Proposition 9.1. The decomposition of the Benders subproblem into each $i \in I$ is solvable by inspection with the following optimal solution.

$$\widetilde{x}_{ip} = 1, \text{where } p \in \underset{j \in F_i \cup H_i}{\operatorname{argmin}} \left(\sum_{k \in k} b_k \lambda_{ik} \left(c_{ij} + c_{j0} \right) - \sum_{k \in K : z^*_{jk} = 1} b_k \lambda_{ik} c_{j0} \right)$$

$$\text{for } j \neq p$$

$$\widetilde{x}_{ij} = 0$$

and

$$\widetilde{\varphi}_{ijk} = \begin{cases} 1, & \text{if } z^*_{jk} = 1 \text{ and } \widetilde{x}_{ij} = 1 \\ 0, & \text{otherwise} \end{cases}$$

Using the solution to the subproblem, one can derive the optimal values of the dual variables. If the solution of the subproblem is infeasible, then the dual variables induce a *Benders feasibility cut* of the form:

$$\alpha_i - \sum_{j \in J} \sum_{k \in K} z_{jk}(w_{ijk} + \varsigma_{ijk}) \leq 0$$

Similarly, a *Benders optimality cut* in the following form can be formed using an optimal solution of the subproblem:

$$\alpha_i - \sum_{j \in J} \sum_{k \in K} z_{jk}(w_{ijk} + \varsigma_{ijk}) \leq \Theta_i$$

where Θ_i is a variable to be minimized in the MP as per Section 9.2.2. As the number of feasibility and optimality cuts is such that it is impractical to generate them *a priori*, an iterative delayed constraint generation algorithm can be used to solve the Benders MP, wherein each iteration $|I|$ subproblems are solved. See Reference 32 for more details and practical enhancements, along with extensive computational results.

In contrast to treating delay or latency as a constraint with an upper bound, one alternative approach is to minimize it along with the more traditional objective of minimizing cost. One of such an approach in Reference 33, where latency is characterized by the number of hops used between a node where request originates and the node where the request is served. Their approach gives way to a biobjective integer linear programming formulation, where one objective minimizes the total number of hops used in the network multiplied by the total volume of content transmitted over the hops, whereas the second objective minimizes total cost of object storage and transmission. The model includes constraints on storage capacities at proxies and guaranteeing that all requests are served. A fuzzy programming approach is used to solve the proposed formulation. The biobjective treatment of the problem allows one to look at trade-offs between latency and operational costs, and whether decrease in one objective is at significant (or not) expense of the other. Of the results that the authors present on synthetic (but realistic) data, it is found that (i) reduction on the number of hops implies an increasingly expensive premium, (ii) storage cost fluctuations have a much greater effect on cost than on hops, and (iii) when storage cost is low, the number of hops are found to be more sensitive to bandwidth charges when their minimization is of secondary concern over minimizing the operational costs.

The efforts mentioned earlier thus far assume that traffic demands are known *prior* to optimizing the problem at hand. Although this information might be available through historical forecasting, the authors of Reference 34 argue that this task is far from trivial as Internet traffic patterns are subject to a high degree of uncertainty. To overcome this within a replica server placement and request-routing problem, they propose to use robust optimization, which is achieved through using two criteria: (i) minimizing the maximum total distribution cost and (ii) minimizing relative regret, which is defined as the difference of a given solution cost from that of the optimal. The authors state that the first criterion minimizes the worst-case scenario, which might be overly conservative or pessimistic, and thus might result in excessively high costs. Also presented in Reference 34 is a bicriteria robust optimization model in which the two criteria are jointly optimized. Empirical results are provided to show that the approach results in substantial improvements in robustness without the cost increases required by the worst-case scenario.

A dynamic and online version of the replica placement and request distribution problem is presented in Reference 35, where models and heuristic algorithms are described for its solution. The model accounts for the limited capacities on the servers, QoS constraints that are characterized by the minimum bandwidth required, and maximum delay tolerated for a client's request.

9.4.6 CDN Pricing

CDNs provide a reliable and robust surrogate caching server infrastructure to the publishers, which covers most of the globe. The origin servers "rent" portions of the caching capacity from the CDN's caching servers and receive varying benefits depending on the surrogates' locations, capacities, and the user request arrival rates. The qualities of service classes differ according to the amount of resource (caching space, bandwidth, etc.) allocated to each service class and the number of publishers subscribing to each class. Optimal pricing problem has several parts: determining the optimal number of service classes given the publisher subscription statistics, finding the optimal resource allocation for each service class, and finally devising an optimal resource tariff. For example, some CDNs provide volume discounts in their pricing policies, whereas others charge significant overages in the event of traffic surges. Similarly, while some CDNs use traditional volume-based pricing (metering bytes delivered in a given time window), others use percentile-based pricing (sampling CP traffic periodically and pricing based on 95th percentile of traffic), and yet others price based on number of deployed assets [36].

A critical problem in developing an optimal pricing problem is to understand when content providers are willing to subscribe to a service from CDN. A monopolistic CDN offering services to content providers receiving requests from users according to Poisson distribution is studies in [37]. The utility of content providers is defined in terms of the value they gain by serving X user requests with faster response times using geographically distributed N CDN servers. The cost of serving X users is then calculated by taking into account the cost of outsourcing content delivery that reflects the cost of losing confidentiality of data outsourced, and the usage-based prices the CDN charges. The net utility in this case is given as

$$u_{\mathrm{CDN}}(X) = V(X) + \tau(N, X) - C_o - P(X)$$

The content provider also has the choice of delivering its content by investing in the infrastructure to process a mean of I requests per unit time. In this case, however, since the demand is random and unknown *a priori*, some requests may not be served due to capacity limitations. The net utility is given as

$$u_{\text{self}}(X) = V(X) - C(I) - c \cdot L(I, X)$$

On the basis of an M/G/1/K*PS queueing system, the authors calculated the expected number of lost requests and the optimal investment cost. Given the optimal investment cost for the content provider, it will subscribe to the service of CDN if utility obtain by this subscription exceeds that of the utility of serving all requests by itself. The CDN announces a single price function $P(X)$ to all content providers. The optimal price function is then obtained by maximizing the expected profit of the CDN by taking into account the probability of a content provider with average request arrival rate λ subscribing to the service. This work shows that it is optimal for CDNs to provide volume discounts, and the optimal price should place lower emphasis on value-based pricing and greater emphasis on cost-based pricing as the relative density of content providers with high outsourcing costs increases.

The prices can also be used to efficiently allocate finite storage resources of CDNs among the content providers. The distributed optimal pricing of caching services of geographically distributed surrogates is investigated in Reference 38, where content providers have finite budgets to invest in the content delivery services. The objective of a content provider is to maximize its net benefit of subscribing to CDNs, where the net benefit is the utility a content provider receives by delivering its content to the users with delay D minus the cost of the resources required to ensure that delay. A cost arises because the CDN has limited resources, such as bandwidth, cache, and so on, and uses prices to induce efficient allocation. The objective of the CDN is to maximize its revenue, while satisfying the requested quality of service (QoS). Content providers act noncooperatively and try to maximize their own benefit, regardless of the state of other content providers.

The content delivery problem can be divided into two subproblems of content distribution and request routing. The objective of the distribution subproblem is to maximize the reduction in average user latency by the use of surrogate caches, while the total cost of service is less than the highest investments the publishers are willing to make for the caching resources. Surrogates have limited cache size and the cache is shared among the publishers. Surrogates charge the publishers for the portion of the cache their content occupies. It is assumed that the surrogates of a CDN do not cooperate—other than announcing their resource prices—due to high messaging and processing overheads associated with cooperative caching. Instead, they act noncooperatively with the objective of maximizing their individual revenues (see Figure 9.3).

Assume that there are I different content provider and J surrogates present in the network. User requests arrive from N different user local area networks (LANs). Let λ_i^j be the total arrival rate to the surrogate j for the content in content provider i. It is assumed that the user interest for the objects of the publishers is distributed according to Zipf distribution. Also, let B_i^j be the investment of the ith publisher in the jth surrogate. Let $B_i = \Sigma_j B_i^j$ be the total investment of the ith publisher. Let p_j denote the price of the unit cache space in surrogate j. Let the pricing policy, $\mathbf{p} = (p_1, p_2, \dots, p_J)$, denote the set

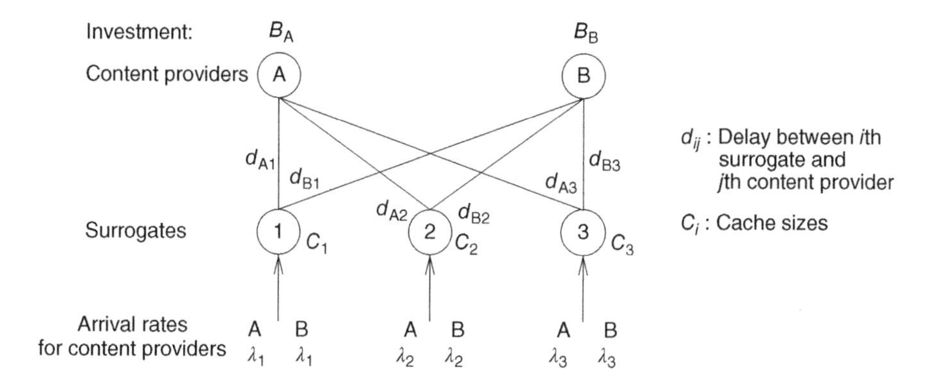

Figure 9.3 Content distribution among surrogates.

of prices for unit cache space of all the surrogates in the network. The additional average delay that a user request forwarded from surrogate j to the origin server of publisher i will experience is denoted by d_{ij}. Let x_i^j be the cache space allocated to publisher i in surrogate j. If ith publisher's investment in jth surrogate is B_i^j, then the total cache space allocated to the content of publisher i in surrogate j is $x_i^j = B_i^j/p_j$. For a given pricing policy $\mathbf{p}$, the content provider optimization problem can be given as

$$\max_{\{x_i^j\}, B_i} \; U_i(\mathbf{X}_i) - \frac{\sum_j x_i^j p_j}{T}$$

$$\text{subject to} \; \sum_j x_i^j p_j \leq B_i$$

where the utility function, that is, the total additional average benefit, $U_i(x_i)$ of publisher i is $U_i(x_i) = \Sigma_{j=1}^{J} \beta_i^j (x_i^j)^{1-\alpha_i}$, β_i^j is the gain factor for content provider i for using surrogate j, α_i is the characteristic of the Zipf distribution of the content of provider i and (S) is a convex optimization problem with a unique solution.

Meanwhile, the optimal pricing strategies of the surrogate maximizing their revenues are obtained when the total publisher demand is equal to the surrogate cache space. Once the optimal distribution and pricing strategies are determined, the authors investigate whether there is an equilibrium at which solutions to both subproblems coincide. For this purpose, the authors defined a surrogate–publisher game and investigated the existence and uniqueness of the Nash equilibrium to this game. The authors determine that a unique equilibrium is reached if the content providers are not willing to pay high amounts and the cache sizes are sufficiently small. These results suggest that a price-directed, market-based distributed algorithm can be used for solving the content provider–surrogate resource allocation problem. For this purpose, an iterative pricing algorithm is proposed as follows.

1. Surrogates announce a set of initial prices $\mathbf{p}^{(0)} = ({p_1}^{(0)}, {p_2}^{(0)}, \dots, {p_J}^{(0)})$.
2. At iteration k, each content provider i calculates its optimal cache demand for surrogate j, $x_i^j(k)$ as the optimal solution to the aforementioned content provider optimization problem. These demands are announced to the surrogates.
3. At iteration k, each surrogate j updates its price according to the publisher demands.

$$p_j^{(k+1)} = \max \left\{ \epsilon, p_j^{(k)} + \gamma \left(x^j \left(\mathbf{p}^{(k)} \right) - C_j \right) \right\}$$

where $x^j = \Sigma_{j=1}^{J} x_i^j$ and γ is the stepsize. Let $\varepsilon > 0$ be a sufficiently small constant preventing prices to approach zero. Thus, if the total demand is greater than the cache capacity C_j, then the new price is increased; otherwise, it is decreased.

9.5 VISIONARY THOUGHTS FOR PRACTITIONERS

The Internet has evolved into a network-supporting generation, sharing and access to content, applications for which its primitive version was not designed for. These new applications have brought with them a number of problems to be resolved, including those pertaining to dynamic or online allocation, storage, and QoS considerations, in order to achieve an efficient running of the network.

We see the techniques and approaches mentioned in this chapter, particularly optimization methods, as being crucial for addressing the issues in the new-generation CDNs. It has been echoed in the literature (see, Reference 12) that commercial CDNs do not exploit optimal replication policies in their planning due to the lack of practical algorithms and instead resort to greedy-type heuristics. Although the need to use fast heuristics in practice is clear, it would be ideal to benchmark the performance of these methods through optimization techniques to have an idea of the potential savings that could be gained.

Compared to the early 2000s, capacity limitations on storage is much less of an issue nowadays, particularly with storage being significantly cheap coupled with the options of leasing space and storing at a cloud level. Such options imply the use of decentralized and collaborative mechanisms. We envisage the future of CDN modeling, therefore, to look at incentives for collaborations and to be able to collectively use the available resource in an efficient way. To achieve this, pricing problems for collaborative and decentralized architectures will need to be solved in greater efficiency and speed. We believe that these issues will be captured within the context of network virtualization, either internal or external, but will bring with it additional issues such as security and scalability.

9.6 FUTURE RESEARCH DIRECTIONS

Although a fair amount of work has been done on CDN modeling, we believe that the rapidly changing nature of the digital world and CDN applications will still give rise to new and exciting research directions. Of particular importance, and as mentioned in the previous section, is the need for dynamic or online algorithms that are able to yield good quality solutions to problems at a network-wide level. Particular attention should be given to request routing, which requires dynamic optimization. Previous research has shown that serving a client request through a single proxy, as opposed to multiple proxies, has provided better results. However, with new technologies such as network virtualization or near-real-time streaming that brings with it stringent QoS requirements, single-server strategies may no longer be viable and multiple-server strategy may be preferred instead. In the latter, one should be aware of the potential privacy issues that may arise, although splitting the content into small enough pieces might overcome such issues. We see this as one interesting research direction to pursue.

Incorporating additional constraints such as delay or SLAs into the existing mathematical models is another research challenge that still needs addressing. Some of the work mentioned in this chapter does include such considerations in the modeling framework, but the functions (e.g., representing delay) are only linear estimations of the actual measure. An explicit account of the true nature of such functions would be needed for a more accurate calculation. An example for this can be seen in Reference 39 who uses functions based on queuing models to represent delay in the network, and a similar treatment is yet to be seen in CDN applications.

The discussions and the models presented in this chapter implicitly assume that various parameters of a CDN are quantifiable and, therefore, can readily be inserted into a modeling framework. We believe that another very interesting research direction is to look at parameters that are not easily quantifiable or those that have not yet been treated as such. One such parameter is user behavior modeling, on which research has only recently started to emerge (e.g., [40,41]). Incorporating user behavior into the optimization models presented here, and particularly within the context of the emerging CDN technologies, would be promising.

9.7 CONCLUSIONS

Mathematical modeling for addressing resource allocation problems in CDNs offers many benefits as well as challenges. The flexibility afforded by these models in incorporating various types of quantifiable objectives or constraints provides for a rich and a transparent framework within which such problems can be represented. These flexibilities range from incorporating single or several objective functions and various constraints on storage, latency, routing, and load balancing. Moreover, the state of the art on the solution of mathematical models, in particular linear and integer programming formulations, is such that several techniques exist that can be customized for CDN allocation problems to give way for efficient algorithms.

However, CDN environment is of an active and dynamic nature, and most applications call for the use of fast and scalable methods that provide good quality solutions. The ever-increasing number of clients and type of contents call for solving resource allocation problems in real time. Online (and fast) optimization or reoptimization approaches, or heuristic solution techniques, seem better suited for this purpose. One should, however, bear in mind that although one may show the superiority of one heuristic method to another, one has no indication of the quality of the solutions obtained with such methods. Our intention through this chapter was to stress the importance of using mathematical models and especially exact solution methods to solve CDN problems and to recognize the benefits of using these approaches. In particular, mathematical modeling and optimization methods are invaluable tools to assess a variety of heuristics, in terms of solution quality, and to guide the choice of the right method to be used in practice.

ACKNOWLEDGMENTS

Some of the work presented here appeared in References 32, 42 and part of Figure 9.1 appeared in Reference 30.

REFERENCES

1. Cisco Systems. 2013. Visual networking index. Available at http://www.cisco.com/en/US/netsol/ns827/networking_solutions_sub_solution.html#~forecast. Accessed 2013 Feb 1.
2. Wolsey LA. *Integer Programming*. Chichester: Wiley; 1998.
3. Chen D-S, Batson RG, Dang Y. *Applied Integer Programming: Modeling and Solution*. New Jersey: Wiley; 2010.
4. Talbi E-G. *Metaheuristics: From Design to Implementation*. Chichester: Wiley; 2009.
5. Benders J. Partitioning procedures for solving mixed-variables programming problems. Numer Math 1962;4:238–252.
6. Geoffrion AM. Lagrangean relaxation for integer programming. In: *Approaches to Integer Programming*. Berlin, Heidelberg: Springer; 1974. p 82–114.
7. Fisher ML. The Lagrangian relaxation method for solving integer programming problems. Manage Sci 2004;50(12):1861–1871.
8. Luss H. Equitable bandwidth allocation in content distribution networks. Nav Res Logist 2010;57:266–278.
9. Ghose D, Kim HJ. Scheduling video streams in video-on-demand systems: A survey. Multimed Tools Appl 2000;11:167–195.
10. Kontothanassis L, Sitaraman R, Wein J, Hong D, Kleinberg R, Mancuso B, Shaw D, Stodolsky D. A transport layer for live streaming in a content delivery network. Proc IEEE 2004;92(9):1408–1419.
11. Andreev K, Maggs BM, Meyerson A, Sitaraman R. Designing overlay multicast networks for streaming. Proceedings of the Fifteenth Annual ACM Symposium on Parallel Algorithms and Architectures (SPAA); 2003 June 07–09; San Diego, CA, USA; 2003.
12. Passarella A. A survey on content-centric technologies for the current Internet: CDN and P2P solutions. Comput Commun 2012;35:1–32.
13. Li B, Golin M, Italiano G, Deng X, Sohraby K. On the optimal placement of web proxies in the Internet. Proceedings of IEEE INFOCOM'99; vol. 3; 1999 Mar 21–25; New York; 1999. p 1282–1290.

14. Qiu L, Padmanabhan V, Voelker G. On the placement of web server replicas. Proceedings of IEEE INFOCOM'01; vol. 3; Anchorage, AK, 2001 Apr 22–26; 2001. p 1587–1596.
15. Krishnan P, Raz D, Shavitt Y. The cache location problem. IEEE/ACM Trans Netw 2000;8:568–582.
16. Salama H, Reeves DS, Viniotis Y. A distributed algorithm for delay-constrained unicast routing. Proceedings of INFOCOM 1997; vol. 1; Kobe; 1997 Apr 7–11; 1997. p 84–91.
17. Oliveira C, Pardalos P. A survey of combinatorial optimization problems in multicast routing. Comput Operat Res 2005;32:1953–1981.
18. Datta A, Dutta K, Thomas H, VanderMeer D. World Wide Wait: a study of Internet scalability and cache-based approaches to alleviate it. Manage Sci 2003;49:1425–1444.
19. Kangasharju J, Roberts J, Ross K. Object replication strategies in content distribution networks. Comput Commun 2002;25:376–383.
20. Khan SU, Ahmad I. Replicating data objects in large distributed database systems: An axiomatic game theoretic mechanism design approach. Distrib Parallel Databases 2010;28:187–218.
21. Fan L, Cao P, Almeida J, AZ Broder. Summary cache: A scalable wide-area web cache sharing protocol; Technical Report 1361; Department of Computer Science, University of Wisconsin-Madison; 1998.
22. Pathan M, Broberg J, Buyya R. Maximizing utility for content delivery clouds. WISE '09 Proceedings of the 10th International Conference on Web Information Systems; Springer-Verlag: Berlin; 2009. p 13–28.
23. Broberg J, Buyya R, Tari Z. MetaCDN: Harnessing 'Storage Clouds' for high performance content delivery. J Netw Comp Appl 2009;32(5):1012–1022.
24. Xu J, Li B, Lee DL. Placement problems for transparent data replication proxy services. IEEE J Select Areas Commun 2002;20:1383–1398.
25. Xuanping Z, Weidong W, Xiaopeng T, Yonghu Z. Data replication at web proxies in content distribution network. In: *Lecture Notes in Computer Science*. vol. 2642. Xian: Springer; 2003. p 560–569.
26. Luss H. Optimal content distribution in video-on-demand tree networks. IEEE Trans Syst Man Cybern A Syst Humans 2010;40(1):68–75.
27. Luss H. An equitable bandwidth allocation model for video-on-demand networks. Netw Spat Econ 2008;8:23–41.
28. Ouveysi I, Sesana L, Wirth A. The video placement and routing problem. In: Kozan E, Ohuchi A, editors. *Operations Research/Management Science at Work: Applying Theory in the Asia Pacific Region*. Kluwer Academic Publishers International Series in Operations Research and Management Science; 2002. p 53–71.
29. Bektaş T, Oguz O, Ouveysi I. A novel optimization algorithm for video placement and routing. IEEE Commun Lett 2006;10(2):114–116.
30. Bektaş T, Oguz O, Ouveysi I. A Lagrangean relaxation and decomposition algorithm for the video placement and routing problem. Eur J Oper Res 2007;182(1):455–465.
31. Andreev K, Maggs BM, Meyerson A, Saks J, and Sitaraman R. Algorithms for constructing overlay networks for live streaming, ArXiv CoRR abs/1109.4114; 2011.
32. Bektaş T, Oguz O, Ouveysi I. Designing cost-effective content distribution networks. Comput Oper Res 2007;34(8):2436–2449.
33. Deane JK, Rakes TR, Agarwal A. Designing content distribution networks for optimal cost and performance. Inform Technol Manage 2012;13:1–15.

34. Ho K-H, Georgoulas S, Amin M, Pavlou G. Managing traffic demand uncertainty in replica server placement with robust optimization. In: Fernando B, Thomas P, Burkhard S, Cedric W, and Edmundo M, editors. *Lecture Notes in Computer Science 3976*. Heidelberg: Springer; 2006. p 727–739.
35. Neves TA, Drummond LMA, Ochi LS, Albuquerque C, Uchuoa E. Solving replica placement and request distribution in content distribution networks. Electron Notes Discrete Math 2010;36:89–96.
36. Rayburn D. 2007. CDN pricing data: What the CDNs are actually charging for delivery. Available at http://blog.streamingmedia.com/the_business_of_online_vi/2007/08/cdn-pricing-dat.html. Accessed 10 June 2013.
37. Hosanagar K, Chuang J, Krishnan R, Smith MD. Service adoption and pricing of content delivery network (CDN) services. Manage Sci 2008;54(9):1579–1593.
38. Ercetin O, Tassiulas L. Pricing strategies for differentiated services content delivery networks. Comput Netw 2005;49(6):840–855.
39. Gavish B. Topological design of computer communication networks—the overall design problem. Eur J Oper Res 1992;58:149–172.
40. Vilas M, Pañeda XG, García R, Melendi D, García VG. User behavior analysis of a video-on-demand service with a wide variety of subjects and lengths. Proceedings of the 31st EUROMICRO Conference on Software Engineering and Advanced Applications; August 2005; Xixon, Spain; 2005. p 330–337.
41. Lobo A, García R, Pañeda XG, Melendi D, Cabrero S. Modeling video on demand services taking into account statistical dependences in user behavior. Simul Model Pract Theor 2013;31:96–115.
42. Bektaş T, Cordeau JF, Erkut E, Laporte G. Exact algorithms for the joint object placement and request routing problem in content distribution networks. Comput Oper Res 2008;35(12):3860–3884.

10

ANALYZING CONTENT DELIVERY NETWORKS

Benjamin Molina[1], Jaime Calvo[2], Carlos E. Palau[1], and Manuel Esteve[1]

[1] *Universitat Politecnica de Valencia, Valencia, Spain*
[2] *Universidad de Salamanca, Escuela Politecnica Superior de Zamora, Zamora, Spain*

10.1 INTRODUCTION

Internet continues growing in terms of users and services. Traditional pages with text and images provide little value to demanding users who increasingly require the integration of services such as multimedia, privacy, and personalization. Thus, it requires some intelligence in the systems or networks that offer such services, which is often implemented at the application layer because of its greater flexibility compared with network-based solutions. Content delivery networks (CDNs) are a scalable solution in providing all types of content to users. CDNs initially emerged as a method of reducing the latency of accessing Web objects, but their applicability has been extended to multimedia (both video and audio streaming) and mobile applications where the temporary requirements and relationships are more demanding and perceptible.

CDNs minimize delay reducing the communication path over the Internet; therefore, multiple servers are placed at the edges of the network and in the vicinity of the

Advanced Content Delivery, Streaming, and Cloud Services, First Edition.
Edited by Mukaddim Pathan, Ramesh K. Sitaraman, and Dom Robinson.

customers, following the concept of CEN (content edge networking). This simple idea represents a technological challenge in the deployment of the network infrastructure, and the necessary intelligence to offer content delivery services has to cope with two relevant aspects:

- *Redirection Model.* The client requesting content should be redirected to a nearby low loaded surrogate that provides such content.
- *Content Model.* The content offered by the CDN is usually not completely replicated in all surrogates for space and efficiency issues.

Deploying and testing CDNs effectively requires a large infrastructure that is difficult to afford and manage. Even though there are some distributed platforms for developing and testing distributed applications, such as PlanetLab [1], there is no full control of the available resources and therefore the performance results are partially limited. Therefore, some models and simulators are frequently used for evaluating CDN configurations.

Regarding simulators, there is hardly any global CDN model in simulation environments, as the diversity in the configurations of various CDNs can be huge. This normally leads researchers to build their own simulation model focusing on some performance parameters and even assuming or simplifying other parameters. The two most common CDN simulators are CDNsim [2], based on OMNeT++ [3], and CDN simulator [4], based on NS-2 [5].

Published CDN studies analyze performance on either real test beds or simulation environments, and there is no basic analytic model that allows analyzing basic and advanced properties of a CDN and how and why such systems reduce user-perceived latency. This chapter provides an advanced analytic model enhanced from a basic model already described by the authors in [6]. The enhanced model overcomes the limitations of the previous model and allows a more in-depth study as it introduces existing delivery and storage mechanisms, such as caching and replication strategies.

10.2 PREVIOUS WORK

CDNs are complex systems that employ different technologies at various levels to provide efficient delivery of content. However, it is rather difficult to model and evaluate such systems unless some simplifications are used or only some aspects of the global system are analyzed. In fact, there are various topics of deep research in CDNs. Caching mechanisms are described and evaluated in References 7–9. Another relevant aspect in large systems is traffic analysis [10–12] for handling properly user redirection [13–15] and efficiently deploying physical systems, which is often referred as replica placement [16–18]. The mechanism of user redirection is normally (tightly) coupled with the policy employed in selecting the optimal surrogate [19–21], which is often a trade-off between low response time and overall load balancing strategy [22–24]. Obviously, a distributed system such as a CDN with a large amount of servers must deal with consistency and content management [19–21]. Once the CDN is deployed, it is important to measure the performance [25–27] and provide latest services, such as streaming [28–30].

To the extent of our knowledge, there is no analytic model that describes a CDN at a basic level (without too much complexity) and in general terms (considering the whole dimension of a CDN). Existing theoretical models restrict to specific aspects, such as resource allocation and management [31], replica placement [32], content externalization [17], pricing models evaluation [33], and redirection mechanisms [31,34]. In some cases, the mathematical models are also used to study specific solutions to real problems appearing in existing CDNs (research and commercial). Thus, in Reference 35, a solution is proposed based on a mathematical model that evaluates the data-clustering effect for the benefit of a CDN provider. In other cases, the mathematical models are used as reference points for evaluating a series of heuristic methods [36]. However, all these previous proposals deal with individual models in a separate way, without considering the potential interaction among subsystems in the CDN. Although the research provided in such topics seems relevant, it is necessary to provide a generic model where such interactions may be studied and analyzed.

A general model is already provided by the authors in Reference 6 that will be introduced in the following section. This model will be enhanced in several ways that provides a more realistic view of a CDN.

10.3 BASIC CDN MODEL

In the previous model [6], the basic configuration assumed an origin server as the central point of the model, around which both customers and surrogates are placed, as depicted in Figure 10.1.

This previous model offers a starting point for basic simulations, but does not cover several issues or they are covered in a limited way, which are as follows:

- *Difficulty in Scalability*. It provides two possible access levels either through the origin server or through the surrogates. The new model provides a hierarchical configuration, which increases flexibility.

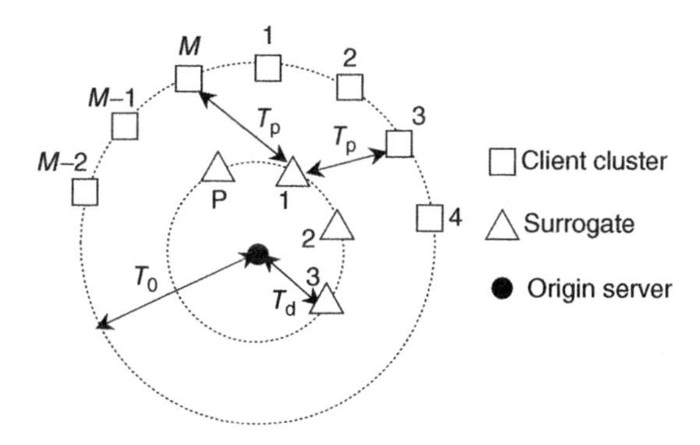

Figure 10.1 Basic CDN model presented in Reference 6.

- *Cluster Concept.* It grouped clients into clusters, but not surrogates, which is targeted in the presented model.
- *Caching.* It assumed content replication in all surrogates. Otherwise, a probability of success was introduced for modeling the hit rate. However, it was rather difficult to study the caching behavior separately on each surrogate. Basic caching parameters include the amount of stored objects, update cache policy, and intracluster and intercluster communication.
- *Redirection Model.* Most current CDNs base their redirection model in the domain name system (DNS). This communication mechanism has not been entirely addressed in the previous model that established a contact probability on the client for the surrogates and the origin server, depending on the hit rate: a certain amount of requests were redirected to some of the surrogates, whereas other requests were served by the origin server. Thus, communication (and delay) involved two different paths: client-surrogate or client-origin. In the hierarchical model, there are more alternatives to be considered, such as intercluster and intracluster communication. This chapter will not provide an analytical model for the DNS; instead, it will be assumed that clients are redirected to their nearest surrogate cluster. Obviously, this implies a certain processing time to be taken into account. In the case of Web objects, this processing time may be significant compared to response times once a server or surrogate is contacted. For streaming objects, we will assume that the initial delay due to the redirection mechanism is worthwhile for serving content from the nearest (optimal) surrogate, as the delay and jitter perceived by the user will be reduced.

10.4 ENHANCING THE MODEL

10.4.1 Representation and Notation

The new model is presented in Figure 10.2. It shows a hierarchical structure centered on the origin server. The origin server is in direct communication with a group of primary surrogate clusters (first level), who act as parents for a small group of secondary surrogate clusters (second level). The model could be easily extended to a larger number of levels, but the paper will only analyze this two-level scenario. Note that adding too many levels may complicate the general management, and the communication delay between various levels may be even greater than contacting directly the origin server.

Surrogates are grouped into distributed clusters, building a special mesh network among them. This is a mechanism to interconnect surrogates that are geographically close together. For small CDNs, the surrogate cluster may be even composed by a server farm building a single datacenter. For large CDNs, on the contrary, a single surrogate in Figure 10.2 may represent a whole datacenter. The model presented here is, therefore, an intermediate approach that may be easily adapted to various CDN configurations.

This new model requires expressing each of the parameters that make up the complete system univocally. The differences from the previous model make it difficult of

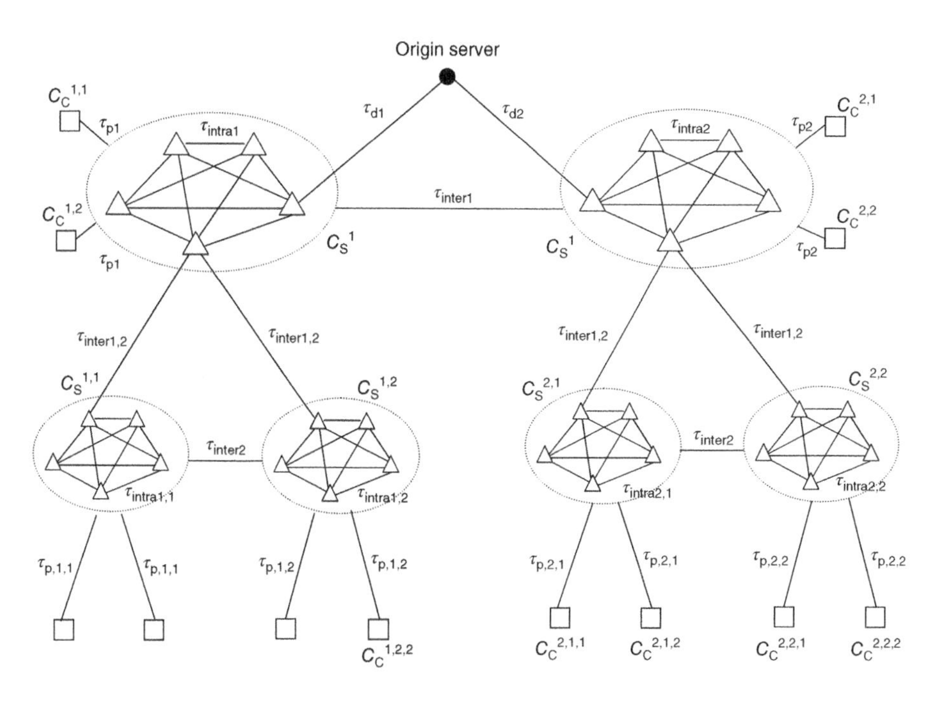

Figure 10.2 Hierarchical CDN model.

retaining the naming (notation) of objects that constitute the infrastructure of the CDN. Each of these parameters is defined as follows:

- $\tau_{\mathrm{d},i}$ represents the mean round trip time (RTT) between the origin server and the ith surrogate from the first level (connected directly to the origin server). It is assumed that all surrogates in the same cluster are approximately at the same distance from the origin server.
- C_{S}^{i} represents the ith surrogate cluster in the first level, whereas $C_{\mathrm{S}}^{i,j}$ denotes the jth surrogate cluster in the second level associated with the ith cluster in the first level.
- Each surrogate cluster, both at first level (C_{S}^{i}) and second level ($C_{\mathrm{S}}^{i,j}$), is formed by a set of surrogates, $N_{\mathrm{S}}^{C_{\mathrm{S}}^{i}}$ (first level) and $N_{\mathrm{S}}^{C_{\mathrm{S}}^{i,j}}$ (second level).
- Individually, each kth surrogate inside a cluster is designated by $S_k^{C_{\mathrm{S}}^{i}}$ y $S_k^{C_{\mathrm{S}}^{i,j}}$ for the first level and second level, respectively.
- Each surrogate forming the cluster is connected in a mesh topology and are so closely connected (topologically) that only one parameter may characterize the mesh; thus, $\tau_{\mathrm{intra},i}$ is defined as the mean RTT between any two surrogates inside the same ith cluster (first level). Analogous for the second level, $\tau_{\mathrm{intra},i,j}$ represents the mean RTT between two surrogates from cluster $C_{\mathrm{S}}^{i,j}$.

- The surrogate clusters have associated with them (due to an intelligent redirection system) a set of clients that are grouped in the form of clusters; $C_{\mathrm{C}}^{i,j}$ represents the jth client cluster from the jth surrogate cluster in the first level (C_{S}^{i}), whereas $C_{\mathrm{C}}^{i,j,k}$ represents the kth client cluster associated with the surrogate cluster in the second level ($C_{\mathrm{S}}^{i,j}$).
- Client clusters associated with the same surrogate cluster have in common the same mean access delay; thus $\tau_{p,i}$ and $\tau_{p,i,j}$ denote the mean RTT to cluster C_{S}^{i} and $C_{\mathrm{S}}^{i,j}$ from any client cluster ($C_{\mathrm{C}}^{i,j}$ and $C_{\mathrm{C}}^{i,j,k}$, respectively) associated with it.
- Surrogate clusters also interact between each; thus, the parameter $\tau_{\mathrm{inter},1}$ is defined as the mean RTT intercluster at level 1, whereas $\tau_{\mathrm{inter},2}$ represents the mean RTT intercluster at level 2; the interaction between surrogate clusters at different levels is defined as $\tau_{\mathrm{inter},1,2}$. These parameters have not been individually defined for each cluster in order to simplify the model and can be considered for further work.
- Related to the number of client and surrogate clusters, it is assumed a number of N_{P1} primary surrogate clusters, N_{P2} secondary surrogate clusters, and N_{C} client clusters.

10.4.1.1 Content Model The content served by the CDN can be seen as a set of O objects that may be requested by any of the clients. If we use vectorial notation:

$$\overrightarrow{o_O} = [o_1 \;\; o_2 \;\; o_3 \quad \cdots \quad o_O] \tag{10.1}$$

Initially, we assume that the access pattern to objects is the same for all client clusters, so an access probability can be associated following any particular distribution (Pareto, Zipf). The access pattern is the same, whereas the probability of accessing each object may vary between clusters. A matrix P_0 can be defined in order to describe access patterns to the client clusters ($C_{\mathrm{C}}^{i,j}$ or $C_{\mathrm{C}}^{i,j,k}$). Note that each row of matrix P_0 holds the properties of a probability density function in a discrete form:

$$\sum_{i=1}^{O} p_{i,C_j} = 1, \; \forall \; j \in [1 \; \ldots \; N_{\mathrm{C}}] \tag{10.2}$$

After defining all the objects and their access patterns, it is time to establish a content distribution policy within all surrogate clusters. We assume that it is impossible to duplicate all content in all surrogate clusters; the set of O objects can be distributed among all surrogates following a particular criterion. We assume that the surrogate clusters store initially the first (most demanded) Q objects:

$$\overrightarrow{o_{\mathrm{C},Q}} = [o_1^* \;\; o_2^* \;\; o_3^* \; \cdots \; o_Q^*], \; Q < O, \; C \in [C_{\mathrm{S}}^{i}, C_{\mathrm{S}}^{i,j}], \; i \in [1 \; \ldots \; N_{\mathrm{P1}}] \, j \in [1 \; \ldots \; N_{\mathrm{P2}}] \tag{10.3}$$

Note that objects in $\overrightarrow{o_{\mathrm{C},Q}}$ (equation (10.3)) have a different order as $\overrightarrow{o_O}$ (equation (10.1)), as the order in $\overrightarrow{o_{\mathrm{C},Q}}$ is determined by the most probability factors in matrix P_0. It is possible that a surrogate cluster has more than one associated client clusters. This leads to the fact that such surrogate cluster (C_S^i or $C_S^{i,j}$) has more than one associated row in matrix P_0. In this case, we take those factors with greater combined probabilities, independent of their associated rows. We assume that the number of clients in each cluster is similar; otherwise, it may affect the perceived requesting rate by the surrogate. With this behavior the most popular content perceived by each surrogate is stored, and the remaining content (less requested) will be stored either in other clusters or in the origin server. Each surrogate ($S_{k^{C_S^i}}$ o $S_{k^{C_S^{i,j}}}$) within the same cluster caches Q_1 objects ($Q_1 < Q$).

Object assignment (externalization) is performed uniformly among all surrogates (initially, all surrogates have the same probability to be contacted by a client of an associated cluster). Even if there is a certain overlap in the objects cached by the surrogates within a cluster, the union of all objects in the cluster results in the set of Q objects. Note that the optimal efficiency in terms of caching is reached when the object overlap within a cluster corresponds to the subset of most popular objects.

The previous structure described represents a static scheme from the point of view of a surrogate and its memory cache management. This may lead to a situation in which a frequently demanded object requested by a client to a given surrogate that does not have (store) it has to continuously fetch it from other surrogates, either in the same cluster or in an adjacent cluster. Note that probabilities are assigned per cluster, not per client, and a client–object traffic pattern may be significantly different from its correspondent client–cluster object traffic pattern (and even from the associated surrogate cluster–object traffic pattern). From another point of view, a completely adaptive scheme may lead to a highly changing scenario, so that after several requests, it may be practically impossible (or extremely difficult) to determine the stored content in each surrogate unless we introduce additional control traffic, which will be excessive for the available network capacity. The presented model introduces an intermediate trade-off approach; therefore, the memory cache comprises a fixed part and a variable part, as represented in Figure 10.3. The fixed section S_f represents the Q_1 objects that are initially cached by all surrogates at start-up. The variable section S_v stores the most recent objects unless they are not already stored in the fixed section S_f. In our model the measurement unit is an object, though in practice physical cache memories are measured in kilobytes. The previous assumption allows a simplified model and does not have a relevant impact unless a cache replacement strategy is used where the size is relevant. For simplicity, we use an LRU cache policy and assume that the cache memory can store and manage efficiently the cached objects.

In this scenario, the objects stored by a surrogate are represented by $\overrightarrow{o_{S,Q1+S_v}}$. The hierarchical level structure does not impact the content model, as client clusters are treated equally. The level separation will only have an impact on the communication delay.

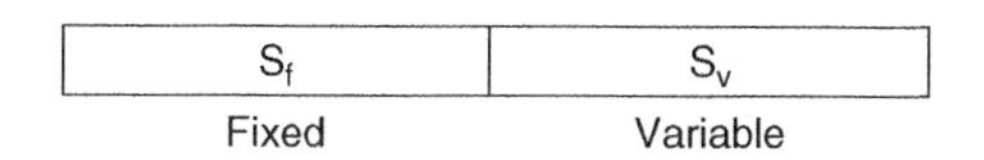

Figure 10.3 Memory cache decomposed of fixed and variable parts.

10.4.1.2 Communication Model The communication model describes the path traversed by the client to achieve the desired content, as well as required intermediate steps to provide such content, if necessary. When a client connects to the CDN, the first step consists in redirecting the client to a nearby surrogate cluster. For simplicity, we assume the existence of a redirection system capable of associating a client with the correspondent surrogate cluster. Following the model in Figure 10.2, a client in cluster $C_{\mathrm{C}}^{2,1,1}$ will be redirected to the surrogate cluster $C_{\mathrm{S}}^{2,1}$. Once this surrogate cluster is contacted, a surrogate is assigned following a round robin mechanism. This server will be denoted as candidate surrogate (S_{C}) for serving the requested content. This is also a simplification but allows obtaining close mathematical expressions, and does work properly in practice if the requests and the server capacity within a cluster are similar. If we denote the requested object as o_{r} ($o_{\mathrm{r}} \in \overrightarrow{o_O}$), we can distinguish various situations:

- If $o_{\mathrm{r}} \in \overrightarrow{o_{S_{\mathrm{c}},Q_1+S_{\mathrm{v}}}}$, then S_{C} serves the content directly to the client, there is a local hit and the hit rate will be denoted as H_{local}. Following the old model [6], the global delay (R_1) can be expressed as the addition of the required process time and transmission time:

$$R_1 = \frac{1}{\mu_{S_{\mathrm{c}}} - \lambda_{S_{\mathrm{c}}}} + N \cdot \tau_{\mathrm{p}}^{\mathrm{C}} \tag{10.4}$$

If we take a process time pattern for each surrogate (μ), we can introduce a factor k, $k \in [1 \ldots N_k]$, so that between the most and less powerful surrogate there is a difference of N_k expressed in units of μ:

$$\mu_{S_{\mathrm{c}}} = k(S_{\mathrm{c}}) \cdot \mu \tag{10.5}$$

As a round robin strategy has been assigned, the inbound flow consumed by the candidate surrogate will be

$$\lambda_{S_{\mathrm{c}}} = \frac{1}{N_{S_{\mathrm{c}}}} \cdot \lambda_{\mathrm{C}},\ C \in [C_{\mathrm{S}}^{i}, C_{\mathrm{S}}^{i,j}],\ i \in [1 \ldots N_{\mathrm{P1}}]\, j \in [1 \ldots N_{\mathrm{P2}}] \tag{10.6}$$

where λ_{C} is the request rate absorbed by the correspondent surrogate cluster. If there is more than one client cluster associated with the same surrogate cluster, the total request rate will be the addition of both. Also note that a surrogate does process not only client requests but also collaborative caching requests, which increases the overall load, as there are more requests to attend. We can reflect such situation with a factor α_1 so that

$$\lambda_{S_{\mathrm{c}}} = \frac{1}{N_{S_{\mathrm{c}}}} \cdot \lambda_{\mathrm{C}} \cdot [1 + \alpha_1(S_{\mathrm{C}})], C \in [C_{\mathrm{S}}^{i}, C_{\mathrm{S}}{}^{i,j}],$$

$$i \in [1 \ldots N_{\mathrm{P1}}],\ j \in [1 \ldots N_{\mathrm{P2}}], \alpha_1 \in [0 \ldots 1] \tag{10.7}$$

Note that α_1 will be typically 0 if there is no caching request directed to such surrogate. Obviously, this is not true in practice, and a feasible value should not exceed 0.25, which represents a quarter of the traffic associated with requests. If $\alpha_1 = 1$, then the correspondent surrogate is absorbing as many client requests as collaborative requests, which represents a bad performance, and a clear hint that either more space should be allocated for storage (caching) or the popular content for this surrogate is not being properly determined. Summarizing previous equations, the final expression for the mean response time in this first situation can be denoted by

$$R_1 = \frac{1}{k(S_c) \cdot \mu - \frac{1}{N_{S_c}} \cdot \lambda_C \cdot [1 + \alpha_1(S_C)]} + N \cdot \left[\frac{1}{N_S^C} \cdot \tau_p^C \right]$$

$$C \in [C_S{}^i, C_S{}^{i,j}],\ i \in [1 \ldots N_{P1}]\ j \in [1 \ldots N_{P2}], \alpha_1 \in [0 \ldots 1] \quad (10.8)$$

- If $o_r \notin \overrightarrow{o_{S_c,Q1+S_v}}$ but $o_r \in \overrightarrow{o_Q}$, then S_C obtains first the content within its own surrogate cluster, and the hit rate will be denoted as $H_{cluster}$. In this situation, the variable parameter is the mean RTT; the surrogate S_C must contact another surrogate in the same cluster (with a time distance of τ_{intra}^C) that has the requested content:

$$\tau_2 = \tau_p^C + \tau_{intra}^C,\ C \in [C_S^i, C_S^{i,j}],\ i \in [1 \ldots N_{P1}]\ j \in [1 \ldots N_{P2}] \quad (10.9)$$

Regarding the process time we may count it twice: once for processing the client request and another for retrieving the cached object; we are assuming here the same process time for serving a request to the client than serving a request to a collaborative surrogate. We cannot group the process and response time for different surrogates, and S_{CO} refers to the contacted surrogate by S_C. Here, note that the perceived arrival rate from the point of view of S_{CO} is different than the one perceived by S_C. The surrogate S_{CO} will be contacted only on cache misses, and therefore we have to consider $H_{cluster}$ as variable.

- If $o_r \notin \overrightarrow{o_{S_c,Q1+S_v}}$ and $o_r \notin \overrightarrow{o_Q}$, then S_C uses an intercluster mechanism to obtain the request content before serving it to the client, and the hit rate will be denoted as H_{global}; the surrogate S_C must contact another surrogate from a different cluster (with a time distance of τ_{inter}^C) or even from the origin server. Two basic possibilities are here envisioned:
- *Case A*. The surrogate S_C contacts only a nearby sibling cluster. On a cache miss, the surrogate S_C contacts the origin server. If we denote $\overrightarrow{o_{Sib}}$ as the set of objects available in the sibling cluster, we have two situations:
- *Case S (Sibling)*. If $o_r \notin \overrightarrow{o_Q}$, but $o_r \in \overrightarrow{o_{Sib}}$, then S_C gets the requested object from a surrogate located in a sibling cluster. In this case, the mean delay is the sum of detecting the intracluster cache miss and the delay from retrieving the object from the sibling cluster. The detection of a cache miss is caused by the remaining surrogates within the cluster not sending the requested content (as they do not have it). This implies the expiration of a timeout in S_C informing that the

content should be fetched from a nearby cluster. The delay for this detection can be expressed similarly as R_2, with two considerations. First, the timeout should be slightly higher as $\tau^{\mathrm{C}}_{\mathrm{intra}}$, but we may fix it with this value in order to have a minimum (best case) value. Second, the value of N should be set to 1, as no retransmissions will be required; the fact of not responding to S_{C} can be virtually modeled as sending one single small message (no segmentation) telling S_{C} that the requested object is not available. Obviously, this will never be implemented for efficiency in real networks in order to avoid unnecessary traffic. Considering both premises, we may express this response time as various addends, considering the time for processing the request, realizing that there is no available object in the cluster and getting it from a nearby cluster. In addition, we may sum the transmission delay.

- *Case O (Origin)*. If $o_{\mathrm{r}} \notin \overrightarrow{o_Q}$ and $o_{\mathrm{r}} \notin \overrightarrow{o_{\mathrm{Sib}}}$, then S_{C} needs to contact directly the origin server to obtain the requested object. The response time is similar to that of the previous expression, but now we have to consider and include the delay between S_{C} and the origin server. We must also note that the origin server is different from that of the surrogates, and therefore it has its own process rate and it obtains all requests from all clusters (on cache misses).
- *Case B*. The surrogate S_{C} contacts a nearby parent cluster. The process is similar as the previous process and will not be explicitly analyzed, as the only relevant aspect may be comparing the effectiveness of caching between siblings and parents. It may be also interesting to investigate the benefits of issuing concurrent requests from the candidate surrogate S_{C} to both sibling and parent cluster, as some implemented proxy cache protocols allow it. However, this will not be studied in this chapter and can be considered as further work.

10.5 PERFORMANCE EVALUATION

After describing the CDN model and the different scenarios involving communication among clients, surrogates, clusters, and origin server, this section shows some results to evaluate its behavior. As can be appreciated, the expression for the main response time corresponds to an n-dimensional function with various parameters, and some (or many) of them must be fixed in order to analyze one (or many) variables at a time; this allows an easier study of its impact on the CDN performance. Taking the model in Figure 10.2, we assume the following basic scenario in order to show some initial results:

- The surrogate distribution hierarchy is two level, with various first-level surrogate clusters, each of them having two second-level surrogate clusters: $N_{C^{i,j}_{\mathrm{S}}} = 2 \cdot N_{C^{i}_{\mathrm{S}}}$. If we denote as $N_{C_{\mathrm{S}}}$ the number of surrogate clusters, it follows that $N_{C^{i}_{\mathrm{S}}} = \frac{1}{3} N_{C_{\mathrm{S}}}$ and $N_{C^{i,j}_{\mathrm{S}}} = 2/3(N_{C_{\mathrm{S}}})$. This assumption allows us to easily model clusters with only one parameter (N_{C}), just for initial results.
- Each cluster (first level and second level) is composed of a set of N_{S} surrogates: ${N_{\mathrm{S}}}^{C^{i}_{\mathrm{S}}} = N_{\mathrm{S}}^{C^{i,j}_{\mathrm{S}}} = N^{\mathrm{C}}_{\mathrm{S}}$.

- As the previous assumptions may result in too much homogeneous scenario, even for initial results, we need to introduce differences in the elements. Thus, each surrogate cluster (first level and second level) will have two associated client clusters ($N_{C_C^{i,j}} = 2N_{C_S^{i,j}}$ $N_{C_C^i} = 2N_{C_S^i}$), but each of them with different traffic patterns. If we denote N_{C_C} as the number of client clusters, it follows that $N_{C_C} = 2N_{C_S}$ and $N_{C_C^{i,j}} = 2N_{C_S^i}$.
- Latencies in this basic scenario can be ordered following the tree hierarchy of the model. We assume therefore that the tree structure guides to a denser structure in the leaf entities (lowest levels). We will take the internal delay as the smallest value in each surrogate cluster because they are supposed to be topologically close together and will only characterize such delay as $\tau_{\text{intra1}} = \tau_{\text{intra2}} = \tau_{\text{intra}}$. The delay between client clusters and the associated attached surrogate clusters can be considered as the second smallest delay. In fact, we can initially assume the same mean delay for all client clusters (first level and second level): $\tau_{\text{p}}^{i,j} = \tau_{\text{p}}^{i} = \tau_{\text{p}}$. The third smallest delay can be considered the one between second-level surrogate siblings $\tau_{\text{inter2,Sib}}$, which we will assume the same value for all second-level surrogate clusters. Then follows the delay between second-level surrogate clusters and first-level surrogate clusters (parents), which we will characterize with only one value denoted by $\tau_{\text{inter,Par}}$. Afterwards, we may find in the hierarchy the delay between first-level surrogate clusters denoted by $\tau_{\text{inter1,Sib}}$. The biggest delay is considered to be the one regarding the origin server: τ_{d}^{1} for the first level and τ_{d}^{2} for the second level. If the origin is far away compared to other entities (clusters), then we may assume the same value for τ_{d}^{1} and τ_{d}^{2}. However, as we are using a tree structure, we will take the following relationship $\tau_{\text{d}}^{2} = 2 \cdot \tau_{\text{d}}^{1}$. In this situation, it makes sense for a surrogate in a second level to retrieve an object from a parent surrogate rather than directly contacting the origin server.
- Another aspect to consider is the object externalization policy, which represents how the overall content composed of O objects is initially distributed among all surrogate clusters. This will be characterized by the following parameters:
- The overall capacity of each surrogate cluster, denoted by $C_{\text{S,cap}}$, indicating how many objects they can store (cache) internally. If $C_{\text{S,cap}} = 0.25$ initially, this means that 25 of the O objects available in the origin server have been already externalized to this particular surrogate cluster. We suppose here that not all O objects can be offloaded to each surrogate cluster, otherwise we may have a mirror. The object's cache in each surrogate cluster depends on the associated client cluster's traffic patterns. This means that, initially, objects from the origin server will be stored in each surrogate starting for those that are more frequently demanded by the associated client clusters. In our model, we have only two client clusters associated with a surrogate cluster. As they have different traffic patterns (otherwise, they may be considered the same client cluster), the set of selected objects to be cached will be taken from the P_0 matrix, described in (equation 10.2).

- The traffic access pattern for each client cluster, determined by the P_0 matrix. Though we may use here any distribution function, we will use a Zipf distribution (the discrete form of a Pareto distribution as the number of web objects is finite). In this scenario, the Zipf's law is expressed as $f(r,e,O) = \frac{1}{r^e} / \sum_{n=1}^{O} \frac{1}{n^e}$ where O is the number of objects, r is the rank ($r \in \{1 \ldots O\}$), and e is the value of the exponent characterizing the distribution. For simplicity, we may take the same value of e for all the client clusters, however, the order of the set of O objects will be different for the client clusters, otherwise it will be the same for all of them. We will just take an index in the vector of O objects indicating the start of the set. This will be based on the number of client clusters. For a set of two clusters, the first set of objects will be taken as the initial set of O objects. The second set (associated with the second client cluster) will start in the middle element of the initial set of O objects, and will have the first half of the initial set of O objects as tail.

Figure 10.4 shows the mean response time as a function of the number of surrogate clusters. As can be observed, the response time diminishes as the number of surrogate clusters increases, and its impact is more significant when the number of surrogates is small. This makes sense in our model as delays have lower values in the lower levels of the communication tree, as commented previously.

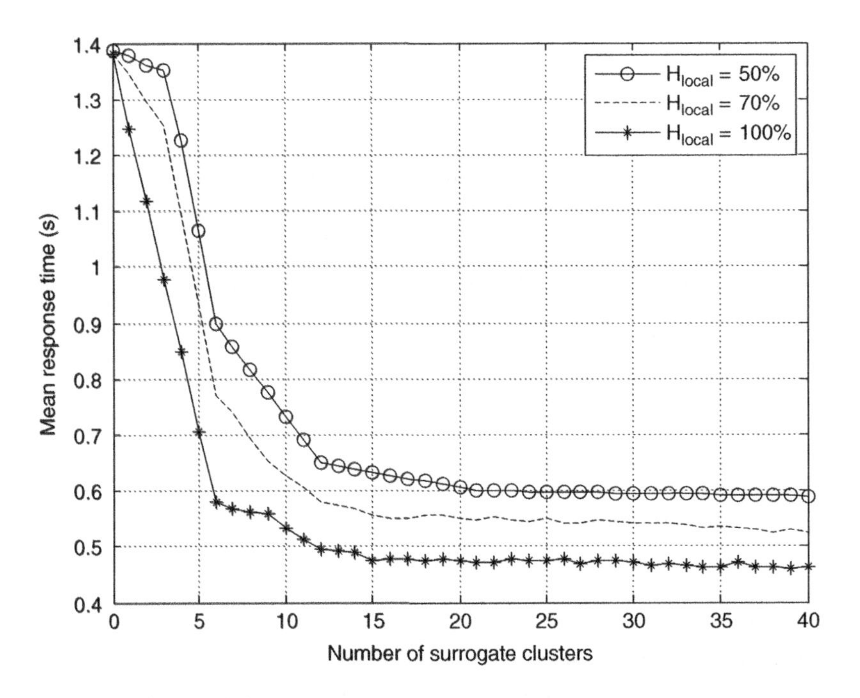

Figure 10.4 Mean response time depending on surrogates.

The effective replication on each surrogate is also shown in Figure 10.4 through the local hit rate. If the local hit rate diminishes the mean response time increases, as more communication messages should be exchanged either for collaborative proxy caching or contacting directly the origin server. A complete replication ($H_{local} = 100\%$) provides the best case scenario, as all client requests are treated within the same local surrogate cluster. There may be some (small) variability here if the contacted surrogate does not have the requested object and has to obtain it internally. A good redirection mechanism would minimize this effect, but it is out of the scope of this model for the moment.

Figure 10.5 shows the mean response time as a function of the local hit rate for various values of the Zipf function. As can be observed, the response time is reduced when objects are effectively cached and there is no need to contact other surrogates. The value of the e parameter in the Zipf also has an impact on the response time; lower values favor a better (more effective) caching mechanism as there is a higher overlap in the common objects between client clusters.

As can be observed, there are several and different probes that can be tested with this enhanced model in order to check whether a particular configuration is useful in terms of latency, caching hit rate, surrogate clustering, client clustering, object externalization policy, and so on. In general terms, one may build their own use cases in order to test a particular network deployment or caching strategy. In the first case (network deployment), the model allows to easily group surrogate and clients into clusters; the

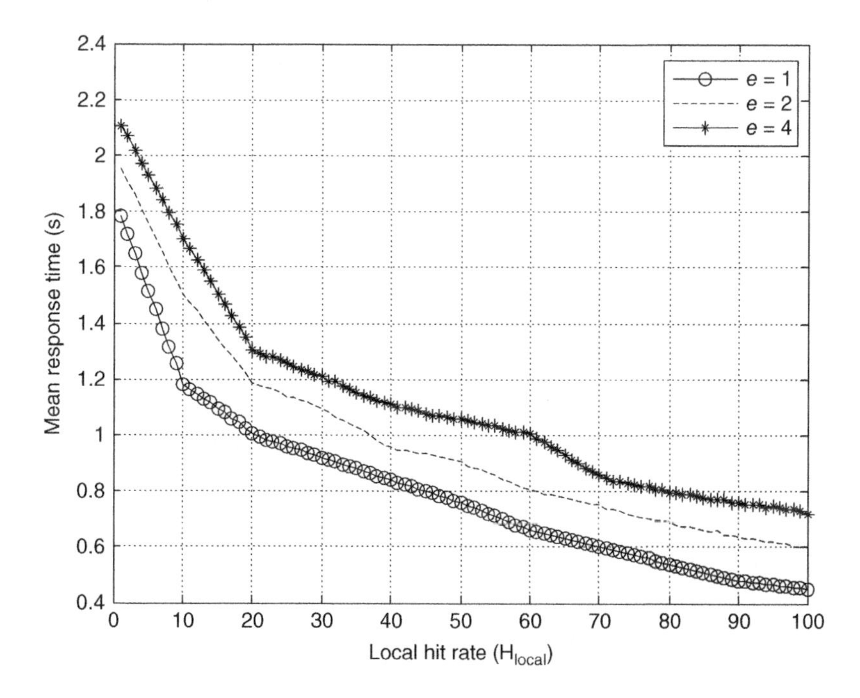

Figure 10.5 Mean response time depending on local hit rate.

clustering policy is a possible research line: how to group clients and surrogates for a given topology in order to minimize the overall client response time. Mapping the given topology into the model may not be trivial, but definitely not difficult. Here, the researcher may start focusing on either the client topology or the surrogate topology and obtain the most appropriate approach, or even simulate both scenarios and compare the obtained results. In the second case (caching strategy), one may vary caching policies at surrogate and cluster levels. For the surrogate caching policy, a researcher may also study the user redirection strategy within the cluster, so that a client is redirected to a surrogate with the desired content available and thus there is no need to obtain it from the cluster.

Other scenarios are also interesting to investigate related to the applications that will be exploited in the CDN. Web and streaming applications could be analyzed separately, but with a similar object model for both. In this case, the streaming service would be simplified by means of large objects whose size is significantly higher than the Web objects. For streaming modeling, the jitter would be also a parameter to study in addition to overall response time.

10.6 CONCLUSIONS

This chapter has presented a new and enhanced model for analyzing content delivery networks in a more realistic way based on a previous analytic model. CDNs typically consist of large networks and many servers; thus, the number of network parameters to model is significant. It is also important how to group network entities into network clusters. Surrogate clusters provide local scalability simulating a local datacenter. Client clusters provide a means for grouping individual clusters and characterizing traffic patterns, which may influence the caching activity in the contacted surrogate clusters. Moreover, a contacted surrogate may cooperate using some proxy collaboration mechanism in order to retrieve a requested object from another surrogate located in either its internal cluster, a sibling cluster, or a parent cluster.

The chapter has described a basic scenario in order to test and evaluate the performance of a CDN. As can be observed according to the results, CDNs allow reducing mean response time as the number of surrogate (surrogate clusters) increases. Moreover, the usage of a collaborative proxy mechanism between siblings and parents allows further reducing the response time compared to contacting the origin server on a cache miss. Client traffic pattern characterization is also a dynamic process that allows an efficient management of the cached objects within a cluster; moreover, using a partial static and dynamic cache within each surrogate also enhances the (local) hit rate.

REFERENCES

1. Ripeanu M, Bowman M, Chase JS, Foster I, Milenkovic M. Globus and PlanetLab resource management solutions compared. High Performance Distributed Computing, 2004. Proceedings. 13th IEEE International Symposium on; 2004 Jun 4–6, 10.1109/HPDC.2004.1323548; 2004. p 246,255.

2. Konstantinos S, George P, Athena V, Dimitrios K, Antonis S and Yannis M. CDNsim: A simulation tool for content distribution networks. ACM Trans Model Comput Simul 2010;20(2)Article 10.
3. Varga A. Using the OMNeT++ discrete event simulation system in education. IEEE Trans Educ 1999;42(4):11. DOI: 10.1109/13.804564.
4. CDNSimulator. 2013. Available at http://cdnsimulator.sourceforge.net/. Accessed 15 June 2013.
5. The Network Simulator NS-2. 2013. Available at http://www.isi.edu/nsnam/ns/. Accessed 15 June 2013.
6. Molina B, Palau CE, Esteve M. *Next Generation Content Delivery Infrastructures: Emerging Paradigms and Technologies*. Chapter 1. IGI Global; 2012. ISBN: 9781466617964.
7. Ronald D, Jeffrey C, Syam G and Amin V. The trickle-down effect: Web caching and server request distribution. Comput Commun 2002;25:345–356.
8. Craig J. Caching hierarchies: Understanding content distribution/delivery networks. Dell Power Solutions; 2001. Issue 1.
9. Syam G, Jeffery C and Michael R. Web caching and content distribution: A view from the interior. Proceedings of the Fifth International Web Caching and Content Delivery Workshop; 2000.
10. Fujiwara K, Sato A and Yoshida K. DNS traffic analysis: Issues of IPv6 and CDN. 2012 IEEE/IPSJ 12th International Symposium on Applications and the Internet (SAINT); 2012. p 129–137, 10.1109/SAINT.2012.26.
11. Coppens J, Wauters T, De Turck F, Dhoedt B and Demeester P. Evaluation of a monitoring-based architecture for delivery of high quality multimedia content. 10th IEEE Symposium on Computers and Communications, ISCC 2005, 10.1109/ISCC.2005.68; 2005. p 611–616.
12. Asano T, Tsuno A and Nishizono T. Quality measuring system for network provisioning and customer service in content delivery. IEEE/IFIP Network Operations and Management Symposium, 2004, Vol. 1, 10.1109/NOMS.2004.1317776; 2004. p 875–876.
13. Turini E. An architecture for content distribution internetworking. Technical Report UBLCS-2004-2. Department of Computer Science, University of Bologna; 2004.
14. Sivasubramanian S, Szymaniak M, Pierre G and van Steen M. Replication of Web hosting systems. ACM Comput Surveys Syst 2004;36(3):291–334NY, USA: ACM Press.
15. Sampath R, Sarit M and Pablo R. A technique for user specific request redirection in a content delivery network. Proceedings of the 8th Workshop on Web Content Caching and Distribution; New York; 2003.
16. Sivasubramanian S, Pierre G, van Steen M and Alonso G. Analysis of caching and replication strategies for Web applications. IEEE Internet Comput 2007;11(1):60–66.
17. Kangasharju J, Roberts J and Ross KW. Object replication strategies in content distribution networks. Comput Commun 2002;25(4):367–383.
18. Cameron C, Low SH and Wei D. High-density model of content distribution network. Proceedings of Information, Decision and Control; 2002.
19. Wu B, Kshemkalyani AD. Objective-optimal algorithms for long-term Web prefetching. IEEE Trans Comput 2006;55(1):2–17.
20. Chen Y, Qiu L, Weiyu C, Nguyen L and Katz RH. Efficient and adaptive Web replication using content clustering. IEEE J Select Areas Commun 2003;21(6):979–994. DOI: 10.1109/JSAC.2003.814608.

21. Fujita N, Ishikawa Y, Iwata A and Izmailov R. Coarse-grain replica management strategies for dynamic replication of Web contents. Comput Netw Int J Comput Telecom Netw 2004;45(1):19–34.
22. Hofmann M, Beaumont LR. *Content Networking: Architecture, Protocols, and Practice*. San Francisco, CA, USA: Morgan Kaufmann Publishers; 2005. p 129–134.
23. Gayek P, Nesbitt R, Pearthree H, Shaikh A and Snitzer B. A Web content serving utility. IBM Syst J 2004;43(1):43–63.
24. Barbir A, Cain B, Nair R and Spatscheck O. Known content network request-routing mechanisms. Internet Engineering Task Force RFC 3568; 2003.
25. Vakali A, Pallis G. Content delivery networks: Status and trends. IEEE Internet Comput 2003;7(6):68–74 IEEE Computer Society.
26. Mirco M, Parravicini E. Impact of request routing algorithms on the delivery performance of content delivery networks. 22nd IEEE International Performance Computing and Communications Conference; Phoenix, USA; 2003.
27. Calo SB, Verma DC, Giles J and Agrawal D. On the effectiveness of content distribution networks. International Symposium on Performance Evaluation of Computer and Telecommunication Systems; San Diego (USA); 2002.
28. Zhuang Z, Guo C. Optimizing CDN infrastructure for live streaming with constrained server chaining. 2011 IEEE 9th International Symposium on Parallel and Distributed Processing with Applications (ISPA), 10.1109/ISPA.2011.44; 2011. p 183–188.
29. ZhiHui L, XiaoHong G, SiJia H and Yi H. Scalable and reliable live streaming service through coordinating CDN and P2P. 2011 IEEE 17th International Conference on Parallel and Distributed Systems (ICPADS), 10.1109/ICPADS.2011.113; 2011. p 581–588.
30. Sheu S-T, Huang C-H. Mixed P2P-CDN system for media streaming in mobile environment. 2011 7th International Wireless Communications and Mobile Computing Conference (IWCMC), 10.1109/IWCMC.2011.5982624; 2011. p 657–660.
31. Bektas T, Ouveysi I. *Mathematical Models for Resource Management and Allocation in CDNs*. Lecture Notes in Electrical Engineering Vol. 9. Berlin: Springer; 2008. p 225–250.
32. Qiu L, Padmanabhan VN and Voelker GM. On the placement of Web server replicas. Proceedings of IEEE INFOCOM; Anchorage, Alaska, USA; 2001. p 1587–1596.
33. Hosangar K, Chuang J, Krishnan R and Smith MD. Service adoption and pricing of content delivery network (CDN) services. Technical Report, Social Science Research Network; 2006.
34. Oliveira CAS, Pardalos PM. A survey of combinatorial optimization problems in multicast routing. Comput Operat Res 2005;32(8):1953–1981 ISSN 0305-0548. DOI: 10.1016/j.cor.2003.12.007Elsevier Science Ltd. Oxford, UK.
35. Thanh VN, Farzad S, Paul B and Chun TC. Provisioning overlay distribution networks. Int J Comput Telecommun Netw Arch 2005;49(1):103–118. DOI: 10.1016/j.comnet.2005.04.001 Elsevier.
36. Nikolaos L, Vassilios Z and Ioannis S. On the optimization of storage capacity allocation for content distribution. Int J Comput Telecommun Netw Arch 2005;47(3):409–428. DOI: 10.1016/j.comnet.2004.07.020 Elsevier.

11

MULTISOURCE STREAM AGGREGATION IN THE CLOUD

Marat Zhanikeev

Kyushu Institute of Technology, Iizuka, Japan

11.1 INTRODUCTION

Content delivery networks (CDNs) have advanced much in recent years [1]. Technology evolved from the traditional one server–one client models [2] to several distinct peer-to-peer (P2P) models. Although P2P models are relatively recent, they already power several real large-scale streaming services such as Uusee [3], Afreeca [4], and NewCoolStreaming [5]. According to several chapters in this book, this list has grown substantially in recent years.

There are two fundamental approaches in streaming—single versus multistream. Traditional CDN is the case of a single-server streaming while P2P streaming methods are multistream [5]. This chapter presents yet another fundamental streaming method—*multisource streaming in the cloud.*

Let us discuss the main distinctions across all the methods discussed in this chapter. Figure 11.1 displays all the methods in a *problem–solution* diagram. The figure shows very primitive models of methods given that the objective is to show only the main features of each method. For more accurate and complete models, refer to Chapters 9 and 10 of this book, which mainly focus on the modeling issue.

Advanced Content Delivery, Streaming, and Cloud Services, First Edition.
Edited by Mukaddim Pathan, Ramesh K. Sitaraman, and Dom Robinson.

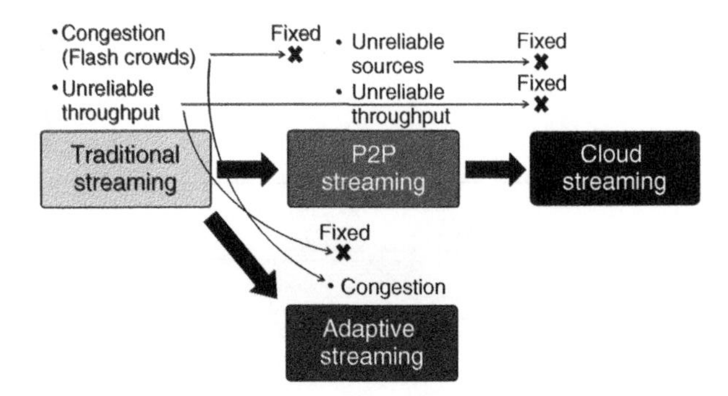

Figure 11.1 Evolution of video streaming technology.

There are two main evolutionary forks where the method discussed in this chapter is the extension of the P2P method. The two main problems are *congestion* and *throughput reliability*. Specifically, congestion of servers or networks along end-to-end paths directly affects streaming quality.

The part about throughput reliability is slightly more tricky. Although *end-to-end delay* and *throughput* are independent (between each other) metrics, longer end-to-end paths may indirectly cause unreliable throughput. Although unreliable throughput cannot be blamed solely on content providers, it can cause degradation of streaming quality.

Returning to Figure 11.1, evolution of streaming technology can be described in the following stages.

Traditional streaming suffers from the two above problems—congestion and unreliable throughput. P2P streaming solves the problem of congestion by moving to a distributed mode of delivery. However, P2P does not particularly solve the problem of unreliable throughput. Research in Reference 4 finds that it may actually make things worse. Moreover, P2P streaming creates a new problem of *unreliable sources* that is closely related to the problem of unreliable throughput.

Adaptive streaming [37] solves the problem of unreliable throughput by adapting to changes in throughput. In fact, adaptive streaming can partially solve the congestion problem as well. Chapter 13 of this book is entirely dedicated to adaptive streaming and should provide better detail.

The cloud streaming is the extension of the P2P streaming model. It solves the problem of unreliable sources by providing reliable sources backed by the cloud. The same solution also covers the problem of unreliable throughput because, as will be shown later, management of sources in the cloud is much more reliable than that in a fully distributed P2P system.

This chapter covers the following topics related to cloud-based multisource streaming:

- a method for cloud-based substream multisource aggregation and the related optimization problem;

- models and simulation analysis of the multisource method versus traditional methods;
- a realistic technology that can make multisource streaming work in cloud environments, specifically the new role of a service provider that can make it work in practice.

The structure of this chapter is as follows. Section 11.2 establishes the necessary terminology. Section 11.3 discusses the background for the topic of multisource streaming and the related research. Section 11.4 introduces the topic of multisource streaming and specifically the cloud-based substream method. Section 11.5 focuses on the unique features of the cloud-based stream aggregation method and compares the feasibility of its implementation by comparing it to the existing P2P streaming models. Section 11.6 presents the results of simulation models, which are discussed in Section 11.7. Section 11.8 contains visionary thoughts for practitioners. Finally, having discussed future research directions in Section 11.9, the chapter is concluded in Section 11.10.

The methods and algorithms described in this chapter can be implemented at today's level of technology. In fact, a prototype of the client side of a multisource stream aggregator has been developed by this author and made publically available at the link present in Reference 6. The prototype is based on the WebSockets [7] functionality implemented in many browsers today. Note that the multistream aggregation methods described in this chapter do not have to be based on P2P networks, and, therefore, are open to alternative platforms such as browsers.

11.2 TERMINOLOGIES

The three fundamental models considered in this chapter are *Traditional*, *P2P*, and *cloud*. They are referred to as methods or models, depending on the context.

The terms *server* and *source* are used interchangeably. The terms *client* and *user* are also interchangeable. The discussed methods can be considered from the viewpoint of the client-and-server model or from the viewpoint of users of a video streaming service.

The term *stream* is only loosely related to the video bitstream. For the method presented in this chapter, the contents, design, and technology of the video stream are not important. Therefore, the *stream* is used as the abstraction for any content streamed over the network. As is common among video streaming methods today, a fixed-rate stream is considered, which would roughly refer to constant bitrate (CBR) video.

Streams in this chapter can be split into multiple *substreams*. Similarly, multiple substreams can be merged to recreate the full original stream. When using substreams, it is important to isolate sources from one another. Each source can form and send its substream over the network when it has its block size (*blocksize*), starting position (*startpos*), and step (*step*) configuration. For example, *blocksize=10kb, startpos=1, and step=3* means that the source should send its substream in the form of 10 kb blocks, starting at the first block and skipping two blocks per each sent block. It is unnecessary to spell out that substreams overlap. More details are provided on the substream design later in this chapter.

The *push* versus *pull* is the old argument in video streaming [8]. The terms denote the two fundamental methods for two-way communication between client and server. The best way to describe these technologies is to replace *pull* with HTTP GET requests and *push* with WebSockets [7]. More details on this argument with some basic math are provided later in this chapter.

It is not enough to aggregate the stream. One also has to play it. *Playback* is traditionally *buffered*. More details on buffer management during playback are provided further in this chapter. Note that *discrepancy* in throughput on substreams can cause discrepancy in per-substream "download" positions, which in turn can cause playback to freeze waiting for a substream to catch up.

The quality of streaming (not the content itself) can be evaluated using several metrics. *Lag* is defined as the difference between the timestamp of receiving a block and its scheduled playback time, where positive *scheduled–received* is good performance. The term *playback continuity* is somewhat related to lag and is defined loosely as the time spans during which playback was not interrupted. The *freeze time* metric is the opposite of continuity because it captures the time during which playback was rendered impossible because of missing (delayed) blocks. All the three metrics are related to each other. This chapter will only use the *freeze time*.

Streaming quality can be viewed in a broader framework of quality of service (QoS) and quality of experience (QoE) where QoS is measured based on measurable performance metrics of the video itself and QoE can be calculated from the QoS measurements [9]. This chapter briefly talks about these technologies but places discussion of these topics out of scope. The justifiable assumption is that QoE is good when QoS is good and QoS is good when streaming has zero *freeze time*.

With respect to the cloud, this chapter deals with two roles—the currently nonexistent role of *service provider* (SP) and the currently available role of *cloud provider*. Today, users are in direct relationship (via SLA, etc.) with CPs. The technology discussed in this chapter calls for the emergence of a new role—that of the SP, whose role is to mediate between users and CPs. In traditional terms, SPs can be referred to as resource brokers.

The specific resource that SPs need to broker is VM *populations* supporting the streaming service. Again, the term itself does not exist in current literature but it is easy to understand. SPs create and manage their VM populations depending on the current size of user base. Such a system would require SPs to have the capability to get more VMs from CPs or return unnecessary VMs back to CPs for recycling. This situation can be academically viewed as an economy where relationships between SPs and CPs as well as those between users and SPs are governed by respective service-level agreements (SLAs) or clearly spelled out policies.

11.3 BACKGROUND AND RELATED WORK

This section discusses the basics of existing streaming methods and the related literature. The cloud model presented further on can then refer and compare to these features. For further details on the existing streaming methods, see other chapters in this book.

Chapter 12 of this book discusses adaptive streaming. Chapter 19 discusses cache strategies in traditional single-source multiple-destination streaming.

11.3.1 Traditional Streaming

Traditional streaming means there is one server and one client, so there is only one substream—the stream itself. The biggest problems of this method are *congestion* and *throughput reliability*, as was already mentioned earlier. Throughput reliability is mostly out of hands of the traditional method, but can be somewhat alleviated by one of the methods mentioned in this section.

Several methods in the literature deal with the congestion problem. The most common method is caching and replication [2]. In this method, several locations for the same content are spread across the network. Once replicas are in place, users have to be redirected to the least congested replica at the time [10]. Today, implementation of request redirection allows for seamless connection where users do not even notice (or know) which particular server they are in contact with. Domain name system (DNS) tweaking is one of practical methods which make it possible.

Several methods similar to that in Reference 11 have attempted to involve the network into the optimization process. Specifically, Reference 11 not only proposes to redirect users to the least busy servers but also optimizes end-to-end paths of the delivery using the traditional Open Shortest Path First (OSPF) problem widely used in traffic engineering.

Although proposal in Reference 11 can partially alleviate the *throughput reliability* problem, it generally remains unsolved by the traditional methods. When redirecting requests, it is common to take into consideration either end-to-end delay or congestion conditions at the destination server, or both. At that time, end-to-end throughput is unknown because delivery itself is yet to start. Once the throughput becomes known, however, it is difficult to rewire the connection in the traditional method.

11.3.2 Adaptive Streaming

Adaptive methods solve some of the problems with traditional delivery. Although the idea itself is relatively old, it has gained new momentum recently as part of the HTML5 standards [12]. In fact, adaptive streaming is a separate standard within HTML5 [13]. The term *adaptive* can mean switching either between different streaming sources or between bitrates, where Reference 13 normally addresses the former.

When variable bitrate is used, it can potentially solve both the congestion problem and the problem of low throughput reliability. In this aspect, the technology has great potential. However, *variable bitrate* is not a common topic in both standards and research today. The default problem in adaptive streaming today seems to be the switching between sources and bitrate. One should not confuse the term variable bitrate here, with the same term used in video coding. In adaptive streaming, *variable bitrate* simply means that the stream can be switched to a lower *CBR* to cope with changes in throughput.

Adaptiveness requires a two-way communication. This is made possible in browsers today through the WebSockets [7] part of the HTML5 standard. This author has tinkered with WebSockets in the past by creating software for benchmarking WebSockets in the Chrome browser. The software is publically available at the link present in Reference 14.

Adaptive streaming has great potential. For example, study in Reference 15 considers using it for TV streaming in browsers.

11.3.3 P2P Streaming

There are two types of P2P streaming. One is based on the traditional BitTorrent Protocol. The other is called the *substream* method and is a clean-slate approach.

BitTorrent is basically a *pull* kind of application because it needs to request every piece separately. The research that proposes to use BitTorrent for streaming needs to overcome yet another limitation known as the *rarest-first* policy. Usually, this is done by modifying the piece selection algorithm that retains the *rarest-first* behavior but adds some real-time compatible constraints [16]. This goal has proven to be hard to achieve [17], so some research on this problem tries to use some other "tricks" to make the real-time piece selection work in practice [18]. There is a considerable number of research works that view BitTorrent piece selection logic from the theoretical viewpoint [19–21].

Substream method is a good alternative to BitTorrent [5]. The method uses the *push* communication paradigm. Also, the stream is split into several substreams, which are downloaded in parallel. In fact, since the substreams overlap each other in terms of the sequence number of media blocks, playing while downloading is possible. There are several successful projects that use the substream method in practice, where Uusee [3], Afreeca [4], and NewCoolStreaming [5] are the most prominent and publicized services.

Substreams have to be kept in sync with each other. Various methods that can be used to optimize for minimum discrepancy across substreams can be found in References 8, 22, 23. More details on the substream method are presented further in this chapter. Note that the main focus in this chapter is on a cloud-based version of the substream method.

Even substream P2P methods are found to be unstable in practice [4]. Practical studies find that server capacity is still important. Also, practice shows that streaming is unstable and can freeze often for a considerable portion of user base.

11.4 THE SUBSTREAM METHOD IN THE CLOUD

Figure 11.2 describes the basic design of the substream model. As was mentioned earlier, each incoming substream is configured with three parameters: *blocksize*, *startpos*, and *step*. The figure shows the example for three substreams.

Client aggregates the blocks it receives into the playback buffer. The playback reads the buffer in the round robin manner, reading each substream in each cycle. It is clear that playback freezes when it encounters a missing block, which happens when download positions are different across substreams. The playback is safe and will not freeze as long

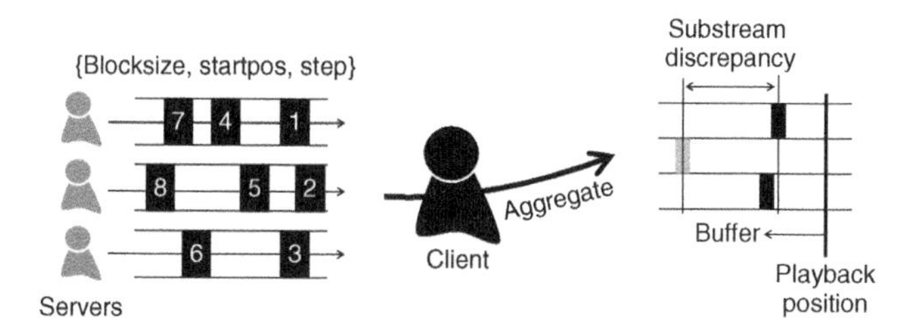

Figure 11.2 The basic idea behind the substream method. Note that the difference in throughput among substreams can potentially cause the playback to freeze.

as the slowest substream is ahead of the playback position, so it is not strictly necessary to align download positions.

However, in practice, substream discrepancy is one of the biggest problems in real P2P streaming services [5], which is why sources have to be optimized. Optimization is covered later in this section.

11.4.1 Presubstream: Push versus Pull

The substream method itself, in part, takes root in the *push* versus *pull* argument. This section briefly explains the trivial reality of the *push* versus *pull* argument, so that it makes sense when the substream method is made default further on. Let d be one-way delay, s the size of a chunk of data (*chunksize*), m the number of chunks, and $S = sm$ the total volume of data. Then, with a given throughput r, the total time T to receive all the data is

$$T = m\left(\frac{s}{r} + 2d\right) = \frac{S}{r} + 2md \tag{11.1}$$

which heavily depends on the value of d. The spirit of the push method is then

$$\lim_{m \to 1}\left(\frac{S}{r} + 2md\right) = \frac{S}{r} + 2d \tag{11.2}$$

Also, for large S, we can ignore the one-time d:

$$\lim_{S \to \infty}\left(\frac{S}{r} + 2d\right) = \frac{S}{r} \tag{11.3}$$

The trivial conclusion is then that *push* is better for streaming than *pull* simply because lowering the number of round trips improves end-to-end throughput of data.

Note that the *pull* method depends on m and by extension on the chunksize s, while the *push* method is freed from this dependency.

11.4.2 Sources

Let us assume that we have a fixed size stream of 500 kbps and 10 substreams for all users, following the practical example in Reference 5.

Each peer then performs the following actions:

- contacts the Tracker of the current streaming service and requests a number of sources, normally more than the number of substreams (double in Reference 5);
- makes substream requests to each elected parent, initial parents are selected randomly from the peer list;
- starts receiving content;
- at regular intervals, the slowest parent is replaced by a fresh parent from the peer list hoping for better performance.

The final stage of this process is part of continuous operation and depends on regular optimization of both the parent and child lists for each peer.

11.4.3 Online Optimization

The action sequence presented earlier applies in equal measure to both parents and children of each peer as well. Because P2P streaming is a peer solution, peers have to optimize not only the incoming upstream but also the downstream [5].

Note that the optimization is necessary primarily because of the initial shortcoming of the Kademlia Protocol [24] used by BitTorrent. Kademlia Protocol judges distance between peers based on the XOR distance between two IDs, where IDs in Kademlia are 160-bit random hash digests. This means that end-to-end delay or throughput is not considered when matching peers. Although this simplifies ID management and content search, it is detrimental for real-time streaming. Imperfections of Kademlia specifically in relation to streaming were previously studied by experiments in Reference 25.

In clouds, some of these problems can be resolved automatically by design. Since in cloud streaming SP provides the sources (on request), SP is free to choose its own metric to evaluate distance between any two locations. Since SP knows exactly where sources are, it can select those that are likely to provide the most reliable throughput. Considerations similar to those in the traditional method can be applied to each source [2].

Specifically, the following two actions are possible in clouds:

- creation of new VMs on demand, that is, when necessary, to make new sources;
- migration of existing sources using existing VM migration techniques [26].

11.5 STREAM AGGREGATION IN THE CLOUD

Figure 11.3 shows the designs of the two services, P2P and cloud, so that features of the latter can be drawn by comparison. The two models are similar only at the level of the

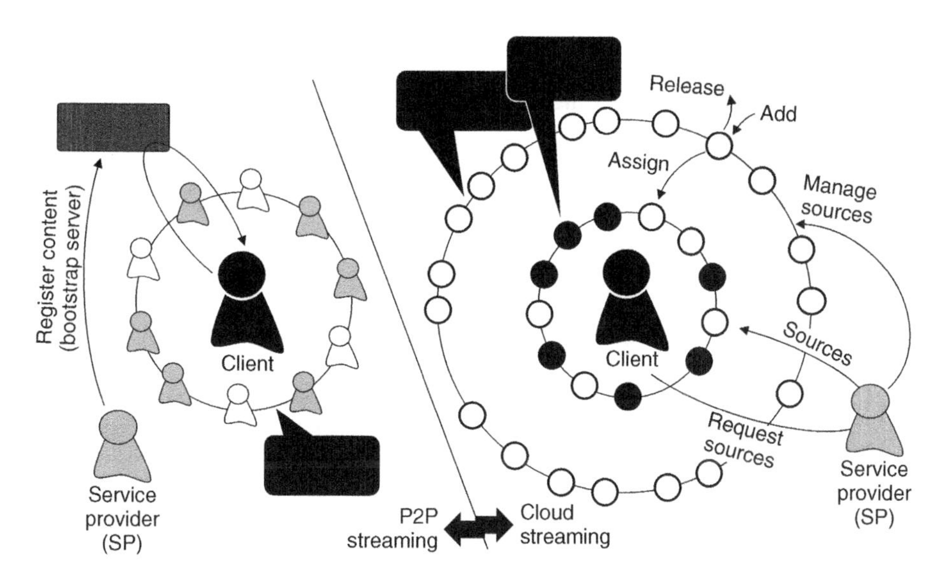

Figure 11.3 The overall designs of traditional P2P and the proposed cloud streaming services. Both designs are client-centric.

source list managed by each user. In P2P, it is the list of parent peers. In the cloud model, it is the list of sources made available to each client.

In P2P model, SP has to register its content with the Tracker. This process is referred to as *bootstrap*. Once bootstrapped, the Tracker can start registering peers for the service as they join the network. Several initial sources, by the current definition—the 10 initial sources, have to be provided by the SP itself, but with more joining users, the streaming will rely on the peers themselves—thus, fulfilling the main purpose of the P2P streaming itself.

In the cloud model, SP replaces the Tracker. This section will further elaborate on the details, but at the level of design it can be represented as the relation between outer and inner circles, as shown in Figure 11.3. SP is in charge of optimizing the outer circle by, as was mentioned earlier, adding or releasing VMs from the VM population of the service, or by triggering VM migrations. Also, SP can provide the most optimal list of VMs as sources for a given user on request.

11.5.1 Content Delivery Virtualization

Sources are handled as VMs in clouds. VMs are not full servers and are normally created with stringent restrictions on performance and load, for example, each VM can only open so many sockets, and so on. With this in view, SP has to keep a surplus of VMs ready for use and distribute (migrate) them to the locations that are most likely to be closest to new users. Naturally, there could be other formulations of VM managements, but this one is within current technological capabilities and is suitable for streaming.

The following practical application can be implemented.

Step 1. User logs into a website and goes through a procedure to start streaming of some content.

Step 2. SP can optionally trigger a network coordination procedure. There are several *network coordinate* methods that are feasible in practice [27–29] and can provide SP with the network coordinates of the user. Alternatively, traditional caching methods can be used to select best serving locations [2]. The outcome of this step is the list of sources that SP sends to the user. Incidentally, SP can create new VMs or migrate VMs from other locations to improve the potential list of sources.

Step 3. User starts the streaming using the substream method explained previously. User does not have to optimize the source list since it has already been optimized by the SP.

11.5.2 Optimization of VM Populations

VM optimization as part of a streaming service is feasible in practice [30], even at the scale of global networks, which is the area of federated clouds [26].

In fact, dynamic VM management as part of a streaming service can be simplified by separating streaming applications from the storage (media files). This separation can substantially decrease cost of VM migration, as was shown by recent research [31]. Another related topic is optimization of VM management in view of variable workloads in VMs on the same physical infrastructure [32]. Discussion of these topics in detail is out of scope of this chapter. It is sufficient to establish the feasibility of streaming-specific VM management.

In large VM populations, the topic of energy efficiency enters the picture. Green clouds is a recent research topic where VM migration has another optimization constraint—the energy efficiency [33]. For lack of a better term, "dormant VMs" might substantially decrease the management cost of VM populations.

11.5.3 Client-Side Optimization

As was mentioned earlier, client-slide optimization of source lists is not necessary, since the list already arrive preoptimized. However, if deemed necessary, the general approach can be borrowed from P2P streaming.

11.6 MODELS

This chapter implements all models except for the adaptive streaming. The latter is left out partially because there are very few practical implementations (meaning real services) at present time, but mostly because adaptive models can have multiple distinct formulations some of which may require distinct format of the video content itself—for example, *variable bitrate* have been proposed as better suited for adaptive streaming. See Chapter 12 of this book for details on adaptive streaming.

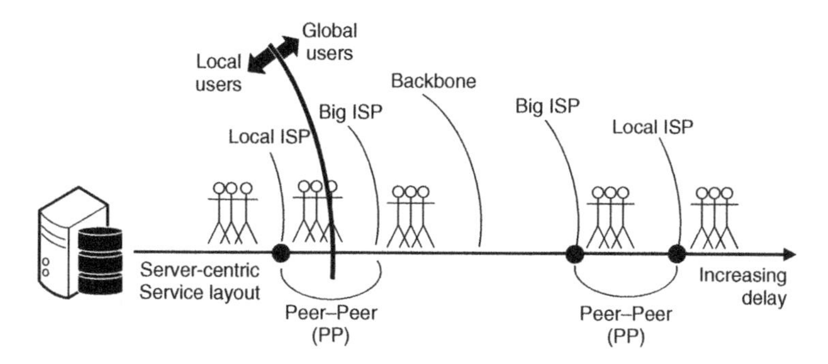

Figure 11.4 A generic service model designed as the world viewed by the primary content server.

11.6.1 A Generic Service Model

A generic service model is shown in Figure 11.4. The layout is server-centric. In cloud streaming, it applies to each VM or, at a bigger unit, to each data center. As far as end-to-end delay is concerned, global Internet can be as big as 6–11 ISPs and 26–31 hops in the diameter between server and user.

It is not unusual for SPs to limit the use of their streaming software to local ISPs, in which case ISP becomes ASP as well. In this case, it might be easier to control throughput volatility and end-to-end delay. With P2P and federated clouds, however, local restriction is no longer necessary; in fact, worldwide distribution seems to me a more common case among streaming services today. For example, at the time of this writing, Hulu—the well-known video streaming service—has announced its entry into the market in Japan.

11.6.2 Service in the Network Graph

It is important to use a realistic topology for analysis. The realistic world-size AS-level topology generated by IGen [34] shown in Figure 11.5 is used in this chapter. One can see the rough shapes of continents in the graph. There are 215 nodes and 900+ links.

The graph is used by all models as follows. Servers (parameter name *servers*) are randomly distributed over the graph. Servers are data centers in the cloud model, in which case multiple VMs are attached to the same node. Users are also assigned to nodes randomly. More details on simulation setup are provided further in this chapter.

11.6.3 Throughput Model

Existing research does not share a common throughput model. In fact, the closest research to a throughput model is the practical observations in Reference 4. However, no traces on dynamics of throughput for streaming users are available today. This author hopes that such traces will become more readily available in the future.

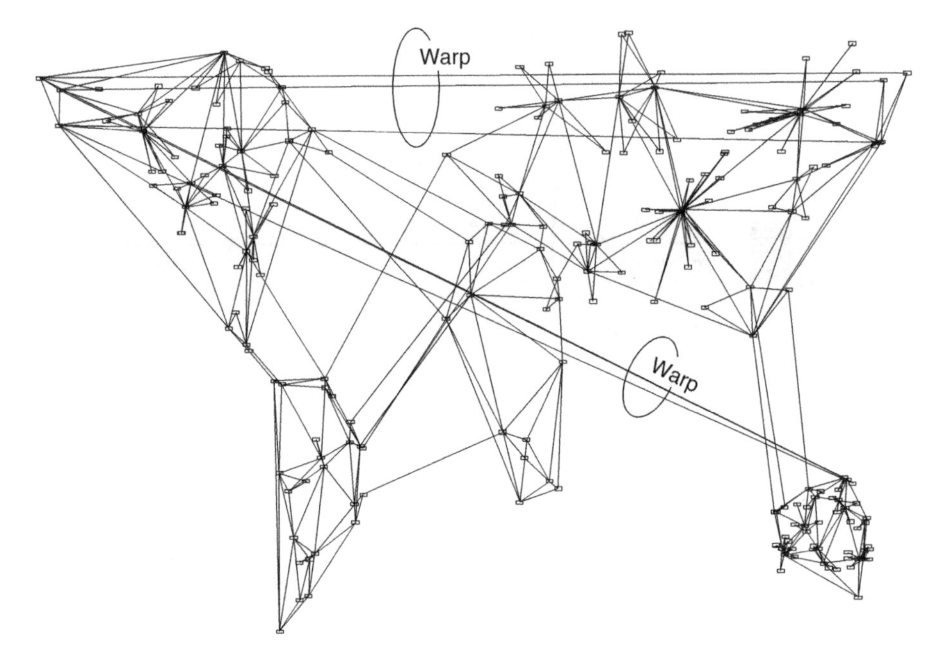

Figure 11.5 A realistic AS-level worldwide network topology.

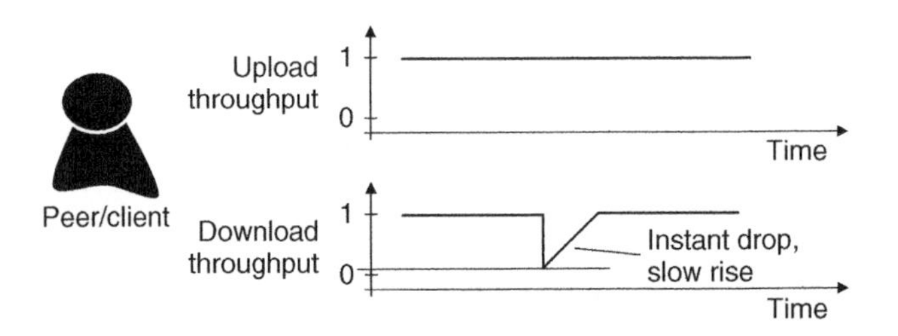

Figure 11.6 The throughput model in which upload throughput is constant while download throughput is affected by end-to-end network performance.

In absence of real traces, the simplistic model in Figure 11.6 is used. The model only describes the download while upload throughput is considered to be perfect. This one-side approach is common in existing research as well [5]. In fact, it should be clear that the effect of capacity on streaming can be studied as long as either upload or download capacity varies over time. If both are used, the effect will be caused by the smallest one at the time. This makes it unnecessary to model both throughputs. In fact, modeling only the download side greatly simplifies modeling in simulation.

The simple model in Figure 11.6 is implemented as follows. Time is split into time slots—10 ms in this chapter, making 100 time slots per second. Probability of throughput drop can then be defined as a discrete (geometric) random variable distribution that can tell the simulation when to drop the throughput. In this chapter, throughput probability is a variable, but the rest is fixed. The rest is specifically the drop depth—drops to 5% in all cases, and rise slope after drop—10% rise in all cases, meaning that throughput recovers in about 100 ms. Drop probability itself is explained further in this chapter.

The use of a partially static configuration simplifies analysis and visualization of results. The issue of complexity is discussed further in this chapter.

11.6.4 Traditional Streaming Model

User contacts the streaming server. The request is captured by SP and redirected to the closest and least busy server. The two constraints are satisfied as follows. First, the servers are listed in an increasing order of hop distance (paths in graph). The group with the shortest distance is selected and the least busy server in that group is selected.

The server does not change during streaming.

11.6.5 P2P Streaming Model

Kademlia is implemented as per description in Reference 24. Each peer issues the *get_peers()* request to the Tracker for 30 peers (three times of the required number of substreams).

The streaming starts as is normal for the P2P model. Parent list is optimized at 1 s intervals, when the current slowest parent is replaced with a randomly selected parent from the reserve list.

For simplicity, only one layer of the distribution tree is simulated. Having multiple layers would mean that each peer would have to dynamically optimize the child list as well. In this case, throughput drops would have to be followed down the stream. Such a model would be very difficult to present visually since performance in lower layers in the distribution tree would gradually degrade. Performance depending on the depth of peers in the distribution tree is left for later publications.

11.6.6 Cloud Streaming Model

Each user contacts SP for the list of 10 sources. SP is assumed to know the location of users and data centers. SH selects 10 closest data centers and lists their VMs in the increasing order of load, defined as the number of current users per VM. The simulation is designed in such a way that a VM would never get more than 10 users, but if shortage of underutilized VMs is found, other underutilized VMs are migrated from other data centers in the order of decreasing degree of underutilization.

In the end, 10 sources each from a different data center are notified to the user. The user then starts the streaming. No further optimization on sources is performed.

11.7 ANALYSIS

Simulation results in this section are analyzed from several distinct viewpoints, most of which have direct practical implications. As was shown above, even simplified models have many parameters. So, to create practical visualizations of performance, some parameters have to be aggregated and presented as averages. However, the analysis below retains the main focus of this chapter and isolates the related parameters for close inspection. The important parameters are *capacity* and its *variation* in time. The main performance metric is *freeze time*.

The following narrative is adopted by this section. The overall performance is presented first, where the advantage of the cloud-based streaming is revealed. Then, detailed performance is analyzed using distributions of freeze time, as in both P2P and cloud-based methods performance varies across users. Distributions help visualize differences in performance across users. Finally, comparative analysis is offered, where results answer practical questions like "how change in available resources affects difference in performance between methods."

The following simulation setup is used.

Server capacity is set to *1 Gbps* so that it is not a limiting factor in downloads, facilitating fair comparison with other models. VM capacity is set to the same value as the servers VMs are attached to, but each VM has a quota of 10 current streaming sessions, beyond which another VMs has to be selected. However, given that SP can trigger VM migrations, shortage of VMs at any given data center is not an issue in this simulation.

Stream rate is 500 kbps. With 10 substreams, substream rate is 50 kbps.

Number of users is selected randomly between 10 and 1000. Nominal download capacity of each user is selected randomly between 1 and 10 times the stream capacity. To model capacity heterogeneity across users (incidentally, P2P peers), capacity can be randomly shifted toward a higher number or a lower number within 90% of the baseline user capacity. When presenting results, this will be referred to as *capacity* and *capacity range*, respectively. Relatively high capacity range causes high variance in capacity across users. Note that download capacity directly affects ability to download the stream, as the stream cannot be downloaded above current user capacity. The throughput model was explained in detail earlier.

Drop probability is based on hop count of end-to-end paths between users and servers (or peers). The following model is used. First, drop probabilities are defined as the fixed value 0.01 (1%) for the minimum value and the random value between 0.05 and 0.2 for the maximum value. Then, exponential function $y = \exp(0.1x)$ is sampled for all hop counts (1 through 18 in the global graph) and the values are mapped to the range between the drop minimum and drop maximum values, thus creating the map between hop count and drop probability. In simulation, drop probability for downloads between two nodes is defined based on that map. Drops are simulated dynamically using geometric function with the drop probability used as the parameter of the random function.

30 s of streaming are simulated for each user. Media blocksize is 10 kbytes.

As was mentioned earlier, *freeze time* is selected as the only metric to evaluate performance. Focusing on a single metric simplifies visualization of results. *Freeze time* is defined as cumulative freeze time during the 30-s stream. Playback in this simulation is unbuffered. However, stream can exceed the playback bitrate (500 kbps) in which case it is stored for future playback. Playback stops if the storage is empty and download rate is below the stream rate.

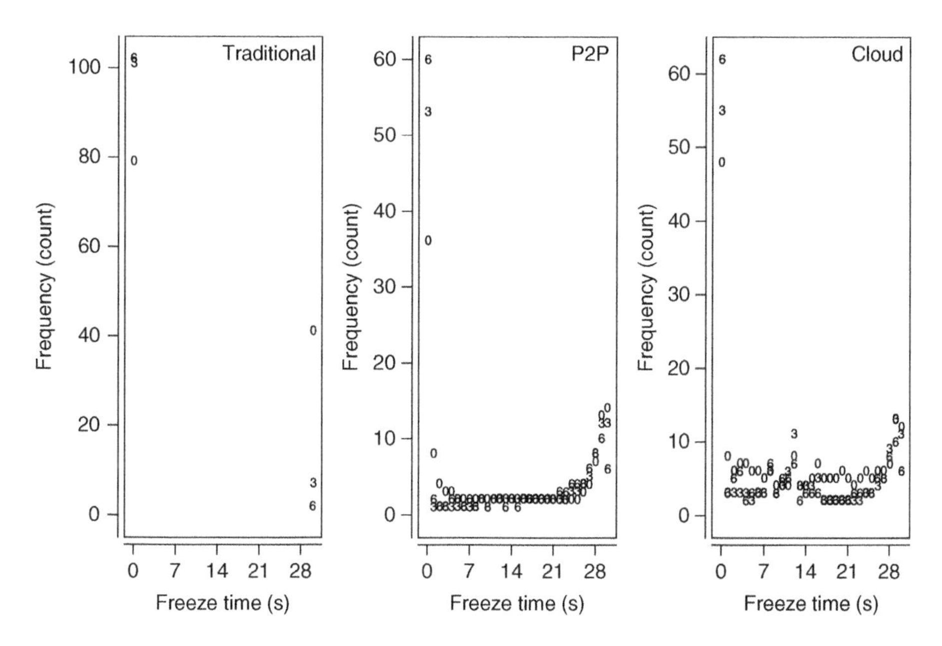

Figure 11.7 Average performance of the three methods over the entire dataset. The bullets denote capacity rounded to the closest 1 Mbps.

11.7.1 Overall Performance

Figure 11.7 shows the overall performance for the entire dataset.

Traditional model shows two clusters at the two extremes. Majority of results show low freeze time while few show the largest possible freeze time. In practice, this means that a large portion of users gets the biggest possible freeze time, which basically indicates a failed streaming session. In fact, both extremes in the traditional model are higher than in the two other models—there are more users with zero freeze times, but there are also more users with the biggest freeze time.

Results for P2P and cloud models are distributions of values. Again, majority of results show good performance, but results with freeze time above zero are gradually becoming more common, finally resulting in a relatively large number of users with the largest freeze time.

Results for the cloud model reveal the main advantage of the model. Remembering that the entire dataset is shared by all models, the cloud model raises the frequency of freeze time across the entire distribution, which means that many users get fewer and shorter freezes compared to the P2P model. Moreover, performance in the cloud model reveals sensitivity to capacity where lower capacity (smaller value in bullets) is more likely to get more freezes (higher count). Note that in the P2P model such effect is not found—bullets for different capacities overlap in most cases.

The lesson to be learned from Figure 11.7 is as follows. Traditional streaming results in extreme performance where between 10% and 40% of user base is likely to get very bad performance. Both P2P and cloud streaming partially alleviate this problem by replacing performance extremes with smooth distributions. Both extreme ends of the distribution are still more likely on average, but there are now users in the medium range of performance. Cloud streaming further improves performance by moving more users to the left of the distribution, thus increasing the number of users with better performance.

11.7.2 Analysis by Curve Aggregates

The term *curve diff* denotes the area between two curves. This can be achieved by calculating integrals. In this chapter, the same job is performed by the Kolmogorov–Smirnov test (*kstest*) readily available in R [35]. Kstest is traditionally used to match two distributions. However, the side effect of the test is the *statistic* that calculates the area between two distributions. As *kstest* is resilient to different length lists and can cope with intersections between distributions, it is preferred to a self-developed algorithm. Positive value of statistic—further referred to as *curve diff*, when measured between curve A and curve B, denotes positive area, that is, the curve A is, on average, above curve B.

Figure 11.8 shows curve analysis for the two cases—traditional versus P2P and P2P versus cloud, expecting positive values in both pairs which would mean that P2P is better than traditional and cloud is better than P2P. To avoid multidimensional analysis, each chart is created by following variations in only one parameter. This way, it is possible to make judgment on how each method performs relative to changes in only one parameter. Remember that each bullet in the plots represents area between two distributions.

The first observation from Figure 11.8 is that all *curve diffs* are positive meaning that P2P is always better than traditional and cloud is always better than P2P. The only exception is for the *drop * capacity/capacity range* metric, in which case all methods perform the same, hence zero curve diffs.

The other lessons from Figure 11.8 are as follows. Increasing the number of *servers* benefits the traditional model (smaller curve diff) but has little effect on the curve diff between cloud and P2P models. Dependence on number of *users* is virtually nonexistent, where the fluctuations are most probably due to randomness in simulation parameters. Increased *capacity* benefits the traditional method but has little effect on the cloud method. Finally, *drop * capacity* and *drop * capacity range* metrics gradually increase the benefit the latter member in both pairs.

The plain outcome of this analysis is as follows. The results clearly show that both cloud and P2P models perform better than the traditional model. Also, the cloud model consistently outperforms the P2P model at a small but consistent margin.

11.7.3 Performance in Realistic Conditions

It should be noted that the *curve diff* analysis shown in the earlier section was not very fair to the cloud model. For example, increasing number of users has positive effect on the P2P model but negative effect on the cloud model. Similarly, increasing number of servers should benefit the cloud method while having no effect on the P2P method's performance. The analysis in this section fixes this problem by visualizing performance boosts relative to difference in values of certain simulation parameters. The cloud model

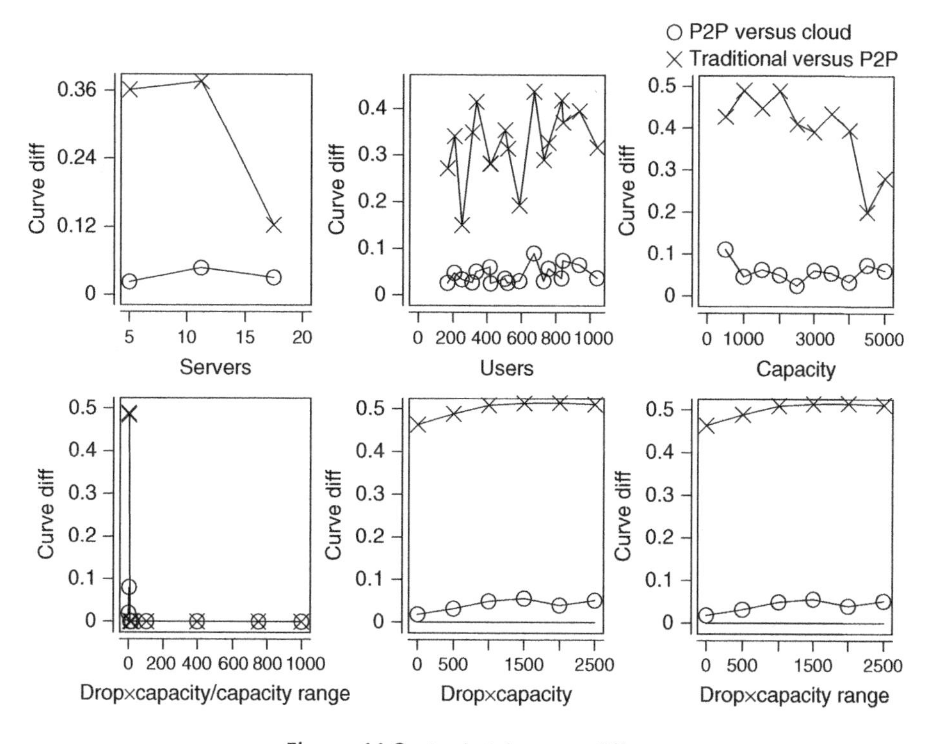

Figure 11.8 Analysis by curve diffs.

with many servers can now be compared with the P2P model with few servers—a more realistic implementation scenario than the one in which the parameter has the same value in both methods.

Figure 11.9 is also based on the analysis of *curve diffs*, but the horizontal scale now represents the gap in simulation setup between the P2P and cloud methods. For example, *Servers (gap)* = *5* means that one of the curves comes from simulations where server count was, say, *1*, and the other from simulations with server count of *6*, thus making the gap of *5*. The gap is always between a higher value in the cloud model and a lower value in the P2P model.

Having established the visualization mechanism, it is now possible to read Figure 11.9. Increasing the number of *servers* results in linearly increasing curve diffs, that is, the cloud method showing increasingly better performance. This is an expected outcome because it literally means that increasing the volume of cloud resources allocated to streaming improves the advantage of the cloud over the P2P model. Note that growing gap in the number of users (second plot) does not generate a consistent trend showing little direct benefit for the P2P method. The cloud method also benefits from increasing gap in *capacity*, as well as two complex metrics *drop * capacity* and *drop * capacity range*. The *drop * capacity/capacity range* complex metric does not reveal a consistent trend, but this gap also benefits the cloud method.

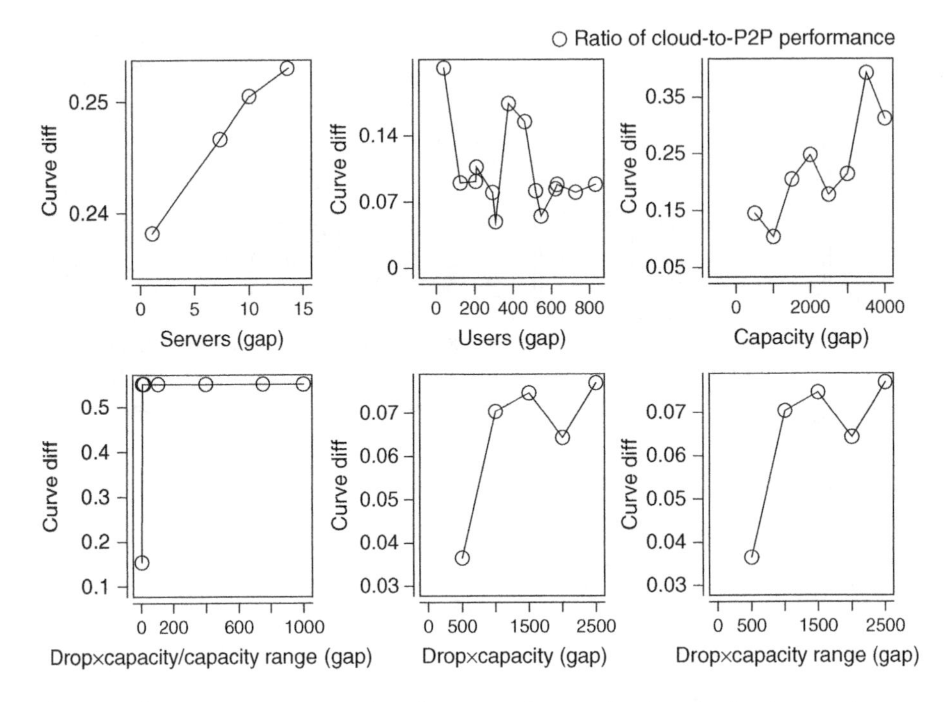

Figure 11.9 Performance comparison between P2P and cloud methods where each curve shows simulation results across a range of gaps in simulation conditions.

11.8 VISIONARY THOUGHTS FOR PRACTITIONERS

11.8.1 Open Cloud

The SP role does not exist today. Instead, cloud providers seem to prefer to work with clients, private, and business, directly. SLAs are written for a given number of machines with a performance description for each VM.

SLA-governed migration is not part of reality today. Instead, CPs develop migration algorithms as part of management of their internal resources.

In order to make cloud streaming work, the role of SP is crucial. The simple logic is as follows.

Future streaming is done in the cloud with possibility that some of today's P2P streaming will convert to the cloud model as well. CPs then have a simple choice: whether to delegate management of per-service VM populations to SPs or do everything by themselves. If CPs choose not to relinquish control over VM populations, they will face very difficult optimization problems for large number of per-service VMs, across multiple locations, across multiple services. Delegating per-service optimizations to SPs can considerably simplify optimization problems.

Services will span multiple CPs in the future, as research on federated clouds has already shown [26]. Clouds will have APIs between each other, which will allow them

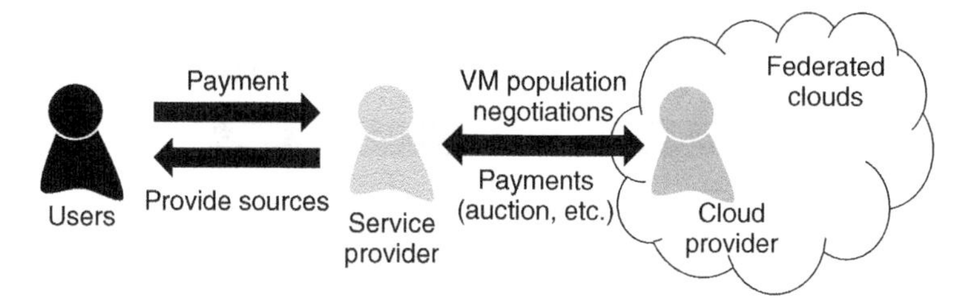

Figure 11.10 The new role of *Service Provider* (SP) and the economy built around it.

to hand VMs over the border. Delegation of some management functions pertaining to VM populations in this case is the practical way to go.

Finally, SPs have content. It is likely that content providers will develop cloud-based streaming applications in the future. Today, the common practice is for SPs to build their own clouds (Amazon, etc.). SPs and CPs joining their forces should create a much richer environment for business growth.

11.8.2 Cloud Economy

Expanding on the subject in the earlier section, relation between users, SPs, and CPs can be studied as two economies, one between users and SPs and the other between SPs and CPs. Figure 11.10 shows the basic concept of such an economy. The term "economy" here loosely refers to a market where prices for VMs are variable and depend on scarcity and current conditions. Scarce resources should cost more, and less scarce should cost less. This economy can reflect onto the relation between SPs and users where difficulty to support a given level of QoS in video streaming may result in stratification of user-ship into high class pay-more users and users who pay less but get a lesser QoS guarantee.

11.8.3 Performance Modeling

Performance modeling is important to large-scale streaming services. All the models found in literature are simplifications [5]. The model presented in this chapter is also a simplification, justified by the focus shifted toward simplicity and clarity.

Realistic simulations have to be much more complex. For example, evaluation of what happens at various depths of distribution tree in P2P streaming is an interesting problem but it is also hard to evaluate and even harder to present results. Note that the cloud model has a shallow distribution tree—only one level of cloud to users, but has other complexities on the part of time dynamics in VM population management.

Adding the buffer as shown in Figure 11.11 also complicates analysis. When using the buffer, it is hard to tell when you have strained the system enough to overflow the buffer versus when the strain was absorbed by the buffer and did not affect the playback. There is no standard approach pertaining to the buffered playback in the literature. Perhaps, such a standard would be useful. For example, buffer effect might be modeled as

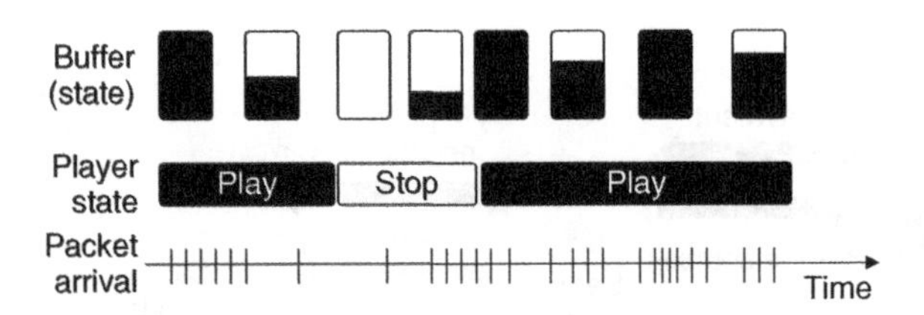

Figure 11.11 A buffer model commonly implemented in software.

a transform on the data where the short-time bursts in capacity are smoothed out by the buffer.

The relation between freeze time and QoE is not fully clear. Research in the literature focuses on much more subtle QoS metrics such as PNSR (peak noise-to-signal ratio) [9], which is not directly related to freeze time. It might be useful to develop models that would calculate QoE based on freeze time. For example, how do people react to a one-time 1 s freeze? How about two times per minute each time a 10 s freeze?

11.8.4 Federated Performance Measurement

Federated network performance measurement is not a new topic. Data centers are at the edges of the network, by definition. Users are also at the edges, but majority is connected to the Internet via a relatively less reliable last-mile connection. When SPs are in charge of VM population management at the scale of the worldwide network, it may be possible that large-scale end-to-end performance measurement can be part of their job description.

End-to-end performance measurement itself, including network coordination, is a well-researched subject [36] with many software implementations.

11.9 FUTURE RESEARCH DIRECTIONS

The main argument of this chapter is that cloud-backed sources are more stable and reliable than peers in a P2P delivery network. However, while the CDN model discussed in this chapter can be implemented using present-day cloud technology, the cost is an open issue. For example, P2P delivery inflicts minimal cost because delivery is done by the users. High performance traditional streaming requires substantial investments. Investments into a cloud streaming service are also necessary, but they can potentially be less than that of a traditional service. For starters, VMs can be traded in and out between SP and CPs, and the number of active VMs can be kept to the required minimum. Provided users pay for the streaming service, highly efficient VM management can help turn in profit at the end of the day.

Also, a hybrid between clouds and P2P delivery methods is a possibility. In this case, only a portion of VMs are injected into the cloud in such a way that the overall integrity of the P2P delivery network is kept feasible for freeze-less delivery. This, however, will require complicated analysis of performance dynamics relative to user depth in the delivery tree, which is a challenging topic in itself.

Open cloud and the related subject of VM populations and the user-SP and SP-CP economies are interesting subjects worth exploring more in the future.

On the practical side, as was mentioned earlier, real traces of throughput dynamics in time are in short supply. Making such traces and releasing them publically (after proper anonymization) should substantially ease modeling in simulations—throughput can now be replayed rather than modeled probabilistically. No such traces are publically available at the time of this writing.

11.10 CONCLUSION

This chapter discussed the next evolutionary step in video streaming. Just like the P2P, streaming model was the logical evolution of the traditional model, the cloud-based streaming presented in this section is the logical evolution of the P2P model.

The main improvements offered by the cloud model are as follows. Homogeneity of streaming sources is facilitated by the cloud, where peers in P2P networks are known to be heterogeneous. Clouds also help facilitate higher availability of streaming sources. Since sources are hosted by VMs, the latter can migrate away from congested data centers. Also, more sources can be created or purged by SP on demand simply by requesting additional VMs or returning unused VMs to the cloud. Since sources are provided to users by a single SP, it is possible to optimize source lists prior to notifying them to users. Note that the P2P model traditionally relies on the Kademlia Protocol where optimization of source lists requires tedious selection and reselection procedures at each P2P client.

Simulation analysis showed that these improvements made the cloud model perform consistently better than the P2P model. Both the cloud and P2P model performed much better than the traditional model, as is expected. In addition, it was shown that realistic conditions consistently favor the cloud over the P2P model with widening gap as conditions become more realistic.

A feasible implementation of the cloud model was discussed at length. In fact, the described example service can be implemented in clouds today. In view of the missing role of SP, the role can be performed by CPs in the interim. The various QoS optimizations performed on VMs, which in turn serve as streaming sources can be implemented at the current level of cloud technology.

REFERENCES

1. Buyya R, Pathan M, Vakali A. *Content Delivery Networks*. Lecture Notes in Electrical Engineering (LNEE) Vol. 9. Springer; 2008.
2. Sivasubramanian S, Pierre G, Steen M, Alonso G. Analysis of caching and replication strategies for web applications. IEEE Internet Comput 2007;11(1):60–66, Vrije Universiteit, Amsterdam.

3. Wu C, Li B, Zhao Sh. Diagnosing network-wide P2P live streaming inefficiencies. IEEE INFOCOM; New York, USA; 2009. p 2731–2735.
4. Park K, Chang D, Kim J, Yoon W, Kwon T. An analysis of user dynamics in P2P live streaming services. IEEE International Conference on Communications (ICC); Beijing, China; 2010. p 1–6.
5. Li B, Xie S, Qu Y, Keung G, Lin C, Liu J, Zhang X. Inside the new coolstreaming: Principles, measurements and performance implications. IEEE INFOCOM; Phoenix, AZ, USA; 2008. p 1031–1039.
6. 2013 e2eprobe: A number of end-to-end available bandwidth measurement tools. Available at https://github.com/maratishe/e2eprobe.
7. Antonio C, Tusa F, Villari M, Puliofito A. Improving virtual machine migration in federated cloud environments. Second International Conference on Evolving Internet; 2010. p 61–67.
8. Li Z, Yu Y, Hei X, Tsang D. Towards low-redundancy push-pull P2P live streaming. 5th International ICST Conference on Heterogeneous Networking for Quality, Reliability, Security and Robustness (QShine); 2008; Article 16; 7 pages.
9. Fiedler M, Hossfeld T, Tran-Gia P. A generic quantitative relationship between quality of experience and quality of service. IEEE Netw J 2010;24(2):36–41.
10. Walkowiak K. QoS dynamic routing in content delivery networks. NETWORKING; Springer LNCS Vol. 3462; Wroclaw, Poland; 2005. p 1120–1132.
11. Chen C, Ling Y, Pang M, Chen W, Cai S, Suwa Y, Altintas O. Scalable request-routing with next-neighbor load sharings in multi-server environments. 19th IEEE International Conference on Advanced Information Networking and Applications; Piscataway, NJ, USA; March 2005. p 441–446.
12. Daoist F Adopting HTML5 for television: Next steps, Report by W3C/ERCIM Working Group; 2011.
13. Lederer S, Muller Ch, Timmerer Ch. Dynamic adaptive streaming over HTTP dataset. Multimedia Systems Conference (MMSys); February 2012. p 89–94.
14. 2013 msplayer: A multi-source media player in Chrome. Available at https://github.com/maratishe/msaplayer.
15. The WebSocket protocol, IETF HyBi Working Group document draftietf-hybithewebsocket protocol-09; June 2011.
16. Charls A, Sharma T, Singh P. Media streaming in P2P networks based on BitTorrent. International Conference on Computer Engineering and Applications (IPSCIT); Vol. 2; Singapore; July 2011. p 158–162.
17. Stais C, Xylomenos G. Realistic media streaming over BitTorrent. Future Network and Mobile Summit; Berlin, Germany; 2012. p 1–5.
18. Erman D. Extending BitTorrent for streaming applications. 4th Euro-FGI Workshop on New Trends in Modelling, Quantitative Methods and Measurements (WP IA.7.1); Ghent, Belgium; May 2007.
19. Erman D, Vogeleer K, Popescu A. On piece selection for streaming BitTorrent. 5th Swedish National Computer Networking Workshop (SNCNW); Karlskrona, Sweden; October 2008.
20. Szkaliczki T, Eberhard M, Hellwagner H, Szobonya L. Piece selection algorithm for layered video streaming in P2P networks. Electron Notes Discrete Math 2010;36:1265–1272.
21. Sandvik P, Neovius M. A further look at the distance-availability weighted piece selection method: A BitTorrent piece selection method for on-demand media streaming. Int J Adv Netw Services 2010;3(3/4):473–484.

22. Zhang M, Zhang Q, Sun L, Yang S. Understanding the power of pull-based streaming protocol. IEEE J Select Areas Commun 2007;25(9):1678–1694.
23. Guofu W, Qiang D, Jiqing W, Dongsong B, Wenhua D. Towards low delay sub-stream scheduling. Int J Comput Commun Control 2010;V(5):727–734.
24. Maymounkov P, Mazieres D. Kademlia: A peer-to-peer information system based on the XOR metric. The 1st International Workshop on Peer-to-Peer Systems (IPTPS); April 2002. p 1–6.
25. Silverston T, Jakab L, Cabellos-Aparicio A, Fourmaux O, Salamatiane K. Large-scale measurement experiments of P2P-TV systems: Insights on fairness and locality. Signal Process Image Commun 2011;26(7):327–338.
26. Wood T, Shenoy P, Venkataramani A, Yousif M. Black-box and gray-box strategies for virtual machine migration. 4th USENIX Symposium on Networked Systems Design and Implementation; 2007. p 229–242.
27. Dabek F, Cox R, Kaashoek F, Morris R. Vivaldi: A decentralized network coordinate system. ACM SIGCOMM; August 2004. p 15–26.
28. Wong B, Slivkins A, Sirer E. Meridian: A lightweight network location service without virtual coordinates. ACM SIGCOMM; October 2005. p 85–96.
29. Tanaka Y, Zhanikeev M. *Active Network Measurement: Theory, Methods and Tools*. ITU Association of Japan; 2009.
30. Antonio C, Tusa F, Villari M, Puliofito A. Improving virtual machine migration in federated cloud environments. Second International Conference on Evolving Internet; March 2010. p 61–67.
31. Stage A, Setzer T. Network-aware migration control and scheduling of differentiated virtual machine workloads. CLOUD; May 2009. p 9–14.
32. Zhanikeev M. Optimizing virtual machine migration for energy-efficient clouds. IEICE Transactions on Information (unpublished, can be made available on request); March 2012.
33. Tang L, Crovella M. Virtual landmarks for the Internet. Internet Measurement Conference (IMC); 2003. p 143–152.
34. 2013 R project for statistical computing. Available at http://www.r-project.org/.
35. 2013 webstreams: A benchmark for HTML5 WebSockets. Available at https://github.com/maratishe/webstreams.
36. 2013 IGen topology generator. Available at http://informatique.umons.ac.be/networks/igen/.
37. Dynamic adaptive streaming over HTTP (DASH), Normative document ISO/IEC DIS 23001-6; 2011.

12

BEYOND CDN: CONTENT PROCESSING AT THE EDGE OF THE CLOUD

Salekul Islam[1] and Jean-Charles Grégoire[2]

[1] *United International University, Dhaka, Bangladesh*
[2] *INRS-EMT, Montréal, QC, Canada*

12.1 INTRODUCTION

Over the years, the content delivery network (CDN) [1,2] has evolved to become a well-established technology for delivering a wide range of content including Web objects (e.g., text, graphics, uniform resource locators (URLs), and scripts), downloadable objects (e.g., media files, software, and documents), applications (e.g., e-commerce and portals), live streaming media, on-demand streaming media, social networks, and so on. Although CDNs deliver content to end-users with high availability and performance, they fail to meet the more recent, quickly increasing demand of multimedia functions on the delivery/server side [3,4]. Multimedia processing includes a large spectrum of multimedia functions including transcoding, delivery of interactive media, mixing different streams, dynamically modifying the resolution according to the user equipment, scaling down/up with respect to the number of customers, custom mashups, and so on.

On the other hand, cloud computing—a collection of applications, hardware, and system software—has become the technology of choice to deliver services to end-users over the Internet [5]. The datacenter that deploys the necessary hardware and software

Advanced Content Delivery, Streaming, and Cloud Services, First Edition.
Edited by Mukaddim Pathan, Ramesh K. Sitaraman, and Dom Robinson.

is the essence of the cloud. Cloud computing offers a wide range of services, including computing, storage and delivery, and different modes of providing services, that is, Software/Platform/Infrastructure-as-a-Service (S/P/IaaS) [6]. Through virtualization technology, it is now possible to run not just an application but also a full server inside the cloud, thereby reducing the cost of the hardware base in a typical one service—one server model, where hardware is often underused. By leveraging the advent of cloud computing, a recent trend has emerged for CDN providers to deploy elements of the cloud infrastructure (e.g., the Netflix architecture [7]), just as some cloud providers offer CDN functions. However, these recent developments have so far failed to exploit the full benefits of cloud computing, that is, the CDN-derived services that are deployed on top of the cloud computing infrastructure remain limited.

According to Cisco's Visual Networking Index tool [8], in 2013 the various forms of video (i.e., TV, video on demand (VoD), Internet video, and peer-to-peer (P2P)) will have started to exceed 90% of the global consumer traffic. Hence, a focus on CDN should not be limited to deliver content only but should rather extend to various forms of processing. More specifically, media processing, which is a fairly generic, but also demanding function, should be the focal point of our investigation. This chapter focuses on the intersection of CDN and cloud computing by exposing a number of trade-offs on the deployment of multimedia processing functions inside the cloud and the complementarity of cloud and CDN deployments. We also show how the popularity of the cloud and its deployment at various scale factors has further blurred the distinction between the different service delivery infrastructures.

12.2 EXISTING CONTENT DELIVERY PLATFORMS

There are three established content delivery platforms—CDN, P2P, and cloud computing—to support large-scale service deployment in the Internet. Next, we discuss how these can be combined for delivering content. This study will later serve to identify how an evolution of cloud computing can combine the benefits of these techniques.

12.2.1 CDN

CDN is a system of caches hosted by Internet service providers (ISPs), but owned by a single company [1,2]. CDNs are an evolution of the client–server architecture introduced early in the emergence of the Web and designed to reduce the overhead of the content server by bringing (parts of) the content to the network edge, closer to the user, similar to a cache memory on a computer. From a resource perspective, a CDN has limited storage and requires only little computing power as it is only a Web server; its main benefit lies in bandwidth scaling through the number of servers. For efficient content delivery, CDN providers deploy thousands of servers at the edge of the Internet. Akamai [9], the market leader, has over 73,000 servers deployed in 70 countries within nearly 1000 networks.

Figure 12.1 illustrates the interworking of different architecture components of CDN [1]. First, the originating server delegates its uniform resource identifier (URI)

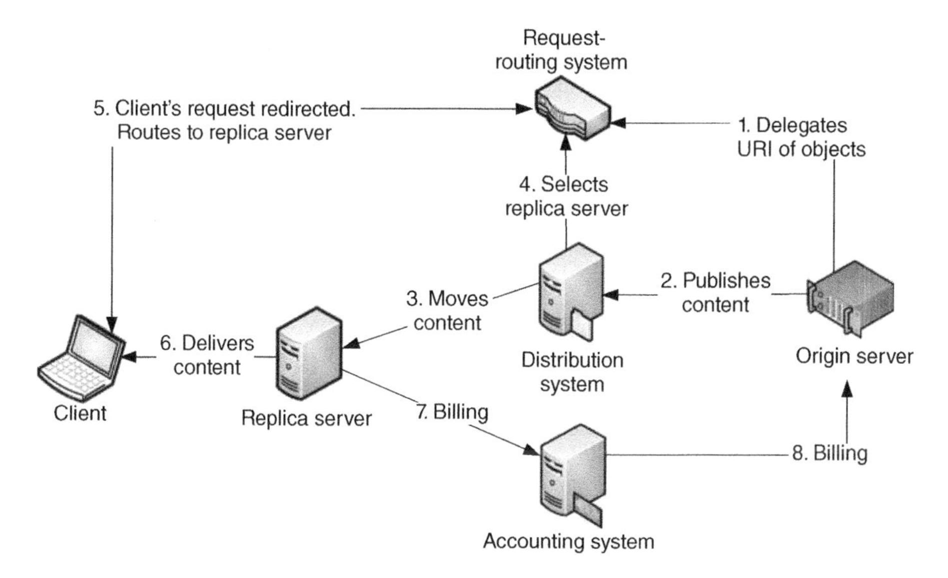

Figure 12.1 Architecture components of CDN.

name space for document objects to be distributed to the request-routing system and publishes content into the distribution system. The distribution system moves the content to replica servers. When a client requests documents because of URI name space delegation, the request is actually directed to the request-routing system. The request is routed to a suitable replica server, close to the client, which delivers the requested content. The accounting system receives billing information for all delivered content, aggregates, and distills the accounting information and forwards this information to the originating server. Note that we have explained the generic architecture of CDN, and many variations of this generic model have been deployed. Interested readers are referred to Reference 10 for a complete list of CDNs and their classification.

Although technically not being associated with a cloud model, the CDN fits with the notion of pooling of resources in the Internet, but for a highly specialized purpose: in this case, the access bandwidth offered to customers is provided by replication of content, closer to customers. Associated with this practical benefit is a semantic model dedicated to Web content access and the related tools to help customers prepare the data suitably for appropriate distribution. For example, while the first request may refer to a remote website, the content of the first page will be extracted from a variety of sites, some chosen dynamically based on the originating domain of the request.

12.2.2 Hybrid of P2P and CDN

CDN and P2P are the two widely used mass content distribution/sharing methods, especially in the case of real-time audio and video distribution or also streaming. A number of studies [11–13] have been carried out to create a hybrid technique that inherits the

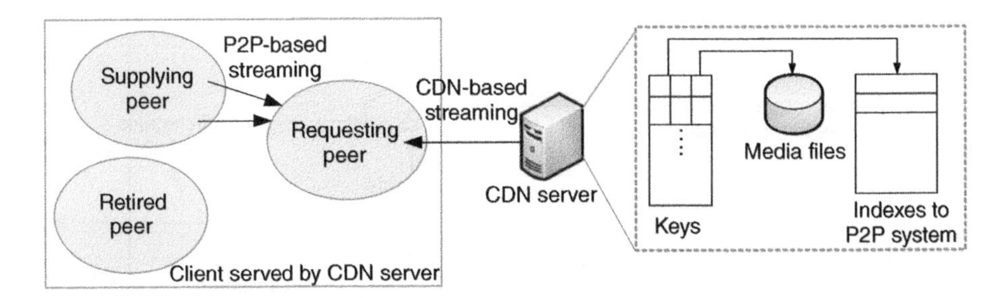

Figure 12.2 A hybrid architecture of P2P and CDN [11].

best of both worlds and mutually offsets each other's deficiencies as well. CDN servers are expensive to deploy and maintain, and also demand a sufficient bandwidth. On the other hand, the P2P architecture needs a good number of reliable seed-supplying peers. Recall that the P2P model is based on a pooling of limited resources, mostly storage and bandwidth, across a large community of users—a model derived from the grid. A single peer offers a fraction of its bandwidth, but multiple peers jointly provide sufficient bandwidth. By considering these factors, the hybrid architecture shown in Figure 12.2 has been developed [11].

The CDN server (e.g., replica server in Figure 12.1) plays different roles of the media streaming server and the P2P index server. Before the CDN-to-P2P handoff, a media request will be served by either the CDN server or a set of supplying peers (whose indexes will be supplied by the CDN server). In this model, the CDN server performs the tasks of a seed and a tracker of a BitTorrent model [14]. Note that the P2P network has been developed by participating clients only. Different from this model, both the core CDN servers and the clients might build separate P2P networks [11].

Although this hybrid has been developed for distributing multimedia content, it also applies to other types of contents.

12.2.3 Cloud-Based CDN

Deploying a number of replica servers around many places is not cost effective for a single distributor. Compared to that cost, using cloud infrastructures is relatively cheap. By leveraging the elasticity and lower price of cloud infrastructure, content providers can lower their operation costs for similar resources. Note that the CDN is still mainly used for content delivery; therefore, we can identify such a model as a cloud infrastructure-assisted CDN. Netflix [7], MetaCDN [15], and ActiveCDN [16] fall in this category.

The video streaming architecture of Netflix is shown in Figure 12.3. The Netflix architecture is custom-tailored to its large-scale streaming service; it uses simultaneously its own datacenter, different service offerings from Amazon's cloud (e.g., EC2, S3), and three leading CDNs (i.e., Akamai, LimeLight, and Level-3). Observe here that each infrastructure element targets specific performance factors for the specific needs of some function of the service. Netflix's datacenter is responsible for the registration of new

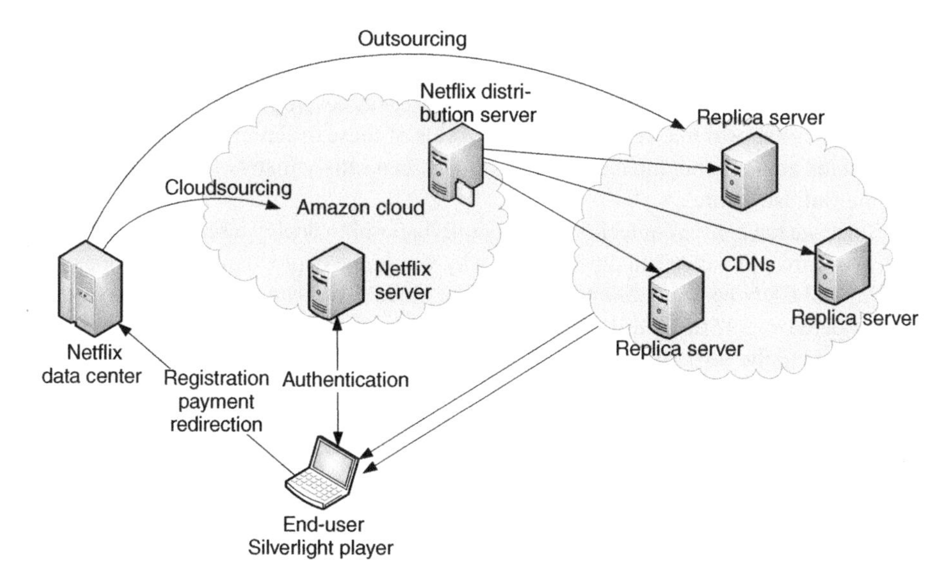

Figure 12.3 Video streaming platform of Netflix [7].

users, payment management, and redirection of users to a Netflix server located inside Amazon's cloud. This latter server performs content origination, log recording/analysis, DRM, CDN routing, user sign-in, and so on. The CDNs deploy a number of replica servers that host the same video content with different quality levels (bitrates) and formats. Depending on the available bandwidth at the client's end, the CDN servers may decide to send the video content with a lower/higher bitrates, typically using the recent dynamic adaptive streaming over HTTP (DASH) protocol [17].

ActiveCDN [16], although primarily based on traditional CDNs, allows on-demand, pop-up content store nodes that are activated on an as-needed basis in the Internet without having to be predeployed. These special purpose nodes can dynamically cache video contents, and consequently, video contents might be provided by these content-storing nodes instead of the CDN replica server. The result is the lowering of streaming traffic at the CDN servers.

MetaCDN [15] proposes an alternate approach by leveraging the existing, less expensive storage cloud infrastructure. It provides an integrated overlay network that removes the complexity of dealing with multiple storage providers. MetaCDN finds out the best possible match from many storage cloud providers based on their quality of service, coverage, and budget preferences.

12.3 COMPARISON OF EXISTING CONTENT DELIVERY PLATFORMS

The different forms of Internet-based infrastructures we have presented so far, cloud, CDN, or P2P, originally emerged to satisfy different objectives in a cost-effective way:

the cloud offers mostly central processing unit (CPU) and storage scalability, the CDN provides bandwidth scalability (as offered to the general public) while P2P offers storage scalability, more specifically for content search. Recent developments (e.g., the Netflix example) have shown that services can be hybrids of these different infrastructures, but also that the cloud, being the most flexible model, can satisfy many of the needs met by the other infrastructures.

Since we have different technologies at our disposal to deploy a service, how do we choose one over the other? In other words, why would the deployment of a service over a traditional CDN be preferable to one over the cloud? And can all cloud deployments be considered equal? In the following, we compare the three content delivery techniques with respect to the resources they need and the performance factors they exhibit.

12.3.1 Resource-Based Comparison

Historically, in the grid computing model, a collection of hosts would register a fraction of their computing resources (i.e., CPU time, memory, bandwidth, and disk space) with resource brokers [18] and were dispatched computations in accordance to their availability. Since cloud computing has evolved from earlier work on parallel and grid computing, we consider that all Internet-based service infrastructures can be characterized in terms of their demands on those four key resources. These resources are either of a static nature, such as the amount of storage used, or dynamic, but typically bounded, which includes processing (CPU), memory, and input/output (I/O) bandwidth. For this last parameter, we can also differentiate between storage I/O and network I/O, depending on the nature of the application.

To achieve a specific performance level would require guaranteeing adequate resources in one or several of these parameters. Furthermore, the operation of these resources presents another cost element, as they are often used sporadically: that is, servers are overdimensioned or subject only occasionally to peak demand, there is an objective to mutualize them to reduce the investment required to provide them. This can be done in two different ways: by acquiring them and sharing them (renting them out), which could be a "corporate" cloud, or by renting them from some other party, that is, a cloud provider. In either case, there is some degree of management overhead and related costs.

Table 12.1 summarizes and compares the three content delivery techniques with respect to three resources: computing, storage, and bandwidth. Note that computing is included for the sake of completeness, although it was originally not a prerequisite of the CDN—beyond running a (cache) Web server. Still, the availability of computing resources enables new functions to be executed, for example, in the form of J2EE servlets.

From this table, we observe that the main difference between the three models consists of a simple transfer of storage and bandwidth resources, while computing, not exactly an essential component of CDNs, could potentially be a factor of difference. Something that could be added with respect to the P2P model is that close availability of a media resource is not guaranteed. P2P traditionally would not make guarantees on the speed of delivery of content nor on widespread availability. To obtain performance

TABLE 12.1 Resources Needed for Existing Content Delivery Methods

Resources	CDN (Akamai)	CDN + P2P (P2P Networks by Clients)	CDN + Cloud (Netflix)
Computing	Not a design goal: availability depends on computing resources provided by the redirected replica server.	By using a fraction of computing from all the peers, a large computing facility might be built, in grid manner.	Computing resources are available by leveraging the cloud computing infrastructure.
Storage	A number of replica servers store the same copy of content. The amount of storage used might be reduced by reducing number of replicas, which may increase the response time for some users.	The CDN server works as a seed and, thus, a large amount of storage is needed for that specific purpose, while content may be spread among P2P participants. Since not so many CDN servers are deployed, unlike the pure CDN model, the size of the total storage is reduced.	In Netflix model, more storage is used while video content is stored in different resolutions. Not so big storage is required for the Netflix server since it does not store any video content.
Bandwidth	The total bandwidth is the aggregation of the bandwidths provided by all replica servers.	The CDN server needs much lower bandwidth, as bandwidth to users consists mostly of the aggregation of the offerings of the peers.	Total bandwidth is the aggregation of the bandwidths provided by all the replica servers plus the (reference) Netflix server.

guarantees, there would be a need to manage the availability of the content, for example, by guaranteeing a balanced distribution of all content across P2P nodes, while the CDN can be restricted to a pure cache behavior.

12.3.2 Performance Factors

A number of performance factors are needed to compare alternate proposals for service deployment and their efficient use of resources. We consider here not only resource utilization but also elements of reaction time: latency, response time, and scheduling time. We also consider the cost of the availability of the resource. We only look at these factors in a general sense, and not specifically as economic and performance models for the cloud are beyond the scope of this article. At this point, suffice to say that, for some applications, some factors may be more critical than others, and this is certainly true of multimedia applications, more demanding in terms of compute power and latency.

Table 12.2 compares three content delivery techniques on the basis of different performance factors.

TABLE 12.2 Comparison of Performance of Existing Content Delivery Methods

Performance Factors	CDN (Akamai)	CDN + P2P (P2P Networks by Clients)	CDN + Cloud (Netflix)
Cost	Services offered by the commercial CDN providers are pricy.	Lower than pure CDN. A lower number of replica servers is required and resources are provided free of charge by peers.	Services offered by the commercial CDN providers are pricy. The addition of cloud computing incurs some additional cost.
Resource utilization	High, only when a good number of users access content from many replica servers.	Resources are spread across many nodes and used more lightly so as not to inconvenience peers.	Low, especially storage and bandwidth are not fully used. Multiple CDNs deploy the same content.
Latency	Low, when there is a replica server with the content looked for in a close proximity.	It varies, depending on the number of peers offering their bandwidth and also on the location of those peers.	Low, when there is a replica server with the content looked for in a close proximity.
Response time	Low, when latency is low and also the replica server is not overwhelmed.	Low to high, when few peers are present or the seed is overwhelmed.	Low, when latency is low and also the replica server is not overwhelmed.
Scheduling	CDN acts like a cache, with some replacement strategy based on available storage.	P2P content is not so easily controllable and more requests may be sent to replica servers.	Acts like a CDN.

Clearly, response time to a request and low latency of delivery are prime considerations in the deployment of media services, and we can expect the different alternatives to perform well in that respect. Using the cloud as an alternative to CDN depends on the large-scale availability of resources and on a reasonable cost, which did not use to be the rule when cloud resources were limited. P2P, while being an interesting alternative, has a degree of randomness in the availability of content or resources, which makes its use more challenging.

What has changed in this picture over the recent years is the "democratization" of the cloud. Technology and competence have become sufficiently affordable, and the market is large enough to support clouds of various sizes and locations. ISPs, for example, have seen a benefit to deploy such infrastructures for providing their own services and also an extension of their offerings to their customers. This in turn leads to the emergence of the game-changing edge cloud, which we expose in the next section.

12.4 AN EDGE CLOUD-BASED MODEL

Before presenting our model for a cloud-based deployment of media functions, we must first introduce a distinction between the traditional, remote, cloud, and the edge cloud.

12.4.1 Remote Cloud

Most commercial cloud infrastructures have been established quite remotely from users, in the "deep core" of the Internet. In this picture, the remote (deep) cloud is unique (up to backup), monolithic, providing a large supply of storage and computing, plus bandwidth to access it. It can offer access to its resources in different ways, in raw or semantic form.

The remote cloud model acknowledges that one size does not fit all. Different needs require different platforms, which can, and indeed in some cases are hosted simply under the same roof. To illustrate this point, it is interesting to observe the diversity of offerings of Amazon [19] in this aspect: different forms of high performance computing (specialized processors, parallel computing), different forms of storage access (database, backups), and raw machines (generic servers based on the most popular operating system (OS)).

12.4.2 Computing at the Edge

Akamai, the largest CDN provider, implements different cloud optimization services including path, communications, and application optimizations that take the CDN beyond a simple role of content caching to also plan an accelerator for cloud computing applications (i.e., the applications deployed inside a cloud infrastructure) [20]. The cloud optimization services rely on edge servers that leverage prefetching, implementing a just-in-time caching technology that retrieves dynamic content the same way that edge caching stores the static content. With the aid of cloud optimization services, the application response time is improved up to twice as fast as an ordinary cloud infrastructure alone. The idea of edge server is further extended by EdgeComputing [21], where the application itself or part of an application can be deployed to the edge of the cloud, close to the end-users. Although EdgeComputing distributes computing at the edge, the functionalities supported by are limited when compared to full-scale cloud computing infrastructure, inasmuch as the technologies supported are strongly tied to the Web.

12.4.3 Cloud from Periphery to Edge

The not-so-remote or periphery cloud can be viewed as a growth of the remote cloud model as demand becomes such that multiple sites are required. Recall that a site can be limited by the aggregated bandwidth it can provide, energy needs, and the related heat dissipation. It is also a massive single point of failure. At some point, it can be economically viable to multiply the cloud sites with similar offerings based on the concentration

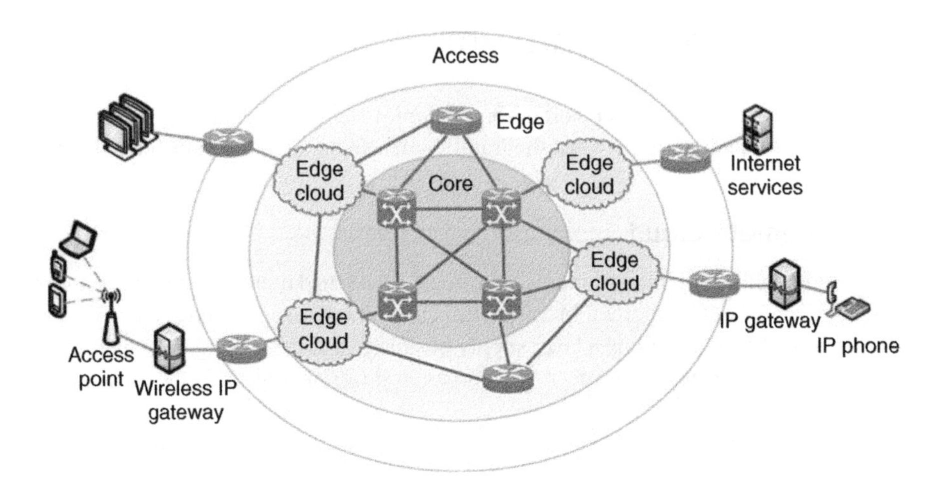

Figure 12.4 Architecture of edge cloud [22].

of customer demand and the opportunity to create sites. One example of bandwidth constraint is a video streaming such as YouTube, which has a much higher demand than normal hosted services, which tend to be more transactional in nature (Web, database, etc.) or generally have more elasticity in terms of user expectation. This leads to a situation with several sites in the periphery of the Internet, where, by periphery, we simply imply that the distance from user to a cloud is more uniform, and smaller on the average than if it were in a single location. Globally, it is also necessary because of growing demand on bandwidth-constrained content. Ultimately, we have cloud sites close to the user, within their ISP's domain, which we call the edge cloud.

Previously, we have introduced the edge cloud model [17,22] as a collection of many small clouds: in addition to mostly application-centric remote cloud infrastructures, core-deep locations, multiple smaller, generic clouds at the edge of the Internet are implemented in partnership with the edge ISPs [23]. The architecture of the edge cloud is shown in Figure 12.4. The edge cloud adds many benefits to the existing cloud computing model including support for desktop virtualization, lower latency for the user, improved data transfer rates and traffic engineering, breaking the service vendor lock-in, fast access to local content, and enhanced security and billing.

12.4.4 Benefits of Introducing the Edge Cloud

The edge cloud follows a similar business model as the remote or periphery clouds. In this case, the initial drive for an ISP is to create a cloud infrastructure for services offered to its customers, with added benefits. Typically, performance factors such as a lowered bandwidth demand on the Internet link and improved response times for customers are the deal makers in such a choice. We also have a normal trend where companies like to deal with providers of a smaller (business) size or are simply under the same legal jurisdiction.

Furthermore, ISPs ever more often require a cloud-like infrastructure for the deployment of their own services to the general public, such as IPTV or other media services, or also services to corporate customers, such as large-scale storage (e.g., backups). Since the ISP is essentially the network connectivity (i.e., bandwidth) provider, it is also in its interest to offer such services to reduce the traffic exchanged with peer operators. But other opportunities, more global, emerge as well.

First, the availability of this cloud infrastructure creates an opportunity for providers such as CDN operators to partner with ISPs to deploy their infrastructure over the ISP's cloud, rather than through dedicated hardware as is typically done now. This directly follows the CDN offers of remote cloud operators, but in this case it preserves the bandwidth scaling capabilities of traditional CDNs. Also, availability of CPU allows the expansion of CDN offerings to more dynamic forms of content creation and processing. Further, such content manipulation can be performed in interaction with the user.

12.4.5 Content Processing

Content processing goes beyond traditional content delivery. Content processing is a set of content-related media services where the services will be built on cloud computing infrastructure in a SaaS manner [3,4]. We name such services content-processing-as-a-service (CPaaS). Figure 12.5 shows the edge cloud-based content processing service model. In this model, important but relatively lesser bandwidth-demanding services (e.g., clients account creation, authentication, billing) will be deployed in the content provider premises and, therefore, inside the core of the Internet (i.e., in the remote cloud). CPaaS and the actual content will be deployed inside the edge cloud, using the storage of these infrastructures.

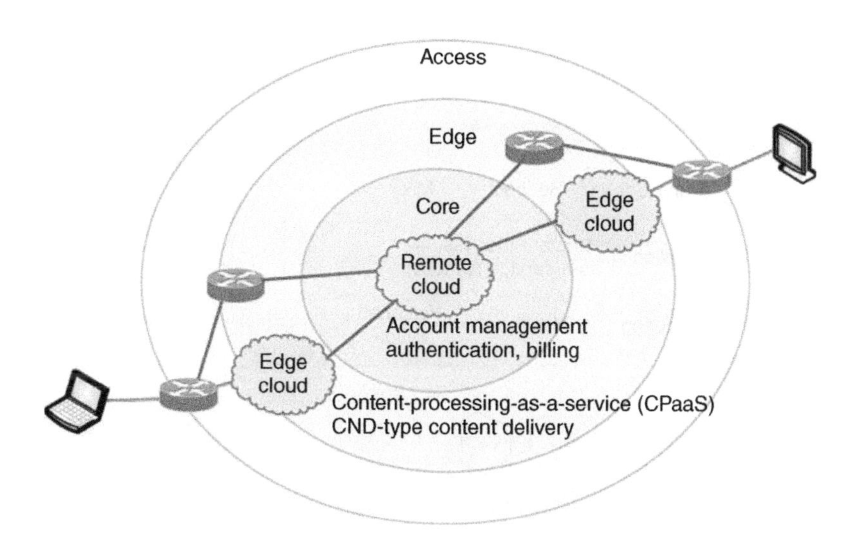

Figure 12.5 Edge cloud-based content processing services.

This however is not the full story. Rather than looking at the services offered as complete in their own right, they can also be seen as building blocks of a larger application. For example, recall that Netflix uses the services of a CDN provider (several in fact) to supply content and hosting services to host its AAA servers, with a custom glue between the two. The CDN provides bandwidth scalability while support services can be hosted in a place offering CPU scalability and security, for example, against denial of service (DoS) attacks. But essentially these different functions can be deployed over a cloud infrastructure.

12.4.6 Detailed Architecture

The cloud-based content processing architecture is shown in Figure 12.6, which explains how different services could be built on top of the infrastructure. Note that not all services shown in this figure would be deployed in the edge cloud, but some account management, redirection, and billing services might rather be deployed in the remote cloud.

The presented model supports a wide range of service models depending on the content provider's needs and the available cloud infrastructure. Some of the open issues and influencing factors that should be addressed in the future are summarized here:

Replica/Cache. Following the CDN model, a number of replica servers can be deployed inside the edge cloud. For noninteractive services, such as VoD, a hierarchy of replica servers, consisting of a number of regional servers, could be deployed. For example, Amazon's datacenter located in the East Coast can serve as the regional replica server for the potential edge clouds located in the

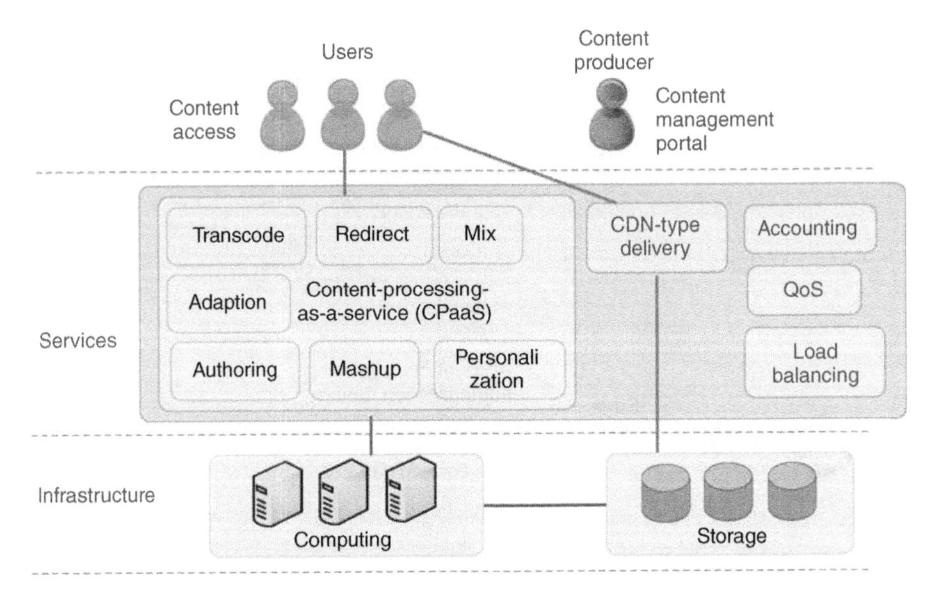

Figure 12.6 Content processing inside the edge cloud.

East Coast of United States and Canada. To support replication, an originating server should be deployed in the remote cloud, possibly maintained by the content provider.

Service and Content Replication. As there is a clear segregation between content and services, it is possible to deploy a single or few copies of the content while deploying a number of replicas of the same service in many edge clouds. Then, the service will access the content from the closest repository. In another possible scenario, the same content might be used by different services. For example, either a photo album or a video might be built from a set of image files.

Redirection. In the CDN model, an end-user usually accesses a website through its publicly known uniform resource locator (URL) and gets redirected to the nearest replica server. In the presented model, the user will be redirected to the nearest edge cloud and the content will be received from there rather than from a remote cloud.

Local Demand-Only Content. Many contents have a geographically bounded demand for local users, for example, a local TV radio/station broadcast to a locality. In such cases, content replication is not required and the edge cloud provides an excellent service model with the minimum response time.

Content Processing. By leveraging the cloud infrastructure, numerous services may be developed that use the content from local cloud storages. Specialized services, especially localized services, in the form of downloadable "apps" might be developed for smartphones and handheld tablets.

Peering of Edge Cloud. The edge clouds may build overlays to replicate content/service without communicating with the remote/core cloud, thus facilitating the sharing of content and supporting resource overflow (elasticity).

Interactive Content. Since the edge cloud has computing resources, it offers the possibility to enrich content with user interaction or even with simply user-specific customization: application of filters, specific menus, and mixing of different contents are examples of possibilities arising from this context.

12.5 RESULTS AND INSIGHTS

Deploying the CDN on the cloud opens opportunities for many added values such as the following:

Unbounded Resources. Elasticity is one of the elegant features of cloud computing. Using this elasticity, access to compute power and greater storage is possible, which is no longer restricted to a simple "cache." As demand for some services grow, it is easy to deploy additional resources, closer to the origin of the demand.

Interactive Content Delivery. It is possible to deliver interactive content (separately from browser control) by processing content inside the CPaaS cloud infrastructure.

Enhanced Security and Billing. An end-user is always authenticated by the ISP to grant access to the network. This authentication can be further extended to access restricted content. User-specific control of access rights or Digital Rights Management (DRM) would be easy to establish. Fine-grained billing for the usage of restricted content also becomes possible.

Local Content Access. Users have strong interests in local content due to cultural and language influence. Hence, a content provider need not have to deal with a cloud/CDN provider with a global presence to replicate its content around the world. Instead of that the regional edge cloud providers provides superior local network coverage and delivers flawless high quality media content throughout.

Simple User Device. Since a lot of computing is offloaded to a CPaaS infrastructure, the resource constraints on the user device are reduced. It is also possible to extend the lifespan of the user device: there is no need to upgrade if the service's implementation is changed. Moreover, legacy applications could still be used in parallel for communications.

The (traditional) CDN model has limits: it is restricted to a cache model. It requires dedicated hardware. It has to support a simple business model, to avoid dispersion of resources. This is probably why there are different CDN providers on the market, more or less specialized in some segments. On the other hand, the CDN model makes the work of the content provider easier—there is only one supplier to deal with, rather than many ISPs. We must acknowledge that the edge cloud-based model may introduce the additional complexity of interacting with multiple providers and ensuing contract negotiations. Still, we must consider that a single provider can cover a whole country or large geographic area, satisfying a single stop solution. We must also take into account the way demand for content would grow, giving time to the content provider to adapt its means of deployment.

The management of this scalable, distributed resource will not be straightforward, and will probably be beyond the care of the content provider. We can still imagine that a CDN provider would take care of providing such a service. It remains a possibility, but not a necessity. This edge cloud-supported CDN provider will be able to play at different levels of cloud availability, depending on the markets—deep, remote, or edge. This is much more flexible in terms of deployment than what has been considered so far. Also, this model benefits from elasticity as excess capacity at one level can be used to compensate excessive demand elsewhere (or distributed caching of less frequently access content).

12.6 FUTURE RESEARCH DIRECTIONS

A proliferation of clouds at different tiers leads to the question of their optimal use. From this premise, we envision three clear avenues for research in the future, which we characterize as *scalability*, *complementarity*, and *infrastructure*.

Scalability means to deploy services where (in the cloud) they will be the most efficiently accessible. It may well be that deep deployments remain the most rational repository for services, but tools are required to analyze alternatives and compare their costs with their respective performance. At the same time, using multiple edge clouds or multitier clouds opens the path to new forms of elasticity, within and between clouds.

Complementarity means affecting different functions on clouds depending on their proximity to—or remoteness from users. Content and computations should be deployed closer to users while support could be centralized, and avoid replication. Essentially, we need to create overlay architectures for services, not unlike Netflix's, to distribute their different support elements—for example, billing, accounting—across different clouds in an efficient and robust way, to minimize unnecessary duplication and coordination efforts. We already see early signs of emerging service-building blocks in support for generic authentication in Amazon. The challenge in the future will be to standardize such building blocks to ease the creation of multicloud, service-tailored overlays.

Infrastructure means the coordination and communication mechanisms required to facilitate the construction of such overlays in a secure and robust way between clouds. Rather than the building blocks we have just mentioned, we mean here the primitives required to facilitate communications between clouds, typically extensions of private networks or public key infrastructure (PKI) services.

12.7 CONCLUSION

The existing, cache-redirection-based CDN model can be reengineered to meet the demands emerging from user-centric, highly interactive, and rich content-based services. The cloud computing infrastructure provides low cost content delivery services along with an open, versatile, and multidimensional platform for developing new content processing services and, consequently, the existing CDNs can move to a cloud computing infrastructure. In this chapter, we have depicted the benefits of this model and illustrated a variant of services that possibly evolve in the future from the emergence of an edge cloud.

REFERENCES

1. Peng G. CDN: Content distribution network. Stony Brook University, Technical Report TR-125; 2008.
2. Pallis G, Vakali A. Insight and perspectives for content delivery networks. Commun ACM 2006;49(1):101–106.
3. Zhu W, Luo C, Wang J, Li S. Multimedia cloud computing. IEEE Signal Proc Mag 2011:59–69.
4. Ranjan R, Mitra K, Georgakopoulos D. MediaWise cloud content orchestrator. J Internet Serv Appl 2013;4(2).
5. Michael A, Armando F, Rean G, Anthony DJ, Randy HK, Andrew K, Gunho L, David AP, Ariel R, Ion S, and Matei Z. Above the clouds: A Berkeley view of cloud computing. University of California at Berkley, USA, Technical Report No. UCB/EECS-2009-28; 2009.
6. Leavitt N. Is cloud computing really ready for prime time? Computer 2009;42:15–20.

7. Adhikari VK et al. Unreeling Netflix: Understanding and improving multi-CDN movie delivery. INFOCOM; 2012. p 1620–1628.
8. Cisco. 2013. Cisco visual networking index forecast. Available at http://www.cisco.com/go/vni/. Accessed 17 July 2013.
9. John D, Bruce M, Jay P, Harald P, Ramesh S, and Bill W. Globally distributed content delivery. IEEE Internet Comput 2002:50–58.
10. Pathan M, Buyya R. A taxonomy and survey of content delivery networks. University of Melbourne; Technical Report; 2007.
11. Xu D, Kulkarni S, Rosenberg C, Chai H. Analysis of a CDN-P2P hybrid architecture for cost-effective streaming. Multimedia Syst 2006;11(4):383–399.
12. Yin H, Xuening L, Zhan T, Sekar V, Qiu F, Lin C, Zhang H, Li B. Design and deployment of a hybrid CDN-P2P system for live video streaming: Experiences with LiveSky, ACM Multimedia; 2009. p 25–34.
13. Huang C, Wang A, Li J, Ross KW. Understanding hybrid CDN-P2P: Why limelight needs its own red swoosh, NOSSDAV; 2008. p 75–80.
14. Pouwelse J, Garbacki P, Epema D, Sips H. The bittorrent P2P file-sharing system: Measurements and analysis, IPTPS; 2005.
15. Broberg J, Buyya R, Tari Z. MetaCDN: Harnessing 'storage clouds' for high performance content delivery. J Netw Comput Appl 2009;32:1012–1022.
16. Srinivasan S, Lee JW, Batni D, Schulzrinne H. ActiveCDN: Cloud computing meets content delivery networks. Columbia University; Computer Science Technical Reports; 2011.
17. Islam S, Grégoire J-C. Giving users an edge: A flexible cloud model and its application for multimedia. Future Gener Comp Syst 2012;28(6):823–832.
18. Alexandrov AD, Ibel M, Schauser KE, Scheiman CJ. SuperWeb: Towards a global web-based parallel computing infrastructure. The 11th International Parallel Processing Symposium; 1997; 100–106.
19. Amazon Elastic Compute Cloud (Amazon EC2). 2013. Available at http://aws.amazon.com/ec2/. Accessed 17 July 2013.
20. Leighton T. Akamai and cloud computing: A perspective from the edge of the cloud, White paper; 2010.
21. Davis A, Parikh J, Weihl WE. EdgeComputing: Extending enterprise applications to the edge of the Internet. 13th international World Wide Web (WWW); 2004. p 180–187.
22. Islam S, Grégoire J-C. Network edge intelligence for the emerging next-generation Internet. Future Internet 2010;2(4):603–623.
23. Islam S, Grégoire J-Ch. Active ISP involvement in content-centric future Internet. The 4th IFIP International Conference on New Technologies, Mobility and Security (NTMS); Paris, France; 2011.

13

DYNAMIC RECONFIGURATION FOR ADAPTIVE STREAMING

Norihiko Yoshida

Information Technology Center, Saitama University, Saitama, Japan

13.1 INTRODUCTION

When a website catches the attention of a large number of people, it gets an unexpected and overwhelming surge in traffic, usually causing network saturation and server malfunction, and consequently making the site temporarily unreachable. This is the *flash crowd* phenomenon on the Internet.

For a content delivery network (CDN) to be scalable in such dynamic situations, the CDN should be dynamically reconfigurable. The network would adaptively change its topology and volume according to the observed traffic load and pattern. For delivery and distribution of static contents such as texts and images, there have been a fair amount of studies and systems, some of which are already employed in industry. However, for delivery and distribution of streaming contents such as voices, music, and videos, further studies are still required and anticipated. The difficulty of a streaming CDN comes mostly from the continuity of the contents and the QoS (quality of service) requirement. A stream is delivered spending a certain period of time, meanwhile a load to the CDN would change. The CDN must adapt dynamically during the period by changing its network in the same way as static content delivery and distribution, and also by adjusting the QoS of the stream dynamically. One of the most promising solution to the issues

Advanced Content Delivery, Streaming, and Cloud Services, First Edition.
Edited by Mukaddim Pathan, Ramesh K. Sitaraman, and Dom Robinson.

is to divide a stream into several slices or segments. The CDN changes and adjusts per segment. However, in this case, the problem is how to manage the integrity of the whole set of segments as a single stream in the dynamic situation.

This chapter presents a dynamic streaming CDN. It is an extension to a dynamic CDN for static contents and incorporates with solutions to the issues mentioned earlier. This scheme copes with situations that a large and varying amount of clients request a sequential stream from a streaming server. The presentation partly uses some examples, FCAN (Flash Crowds Alleviation Network) as a CDN and HyperText Transfer Protocol (HTTP) live streaming for stream segmentation; however, it is supposed to be generally applicable.

Dynamic load distribution or load balancing is one of the most important issues in both CDNs for static and stream contents. To handle increase and decrease in surrogates, most systems use virtual machine replication and migration. This scheme incurs much overhead on transferring the image of a virtual machine. There is another promising technique for dynamic replication and migration using mobile threads which leads to more lightweight, low overhead. This chapter shows an overview of this technique as well.

This chapter is organized as follows. After summarizing backgrounds and some related works including an overview of FCAN in Section 13.2, Section 13.3 presents dynamic server deployment based on mobile threads. Then, Section 13.4 describes adaptive dynamic CDN for streaming. Section 13.5 gives some future research directions, and Section 13.6 contains concluding remarks.

13.2 BACKGROUND AND RELATED WORK

13.2.1 Flash Crowds and Adaptive CDNs

The term *flash crowd* was coined in 1973 by a science fiction writer Larry Niven in his short novel *Flash Crowd* [1]. In the novel, cheap and easy teleportation enabled tens of thousands of people worldwide to flock to the scene of anything interesting almost instantly, incurring disorder and confusion. The term was then applied to similar phenomena on the Internet in the late 1990s. When a website catches the attention of a large number of people, it gets an unexpected and overwhelming surge in traffic, usually causing network saturation and server malfunction, and consequently making the site temporarily unreachable. This is the *flash crowd* phenomenon on the Internet.

Researches to alleviate flash crowds are divided into three categories: server-layer, intermediate-layer, and client-layer solutions, according to typical architectures of networks.

Server-Layer Solution. Systems in this category form delivery networks on server side similar to that of conventional CDNs. This is an expensive approach. The systems are inefficient on and difficult to deal with short-term Internet congestion. For example, CDN with Dynamic Delegation [2] and DotSlash [3] are in this category.

Intermediate-Layer Solution. Systems in this category let proxy servers work together for load balancing. Proxies in the system are mostly volunteers, and often less powerful than servers in the server-layer solution; however, caching techniques help to alleviate server load during flash crowds by filtering out repeated requests from groups of clients that share a proxy cache. Multilevel caching [4], BackSlash [5], and CoralCDN [6] are included in this category.

Client-Layer Solution. Systems in this category make clients help each other in sharing contents so as to distribute the load burden from a centralized server. Clients form peer-to-peer (P2P) overlay networks and use search mechanisms to locate resources. This is a costless approach. However, it is difficult to manage and control the clients, and to make the system reliable, secure, and transparent to users. CoopNet [7] and PROOFS [8] are included in this category.

13.2.2 FCAN

FCAN is an adaptive CDN that takes the form of client/server (C/S) or CDN depending on the amount of accesses from clients. Specifically, in the C/S mode, a server provides contents to clients as in a traditional C/S. In the CDN mode, when the server detects a coming of a flash crowd, volunteer cache proxies in the Internet construct a temporary P2P network and provide the content on behalf of the server. These volunteer proxies are recruited in advance out of providers and organizations. In case servers in such providers and organizations suffer from flash crowds, they will be helped by other volunteer proxies. FCAN is built on this mutually aiding policy. Figure 13.1 shows an overview of FCAN. FCAN is an intermediate-layer solution, which employs an Internet infrastructure of cache proxies to organize a temporal P2P-based proxy cloud for load balancing. However, FCAN has some extensions with some dynamic and adaptive features. Our FCAN studies achieved very promising results regarding static content delivery on the real Internet [9–11].

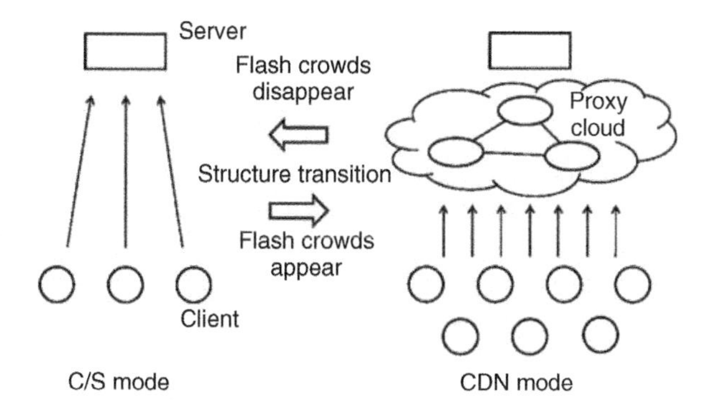

Figure 13.1 FCAN overview.

13.3 DYNAMIC SERVER DEPLOYMENT

To enable dynamic reconfiguration of networks, we must have some mechanism for dynamic server deployment. Without it, we must distribute the identical images of the server to all the hosts in the network beforehand even if the image would be actually not used after all. It would possibly cause wastes of computing and networking resources. There are two types of techniques for server deployment: virtual machine migration and mobile thread migration.

13.3.1 Virtual Machine Migration

One approach is migration and duplication of server virtual machines between hosts. Below are some examples.

vMatrix [12] focuses on distribution and replication of dynamic contents. vMatrix is a network of real machines running virtual machine monitor software, allowing server virtual machines to be moved among the physical machines. vMatrix can improve the response time perceived by end-users, overall availability and uptime, and on-demand replication to absorb flash crowd requests by optimizing the virtual server placement. vMatrix can also reduce the overall bandwidth consumed in the network in the same manner.

XenoServer [13] focuses on service deployment. A user first selects the number of XenoServers on which services are to be deployed, then configures the XenoServers so that the virtual machines accommodate the service components, and launches the service components.

In server proliferation [14,15], services (e.g., Web server, streaming server) are executed on virtual machines. When a new virtual machine is required, a disk image of the virtual machine is delivered from the distribution server (DS) to one of the execution servers (ESs), and the delivered virtual machine runs on the ES. In this architecture, DS may be multiplied so as to avoid load concentration.

13.3.2 Mobile Thread Migration

The other approach is migration and duplication of mobile threads between server hosts. It is more difficult to deploy, but would be much more efficient than the above.

A mobile thread, or sometimes called mobile agent, is a kind of thread that can move among computers over a network while keeping its program code and execution state [16]. Its idea came to help development and operation of large-scale network applications. Conventional network applications are designed based on communication. The idea of mobile threads separates computation from a physical platform and unifies computation and communication instead. Communication is now enclosed within computation, and this encapsulation is also expected to reduce network traffic practically. There are some mobile thread systems such as Aglets [17], MOBA [18], AgentSpace [19], and JavaGo [20].

Thread migration applied to dynamic network reconfiguration has been found mostly in grid computing based on distributed shared memory [21–23] and mobile threads [24], however not found in CDNs thus far.

13.4 FROM CONTENT DELIVERY TO STREAMING

13.4.1 HTTP Live Streaming

Conventional standard streaming protocols such as progressive download and real-time streaming do not allow switching of stream source servers on the client side dynamically. Therefore, massive accesses from clients concentrate on a particular site on the Internet. Accordingly, the server and its surrounding network choke up, and a flash crowd occurs.

In order to resolve this problem of conventional protocols, Apple has introduced a new protocol for video streaming, HTTP live streaming (also known as "HLS") [25]. It is defined in a standard draft for the Internet Engineering Task Force (IETF) [26]. Figure 13.2 shows an overview of this protocol. It would be beneficial for the readers to refer to Chapter 2 in this book as well for the details of HLS.

The server starts providing a video stream with the following procedure: (i) encodes an audio/video inputs; (ii) divides the encoded stream into a set of media segments and makes an index which refers them; (iii) delivers them to clients using HTTP on the Internet.

This protocol enables a client to switch the source server dynamically as opposed to the conventional streaming protocols. The delivery system archives load distribution easily with additional servers.

As another key feature, HLS supports "adaptive bitrate." The server provides alternative streams with different quality levels of bandwidths, so as to enable a client to optimize the video quality according to the network situation, as the load on the network and central processing unit (CPU), on both the server side and the client side, fluctuates on a frequent basis.

13.4.2 Adaptive Streaming on FCAN

Enhancing the HLS technology over FCAN, we built a dynamic CDN for both live and on-demand streaming [27]. The network changes its structure and configuration adapting to request loads. Its design, implementation, and experiment results are shown.

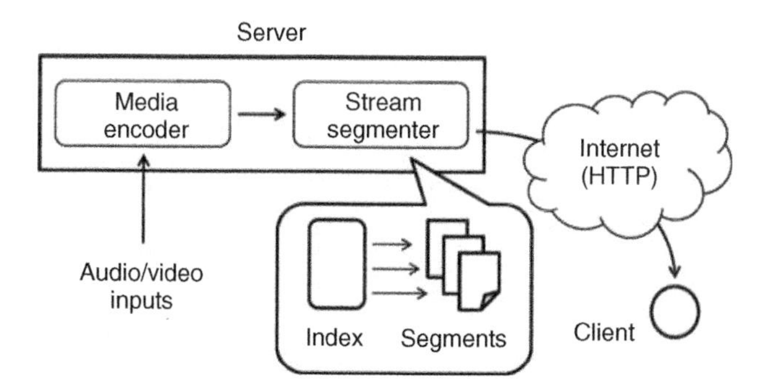

Figure 13.2 HTTP live streaming overview.

13.4.3 Structure Transition

The server and the cache proxies in the proxy network always monitor the amount of accesses they receive from clients and evaluate the load on the network. The system switches to the CDN mode if all nodes' loads are higher than a certain threshold, and switches back to the C/S mode if lower. Each cache proxy sends its own load information to the server periodically, and the server determines whether to perform structure transition.

In peaceful times, the conventional C/S architecture satisfies most of the client requests. A server and cache proxies, both of which comprise FCAN, do little more than what normal ones do. When a flash crowd comes, the server detects the increase in traffic load. It triggers a subset of the proxies to form an overlay, through which all requests are conducted. All subsequent client requests are routed to this overlay.

The server-side procedure is outlined as follows: (i) selects a subset of proxies to form a CDN-like overlay of surrogates and builds a distribution tree; (ii) pushes the index file and stream segments to the node of the distribution tree, so as to meet the real-time constraint of the video streaming; (iii) prepares to collect and evaluate statistics for the object from the involved proxies, so as to determine dynamic reorganization and release of the overlay.

The proxy-side procedure is outlined as follows: (i) changes its mode from a proxy to a surrogate (or, in the strict sense, a mixed mode of a forward proxy and a surrogate); (ii) stores flash-crowd objects (except the index file) permanently, which should not expire until the flash crowd is over; (iii) begins monitoring the statistics of request rate and load and reporting them to the server periodically.

In the live streaming, the index file is updated periodically; therefore, the server monitors its composition and pushes the segments at an appropriate time. Meanwhile, when the proxy is released by the server, it discards the index so as to keep the consistency among the nodes.

When the member server detects the leaving of the flash crowd, the involved proxies are dismissed one by one with the following procedure: (i) the server notifies the proxy to be dismissed; (ii) the server requests the related proxies to modify the relation of connection; (iii) the proxy changes its mode from a surrogate to a proxy.

The CDN-like overlay transits back to the normal C/S mode when all the proxies are dismissed. They are not all dismissed at once, since the low load may be just temporary, and the system should therefore remain in the anti-flash-crowd mode for a while.

13.4.4 Dynamic Resizing and Quality Restriction

The proxy network is a pure P2P network. Therefore, it is highly fault tolerant and scalable. Contrary to traditional P2P systems, it does not include clients into the network itself in order to assure reliability and security.

FCAN resizes a scale of the proxy network depending on a load fluctuation adaptively in order to avoid troubles such as server down by massive access concentration. Figure 13.3 shows how the system works in the CDN mode.

When the server detects a coming of flash crowds, it forms a temporary proxy network as shown in the left-hand side of Figure 13.3. If the initial network cannot handle

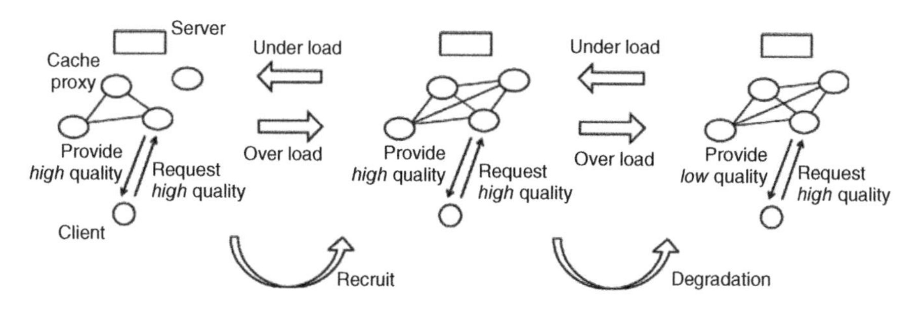

Figure 13.3 Adaptation to load change in the CDN mode.

increasing an amount of accesses, the server recruits a new member proxy one by one as shown in the middle of Figure 13.3.

If the server cannot recruit temporary proxies any more, it degrades the quality of the video stream as shown in the right-hand side of Figure 13.3. For example, in the situation that the system provides video streams of two different qualities, high and low, it delivers the low quality content as substitute for the high quality content under this quality restriction. The network occupancy per client decreases so that the server can alleviate the load of whole delivery network.

If the proxy network can easily handle all the incoming loads, the server may lift the restriction of the stream quality at first. After the derestriction, it releases temporarily recruited proxies one by one until all proxies are dismissed. Finally, the system all turns back to the normal condition.

13.4.5 Results

We conducted some preliminary experiments on a real network with a prototype of the system. Figure 13.4 shows an overview of the experiments. In our experiments, we use some hosts in Saitama University and Kyushu Sangyo University for a server, proxies, a pseudo client, and a client node. We use Apple's stream segmenter (mediastream-segmenter) for the segmenter and QuickTime Player for QuickTime X in the client.

The pseudo client is to trigger the FCAN's functions against flash crowds. It submits requests for randomly chosen segments to the server following the pattern shown in Figure 13.5. In the rest of this section, CP1, CP2, and CP3 are the proxies shown in Figure 13.4.

We made two experiments with two delivery methods, live and on-demand. In the on-demand streaming experiment, the server provides high and low quality contents with adaptive bitrates. We use segment samples of HLS in the Apple Developer's site [28]. The client software, the client has a master index file indicating these two quality contents, and the QuickTime Player requests segments of an adequate quality depending on load fluctuation following the master index. On the other hand, in the live streaming experiment, the server provides contents in a single quality in real time.

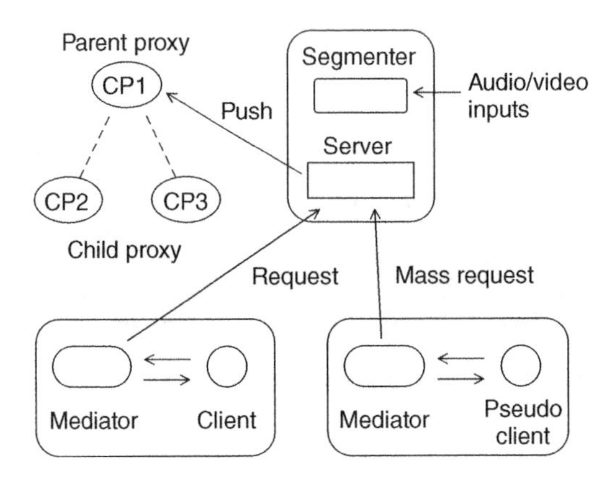

Figure 13.4 Experimental setup.

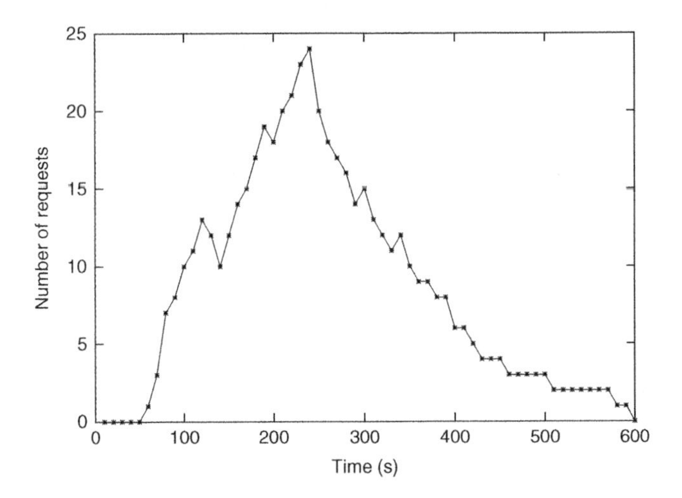

Figure 13.5 Sample request pattern of clients.

The server computes a load value regarding the size of requested segments. In the experiments, thresholds for load detection are defined beforehand based on some experiences. Workloads on the real Internet vary, and automatic and dynamic configuration of the thresholds is difficult. We suppose they may be configured based on the server capacity and the network bandwidth around the server.

Figure 13.6 shows the load transitions of the server with FCAN and without FCAN in the on-demand experiment. The case of the server with FCAN shows the average loads of member nodes in the distribution network. The first peak at the 40th second shows that "buffering" was done when the client started the playback as mentioned earlier. The load on the server exceeded the higher threshold at the 130th second, and then structure

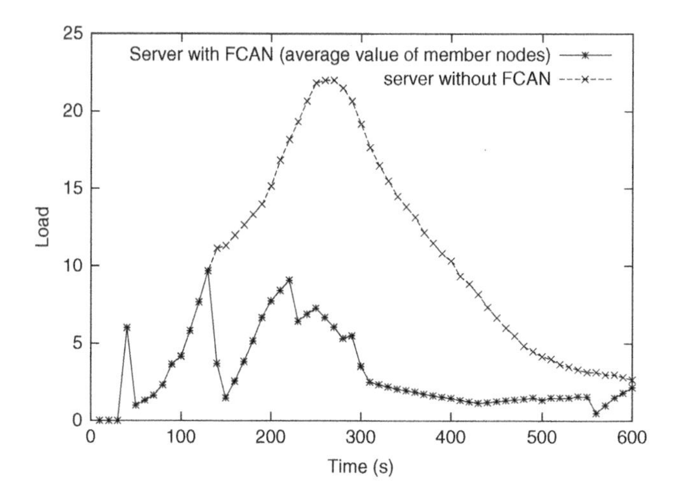

Figure 13.6 Comparison of load transitions in on-demand streaming.

transition to the CDN mode occurred so as to alleviate load concentration. However, the load on the member nodes continued to increase even after the transition, a new member proxy was recruited at the 220th second, and in addition the quality of segments was degraded at the 250th second, and consequently the server withstood the heavy load condition.

On the contrary, in the case of the server without FCAN, we observed that load values were far exceeding values of the server with FCAN consistently. We, therefore, confirmed that FCAN's features against flash crowds were performed as a result of the increase in client requests, and FCAN archived dynamic load balancing. We obtained equivalent results also in the live streaming experiment.

13.5 FUTURE RESEARCH DIRECTIONS

In order to fully deploy dynamic CDN for stream delivery and distribution, this chapter presents (i) dynamic network reorganization, (ii) load distribution and balancing, (iii) stream segmentation, and (iv) QoS assurance.

1. *Dynamic Network Reorganization*. There are two sides in network reorganization: the technical side on how to reorganize, and the management side on whom to reorganize with. Issues related to the former are studied intensively in the area of P2P networks. In large CDNs or global CDNs where voluntary hosts or providers are sometimes involved, centralized management is not effective, and autonomous management of join and leave of surrogates would be preferable. Research results on P2P networks must be helpful for that.

2. *Load Distribution and Balancing*. As stated earlier, the mobile thread technique is promising from the overhead point of view. However, it has not been fully investigated in CDN-related studies compared to the virtual machine migration technique. It may be because the mobile thread technique is rather new and is more difficult to implement especially in heterogeneous environments. The virtual machine technique, VMware [29] and Xen [30] for example, has been well developed to cope with such environments. Large CDNs or global CDNs would be heterogeneous in OS, middleware, and so on, and techniques to integrate them are essential. The mobile thread technique is no exception. Implementation and deployment in heterogeneous environments must be studied for the fruitful future of the mobile thread technique not only in CDN contexts but also in general actually.
3. *Stream Segmentation*. HLS is defined in an IETF draft and now being vigorously discussed. As a possible extension, a segmentation method adaptive to network environments and QoS requirement should be interesting.
4. *QoS Assurance*. QoS in stream delivery and distribution has already been thoroughly investigated in many related research areas. However, QoS assurance assisted by network reorganization is a new challenge. It must have a close correspondence with the research topic, dynamic resilient streaming [31]. The word "resilience" is defined as the persistence of avoiding failures and malfunctioning in the presence of changes. In the case of QoS, the aim is at persistence of acceptable quality in the presence of changes. The technique presented in this chapter is a basic one, and various extensions are possible including multipath streaming, network coding, and underlay awareness.

   There are also some other issues that should be studied for dynamic streaming CDN, such as (5) load estimation and (6) request redirection.
5. *Load Estimation*. A dynamic CDN requires any estimation method for host load and network load to determine how to reorganize its network and how to distribute loads. The most accurate method to estimate throughput and latency would be to measure round trip time of a small dummy message. However, if all the clients applied this, it would cause an equivalent to distributed denial of service (DDoS) attacks to the server. A practical estimation measurement for host load is the number of connections, and a measurement for network load is the number of messages.
6. *Request Redirection*. The system should provide a function for clients or networks to switch the direction of requests to an appropriate surrogate server dynamically. It is because the number and allocation of surrogate servers, as well as the (estimated) load of each surrogate, change dynamically. Some of the practical methods for redirection are using browser cookies, switching DNS records at the client side, and applying anycast infrastructure.

Recently, a technology named "Software Defined Networking" (SDN) [32] has emerged, and is now attracting much attention in the field of Internet routing. The technology introduces a programmable router that can change its routing behavior dynamically according to its programs. Therefore, if we would install a program on an SDN router to change the destination of network packets different from the original destination, the packets could be redirected. Actually, it has been reported that SDN can be used to implement anycast in a more flexible manner than before [33]. SDN-based request redirection for dynamic CDN must be a promising research topic to explore as well.

Actually, there is no best method for estimation and redirection practically, and system designers must choose any appropriate method according to individual requirements and constraints.

13.6 CONCLUSION

This chapter presents a dynamic CDN for streaming. It is an extension to a CDN for static contents and optimizes its organization and size adaptively to a varying amount of requests from clients. The system comprises techniques for dynamic network reorganization and for load distribution and balancing to realize dynamicity, as well as techniques for stream segmentation and reconstruction and for QoS assurance. Live streaming and on-demand streaming require different treatments, which are also illustrated using a system prototype example. Although the example shows a success in adaptive optimization to the varying load, much space still remains to explore and investigate, as discussed in Section 13.5. In particular, HLS is the only established solution to stream segmentation and reconstruction at this moment, and any improvement and sophistication to the technology will bring a wider perspective to streaming CDNs. It is expected that continual advancement of the related technologies will address the future issues and pave the way for smooth deployment in real-world applications.

ACKNOWLEDGMENTS

This chapter is partially based on the author's joint works with Mr. Yuta Miyauchi, Mr. Masaya Miyashita, Prof. Noriko Matsumoto (Saitama University, Japan), Dr. Yuko Kamiya, and Prof. Toshihiko Shimokawa (Kyushu Sangyo University, Japan).

REFERENCES

1. Niven L. Flash Crowd. In: *The Flight of the Horse*. Ballantine Books; 1973. p 99–164.
2. Jung J, Krishnamurthy B, Rabinovich M. Flash crowds and denial of service attacks: Characterization and mplications for CDNs and web sites. Proceedings of 11th International World Wide Web Conference; 2002. p 252–262.
3. Zhao W, Schulzrinne H. DotSlash: A self-configuring and scalable rescue system for handling web hotspots effectively. Proceedings of International Workshop on Web Caching and Content Distribution; 2004. p 1–18.

4. Ari I, Hong B, Miller EL, Brandt SA, Long DE. Managing flash crowds on the Internet. Proceedings of 11th IEEE/ACM International Symposium on Modeling, Analysis, and Simulation of Computer and Telecommunication Systems; 2003. p 246–249.
5. Stading T, Maniatis P, Baker M. Peer-to-peer caching schemes to address flash crowds. Proceedings of 1st International Workshop on Peer-to-Peer Systems; 2002. p 203–213.
6. Freedman MJ, Freudenthal E, Mazieres D. Democratizing content publication with coral. Proceedings of 1st USENIX/ACM Symposium on Networked Systems Design and Implementation; 2004.
7. Padmanabhan VN, Sripanidkulchai K. The case for cooperative networking. Proceedings of 1st International Workshop on Peer-to-Peer Systems; 2002. p 178–190.
8. Stavrou A, Rubenstein D, Sahu S. A lightweight, robust P2P system to handle flash crowds. IEEE J Select Areas Commun 2002;22:6–17.
9. Pan C, Atajanov M, Hossain MB, Shimokawa T, Yoshida N. FCAN: Flash crowds alleviation network using adaptive P2P overlay of cache proxies. IEICE Tr Comm 2006;E89-B(4):1119–1126.
10. Yoshida N. Dynamic CDN against Flash Crowds. In: Buyya R, Pathan A-MK, Vakali A, editors. *Content Delivery Networks*. Springer; 2008. p 277–298.
11. Miyauchi Y, Matsumoto N, Yoshida N, Shimokawa T. Preliminary study on world-wide implementation of adaptive content distribution network. Proceedings of Workshop on Self-Organising, Adaptive, Context-Sensitive Distributive System; 2011. 11 pages.
12. Awadallah AA, Rosenblum M. The vMatrix: A network of virtual machine monitors for dynamic content distribution. Proceedings of 7th International Workshop on Web Content Caching and Distribution; 2002.
13. Kotsovinos E, Moreton T, Pratt I, Ross P, Fraser K, Hand S, Harris T. Global-scale service deployment in the XenoServer platform. Proceedings of 1st Workshop on Real, Large Distributive Systems; 2004.
14. Kamiya Y, Shimokawa T, Tanizaki F, Yoshida N. Scalable contents delivery system with dynamic server deployment. Int J Comp Sci Issues 2010;7(6):81–85.
15. Kamiya Y, Shimokawa T, Tanizaki F, Yoshida N. Dynamic wide area server deployment system with server deployment policies. Int J Comp Sci Netw Sec 2010;10(10):92–96.
16. Chess D, Harrison C, Kershenbaum A, Watson TJ. *Mobile Agents: Are They a Good Idea?, Mobile Object Systems Towards the Programmable Internet, LNCS 1222*. Springer; 1997.
17. Lange DB, Oshima M. *Programming and Deploying Java Mobile Agents with Aglets*. Addison-Wesley; 1998.
18. Shudo K, Muraoka Y. Asynchronous migration of execution context in Java virtual machines. Fut Gen Comput Syst 2001;18(2):225–233.
19. Satoh I. A mobile agent-based framework for active networks. Proceedings of IEEE Conference on Systems, Man, and Cybernetics 1999; 1999; 2. p 71–76.
20. Sekiguchi T, Masuhara H, Yonezawa A. *A Simple Extension of Java Language for Controllable Transparent Migration and Its Portable Implementation, Coordination Models and Languages*. Springer; 1999. p 211–226.
21. Thitikamol K, Keleher P. Thread migration and load balancing in non-dedicated environments. Proceedings of 14th International Parallel and Distributed Processing Symposium; 2000. p 583–588.
22. Hai J, Chaudhary V. MigThread: Thread migration in DSM systems. Proceedings of International Conference on Parallel Processing; 2002. p 581–588.

23. Chen P, Chang J, Liang T, Shieh C, Zhuang Y. A multi-layer resource reconfiguration framework for grid computing. Proceedings of 4th International Workshop on Middleware for Grid Computing; 2006. 13 pages.
24. Miyashita M, Haque ME, Matsumoto N, Yoshida N. Dynamic load distribution in grid using mobile threads. Proceedings of IEEE 3rd International Workshop on Internet and Distributive Computing Systems; 2010. p 629–634.
25. Apple Developer. 2011. HTTP live streaming. Available at http://developer.apple.com/resources/http-streaming/. Accessed 18 August 2013.
26. Pantos R. 2011. IETF Internet draft: HTTP live streaming. Available at http://tools.ietf.org/html/draft-pantos-http-live-streaming. Accessed 17 August 2013.
27. Miyauchi Y, Matsumoto N, Yoshida N, Kamiya Y, Shimokawa T. Adaptive content distribution network for live and on-demand streaming. Proceedings of Workshop on Architecture for Self-Organizing Private IT-Spheres; 2012. p 27–37.
28. Apple Developer. 2011. Bip Bop All. Available at http://devimages.apple.com/iphone/samples/bipbopall.html. Accessed 18 August 2013.
29. VMware. 2013. Available at http://www.vmware.com/. Accessed 17 August 2013.
30. Xen Project. 2013. Available at http://www.xenproject.org/. Accessed 17 August 2013.
31. Abboud O, Pussep K, Kovacevic A, Mohr K, Kaune S, Steinmetz R. Enabling resilient P2P video streaming: Survey and analysis. Multimedia Syst 2011 Springer;17(3):177–197.
32. Software Defined Networking. 2013. Available at https://www.opennetworking.org/. Accessed 18 August 2013.
33. Othman OMM, Okamura K. Design and implementation of application based routing using OpenFlow. Proceedings of ACM 5th International Conference on Future Internet Technologies; 2010. p 60–67.

14

MINING DISTRIBUTED DATA STREAMS ON CONTENT DELIVERY NETWORKS

Eugenio Cesario[1], Carlo Mastroianni[1], and Domenico Talia[1,2]

[1]*ICAR-CNR, Rende (CS), Italy*
[2]*DIMES, University of Calabria, Rende (CS), Italy*

14.1 INTRODUCTION

Mining data streams (DSs) is a very important research topic and has recently attracted a lot of attention, because in many cases data is generated by external sources so rapidly that it may become impossible to store it and analyze it offline. Moreover, in some cases streams of data must be analyzed in real time to provide information about trends, outlier values or regularities that must be signaled as soon as possible. Important application fields for stream mining are as diverse as financial applications, network monitoring, security problems, telecommunication networks, Web applications, content delivery networks, sensor networks, analysis of atmospheric data, and so on.

The mining DS process becomes more difficult when streams are distributed, because mining models must be derived not only for the data of a single stream, but for the integration of multiple and heterogeneous DSs. This scenario can occur in all the application domains mentioned before and specifically in a content delivery network. In this context, user requests delivered to a Web system can be forwarded to any of several servers located in different and possibly distant places, in order to serve requests

Advanced Content Delivery, Streaming, and Cloud Services, First Edition.
Edited by Mukaddim Pathan, Ramesh K. Sitaraman, and Dom Robinson.

more efficiently and balance the load among the servers. The analysis of user requests, for example, to discover frequent patterns, must be performed with the inspection of the DSs detected by different servers. The discovery of popular items can suggest the data and the Web pages that are more interesting (i.e., frequently requested) for the users: such data can be prefetched in the Web cache of servers, in order to serve future demands more efficiently. In order to improve the efficiency and scalability of the whole process, the content delivery network can be hosted on and exploit the facilities of a cloud infrastructure.

Two important and recurrent problems regarding the analysis of DSs are the computation of frequent items and frequent itemsets from transactional datasets. The first problem is very popular both for its simplicity and because it is often used as a subroutine for more complex problems. The goal is to find, in a sequence of items, those whose frequency exceeds a specified threshold. When the items are generated in the form of transactions—sets of distinct items—it is also useful to discover frequent sets of items. A k-itemset, that is, a set of k distinct items, is said to be frequent if those items concurrently appear in a specified fraction of transactions. The discovery of frequent itemsets is essential to cope with many data mining problems, such as the computation of association rules, classification models, data clusters, and so on. This task can be severely time consuming, since the number of candidates is combinatorial with their allowed size. The technique usually adopted is to first discover frequent items, and then build candidate itemsets incrementally, exploiting the Apriori property, which states that an i-itemset can be frequent only if all of its subsets are also frequent.

While there are some proposals in the literature to mine frequent itemsets in a single pass, it is recognized that in the general case, in which the generation rate is fast, it is very difficult to solve the problem without allowing multiple passes on the DS. In this chapter, we elaborate a distributed architecture, firstly introduced in Reference 1, for mining DSs generated from multiple and heterogeneous data sources, with specific focus on the case of content delivery networks.

More in detail, the architecture exploits the following main features:

- The architecture combines the parallel and distributed paradigms, the first to keep the pace with the rate of a single DS, using multiple miners, the second to cope with the distributed nature of DSs. Miners are distributed among the domains where DSs are generated, in order to keep computation close to data.
- The computation of frequent items is performed through sketch algorithms. These algorithms maintain a matrix of counters, and each item of the input stream is associated with a set of counters, one for each row of the table, through hash functions. The statistical analysis of counter values allows item frequencies to be estimated with the desired accuracy. Sketch algorithms compute a linear projection of the input: thanks to this property, sketches of data can be computed separately for different stream sources, and can then be integrated to produce the overall sketch.
- The approach is hybrid, meaning that frequent items are calculated online, with a single pass, while frequent itemsets are calculated as a background activity by a further analysis. This kind of approach allows important information to be

derived on the fly without imposing too strict time constraints on more complex tasks, such as the extraction of frequent k-itemsets, as this could excessively lower the accuracy of models.

- To support the mentioned hybrid approach, the architecture exploits the presence of data cachers (DCs) on which recent data can be stored. In particular, miners can turn to DCs to retrieve the statistics about frequent items and use them to identify frequent sets of items. To avoid excessive communication overhead, DCs are distributed and placed close to stream sources and miners.

The major advantages of the presented architecture are its scalability and flexibility. Indeed, the architecture can efficiently exploit the presence of multiple miners, and can be adapted to the requirements of specific scenarios: for example, the use of parallel miners can be avoided when a single miner can keep the pace of a single stream, and the use of DCs is not necessary if mining frequent itemsets is not required or if the stream rate is so slow that they can be computed on the fly. Such behavior can be naturally obtained if the miners are running on cloud servers, because the number of active machines is dynamically adapted and scaled to the computational load.

Beyond presenting the architecture, we describe an implemented prototype and discuss a set of experiments performed in a distributed environment composed of two domains each one handling a DS.

14.2 BACKGROUND AND RELATED WORK

The analysis of DSs has recently attracted a lot of attention owing to the wide range of applications for which it can be extremely useful. Important challenges arise from the necessity of performing most computation with a single pass on stream data because of limitations in time and memory space. Stream mining algorithms deal with problems as diverse as clustering and classification of DSs, change detection, stream cube analysis, indexing, forecasting, and so on [2].

For many important application domains, a major need is to identify frequent patterns in DSs, either single frequent elements or frequent sets of items in transactional databases. A rich survey of algorithms for discovering frequent items is provided by Cormode and Hadjieleftheriou [1]. Some of these algorithms, for example, CountSketch [3], compute a sketch, that is, a linear projection of the input, and provide an approximated estimation of item frequencies using limited computing and memory resources. Advantages and limitations of sketch algorithms are discussed in Reference 4. Important advantages are the notable space efficiency (required space is logarithmic in the number of distinct items), the possibility of naturally dealing with negative updates and item deletions, and the linear property, which allows sketches of multiple streams to be computed by overlapping the sketches of single streams. The main limitation is the underlying assumption that the domain size of the DS is large; however, this assumption holds in many significant domains.

Even if modern single-pass algorithms are extremely sophisticated and powerful, multipass algorithms are still necessary either when the stream rate is too rapid, or when the problem is inherently related to the execution of multiple passes, which is the case, for example, of the frequent itemsets problem [3]. A very promising avenue could be to devise hybrid approaches, which try to combine the best of single- and multiple-pass algorithms [5]. A strategy of this kind is adopted in the mining architecture presented in this chapter.

The analysis of streams is even more challenging when data is produced by different sources spread in a distributed environment, as happens in a content delivery network. A thorough discussion of the approaches currently used to mine multiple DSs can be found in Reference 6. The chapter distinguishes between the centralized model, under which streams are directed to a central location before they are mined, and the distributed model, in which distributed computing nodes perform part of the computation close to the data, and send to a central site only the models, not the data. Of course, the distributed approach has notable advantages in terms of degree of parallelism and scalability.

An interesting approach for the continuous tracking of complex queries over collections of distributed streams is presented in Reference 7. To reduce the communication overhead, the adopted strategy combines two technical solutions: (i) remote sites only communicate to the coordinator concise summary information on local streams (in the form of sketches); (ii) even such communications are avoided when the behavior of local streams remains reasonably stable, or predictable: updates of sketches are only transmitted when a certain amount of change is observed locally. The success of this strategy depends on the level of approximation on the results that is tolerated. A similar approach is adopted in Reference 8: here stream data is sent to the central processor after being filtered at remote data sources. The filters adapt to changing conditions to minimize stream rates while guaranteeing that the central processor still receives the updates necessary to provide answers of adequate precision. In Reference 9, the problem of finding frequent items in the union of multiple distributed streams is tackled by setting a hierarchical communication topology in which streams are the leaves, a central processor is the root, and intermediate nodes can compress data flowing from the leaves to the root. The amount of compression is dynamically adapted to make the tolerated error (difference from estimation and actual frequency of items) follow a precision gradient: the error must be very low at nodes close to the sources, but it can gradually increase as the communication hierarchy is climbed. The objective of this strategy is to minimize load on the central node while providing acceptable error guarantees on answers.

Stormy [10] is a distributed multitenant streaming service designed to run on Cloud infrastructures. Like traditional data stream processing engines (SPEs), it executes continuous queries against ongoing streams of incoming data and eventually forwards the results to a designated target. To achieve good scalability, the system exploits distributed hash tables (DHTs) for an efficient distribution of the queries across all nodes. Moreover, it guarantees good level of fault tolerance by replicating query executions on several nodes, avoiding in this way a single point-of-failure. The system combines standard techniques of Cloud computing with stream processing to guarantee elasticity, scalability, and fault tolerance.

The *Prism* system [11], the Portal Infrastructure for Streaming Media, is a content delivery network architecture for distributing, storing, and delivering high quality streaming media over the Prism IP network infrastructure. A first service developed on this architecture was the Prism-based stored-TV (STV) service, aimed at allowing users to select content based on the program's name. It is composed of three main modules: *live sources*, *portals*, and *clients*. *Live sources* receive content from a content provider, encode and packetize it, and then stream it into the Prism IP network. *Portals* receive multimedia content from live sources and other portals, and transmit it to Prism clients. Portals can store and archive live content, thus allowing it to be viewed on demand, and provide functions such as fast-forward and rewind. *Clients* receive content from a portal and display it to end users. They are connected to the backbone using broadband access.

14.3 A HYBRID MULTIDOMAIN ARCHITECTURE

This section presents the stream mining architecture that aims at solving the problem of computing frequent items and frequent itemsets from distributed DSs, exploiting a hybrid single-pass/multiple-pass strategy. We assumed that stream sources, though belonging to different domains, are homogenous, so that it is useful to extract knowledge from their union. Typical cases are the analysis of the traffic experienced by several routers of a wide area network, or the analysis of client requests forwarded to multiple Web servers of a content delivery network (CDN). Miner nodes are located close to the streams, so that data transmitted between different domains only consists of models (sketches), not raw data.

The architecture includes the following components, depicted in Figure 14.1:

- *DS*. They are located in different domains.
- *Miners (M)*. They are placed close to the respective DS, and perform two basic mining tasks: the computation of sketches for the discovery of frequent items,

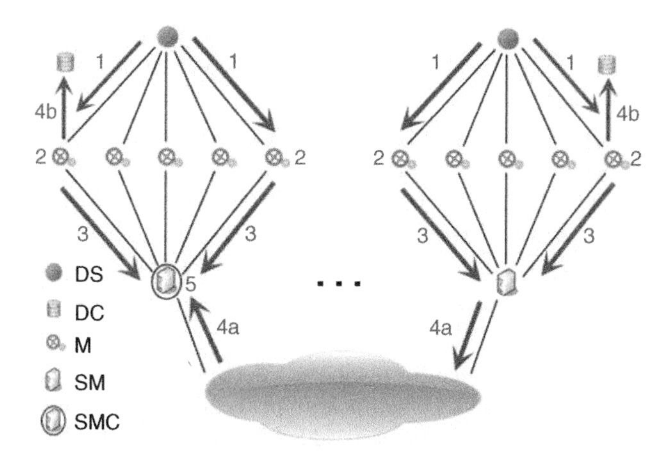

Figure 14.1 Architecture and schema of the algorithm for mining frequent items.

and the computation of the support count of candidate frequent itemsets. If a single Miner is unable to keep the pace of the local DS, the stream items can be partitioned and forwarded to a set of Miners, which operate in parallel. Each Miner computes the sketch only for the data it receives, and then forwards the results to the local Stream Manager (SM). Parallel Miners can be associated to the nodes of a cluster or a high speed computer network, or to the cores of a manycore machine.

- *SM*. In each domain, the SM collects the sketches computed by local miners, and derives the sketch for the local DS. Moreover, each SM cooperates with the Stream Manager Coordinator (SMC) to compute global statistics, valid for the union of all the DSs.
- SMC. This node collects mining models from different domains and computes overall statistics regarding frequent items and frequent itemsets. The SMC can coincide with one of the SMs, and can be chosen with an election algorithm. In Figure 14.1, the SM of the domain on the left also takes the role of SMC.
- *DCs*. They are essential to enable the hybrid strategy and the computation of frequent itemsets, when this is needed. Each DC stores the statistics about frequent items discovered in the local domain. These results are then reused by Miners to discover frequent itemsets composed of increasing numbers of items.

The algorithm for the computation of frequent items, also outlined in Figure 14.1, is performed continuously, for each new block of data generated by the DSs. A block is defined here as the set of transactions that are generated in a time interval P. The algorithm includes the following steps, also shown in the figure.

1. A filter is used to partition the block into as many miniblocks as the number of available Miners.
2. Each Miner computes the sketch related to the received miniblock.
3. The Miner transmits the sketch to the SM, which overlaps the sketches, thanks to the linearity property of sketch algorithms, and extracts the frequent items for the local domain.
4. Two concurrent operations are executed: every SM sends the local sketch to the SMC (step 4a), and the Miners send the most recent blocks of transactions to the local DC (step 4b). The last operation is only needed when frequent itemsets are to be computed, otherwise it can be skipped.
5. The SMC aggregates the sketches received by SMs and identifies the items that are frequent for the union of DSs.

Frequent items are computed for a window containing the most recent W blocks. This can be done easily, thanks to the linearity of the sketch algorithm: at the arrival of a new block, the sketch of this block is added to the current sketch of the window, while the sketch of the least recent block is subtracted. The window-based approach is common because most interesting results are generally related to recent data [12].

Sketch-based algorithms are only capable of computing frequent items. To discover frequent itemsets, it is necessary to perform multiple passes on data. Candidate k-itemsets are constructed starting from frequent $(k-1)$-itemsets. More specifically, at the first step candidate two-itemsets are all the possible pairs of frequent items: Miners must compute the support for these pairs to determine which of them are frequent. In the following steps, a candidate k-itemset is obtained by adding any frequent item to the frequent $(k-1)$-itemsets. Thanks to the Apriori property, candidates can be pruned by checking if all the $k-1$ subsets are frequent: a k-itemset can be frequent only if all the subsets are frequent.

The approach allows us to compute both itemsets that are frequent for a single domain and those that are frequent for the union of distributed streams. Figure 14.2 shows an example of how frequent three-itemsets are computed. The top part of the figure reports items and two-itemsets that are frequent for the two considered domains and for the whole system. The candidate three-itemsets, computed by the two SMs and by the SMC, are then reported, before and after the pruning based on the Apriori property. In the bottom part, the figure reports the support counts computed for the two domains and for the whole system. Finally, the SMs check which candidates exceed the specified threshold (in this case, set to 10%): notice that the {abc} itemset is frequent globally though it is locally frequent in only one of the two domains. In general, it can happen that an itemset occurs frequently for a single domain and infrequently

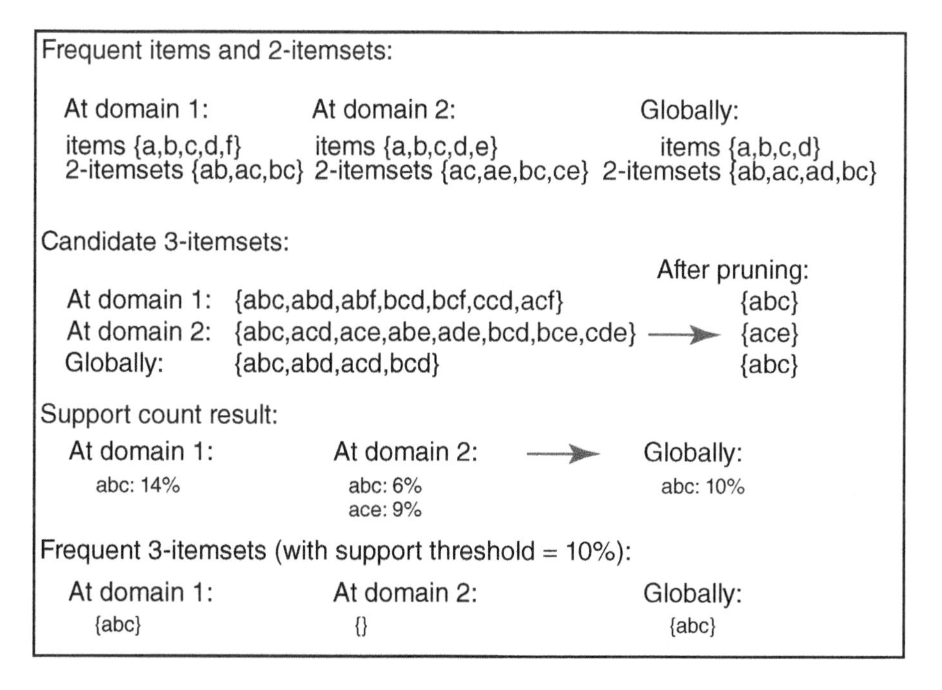

Figure 14.2 Example of the computation of frequent three-itemsets.

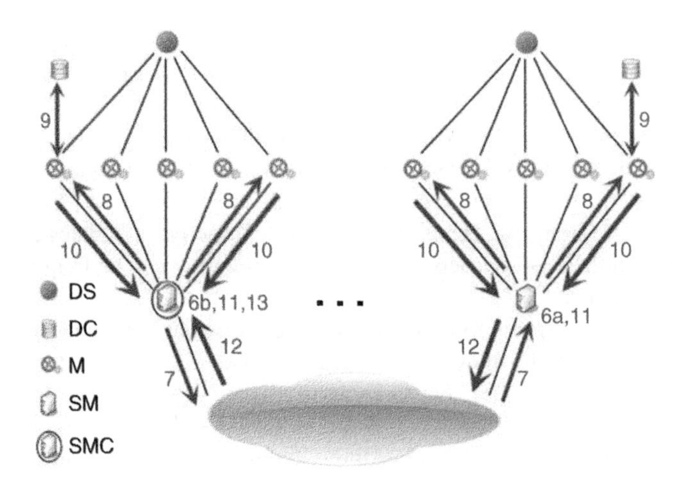

Figure 14.3 Schema of the algorithm for mining frequent itemsets.

globally, or vice versa: therefore, it is necessary to separately perform the two kinds of computations.

The schema of the algorithm for mining frequent itemsets is illustrated in Figure 14.3, which assumes that the steps indicated in Figure 14.1 have already been performed.

The successive steps are the following:

6. Each SM builds the candidate k-itemsets for the local domain (6a), and the SMC also builds the global candidate k-itemsets (6b);
7. The SMC sends the global candidates to the SMs for the computation of their support at the different domains.
8. SMs send both local and global candidates to the Miners.
9. Miners turn to the DC to retrieve the transactions included in the current window (this operation is performed only at the first iteration of the algorithm). Then, the Miners compute the support count for all the candidates, using the window-aware technique presented in Reference 13.
10. Miners transmit the results to the local SM.
11. The SM aggregates the support counts received by Miners and selects the k-itemsets that are frequent in the local domain.
12. Analogously, the SMs send the SMC the support counts of the global candidates.
13. The SMC computes the itemsets that are frequent over the whole system. At this point, the algorithm restarts from step 6 to find frequent itemsets with increasing numbers of items. The cycle stops either when the maximum allowed size of itemsets is reached or when no frequent itemset was found in the last iteration.

14.4 A PROTOTYPE FOR STREAM MINING IN A CDN

The architecture described in the previous section was implemented starting from *Mining@Home*, a Java-based framework partly inspired by the Public Computing paradigm, which was adopted for several classes of data mining computations, among which the analysis of astronomical data to search for gravitational waves [14], and the discovery of closed frequent itemsets with parallel algorithms [15]. The main features of the stream mining prototype inherited from *Mining@Home*, are the *pull* approach (Miners are assigned jobs on the basis of their availability) and the adoption of DCs to store reusable data. Moreover, some important modifications were necessary to adapt the framework to the stream mining scenario. For example, the selection of the Miners that are the most appropriate to perform the mining tasks is subject to vicinity constraints, because in a streaming environment it is very important that the analysis of data is performed close to the data source. Another notable modification is the adoption of the hybrid approach for the single-pass computation of frequent items and the multipass computation of frequent itemsets.

Experiments were performed on the ICAR-CNR distribution infrastructure. We used two networks, connected by a router to test a scenario with two domains and two DSs. Each network has a cluster: the first cluster has 12 CPU Intel Xeon E5520 nodes with four 2.27 GHz processors and 24 GB RAM; the second cluster has 12 Intel Itanium nodes with two 1.5 GHz CPU and 4 GB RAM. The Miners and the SMs were installed on the nodes of the clusters while the Data Sources and the DCs were put on different nodes, external to the clusters. The average interdomain transfer rate measured in the experiments was 197 KB/s, while the average intradomain transfer rates were 918 and 942 KB/s in the two networks. All the nodes run Linux, and the software components are written in Java.

To assess the prototype, we used "webDocs," a stream dataset published by the *FIMI Repository* [13]. The dataset is generated from a set of distributed Web hosts belonging to a CDN. Each Web page, after the application of a filtering algorithm, is represented with a set of significant words included in it. The analysis of most frequent words, or sets of words, can be useful to devise caching policies, indexing techniques, and so on. Some basic information about the dataset is summarized in the following table.

Dataset	MB	Number of tuples	Number of distinct items	Size of tuples (number of items)		
				Min	Med	Max
webDocs	1413	16,92,082	52,67,656	1	177	71,472

The parameters used to assess the prototype are listed in the following:

- N_t is the average number of transactions generated within a block.
- N_M is the number of available miners per domain, assuming that this number is the same for the two domains.

As the webDocs dataset contains representative words of Web pages filtered by a search engine, the data rate was set to typical values registered by servers of Google and Altavista, again using the site http://www.webtraffic24.com to do the estimation. The considered values for N_t were 500, 1500, and 3000 transactions, with the time interval P set to 15 s. It is assumed that the cache of a miner can contain one complete block of data.

The support threshold used to determine frequent items and itemsets is set to 0.02. The size of the sliding window, that is, the number of consecutive blocks of data on which computation is performed, is set to five blocks. The accuracy parameters of the sketch algorithm, ϵ and δ [13], are both set to 0.01. Another peculiar parameter of the sketch algorithm, maximum size of candidate itemsets, is set to 8.

The main performance index assessed during the experiments is the average execution time, that is, the time necessary to compute frequent items and frequent itemsets at the arrival of a new block of data. If this value is not longer than the time interval P, it means that the system is able to keep the pace with data production.

Figure 14.4 reports the average execution time experienced for the computation of frequent items exclusively (I), and for the computation of both frequent items and itemsets (I + IS) versus the number of miners per domain N_M. The execution time is calculated from the time at which a block of data is ready to be delivered from the Data Sources to the time at which the Stream Miner Controller terminates the computation. The dashed line corresponds to the time period P, and it is shown to easily check in which cases the system is stable. Results, for three different values of N_t, show that a single miner per domain is not sufficient: depending on the generation rate, at least 4, 6,

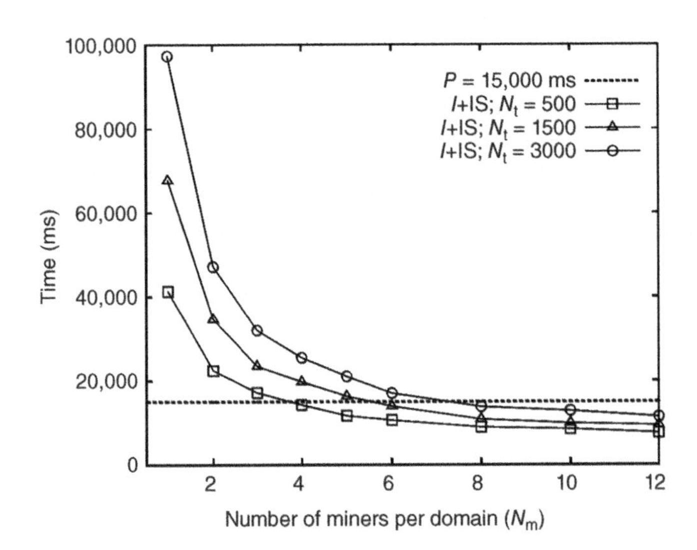

Figure 14.4 Analysis of webDocs streams: average execution time for the computation of frequent items and itemsets (I + IS) versus the number of miners per domain, for different values of the number of transactions per block, N_t.

or 8 miners are needed to keep the processing time below the period length P. In general, the processing time decreases as the number of miners increases, which is a sign of the good scalability of the architecture. Scalable behavior is ensured by two main factors: the linearity property of the sketch algorithm, and the placement of DCs close to the miners.

In this evaluation, the case of $N_M = 1$ corresponds to a nonparallel architecture in which the mining computation is performed on a single node. Therefore the results in Figure 14.4 can also be seen as a comparison between a parallel and a sequential solution. It is worth noting that in this scenario any centralized architecture would have few chances to keep pace with data, which means that the computation of frequent itemsets would have to be done offline, while the architecture presented here can achieve the goal of online processing using an appropriate degree of parallelism.

To better assess the system behavior, it is useful to analyze the data traffic involved in the computation. Figure 14.5 shows the amount of data transmitted over the network at the generation of a new block of data, for different numbers of miners per domain. In these experiments, N_t was set to 15,000. The first three groups of bars show the overall amount of data transferred between nodes of type A to nodes of type B in a single domain, denoted as A→B. For example, DS→M is the amount of data transmitted by the Data Source to the Miners of a single domain at every time period. The values of DS→M and M→DC are equal since each Miner sends to the local DC the data received from the DS. The fourth group reports DT_{Domain}, the overall amount of data transferred within a single domain, computed as the sum of the contributions shown in the first three groups (the contribution of the first group is considered twice). The contribution SM→SMC is the amount of data transferred between the SMs and the Stream Manager Controller. Finally, the last group of bars reports the amount of data transferred over the whole network, DT_{Net}. This is computed as the term DT_{Domain} times the number of domains (in this

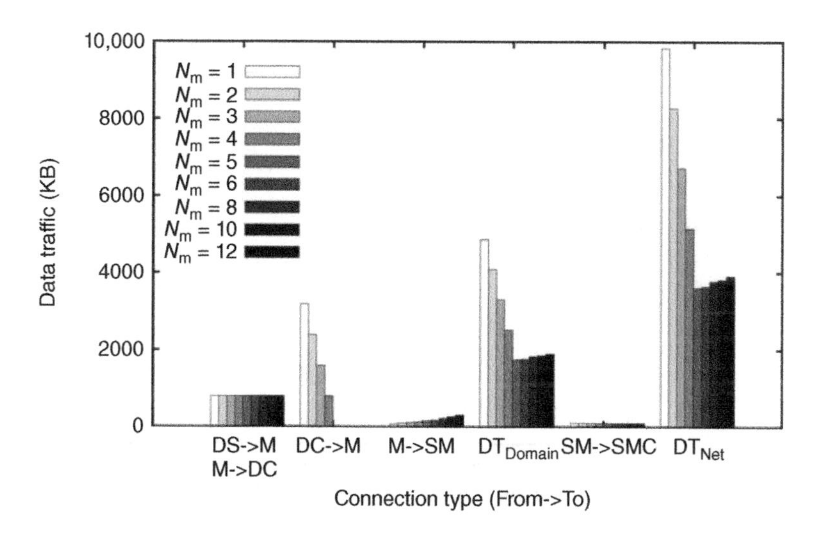

Figure 14.5 Data traffic per each block of data generated by the data streams. In this experiments, $N_t = 1500$.

case 2) plus the term SM→SMC. It is interesting to notice that the contribution DC→M decreases when the number of miners per domain increases, and becomes null when N_M is equal or greater than 5. This can be explained by considering that each miner must compute the frequency of items and itemsets over a window of five time periods, and possesses a cache that can contain one complete block of data. If there is only one miner per domain, this miner can store one block and must retrieve the remaining four blocks from the local DC. As the number of miners increases, each Miner needs to request less data from the DC, and needs no data when N_M equals or exceeds the window size.

Conversely, the component M→SM slightly increases with the number of miners, because every miner sends the local SM a set of results (the sketch and the support counts of itemsets) having approximately the same size. However, as the contribution of DC→M has a larger weight than M→SM, the overall amount of data exchanged within a domain decreases as N_M increases from 0 to 5. For larger values of NM, DT_{Domain} starts to increase, because M→SM slightly increases and DC→M gives no contribution. A similar trend is observed for the values of DT_{Net}. Therefore, not only the presence of multiple miners allows the computation to be distributed, but also leads to a decrease of the overall amount of transferred data.

A study has been performed in order to evaluate the efficiency of the approach. In particular, the efficiency analysis was performed in accordance with the study of parallel architectures presented in Reference 16. Specifically, we extracted the overall computation time T_C, that is, the sum of the computation times measured on the different miners, and the overall overhead time T_O, defined as the sum of all the times spent in other activities, which practically coincide with the transfer times. The efficiency of the computation can be defined as the fraction of time that the miners actually devote to computation with respect to the sum of computation and overhead time: $E = T_C/(T_C + T_O)$. Figure 14.6 reports efficiency values obtained in our experiments. It is observed that

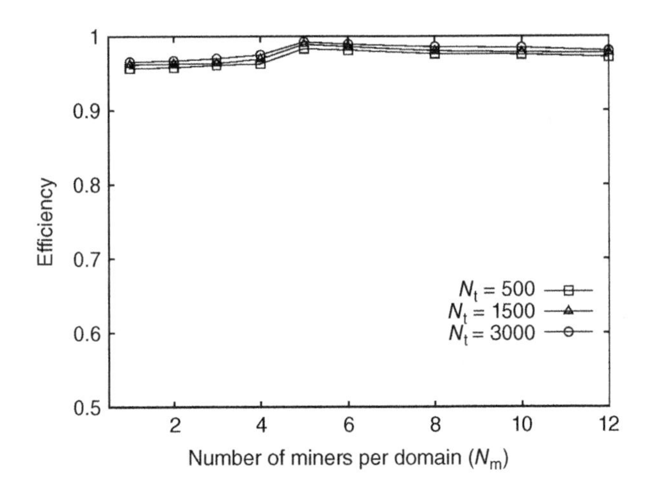

Figure 14.6 Efficiency versus the number of miners per domain, for different values of the number of transactions per block, N_t.

efficiency increases as the number of miners increases up to 5, that is, the window size. This effect is induced by the use of caching, as explained in Reference 16. Specifically, when N_M increases up to the window size, each miner needs to request less data from the DC, because more data can be stored in the local cache: this leads to a higher efficiency. For larger values of N_M, this effect does not hold anymore and the efficiency slightly decreases, being still very high. Moreover, it is noticed that the efficiency increases with the rate of DSs. This means that the distributed architecture is increasingly convenient when the problem size increases, which is a another sign of good scalability properties.

14.5 VISIONARY THOUGHTS FOR PRACTITIONERS

A walk through CDN's evolution, discussed in Reference 17, reveals a smooth transition from preevolution period to the current generation of CDNs. Two of the identified evolutionary avenues are: a more massive deployment of servers with improved caching techniques, and a major boost for delivering streaming rich media content: video, audio, and associated data. The analysis of streaming data is essential to improve the efficiency and effectiveness of CDN architectures and the value of CDN services to the user.

Data mining algorithms can provide an efficient way to perform this analysis, but they must be adapted to the distributed and heterogeneous environment that is peculiar to modern CDN solutions. The architecture and prototype presented in this chapter are, to the best of our knowledge, one of the first attempts to devise a stream mining solution that offer the following characteristics, which can be particularly profitable in the context of CDNs: parallel/distributed architecture, use of sketch algorithms, hybrid online/offline approach, and use of distributed DCs. In particular, we are not aware of attempts to combine the parallel and distributed paradigms in stream mining, nor of implemented systems that adopt the hybrid single-pass/multipass approach, though this kind of strategy is suggested and fostered in the recent literature [5].

The major advantages of the presented architecture are its scalability and flexibility. Indeed, the architecture can efficiently exploit the presence of multiple miners, and can be adapted to the requirements of specific scenarios: for example, the use of parallel miners can be avoided when a single miner can keep the pace of a single stream, and the use of DCs is not necessary when mining frequent itemsets is not required or when the stream rate is so slow that they can be computed on the fly.

14.6 FUTURE RESEARCH DIRECTIONS

The accurate analysis of stream data is becoming increasingly important as more and more companies are exporting not only their public Web pages but also their data and applications to the Cloud. The success of Cloud has a strong impact on the building and usage of CDNs, and future research activities should investigate the advantages that can derive from the integration of Cloud and CDN. Indeed, the services offered by main CDN players, such as Akamai and Mirror Image, are priced out of reach for all but the largest enterprise customers. An alternative approach to content delivery is presented in Reference 18 and consists of leveraging the infrastructure provided by "Storage Cloud" providers, who offer internet accessible data storage and delivery at a much lower cost. The devised architecture, MetaCDN, exploits Cloud storage facilities and creates an integrated overlay network that provides a low cost, high performance CDN for content creators. In this or similar architectures the analysis of stream data coming from distributed and possibly heterogeneous Cloud facilities is of much interest.

On the other hand, Cloud resources can be used to help the analysis of distributed DSs. For example, modern commercial analytics tools such as Gomez, Keynote, and Cedexis, produce large amounts of stream data that is used to drive Web platform analysis and optimize Web business performance. Cloud environments provide efficient computing infrastructures and offer effective support to the analysis of stream data and the implementation of data mining and knowledge discovery systems. Cloud infrastructures can be efficiently exploited to archive, analyze, and mine large distributed data sets. An interesting approach is described in Reference 19. The authors present a combined stream processing system that, as the input stream rate varies, adaptively balances workload between a dedicated local stream processor and a Cloud stream processor. The approach only utilizes Cloud machines when the local stream processor becomes overloaded.

14.7 CONCLUSION

In recent years, the progress in digital data production and pervasive computing technology has made it possible to produce and store large streams of data. Data mining techniques became vital to analyze such large and continuous streams of data for detecting regularities or outlier values in them. In particular, when data production is massive and/or distributed, decentralized architectures and algorithms are needed for its analysis. CDNs have recently experienced a gradual but continuous evolution from the publication of static and semistatic content to the provisioning of multimedia DSs. Therefore, CDNs are a privileged scenario for the application of distributed stream mining techniques.

The architecture in this chapter is a contribution in the field and it aims at solving the problem of computing frequent items and frequent itemsets from distributed DSs by exploiting a hybrid single-pass/multiple-pass strategy. Beyond presenting the system architecture, we described a prototype that implements it and discussed a set of experiments performed in the ICAR-CNR distributed infrastructure. The experimental results confirm that the approach is scalable and can manage large data production using an appropriate number of miners in the distributed architecture.

REFERENCES

1. Cesario E, Grillo A, Mastroianni C, Talia D. A sketch-based architecture for mining frequent items and itemsets from distributed data streams. Proceedings of the 11th IEEE/ACM

International Symposium on Cluster, Cloud and Grid Computing (CCGrid 2011); Newport Beach, CA, USA; 2011. p 245–253.

2. Aggarwal C. An introduction to data streams. In: Aggarwal C, editor. *Data Streams: Models and Algorithms*. Springer; 2007. p 1–8.
3. Charikar M, Chen K, Farach-Colton M. Finding frequent items in data streams. Proceedings of the International Colloquium on Automata, Languages and Programming (ICALP); 2002.
4. Aggarwal C, Yu PS. A survey of synopsis construction in data streams. In: Aggarwal C, editor. *Data Streams: Models and Algorithms*. Springer; 2007. p 169–207.
5. Wright A. Data streaming 2.0. Commun ACM 2010;53(4).
6. Srinivasan Parthasarathy AG, Otey ME. A survey of distributed mining of data streams. In: Aggarwal C, editor. *Data Streams: Models and Algorithms*. Springer; 2007. p 289–307.
7. Cormode G, Garofalakis M. Approximate continuous querying over distributed streams. ACM Trans Database Syst 2008;33(2):1–39.
8. Olston C, Jiang J, Widom J. Adaptive filters for continuous queries over distributed data streams. SIGMOD '03: Proceedings of the 2003 ACM SIGMOD International Conference on Management of Data; San Diego, California; 2003.
9. A. Manjhi, V. Shkapenyuk, K. Dhamdhere, C. Olston: Finding (recently) frequent items in distributed data streams. ICDE '05: Proceedings of the 21st International Conference on Data Engineering; Tokyo, Japan; 2005. p 767–778.
10. Loesing S, Hentschel M, Kraska T, Kossmann D. 2012. Stormy: An elastic and highly available streaming service in the cloud. Proceedings of the 2012 Joint EDBT/ICDT Workshops (EDBT-ICDT '12); 2012. p 55–60.
11. Cranor CD, Green M, Kalmanek C, Shur D, Sibal S, Van der Merwe JE, Sreenan CJ. Enhanced streaming services in a content distribution network. IEEE Internet Comput 2001;5(4):66–75.
12. Datar M, Gionis A, Indyk P, Motwani R. Maintaining stream statistics over sliding windows. SIAM J Comput 2002;31(6):1794–1813.
13. 2013. Frequent itemset mining dataset repository. Available at http://fimi.cs.helsinki.fi. Accessed 23 September 2013.
14. Mastroianni C, Cozza P, Talia D, Kelley I, Taylor I. A scalable super-peer approach for public scientific computation. Future Gener Comput Syst 2009;25(3):213–223.
15. Lucchese C, Mastroianni C, Orlando S, Talia D. Mining@home: Toward a public resource computing framework for distributed data mining. Concurr Comput Pract Exp 2009; 22(5):658–682.
16. Grama AY, Gupta A, Kumar V. Isoefficiency: Measuring the scalability of parallel algorithms and architectures. IEEE Concurr 1993;1(3).
17. Pathan M 2011. Ongoing trends and future directions in content delivery networks (CDNs). Available at http://amkpathan.wordpress.com/article/ongoing-trends-and-future-directions-in-3uxfz2buz8z1w-2/. Accessed 23 September 2013.
18. Broberg J, Buyya R, Tari Z. MetaCDN: Harnessing 'Storage Clouds' for high performance content delivery. J Netw Comput Appl 2009;32(5):1012–1022.
19. Kleiminger W, Kalyvianaki E, Pietzuch P. Balancing load in stream processing with the cloud. Proceedings of the 2011 IEEE 27th International Conference on Data Engineering Workshops (ICDEW '11); Washington, DC, USA: IEEE Computer Society; 2011. p 16–21.

15

CDN CAPACITY PLANNING

Phil Davies[1] and Mukaddim Pathan[1]

[1] *Telstra Corporation Limited, Melbourne, Victoria, Australia*

15.1 INTRODUCTION

The Content delivery network (CDN) marketplace [1–3] consists of a handful of pure-play providers who continuously seek to simultaneously scale up and diversify their current offerings and customer bases in order to maintain and grow market share.

Services to CDN customers are delivered by way of the provider's infrastructure inventory, so it is critical that this infrastructure is able to scale both vertically and horizontally. That is, be able to support demand for existing services along with improvements in those services required to facilitate ongoing growth in the volume of demand. Also, the infrastructure must have the ability to support rapid innovation in new value-added services, to allow the provider to continue to present relevant and up-to-date offerings, which will be attractive in a competitive marketplace.

CDN infrastructure inventory can be considered as the providers' means of production [4]. To ensure continued long-term business success for the provider, ongoing inventory investments are required, ensuring that a sustainable return for the providers' business is realized. A wide range of factors must be considered when these investments

Advanced Content Delivery, Streaming, and Cloud Services, First Edition.
Edited by Mukaddim Pathan, Ramesh K. Sitaraman, and Dom Robinson.

are being made. These factors include a mixture of customer demand, market conditions, corporate business strategy, and regulatory requirements, along with the associated technology and operational drivers.

Capacity Planning [5–7] is the process that takes into account all the diverse factors, and from these factors, a cost-effective infrastructure investment program develops to support the ongoing business success for the provider. Owing to the wide range of activities involved in undertaking the capacity planning process, it is often helpful to consider these factors in their natural combinations and interactions to get a clearer overall picture at a high level of the components of the process. The key areas that can be addressed as discrete activities are listed in the following:

- The procurement and management of capital funding to facilitate any investment requirement.
- The interpretation and translation of strategy relevant to infrastructure development to support the business.
- The development of rules, which ensure expected service levels required of the infrastructure to best serve target markets that can be delivered in the most cost-effective manner for the provider.
- The composition of specific capacity requirements in terms of locations, timing, funding, equipment, and labor.
- The physical deployment of such capacity requirements to realize investment.

This chapter presents the capacity planning process and demonstrates how it can be used in a CDN context. We start with a brief coverage of the capacity planning process and identify the key components. It is followed by the practical implications and realization of capacity planning. A case study of CDN capacity planning is included to demonstrate the approach required to perform effective capacity planning. The chapter is concluded with recent developments in the field of capacity planning and the associated challenges faced by the industry.

15.2 CAPACITY PLANNING PROCESS

A pictorial representation of the capacity planning process and the separate discrete activities that comprise it is shown in Figure 15.1.

To achieve the required outcome, the basic flow of the overall process shown previously can be considered as being from left to right. However, it should be noted that there is necessary interaction between each of the separate activities, which may be in both directions.

Inputs in terms of capital, strategy, service levels, and demands are taken in, analyzed, and processed with the resultant output being a quantifiable infrastructure investment that can be deployed and utilized by the provider to support services and growth.

A brief description of each of the activities and their interactions is outlined in the following.

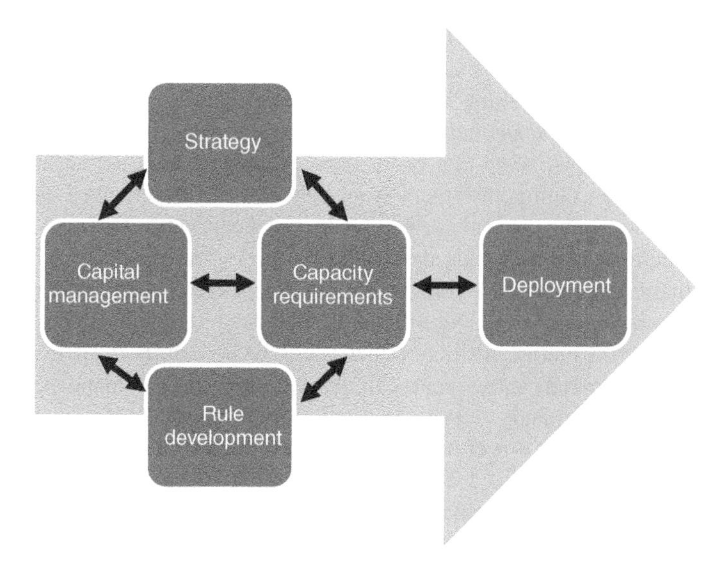

Figure 15.1 Capacity planning process.

Capital Management. This refers to the procurement and management of the capital required to invest in the providers' infrastructure. This activity may typically involve the development and approval of business cases for funding, along with the reporting of capital expenditure and communication of changes (up or down) in the providers' capital availability.

Strategy. The strategy activity for capacity planning is focused around being able to quantify a providers' strategic direction to a point where it becomes a tangible, useful input for the capacity planning process. It may or may not involve the development of strategies for the provider, for example, expansion of product range, markets served, and discontinuation of certain product lines. At a minimum capacity, planning should be able to interpret such strategies so that their required outcomes can be incorporated into discrete and deliverable infrastructure investments.

To assist in undertaking this activity, it may be useful to develop a set of checklist items to ensure that all relevant items are taken into consideration. An example set of checklist items may include financial constraints, capacity requirements, functionality requirements, security, resiliency, performance obligations, plus any lifecycle management, customer reporting, and regulatory impacts.

Each individual relevant checklist item should then be analyzed, queried, and commented on with regard to its likely effect. Areas noted as requiring clarification arising from using such a checklist could then be possibly used as a trigger for discussions with appropriate stakeholders such as infrastructure suppliers and customers.

Rules Development. Typically, a service providers' product set will have associated with it service-level guarantees that outline figures for availability, delay, time to restore, maximum amounts of customers impacted under fault conditions, along with other performance metrics that are relevant to specific or multiple product offerings. The published figures for these guarantees are often utilized by both potential and existing customers as an indicator of the perceived quality of the provider's offering, and subsequently whether the product offering is suitable for their requirements. The rule development activity translates these metrics into pragmatic, implementable rules, which can be utilized for infrastructure deployment. For example, an availability metric will need to be converted into a certain level of infrastructure resiliency that needs to be incorporated in both initial deployment and any subsequent scaling to cater for future demand increases. The definition of resiliency at an infrastructure level may include equipment specification, associated power and connectivity schemes, utilization levels that the equipment is not to exceed, and other tangible, observable items that contribute to the infrastructures' performance in relation to the published availability requirement.

From the provider's perspective, these rules also provide guidance as to how to deploy their infrastructure in the most cost-effective manner to meet expected demand from the targeted market segments. Areas need to be addressed include, but are not limited to, how much capacity excess to existing demand is to be maintained (e.g., to cater for bursts in utilization), how much site infrastructure (power, racking, etc.) is required to be preprovisioned, how much capacity is to be deployed when a build is required, and for how long this capacity should meet expected demand requirements when it is deployed. The values of the variables required to be considered may differ depending on whether the infrastructure being deployed is supporting premium service offerings or those with a lower expected performance criteria.

Another consideration which may require addressing in any rule development is that it is not uncommon for a providers' infrastructure to be a shared resource, simultaneously utilized by multiple customers. The extent and degree of how customers are likely to have their service performance impacted in case of any capacity shortfall should also be a key guideline as this may represent an area of risk to the providers' business that must be mitigated.

Capacity Requirements. The determination of capacity requirements for an infrastructure is at the core of the capacity planning process. This is where the detailed equipment build quantities, locations, timings, and costing are determined. At first glance, this may seem like a complex and potentially iterative task as there are a wide range of differing inputs to consider; however, it can be simplified to some extent by breaking the activity down into three distinct steps.

The structure of these steps attempts to consolidate different inputs required to be analyzed into logical groupings, which are then actioned in a hierarchical sequence. Put simply, the sequence to follow is to firstly determine *What* infrastructure is existing, then *Why* there may be a need to undertake action related to the infrastructure, and if this is the case, *How* exactly to go about it (Figure 15.2).

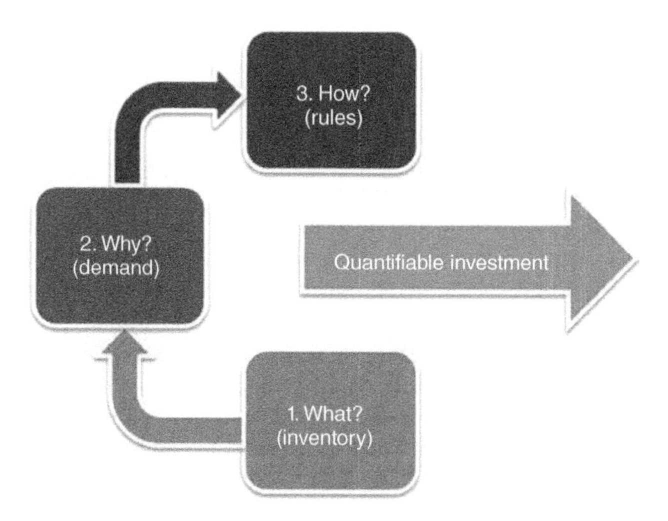

Figure 15.2 Capacity requirements activity.

The first step, the *What*, looks at what the current inventory level of the infrastructure is. This consists of two components, existing inventory and upcoming inventory.

- Existing inventory—in place and in use inventory that can be determined from appropriate records, such as those contained in inventory management systems.
- Upcoming inventory—inventory that is committed to be utilized but is yet to be placed in use. An example of this is the equipment that may have been ordered but not yet installed. Typically, the tracking of such inventory may be more difficult; however, it is critical that its quantity is known so that unnecessary doubling up of investment does not occur.

The next step, the *Why*, is to identify the underlying reason to invest in infrastructure. This essentially means quantifying demands placed on the infrastructure, which typically consists of the following main components:

- Trends that are forward extrapolations of existing historical infrastructure utilization data. The data for this infrastructure utilization may be obtained from the same or similar sources as the inventory; however, this may not always be the case.
- Forecasts that are typically hypotheses of what is expected to happen to the infrastructure or service utilization in the future. These may be different from trends as they could contain information from a market, technology, or strategy development perspective.
- Another component to be considered when determining demand would be incorporating step increases, which may be the result of actions such as on-boarding

new customer bases, as could potentially occur in the case of a company takeover or whole-of-business deal being on-boarded onto the provider's infrastructure.

These components are then analyzed and combined, with the resultant output being a picture of expected demand for which necessary infrastructure will be required to support over a specified period of time. Often this may end up being a range of values.

The final step is to utilize the information obtained from the previous two steps in order to produce a quantifiable infrastructure investment. This is done by applying the rules that govern how investment in the infrastructure is to be undertaken to the previously determined expected demand values. These rules should incorporate guidelines, allowing the most cost-effective investment result to be determined for the infrastructure, given the scenario presented. These rules may include but are not limited to areas such as utilization levels not to exceed, required architecture and relevant standards to adhere to, specification of equipment (hardware and software), and any regulatory considerations that may be required for the infrastructure.

By applying the relevant rules (the *How*) to the information obtained in the first two steps (the *What* and the *Why*), a quantifiable infrastructure investment should be produced. Some of the likely outputs from the final step may include the following.

- Do nothing, for example, existing inventory will meet demand requirements for the appropriate timeframe.
- Do something, for example, build/upgrade/augment a specific quantity.
- The rules do not fit the situation, for example, there is not a solution for the scenario being examined. This case will require the rules to be reevaluated and possibly rewritten to allow the presented scenario to be assessed.

After determination of the capacity requirements, a typical output could be a list of locations where equipment is to be installed, what type and quantity of equipment is to be installed, how it is to be connected to both existing and new infrastructure, and a proposed timing for the installation and an estimated cost.

The deployment activity is the part of the capacity planning process that converts the list of location-specific infrastructure requirements derived from the previous step into detailed site build requirements that can be integrated into the providers' existing infrastructure. This activity ensures that any investment undertaken by the provider can be realized into an active, usable resource, capable of producing a return on that investment by being available to cater for both existing and future expected customer demand.

This activity may include the preparation of detailed installation and configuration instructions for both existing and additional equipment, the procurement, managing and scheduling of resources to undertake both physical and logical tasks that are required to allow completion to occur, and the updating of necessary records, for example, notifications of progress to appropriate parties, site inventory, and costs incurred.

As described at the beginning of this section, there is also a necessity to have interactions between each of the activities where outputs from one activity may be required as an input to another, and vice versa. One obvious example of this is the interaction between determining the level of funding available to the provider (from the capital management activity) and the amount of infrastructure expenditure required to meet demand (from the capacity requirement activity). Often the case may be that these levels are different and a compromise between the two may be required to be developed by the provider.

The degree to which the capacity planning process may be more or less detailed or intensively undertaken does depend on the specific scenario being examined by the individual service provider. For example, a provider investigating a strategy to consolidate multiple, discrete platforms, each of which supports a single product offering into a single platform to support these multiple products will likely require a higher level of capacity planning resource to produce an optimum investment compared to a provider who offers a single product on a single platform.

In summary, CDN capacity planning is carried out to successfully address the following aims listed, and benefits are obtained by regularly undertaking the process to ensure that the provider obtains the most cost-effective infrastructure investment result.

- Meet customer demand for traffic and sessions.
- Minimize capital expenditure.
- Maximize utilization of the existing infrastructure.
- Minimize risk of customers being impacted by the lack of capacity during fault conditions.
- Ensure infrastructure capability and the associated scalability requirements are identified within adequate timeframes and trigger appropriate action.
- Provide adequate flexibility to cater for any future unforeseen requirements.
- Identify opportunities for leveraging existing infrastructure.
- Identify optimal timing for exits to maximize return on investment (ROI).

The process can be broken down into logical activities to assist with both understanding and analyzing the impact of key inputs to and outputs from the process, and that some of these inputs and outputs may be from activities internal to the process. Examples of typical key inputs and outputs—and whether they are from outside the process (external) or within the process (internal) are listed as follows (note that this is not an exhaustive list; Figure 15.3 and Table 15.1):

15.3 UNDERTAKING THE CAPACITY PLANNING PROCESS

15.3.1 Overview

The previous section has provided a description of the overall capacity planning process and each of its individual component activities. As the definition of a process can be considered a task, which takes inputs and converts them into outputs [8], each activity within the process will need to be addressed to ensure the completion of a cycle of the capacity planning process. The completion of a capacity planning process cycle assumes that inputs are fixed at the point in time they are utilized by the process, and that the resultant output produced can be correlated to the relevant inputs.

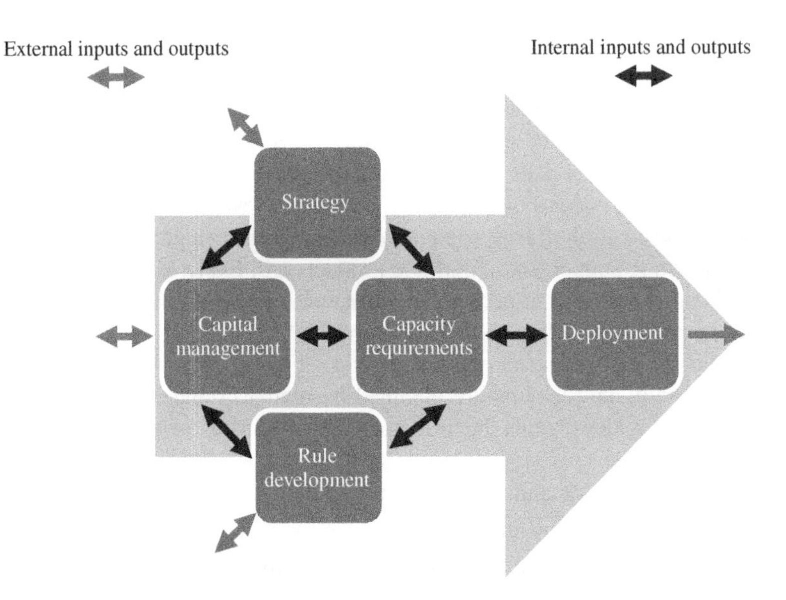

Figure 15.3 Capacity planning inputs and outputs.

As a number of the inputs to the capacity planning process are expected to change over time (e.g., levels of capital available, customer demand forecasts), the cycle of completing the process may need to be undertaken multiple times during the lifespan of a providers' infrastructure [6]. The frequency and timing of each pass of the process may depend on several factors such as the level of variation in customer demand compared to what was expected, triggers that require the reexamination of existing investment plans (e.g., new strategic directions or capital funding constraints), or unforeseen impacts of new technology introduction, for example, new software behaving unexpectedly compared to previous versions.

15.3.2 Benefits

As each additional cycle is performed, this presents an opportunity to refine and improve the process, and as more cycles are performed, typically a greater level of understanding of the infrastructures' operation, behavior, and likely investment requirements to meet ongoing expected demand can be garnered. This point is worth reinforcing as it represents an opportunity for the provider over time to develop tools and resources that allow quite mature, refined, and robust infrastructure investment profiles to be produced. The development of more accurate, appropriate investment plan helps to ensure that a provider is neither under- nor overinvesting in their infrastructure, thus maximizing returns and opportunities and minimizing risk. If a provider is unable or unwilling to leverage the opportunities for improvement that may present during the undertaking of capacity planning process activities, it is likely that the provider's ability to produce investment plans to meet expected customer demand and maximize returns as the end-result is less than optimal.

TABLE 15.1 List of Inputs and Outputs for Capacity Planning

Activity	Key Inputs (External)	Key Inputs (Internal)	Key Outputs (Internal)	Key Outputs (External)
Capital management	Available capital	Capital required to build infrastructure	Capital available to build infrastructure	Expenditure amounts and timings
	Business cases			
Strategy	Business strategy	Plans for infrastructure deployment (sites, costs, and timings)	Considerations for current infrastructure builds	Opportunities to leverage current infrastructure
	Regulatory requirements			
Capacity requirements	Inventory levels	Rules/guidelines	Capital required to build infrastructure	Plans for infrastructure deployment (sites, costs, and timings)
	Trends	Considerations for current infrastructure builds	Scalability limits being reached	
	Forecasts			
Rule development	Service-level agreements Strategies	Scalability limits being reached	Rules/guidelines	Platform architectures
Deployment	Installation resource availability	Plans for infrastructure deployment (sites, costs, timings)	Build progress (sites, costs, timings)	Builds

A disciplined and well-resourced approach to capacity planning also presents the opportunity to develop additional flow on benefits for a provider. It allows for a greater level of responsiveness to rapidly fluctuating market conditions by encouraging a more rigorous approach to examining existing and future demand. Capacity planning also promotes improved effectiveness in areas of the business that are impacted by investment decisions as it provides these areas of the business with better clarity for their future needs, for example, accurate forward knowledge of the volumes of physical site build work required may allow the more efficient procurement of resources to undertake this work.

If the provider commits to refining and improving the capacity planning process, they will be in a better position to enable the development of more accurate and effective infrastructure investment plans. This improved outcome can be achieved by ensuring that a wide ranging and sometimes complex set of inputs and scenarios are tackled in a logical, comprehensive manner. As the development of any future plan entails dealing with an inherent level of uncertainty, the more rigor that is placed on undertaking these activities, the better the long-term result is for the provider. For example, at first glance, there may be two equally valid infrastructure investment alternatives for a provider to satisfy an expected scenario. However, without undertaking the appropriate capacity planning steps, longer term benefit to the provider from an alternative may not become apparent. This is not to say that either alternative is incorrect, one simply may be more appropriate than the other when all relevant factors are taken into account.

Having a mature, well-designed, and well-resourced capacity planning process helps to ensure that the most appropriate investment decisions can be made, as it provides a structured mechanism via which future uncertainty and complexity can be readily converted into sensible outcomes.

15.3.3 Carrying Out the Process

Capacity planning in a CDN context can be performed either on a regular, scheduled basis or triggered due to changes in inputs that impact activities within the process. As the individual activities that comprise the process are likely to have inputs with varying degrees of timing and confidence, the scheduled process undertaking may benefit from aligning with timeframes where there is an expected high degree of validity of the relevant inputs. For example, timeframes for capital procurement may be on biannual basis, whereas customer demand forecasts may be updated on a more regular basis. It may be prudent in such a case to schedule process undertaking immediately after such events where the inputs are expected to be at their most valid state (Figure 15.4).

The validity and hence immediate relevance of the inputs also provide a guide as to which activity to start the process from. For example, if the customer demand forecast has recently changed but the capital position of the provider is relatively stable, then it makes most sense to begin the process undertaking by carrying out the capacity requirements activity first. As all activities are required to be undertaken to complete the process, it is more logical and effective to begin with the activity that has the most recent and relevant inputs, and, therefore, is more likely to be the source of any changes impacting the provider's investment decisions. Working from the source of changed inputs also

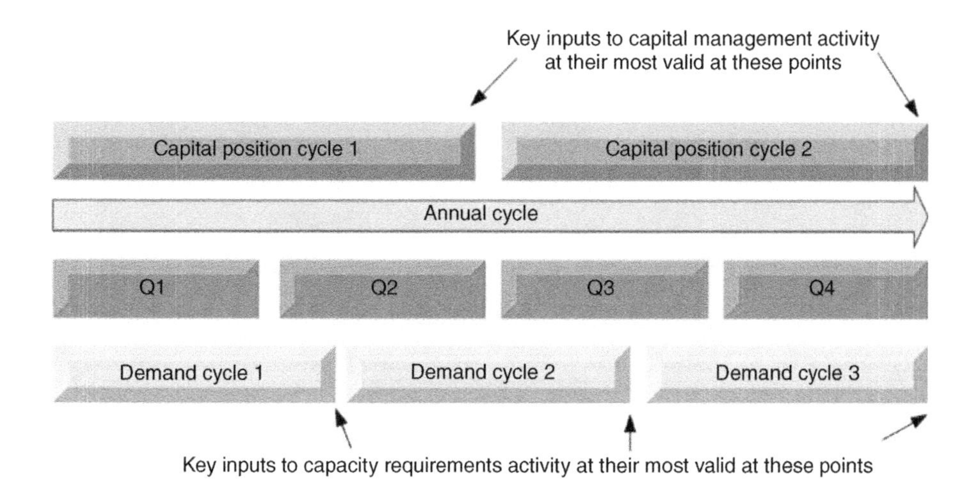

Figure 15.4 Validity of key inputs.

assists in reducing iteration and promotes a more efficient approach to undertaking the process.

For example, if the key inputs to the capacity requirements activity are at their most relevant (such as a customer demand forecast indicated in Figure 15.4 in the early part of Q2), it would be more prudent to begin there, noting that outputs from this activity will flow into other activities within the process (Figure 15.5).

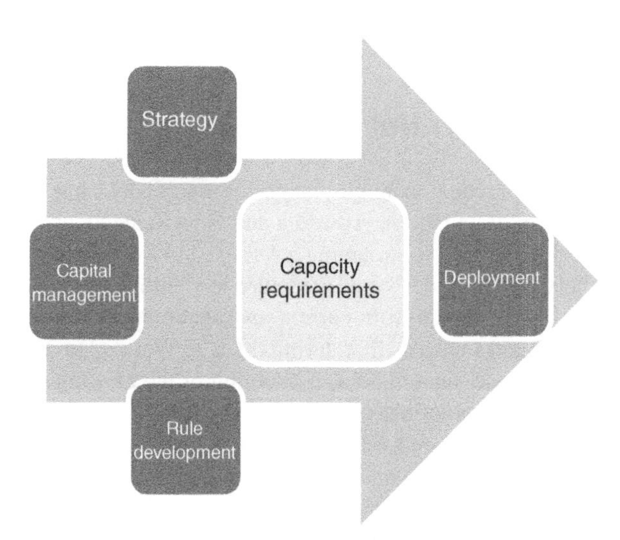

Figure 15.5 Where to begin undertaking the process if key inputs to the capacity requirements activity change.

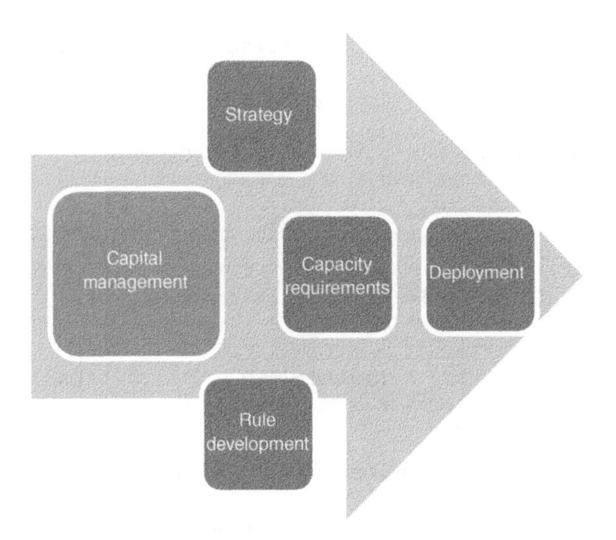

Figure 15.6 Where to begin undertaking the process if key inputs to the capital management activity change.

However, if the key inputs to the capital management activity are at their most valid (e.g., at the end of Q2 in Figure 15.4), it may be prudent to begin undertaking the process by starting with the capital management activity (Figure 15.6).

Note that what has been presented here is a guide and the exact methodology utilized to provide the optimal outcome will depend on the level of skills, experience, and resources available to perform capacity planning. The key point which is worth reinforcing here is that persistent, sustained application of effort into the area of capacity planning will produce the best results for the provider.

15.4 CDN CAPACITY PLANNING CASE STUDY

Let us now consider a practical example for performing CDN capacity planning. The ICC World Cup 2015 will be held in Australia and New Zealand, and a regional CDN provider is responsible for the live streaming coverage and website content delivery during the event. The CDN provider's infrastructure has sufficient capacity to cater for existing and forecasted demands from current and upcoming customers. However, its business intelligence team forecasts that during the ICC World Cup 2015, there will be a significant traffic increase, quantified as a doubling of the existing capacity demands in terms of bandwidth consumption, content delivery transactions, and volumes delivered. In order to provide coverage for the upcoming major event while still maintaining adequate customer service performance, the CDN provider will need to investigate the impact of this event on its infrastructure with a view to determining whether an augmentation with additional capacity is likely to be required.

At the first glance, this may appear to be a scenario that could result in multiple likely outcomes, each with varying degrees of complexity and requiring vastly differing levels of resource to address adequately. From a "do nothing" requirement at one end of the scale to a wholesale redesign of the provider's infrastructure at the other, multiple options in between the investigation of the impact of this event on the providers' infrastructure may at first appear somewhat iterative and confusing. However, undertaking the capacity planning process in a logical manner will assist in determining the required outcome for the provider in the most effective and efficient manner.

The first issue to address here is to determine which activity within the process is directly impacted by the triggered event. Doing this will allow for the most effective analysis of the scenario presented, as it minimizes any potential iteration by addressing the source of the changed input to the process.

As the impact of this event is triggering a change in demand, and demand is an input to the capacity requirements activity, it is appropriate that the provider's analysis begins there, for example:

- Determine existing inventory (the *What*), as mentioned in Section 15.2.
- Determine existing demand (bandwidth, transactions, volumes).
- Then double it. It can be noted that it is worthwhile doing a sanity check with the forecasters at this point, as few forecasts have 100% confidence so it may be more pragmatic to look at a range, or at least get some feedback from the forecaster.
  - At this point you have the *Why*, that is, can it be accommodated within existing capacity?
- A couple of options may present themselves here, for example, the doubling of demand may or may not be able to be accommodated.
- If the doubling of demand is able to be accommodated, then no further action may be required.
- If the doubling of demand cannot be accommodated within the existing capacity, how much more is required and where? (presuming we already have an idea of the when).
  - This can be determined by applying the rules (the *How*) to the inventory and demand figures.
- At this point, a plan can be formed for infrastructure deployment, which will contain information on sites, costs, and timings.
  - Also, indications that scalability limits will or will not be breached will also be known. Note that at this point feedback in terms of build costs to capital requirements can also be provided, along with the feedback to rules development, if scalability limits for the existing platform will be breached by the amount of augmentation required.
- If scalability limits are not breached by the required level of augmentation (and assuming funding along with resources to implement augmentation are available), then the necessary deployments would proceed to meet the demand requirement for the event.
- If scalability limits for the provider's infrastructure are going to be breached by the level of augmentation required for the event, then other alternatives will need to be examined, such as

- Upgrade/improve the existing platform architecture to allow the appropriate level of scaling.
- Utilize another platform, for example, procure capacity from another provider via a purchase or leasing arrangement.
- Note that cost and time analysis for such alternatives will be required, and cost to the business for such alternatives will be key feedback provided from the capacity planning process.

Note that there are numerous other examples whereby a scenario being presented will require analysis via the capacity planning process to determine the impacts to the provider's infrastructure investment plans. Such examples include the following:

- A change to corporate strategy, such as targeting new market segments. Again, this is best tackled by beginning at the strategy activity and working from there.
- A change to available installation resources because of these being diverted to repair activities after a natural disaster. By beginning at the deployment activity, reprioritization of infrastructure installation to align with scarcer deployment resources, it is possible to meet the provider's most urgent needs.

Note that these are only a few examples whereby a complex, wide-ranging set of inputs is required to be tackled to ensure that the provider continues to make optimal infrastructure investment decisions.

15.5 RECENT DEVELOPMENTS AND CHALLENGES

Capacity planning has been in and around the industry since the inception of first network technology. Although some aspects of capacity planning have evolved over time, the core capacity planning remains the same. Nevertheless, the capacity planning process now faces additional challenges resulted in from technological advancements, of which key processes are as follows:

- *Lack of Longer Term Focus*. Modern companies tend to be transactional, rather than transformational. Current revenue is considered as the major driver of activities; thus, there is a major focus on short-term planning and infrastructure development for the capacity that is required to meet near-term demands. Often longer term strategic capacity planning is not featured with a high degree of importance in the company's program of work. Moreover, in the post-GFC time, companies are less willing to make longer term investment as the financial situation can fluctuate, leading to potential loss for the company.
- *Automation*. It is becoming a widely adopted practice to move toward automated tools for capacity planning; however, there is a significant upfront cost for the development of these automation tools. Companies now need to focus on allocating additional budget for capacity planning not only for the core process itself but also for automating it.

- *Tailored Monitoring*. As infrastructure is grown, it is expected that capacity planning is performed at a more granular level. As for example, for a nationwide CDN infrastructure, it may be expected by the business to perform separate and segmented capacity planning at each Point of Presence (PoP), datacenter, and core network backbone level.
- *Technology Diversification*. With the deployment of a new technology, it is difficult to perform capacity planning as in many instances there may not be an existing standard or benchmark to compare to. Moreover, unavailability of relevant forecasting tools and data would mean a capacity planner needs to make planning decisions with little available information. Having a well-resourced and skilled capacity planning, team can assist in scenarios like the experience in decision making is able to be leveraged by the provider.
- *Reduced Predictability*. There has been decreasing predictability on demands and system behavior. This is due to two major reasons: (i) it may not be often clear how customer demand profiles and impact on systems may change with new technologies; and (ii) system and inventory behavior is becoming more complex with addition of new services by augmenting (or repurposing) existing infrastructure.

15.6 SUMMARY AND CONCLUSION

Capacity planning is an integral part of a modern technology organization. Instead of working in a reactionary manner, analyzing and correcting performance problems as users report them, capacity planning helps providers to prepare in advance to avoid performance bottlenecks altogether. With the use of appropriate capacity planning tools, a CDN organization could better handle short-lived, spatiotemporal events (e.g., flash crowds) and meet requirements for a growing business. In this chapter, we have provided a coverage of the capacity planning process, with description of methodology, and how it is conducted in practice. A real-life case study is presented to demonstrate to aid the readers. While capacity planning has been around in the industry for a long time, recent technology advancements demand the adaptation of tools and approaches. We have provided a brief coverage of relevant issues and the associated challenges that are faced in the capacity planning domain. It is expected that capacity planning can be the best weapon for optimal infrastructure investment in order to satisfy service-level requirements of the business.

REFERENCES

1. Buyya R, Pathan M, Vakali A. *Content Delivery Networks*, Vol. 9. Springer; 2008.
2. Lazar I, William T. Exploring content delivery networking. IT Prof 2001;3(4):47–49.
3. Pallis G, Vakali A. Insight and perspectives for content delivery networks. Commun ACM 2006;49(1):101–106.
4. Saul JR. *Voltaire's Bastards: The Dictatorship of Reason in the West*. 1st ed. Penguin; 1993. p 386.

5. Rich J, Hill J. How to do capacity planning, Whitepaper, TeamQuest, USA; April 2010.
6. Huston G. *ISP Survival Guide: Strategies for Running a Competitive ISP*. NY, USA: John Wiley & Sons; 1999. p 297–300.
7. Menasce DA, Almeida VAF. *Capacity Planning for Web Services: Metrics, Models, and Methods*. Upper Saddle River: Prentice Hall; 2002.
8. SAI Global Limited. Demystifying ISO 9001:2000; 2001. p 38.

PART III

CASE STUDIES AND NEXT GENERATION CDNs

16

OVERLAY NETWORKS: AN AKAMAI PERSPECTIVE

Ramesh K. Sitaraman[1,2], Mangesh Kasbekar[1], Woody Lichtenstein[1], and Manish Jain[1]

[1]*Akamai Technologies, Inc. Boston MA, USA*
[2]*University of Massachusetts Amherst MA, USA*

16.1 INTRODUCTION

The Internet is transforming every aspect of communication in human society by enabling a wide range of applications for business, commerce, entertainment, news, and social interaction. Modern and future distributed applications require high reliability, performance, security, and scalability, and yet need to be developed rapidly and sustainably at low operating costs. For instance, major e-commerce sites require at least "four nines" (99.99%) of reliability, allowing no more than a few minutes of downtime per month. As another example, the future migration of high quality television to the Internet would require massive scalability to flawlessly transport tens of petabits per second of data to global audiences around the world.

However, the Internet was never architected to provide the stringent requirements of such modern and futuristic distributed applications. It was created as a heterogeneous network of networks, and its design enables various entities to interact with each other

Advanced Content Delivery, Streaming, and Cloud Services, First Edition.
Edited by Mukaddim Pathan, Ramesh K. Sitaraman, and Dom Robinson.

in a "best effort" manner. Guarantees on high performance, availability, scalability, and security are not inherently provided on the Internet in accordance with its best effort design principle. Today's Internet is a vast patchwork of more than 13,000 autonomous networks that often compete for business. Failures and performance degradation in transporting information across this patchwork are routine occurrences.

So, how would we bridge the gap between what modern Internet-based services need and what the Internet actually provides? A complete clean-slate redesign of the Internet is appealing but would be hard to implement, given the widespread adoption of the current technology. A novel idea to bridge the gap is *overlay networks*, or just *overlays* for short. The fundamental idea of overlays is rooted in the age-old computing paradigm of *virtualization* that states that if you do not have what you want you can *virtually* create what you want with what you have. This principle is at foundation of a number of computing innovations over the decades. For instance, virtualization is why a computer with finite and fragmented storage can be made to appear to the programmer as if it had a single, contiguous *virtual* memory space. Otherwise a Linux server can be made to emulate a *virtual* machine to provide the abstraction of a Windows desktop to a user. Along the same lines, we can build a *virtual* network (the overlay) over the existing Internet (the underlay) to provide the stringent requirements of modern Internet-based services (cf. Figure 16.1).

Overlays use the functional primitives that the underlay has to offer. In turn, the overlay provides richer functionality to services that are built on top of it.

16.1.1 Types of Overlays

Different overlays offer different enhanced functionalities for online services. As such, there are as many types of overlays as there are service requirements. However, our aim is not to be comprehensive. Rather, we only review the following three types of overlays that are illustrative and important for meeting the needs of Internet-based services:

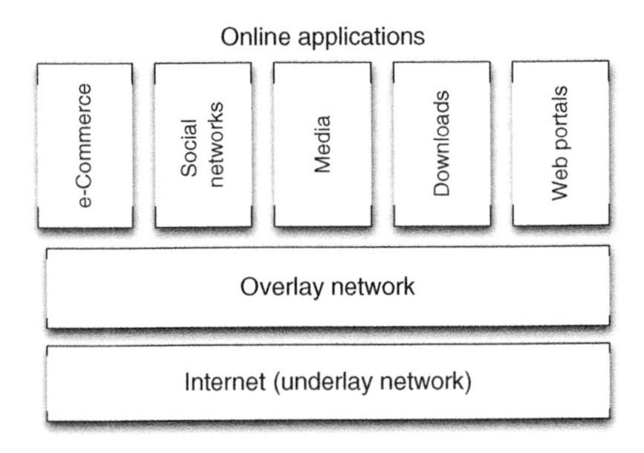

Figure 16.1 An overlay network is built on top of the public Internet to provide the stringent requirements that rich Internet-based services need.

1. The ubiquitous *caching overlay* aims to deliver websites, on-demand videos, music downloads, software downloads, and other forms of online content. Such overlays are applicable for content that does not change over extended periods of time and is hence *cacheable*. The key benefits that a caching overlay provides are greater availability, performance, origin offload, and scalability (cf. Section 16.3).
2. The *routing overlay* provides wide-area communication with more reliability, lesser latency, and greater throughput than the public Internet can. Such overlays could be used to deliver dynamic Web content or live stream content that normally cannot be cached (cf. Section 16.4).
3. The *security overlay* increases the security and mitigates distributed denial of service (DDoS) attacks on websites and other online services (cf. Section 16.5). Such overlays are at the core of some of the most sought after Akamai services, while at the same time they are a classic example of the overlay philosophy for enhancing the underlay by providing new functionality.

As we discuss the architecture and techniques of each type of overlay, it is important to remember that these overlays are often utilized together as a part of a single solution. For instance, an e-commerce application provider would use the caching overlay for the static Web content such as images, the routing overlay for the dynamic content such as the Hypertext Markup Language (HTML), and use the security overlay to ensure security and business continuity.

16.2 BACKGROUND

First, we provide background information on the Internet (the underlay) and what it can or *cannot* provide an overlay designer. In fact, the shortcomings of the Internet architecture are the very reason why overlays are required. Next, we trace the evolution of overlays from both an industry and academic research perspective. Finally, we describe the high level architecture of the overlays, which we study in more detail later. For a more comprehensive treatment of fundamental overlay architecture, the reader is referred to Reference 1.

16.2.1 Deficiencies of the Internet

The Internet is not a single entity but a network of thousands of autonomous networks. No single network dominates the Internet traffic, with the largest controlling less than 5% of the access traffic. This implies that most users accessing a popular website that is centrally hosted must traverse multiple networks to obtain that content. The networks that make up the Internet are autonomous business entities and often compete with each other. Not surprisingly, the end-to-end communication on the Internet is governed largely by business rules rather than on a notion of maximizing the performance perceived by the user.

We outline the major shortcomings of Internet that make it unsuitable for *directly* supporting the stringent requirements of Internet-based services *without an overlay*.

As we will see in the succeeding sections, an appropriately architected overlay can alleviate some of the shortcomings in the following.

1. *Outages.* Partial network outages are common on the Internet caused by misconfigured core routers, DDoS attacks, cable cuts, power disruptions, natural calamities, and depeering because of a business conflict. For instance, a number of major incidents have occurred recently with an important submarine cable system (the SE-ME-WE-4) leading to outages for millions of users in the Middle East, South, and South East Asia several times in each year between 2008 and 2013 [2]. More recently, a power outage in a Virginia data center took down a number of online services [3]. As another example, Sprint and Cogent decided to depeer resulting in a partial loss of connectivity for over 3500 networks in 2008 [4].
2. *Congestion.* When the capacity of routers and links on the Internet is insufficient to meet the traffic demand, congestion occurs resulting in packet loss. Packet loss manifests itself as performance problems perceived by the user, including slow downloads of Web pages and freezing of videos during playback. Peering points where individual networks exchange packets are particularly susceptible to packet loss. Part of the reason for the peering bottleneck is economics. Networks are incentivized to provision surplus capacity on the "first mile" that serves their hosting customers and the "last mile" that serves their paying end-users. However, there is no significant economic incentive to exchange traffic with other potentially competing networks over peering points in the "middle mile." Shared mediums such as wireless links are also particularly susceptible to packet loss.
3. *Lack of Scalability.* Online services require provisioning server and network resources to meet the demand of users at all times, even during unexpected periods of peak demand and flash crowds. Without the existence of overlays, an enterprise may deploy their online services in a centralized manner within a single data center and expect to serve their users from that centralized origin infrastructure. However, such a centralized solution falls significantly short of meeting the requirements of mission critical services. The data center itself is a single point of failure where outages or congestion can adversely impact the availability and performance of the service. Further, the enterprise would have to provision resources for peak demand that is significantly higher than the average. Such overprovisioning for peak means wasting money on infrastructure that is seldom used. On the other hand, underprovisioning to save cost has dire consequences as well, as the service may be unavailable at critical moments of high demand, leading to loss of revenue. To some degree, one can alleviate some of these issues by mirroring the origin in multiple data centers or by multihoming on multiple networks [5]. However, these solutions increase the cost and complexity of the origin infrastructure and by themselves are unlikely to provide the stringent requirements of modern online services. Therefore, an overlay solution that allows for massive replication and automatic scaling is required to meet the user demand that could vary significantly over time.

4. *Slow Adaptability.* Online services and their requirements evolve rapidly. However, the fundamental architecture and protocols of the Internet are slow to change or accommodate new primitives. For instance, the rollout of a new version of Internet Protocol (IP) called IPv6 was first proposed in 1998, but has just started to gain some adoption about 15 years later. IPv6 traffic currently accounts for around 1% of the Internet traffic [6]. The large investment by networks and enterprises in the current Internet technology is often a barrier to change. The complexity in business relations between networks is also a stumbling block. In some cases, the highly decentralized architecture of the Internet means that a large number of autonomous networks must all adopt the change to reap its full benefits, which rarely happens. However, contrary to the Internet architecture, an overlay is often owned and operated by a single centralized entity and can be deployed rapidly with no changes to the underlying Internet. Thus, overlays are an attractive alternative for adapting to the fast changing requirements of online services in a way that the Internet underlay cannot.
5. *Lack of Security.* Modern online services require protection from catastrophic events such as DDoS attacks. A recent report by Arbor Networks found that the average size of a DDoS was 1.77 Gbps in the first quarter of 2013, an increase of almost 20% from a year ago [7]. Aside from the size, the number of DDoS attacks is also growing year after year. Each attack could cause a significant amount of revenue loss if the online business, such as an e-commerce site, is unable to withstand the attack and becomes unavailable to users. The Internet architecture provides no protection against DDoS attack modes such as syn floods and packet floods. For each individual business to provision for additional server and bandwidth capacity to tackle the possibility of a DDoS attack can be prohibitively expensive, making the overlay approach to defend against such attacks a more attractive option.

16.2.2 A Brief History of Overlay Networks

The idea of building one network (overlay) over the other (underlay) is at least a few decades old. In fact, the early Internet was itself initially built as an overlay on top of the telephone network that was the predominant network of the day. Much of the early overlays were build to provide functionality that the underlay natively lacked. A classic example is Mbone [8] that can be viewed as an overlay on the Internet providing multicast functionality. Further, the Internet was not the only underlay studied in the context of overlays. In fact, the concept of overlays found simultaneous and in some cases earlier developments in the domain of interconnection networks for large parallel computers.[1] For instance, early work from the late 1980s [9] showed how to effectively emulate one type of network (say, a 2D mesh) as an overlay on a different type of underlay network (say, a butterfly).

[1] Rather than using the terms overlay and underlay, the terms guest and host network, respectively, were used in the parallel network literature.

Besides creating new functionality, the potential for improving reliability and performance by building a virtual overlay network also has a long history. In the domain of parallel networks, work from the early 1990s showed that it is theoretically possible to build an overlay that provides an abstraction of a failure-free network over a failure-prone underlay network of the same size and type without significant loss in performance [10–12]. Subsequently, in the late 1990s, overlays to enhance the availability, performance, and scalability of the Internet came to prominence within both industry and academia, even as the Internet became crucially important for business, commerce, and entertainment. Seminal research advances were made in academia with systems such as Resilient Overlay Network (RON) [13] that showed that routing overlays can effectively use the path diversity of the Internet to improve both end-to-end availability and performance. Among the earliest to appear in industry were the caching overlay and the routing overlay. By the end of the 1990s, companies such as Akamai had full fledged offerings of caching overlays for delivering web and on-demand videos [14] and routing overlays that relied on the path diversity of the Internet to provide higher quality live streaming [15,16].

Another major development over the past decade is the emergence of peer-to-peer (P2P) overlays that use (nondedicated) computers of the end-users themselves to form overlays that can be used for downloading content. Early systems that used P2P principles include the now-defunct services, such as Kazaa [17] and Gnutella [18], and innovative experimental systems, such as Chord [19], content-addressable networks [20], Tapestry [21], and Pastry [22]. Although pure P2P systems have had less adoption among enterprise customers who demand higher levels of availability, performance, and content control than such systems typically offer, hybrid approaches that combine P2P principles with a dedicated overlay infrastructure are widely used in services such as Akamai's client-side delivery [23]. While P2P overlays are an important type of overlay, we do not describe them in more detail here. All the three types of overlays described here use a dedicated server infrastructure owned and operated by the overlay provider, rather than computers belonging to users.

16.2.3 Overlay Architecture

The overlays that we study in detail are used to deliver content, applications, and services to users on behalf of content providers.[2] Content providers include news channels, e-commerce sites, social networks, download services, Web portals, banks, credit card companies, and authentication services. The end-users access content, applications, and services from across the globe from a variety of devices, such as cell phones, tablets, and desktops, using a variety of client software including browsers, media players, and download managers.

An overlay capable of delivering content, applications, and services to a global audience is a large distributed system consisting of hundreds of thousands of globally

[2]For simplicity, we use the term content provider to denote traditional content owners such as websites, application providers such as a software-as-a-service provider, and service providers such as an authentication service provider.

deployed servers that run sophisticated algorithms. Independent of the specific type of overlay, they share a similar system-level architecture as shown in Figure 16.2. However, the detailed design and implementation of each system component differs depending on the precise functionality that the overlay provides. The content, application, or service is hosted by the content provider in one or at most a few *origin* locations on the Internet. Users who access the content, application, or service interact directly with a wide deployment of *edge servers* (ESs) of the overlay, rather than directly with the origin. For instance, a large-scale Akamai overlay consists of over 100,000 ESs that are physically located in over 80 countries and in over 1150 networks around the world. Contrary to origin that is located in the core of the Internet, the ESs are located at the "edges" of the Internet. The ESs are deployed very close to users in a network sense so as to provide low latencies, low packet loss, and higher bandwidth, resulting in better performance. To understand how the different systems of an overlay interact to deliver content to the user, it is instructive to consider the simple example of a user entering a uniform resource locator (URL) into his/her browser and receiving a Web page through the overlay. The important control or data flow at each step is shown with arrows in Figure 16.2.

- The domain name of the URL is translated by the *mapping system* into the IP address of an edge server that can serve the content (arrow 1). There are a number of techniques for implementing the domain translation. A simple but

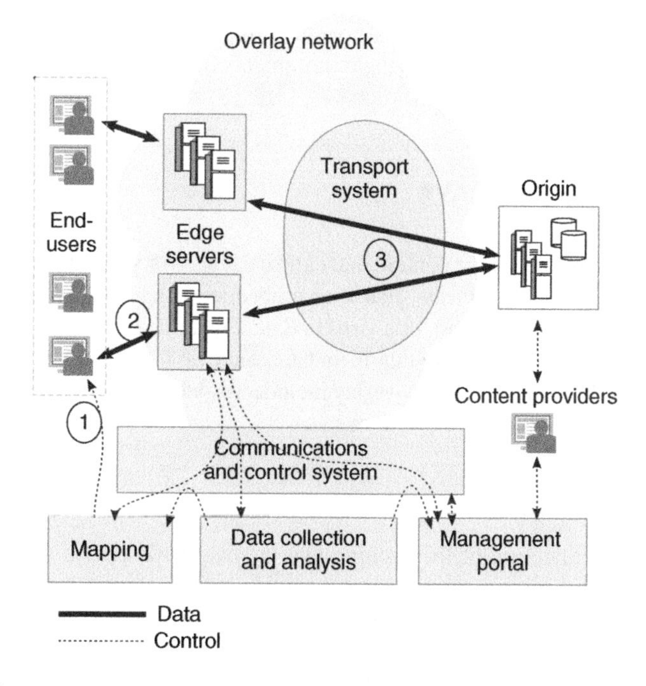

Figure 16.2 The system-level architecture of an overlay network [1].

rudimentary mechanism is to assign the same IP address to all the ESs and rely on network-level anycast for determining the "right" ES for a given end-user. A more robust and scalable mechanism is to use the domain name system (DNS) for the translation and is used in Akamai's overlays. The mapping system bases its answers on large amounts of historical and current data that have been collected and processed regarding the global Internet and server conditions. This data is used to choose an ES that is located "close" to the end-user so as to maximize performance.

- The browser sends the request to the chosen ES that is now responsible for serving the requested content (arrow 2). The ES may be able to serve the requested content from its local cache. Otherwise it may need to communicate with an *origin* server to first obtain the content that is then placed in its local cache and served to the user.
- The *transport system* is used to transfer the required content between the origin and the ES (arrow 3). The transport system is at the heart of the overlay, which moves content over the long-haul Internet with high reliability and performance.
- The *origin* is typically operated by the content provider and consists of Web servers, application servers, databases, and other back-end infrastructure that serve as the source of the online content, application, or service. In the case of live streaming, the origin also consists of encoders and media servers that originate the stream and transmit it to the overlay network via "entry points" [16].

The different types of overlays that we study implement each of these systems differently in accordance with the differing requirements of the overlay. We will delve into those differences in the following sections.

16.3 CACHING OVERLAYS

The caching overlay is used for content that can be cached for some time period. Canonical examples include static objects such as an embedded image on a Web page, a movie, a music file, a software download, or a virus update. Dynamic objects that do not change very often, such as a weather map that is updated once an hour, can also benefit from a caching overlay. The benefits of the overlay include availability, performance, and origin offload, each of which we address in turn.

16.3.1 Architecture

The high level architecture of the caching overlay can be described with reference to Figure 16.2. The mapping system directs each user to the closest server in a network sense, identified by using recent knowledge of the network latency and packet loss rate along the path between the user and the ESs. The ESs provide the functionality of caching HyperText Transfer Protocol (HTTP)/HTTP Secure (HTTPS) proxy servers. If the content requested by the user is found in local cache, the ES serves it to the user. If the content is not found in its own local cache or in any other ES in its cluster, then the ES uses the transport system to download the content from the origin. The transport system may itself have another layer of servers called *parent servers* that are capable of caching content. In this case, the ES requests a parent server for the content. If the parent server has the content, it is served to the ES. Otherwise, the parent server requests the content from origin, and caches the content itself, and then it is forwarded to the ES. In some cases, one could have multiple layers of parent caches that are requested before the request is forwarded to origin. A transport system with one or more layers of parent servers is called a *cache hierarchy*.

16.3.2 Performance Benefits

It is easy to see why the caching overlay improves the performance of cacheable Web content. The *edge hit rate* is defined to be the probability of finding the requested content in the ES without having to download it from a parent server or the origin. If the edge hit rate is high, instead of traversing a major part of the Internet to get content from the origin servers, users fetch their required content from a nearby ES that is reachable on a path that is known to have low latency and low packet loss. For a typical popular website, the edge hit rate can be very high at 90+%, resulting in almost all user requests being served from an ES's cache leading to significant performance improvements.

To better quantify the performance benefits for cacheable content, we used a *performance testing platform* that uses a large collection of "agents" installed on end-user desktop machines located around the world. These agents are capable of performing periodic downloads of Web pages and reporting fine-grain performance measurements about each download. For our experiment, we used 30 agents located in Asia, Europe, and North America, each with broadband connectivity to the Internet. The agents did hourly downloads of two versions of a cacheable 32 KB file. The first version of the file was delivered directly from our origin servers in Dallas where we hosted the file. Thus, this version did not use the caching overlay. The second version of the file used an Akamai caching overlay to deliver the file.

A natural measure of the performance benefits of the overlay is *speedup* that is defined to be ratio of the time to download the file directly from the origin to the time to download the same file using the overlay. In Figure 16.3, we show the speedup aggregated by the continent where the user (i.e., agent) was located. The caching overlay provides large speedups between 1.7 and 4.3 in all continents. Further, users farther away from the Dallas origin experience larger speedups. The reason is that the download time using the overlay remains roughly the same independent of geography, since the caching overlay can find an ES close to the user in all geographies. However, the download time from the origin deteriorates as users move farther away from it.

16.3.3 Origin Offload Benefits

An important benefit of using a caching overlay is the decrease in traffic served by the origin and is measured by a metric called *origin offload* that equals the ratio of the volume of traffic served by the origin without the overlay to the volume of traffic served by the

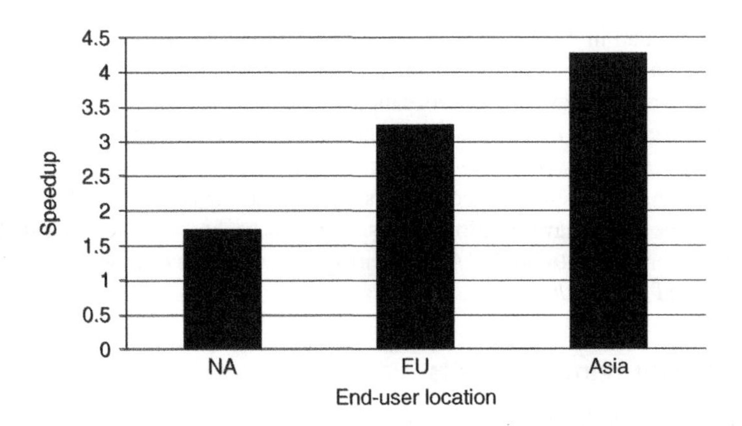

Figure 16.3 Performance benefits of a caching overlay.

origin with the overlay. When an overlay is used, only the traffic due to cache misses at the ESs is served by origin. If cache hit rates are high, as they are with popular Web content, the origin offload could be anywhere from 10 to 100. Note that origin offload is very beneficial to the content provider as they only have to provision their origin to serve a small fraction of the traffic that they would have had to serve without the overlay, resulting in a large decrease in server, bandwidth, colocation, and operational expenses. However, not all Web content is popular and most websites have a "long tail" of less popular content. Take for instance an e-commerce site. Although a few product pages may be "hot" and will yield a high edge hit rate and a high origin offload, most of the other product pages may at best be "warm" or even "cold," yielding much smaller edge hit rates and much less origin offload. Providing better origin offload for "warm" and "cold" traffic requires more sophisticated architectures that we describe later.

16.3.3.1 Cache Hierarchy The basic idea in a cache hierarchy is to add a layer of *parent servers* that serves the requests that experience cache misses at the ESs. If an ES fails to find the requested content in its cache, it forwards the request to a chosen parent server. If the parent server has the content in its cache, it serves it to the ES. Otherwise, the parent obtains the content from origin. Adding a layer of parent servers decreases the traffic to the origin, increasing the origin offload. Typically, the requests that are cache misses at the edge are "funneled" to a smaller number of parent servers, so as to increase the hit rate at the parent, thereby increasing the origin offload.

Cache hierarchy is relatively easy to implement as a parent server uses similar software components as an ES for caching and serving the content. Further, the mapping system can be used to find a parent server that is "proximal" to the ES. Finding a parent close to the ES decreases the parent-to-edge download and improves performance.

To illustrate the origin offload benefit, Figure 16.4 shows the traffic and hit rates over a 2-month period for a content provider with a mixture of popular and less popular pages. The content provider uses a cache hierarchy and has an edge hit rate in excess of 85%. The requests that "miss" at the edge are for less popular content and the probability

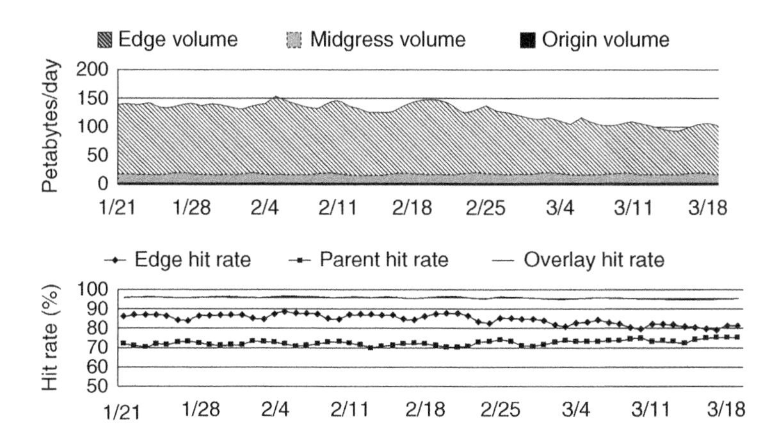

Figure 16.4 Traffic volumes and cache hit rates for a content provider with a mix of popular and less popular content.

that such a request can be served from the parent's cache, known as the *parent hit rate*, is over 70% hit rate. Cumulatively, the *overlay hit rate* which is the probability that a request is served from a parent or ES of the overlay is 96%, that is, only 4% of the total requests are served from origin. Note that without the cache hierarchy about 15% of the requests would have been served by origin.

In Figure 16.4, we plot the *edge traffic* that is served to user, the *midgress traffic* that is served by parents to edges, and the *origin traffic* that is served by the origin to the parents. Observe that without the cache hierarchy, all midgress traffic would be served from the origin. Thus, the origin offload without the cache hierarchy is simply the ratio of the edge traffic to the midgress traffic, that is, only 6.7. While the origin offload with the cache hierarchy is the ratio of the edge traffic to the origin traffic, that is, much larger at about 24.3. Thus, a cache hierarchy significantly increases the offload factor.

Note that while we have described a cache hierarchy with a single layer of parent servers, one can easily extend the hierarchy to consist of multiple levels of parents where a lower level parent fetches from a higher level parent, and the parent at the highest level fetches from origin. Although multiple levels of cache hierarchy can further increase the origin offload, it can be detrimental to performance for the fraction of requests that get forwarded to multiple parents before a response.

16.3.3.2 Dealing with Unpopular Content

In recent years, with social networking sites that carry user-generated content such as photos and videos, the amount of unpopular "cold" content on the Internet has exploded. For such types of content, providing origin offload can be challenging since there is a large footprint of objects that are each accessed only a few times, requiring an "offload oriented" overlay to provide higher overlay hit rates. Although popular content can be served by making several copies of content and placing them close to the end-users, to achieve high origin offload for unpopular content we must make a very small number of copies of the content, thus

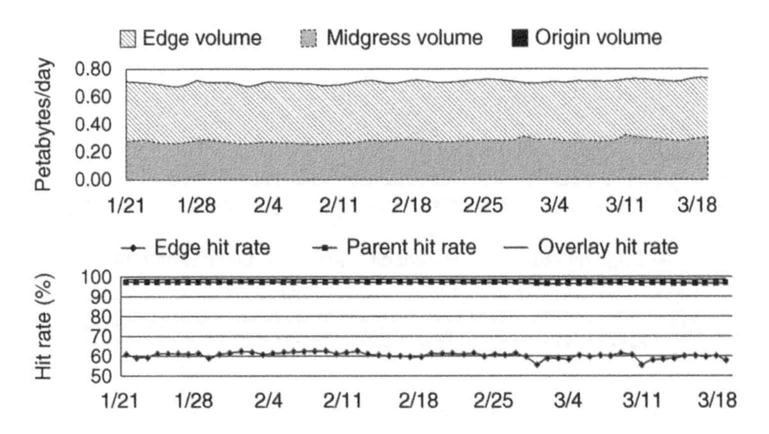

Figure 16.5 Traffic volumes and cache hit rates for a social networking site with predominantly unpopular content. The overlay uses more sophisticated content location techniques to increase the overlay hit rate and origin offload.

using the available cache space to cache the most number of unique objects within the overlay.

Since any given piece of content is available at fewer locations in the overlay, a content location service is needed as part of the caching overlay. It can be implemented as a distributed hash table or with a more advanced directory service. The content placement and location techniques can be designed to automatically adapt to content popularity so that several copies are made of the popular content for enhancing performance, while the number of copies of unpopular content is kept small to conserve storage space.

Figure 16.5 shows the hit rates and traffic for unpopular social networking content using more sophisticated content location techniques. It can be seen that the content has relatively poor edge hit rate of 60.2% in comparison with the more popular content depicted in Figure 16.4. However, the content location techniques provide an extremely high overlay hit rate of 98.9%, that is, only 1.1% of the total request is served by origin. The origin offload is only 2.5 without the content location techniques, but rises to 89 with it.

16.4 ROUTING OVERLAYS

Not all content on the Internet is cacheable for long periods of time. For Internet-based applications such as shopping, banking, and gaming, the downloaded content is dynamic and uncacheable in the sense of being generated based on the user's interaction with the application in real time. Such requests and responses therefore must traverse the Internet between the user and the origin. Another key example of dynamic content is live streams that cannot be cached and must be routed in real time from the origin, that is, the source of the stream to the users who are watching it. Despite the fact that caching cannot be used, a routing overlay improves performance and availability by discovering better "overlay paths" from the origin to the user.

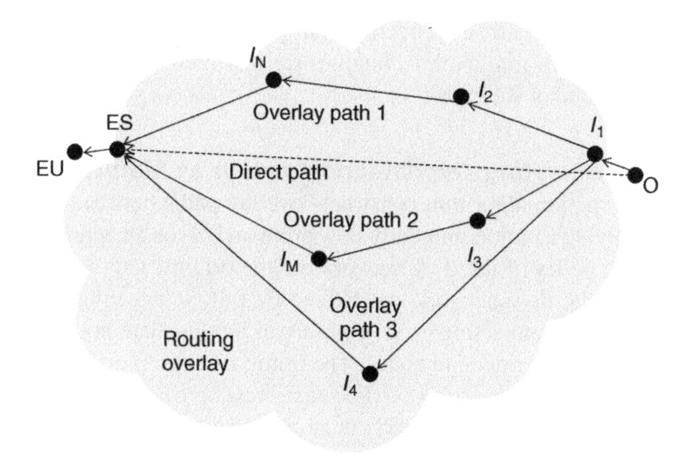

Figure 16.6 An [end-user] ([EU]) downloads dynamic content from an edge server (ES). The ES in turn can download directly from origin (O) or through a set of alternate overlay paths constructed between the ES and a reverse proxy (I_1).

16.4.1 Architecture

The architecture of a routing overlay is shown in Figure 16.6. An *overlay construction algorithm* is used to compute a set of *overlay paths* that each ES can use to reach each of the origins. The overlay construction algorithm takes as input real-time latency, loss, and available bandwidth of the Internet paths connecting all the overlay nodes, and determines a ranked set of alternate overlay paths for requests and responses that travel between each ES and each origin. The overlay paths are chosen to provide multiple high performance paths for origin-to-edge communication and are frequently updated to account for real-time changes in the Internet. Once the overlay is constructed, each ES receives a real-time feed of the set of overlay paths that it could use to connect to each relevant origin in the world.

The overlay paths are used for routing communication as follows. When a request arrives at an ES, it deduces which origin serves the requested content, and forwards the request along a chosen overlay path(s) to that origin. Each intermediate overlay node acts as a forwarder of the request to the next node in the path toward the origin. The response from the origin traverses the same path in the opposite direction to arrive at the ES, which is then forwarded to the end-user. Note that the overlay construction process provides multiple alternate paths for each ES and each origin. In addition, the ES always has the option of the using the *direct path* from the edge to origin that passes through no other intermediate node. The choice of which of these alternate paths to use can depend on real-time testing of the different path options. In some overlays, each edge-to-origin communication uses a single chosen path. However, other routing overlays use multiple overlay paths simultaneously. In this case, the content can be encoded and sent across multiple paths that is then decoded by the ES and sent to the end-user. Using multiple paths is useful if one of the chosen overlay paths experiences transient packet loss.

The data lost in one path can be recovered from packets received on the other path. A more detailed account of multipath techniques for loss recovery in the context of live streaming can be found in Reference 16.

16.4.1.1 Formulating Overlay Construction as Multicommodity Flow

The overlay construction algorithm constructs overlay paths between each origin (O) and each ES by solving a multicommodity flow problem [24] on an appropriately defined flow network. The nodes of the flow network are the origins, intermediate nodes, and ESs. The edges of the flow network are defined by fully connecting the intermediate nodes with each other, connecting each origin to all intermediate nodes, and also connecting each ES to all intermediate nodes. The traffic from each origin O to each ES is represented as a distinct commodity $\langle O, \mathrm{ES}\rangle$ that must be routed on the flow network. The demand of each commodity $\langle O, \mathrm{ES}\rangle$ is an estimate of the traffic that needs to be sent from origin O to ES. Demand needs to be measured over short-time windows, for example, 10 s, since demand (i.e., traffic) can change rapidly over time. The capacity of each link in the flow network is an estimate of the maximum traffic that can be sent across that link. Finally, the cost of each link of the flow network can be modeled so that "good" links have lower cost, where goodness can be defined using a function that captures the objective of the routing overlay. Solving the multicommodity flow problem on the appropriately defined flow network yields a set of low cost paths for each commodity that correspond to the required overlay paths.

The power of the multicommodity flow formulation is the ability to define link costs in different ways to construct different types of routing overlays. For instance, a routing overlay for live streaming may define link costs in a different manner than a routing overlay for dynamic Web content, resulting in different overlay networks being constructed. We list a few ways of defining the link costs following.

1. *Latency Versus Bandwidth Price.* If path latency is considered to be the link cost, then the solution finds the fastest overlay paths from O to ES. If bandwidth price is considered to be the link cost, then the solution finds the cheapest overlay paths from O to ES. Combining the two types of costs could allow one to find different solutions, for example, finding the fastest overlay routes while avoiding links that are too expensive, or finding the cheapest overlay paths while avoiding paths that are too slow.
2. *Throughput.* In the Context of Web content delivery, two different notions of performance apply. Minimizing latency is important when delivering small-size responses. Maximizing throughput is important for large responses. These two are closely related since steady-state Transmission Control Protocol (TCP) throughput is inversely proportional to the round trip time (RTT) and the square root of the packet loss rate. If the loss rate is close enough to zero, it is possible to sustain high throughput over a long connection if TCP buffers are sufficiently large. In general, overlays with different properties can be obtained differently by weighting path metrics such as latency and loss in the derivation of the link costs.

3. *TCP Performance.* A vast majority of Internet traffic is served over the connection-oriented transport protocol TCP. Establishing a new TCP connection penalizes performance in two ways. Establishing a new connection requires additional round trips, adding to the latency. Further, new connections have smaller congestion windows impacts the number of packets you can have "in flight." An overlay routing algorithm that maximizes performance after accounting for TCP penalties must also attempt to reuse existing TCP connections rather than start new TCP connections. Thus, the overlay paths must remain "sticky" over longer periods of time and must change only when the cost parameters have changed sufficiently to overcome the potential penalties for a new TCP connection.

16.4.1.2 Selecting the Reverse Proxy An important aspect of overlay construction is selecting the reverse proxy (node I_1 of Figure 16.6) for each origin. The overlay construction algorithm typically chooses a reverse proxy close to or even colocated with the origin. It is important that the "first hop" from the origin to the reverse proxy has the smallest possible latency and loss, since the first hop is shared by all overlay paths from the origin to the ES. Further, while the overlay provider can ensure persistent TCP connections between any two nodes of the overlay, the first hop is partially dependent on the origin that is controlled by the content provider. For this reason, the first hop is more likely to lack persistent TCP connections and is more likely to incur TCP penalties. Making the first hop as low latency as possible reduces the potential penalty for establishing a new TCP connection. Similarly, while TCP optimizations that speedup downloads are implemented between any two overlay nodes, they are not guaranteed to be available on the first hop. For these reasons, choosing a reverse proxy close the origin is often desirable.

16.4.1.3 Fast Algorithms for Overlay Construction Computing the overlay paths efficiently requires heuristics as the multicommodity flow problem as formulated in Section 16.4.1.1 is, in general, nondeterministic polynomial (NP)-hard. However, the overlays need to be kept updated in real time responding to the ever-varying latency and loss conditions. Thus, using a scalable algorithm for solving the overlay problem is an important requirement. One approach is to use an optimal Lagrangian relaxation scheme that routes traffic through the lowest cost paths as determined by a modified Floyd–Warshall All-Pairs-Shortest-Path (APSP) algorithm. APSP itself can be modified to take advantage of the structure of the flow network that has a nearly fully connected "middle" with origins and ESs attached to the middle.

In addition to Lagrangian relaxation, a different approach is to write a mixed integer program (MIP) whose constraints can be "relaxed" to create linear program. The fractional solution for the linear program can then be rounded using generalized assignment problem (GAP) rounding techniques to produce an integral solution that corresponds to choosing the overlay paths. This approach yields provable guarantees for the approximate solution and is described in the context of constructing routing overlays for live streaming in [15,25].

16.4.1.4 Performance Benefits To illustrate the benefits, we measured the performance of a routing overlay during a recent large-scale Internet outage when a submarine communications cable system (called SEA-ME-WE 4) that links Europe with the Middle East and South Asia was cut on Wednesday, April 14, 2010 in the Mediterranean Sea and became inoperable due to a shunt fault, approximately 1886 km from Alexandria toward Palermo, Italy in the S4 section of the cable. It underwent further repairs from April 25 to April 29 to fix the affected fiber pair in the Mediterranean Sea. The repair work affected several cable systems, severely impacting Internet connectivity in many regions across the Middle East, Africa, and Asia.

Figure 16.7 shows the download time experienced by end-users in Asia using a routing overlay during the outage in comparison with the download time experienced by the same end-users without the overlay. For this experiment, we used a dynamic (i.e., uncacheable) Web page approximately 70 KB in size, including all page objects. The Web page was measured using agents located in India, Malaysia, and Singapore.

The agents downloaded two versions of the Web page—one directly from the origin in Boston and the other through the routing overlay. It can be seen that during the SE-ME-WE 4 disruption, the performance of the download directly from the Boston origin to Asia suffered severe slowdowns. However, downloads for the same Web page from the same Boston origin to the same Asian end-users using the routing overlay experienced minimal performance degradation. The significantly greater performance is due to the ability of the routing overlay to find alternate paths that avoid the failed links between different parts of Asia and the Boston origin. However, without the benefit of the routing overlay, the direct Internet path that exhibits significant degradation due to the cable cut must be used for the downloads. The routing overlay speeded up the downloads for Asian end-users by over eight times at the peak of the outage.

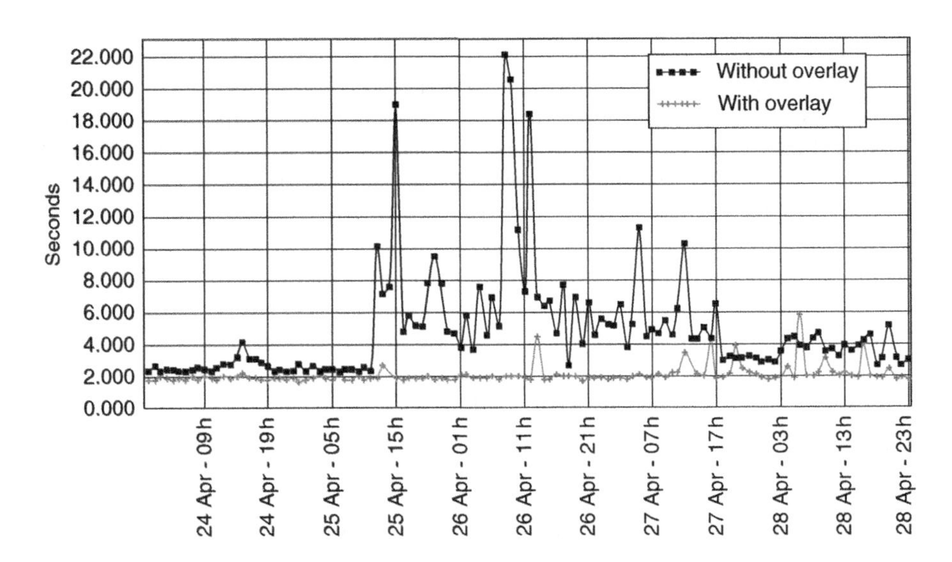

Figure 16.7 Performance of the routing overlay during a cable cut.

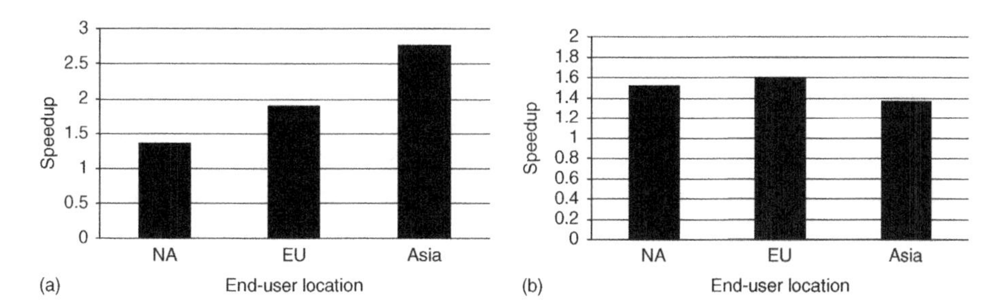

Figure 16.8 A routing overlay provides significant speedups by choosing better performing paths from the origin to the end-user. (a) A large-sized overlay using a single optimized path and (b) a medium-sized overlay using multiple optimized paths and error-correcting codes for loss recovery.

Even when there is no major Internet outage, the routing overlay provides a significant performance benefit by discovering and using better performing overlay paths for communication. In Figure 16.8(a), we show the performance benefit of a large routing overlay that uses a single optimized path chosen from a set of alternate overlay paths to the origin. For the experiment, we stored a dynamic (uncacheable) file of size 38 KB in an origin in Dallas and tested it from agents around the world. Speedup is the ratio of the download time of the file downloaded directly from the origin to the download time of the same file downloaded using the routing overlay. Speedup was significant for all geographies, though the speedups increases as the end-user moves farther away from the origin. The performance benefits seen are due to a combination of the routing overlay finding a path with shorter latency between the origin and the ES, the use of an optimized TCP between the overlay nodes, and a reduction in TCP penalties. The net effects of these benefits are expected to be higher on longer haul paths, resulting in greater speedups for users farther away from the origin.

In Figure 16.8(b), we show the performance benefits for a different smaller routing overlay that uses multiple optimized paths for each communication and uses network coding across the paths for loss recovery. For the experiment, we stored a complete dynamic (uncacheable) webpage 88 KB in size in an origin server located in Japan. The page was tested by agents around the world. The speedups from all geographies are significant with the continent of origin (Asia) having the smallest speedup. However, the speedup of this routing overlay is generally smaller than that of Figure 16.8(a) largely due to that fact that smaller overlays have fewer choices for alternate paths and the ESs are less proximal to the user.

16.5 SECURITY OVERLAYS

An Internet-based service needs to defend itself from DDoS attacks that aim to take it out of commission and hackers who want to steal sensitive user information from the service. The Internet architecture by itself does not provide security, and it is desirable that security be automatically provided by the platform on which Internet services are offered. There are good reasons why security should be dealt with in the underlying platform and not as part of every individual website or service. Defending against DDoS and hacker attacks requires a vast amount of spare capacity and up-to-date expertise in security vulnerabilities. If a website experiences DDoS attacks few days each year, then maintaining the excess capacity all year round is a costly proposition. Short-duration DDoS attacks happen frequently against some website or the other, and maintaining excess capacity and security expertise in a shared platform is more cost effective than individual measures taken by the owner of every Internet-based service.

For websites and services that already use the caching and/or routing overlays, most Internet traffic is already being handled by the servers of the overlay. If the security policy needs user requests to be examined, performing that task at the first ES to receive the request is more efficient from a performance and workload perspective than having to forward it someplace else. Thus, for Internet-based services that already use a caching and routing overlay, a security overlay is a synergistic addition.

16.5.1 Architecture

The design of a security overlay is not just about having spare capacity and security features but also about how best to use the capacity and the features to protect against a security attack at short notice. A security overlay incorporates the following described architectural elements.

1. *Shared Spare Capacity.* The overlay has a very large number of servers, each with high capacity network connections. The operating model of the overlay is flexible enough to divert any fraction of traffic to any server and increase the network bandwidth capacity at some locations on demand as needed. The spare capacity that can be made available should be large enough that it is unlikely to be overwhelmed even by the largest DDoS attacks that are likely to occur. Thus, the spare capacity can be used to hold a high volume attack away from the origin servers. Having the excess capacity shared across many content providers is a cost-effective way of reserving a large pool of resources as attack capacity.
2. *Shared Expertise and Lower Costs.* Hackers often exploit new and known vulnerabilities in operating systems and applications. In the scenario without a shared security overlay, every Internet-based service has to look after its own security by keeping current with new and known vulnerabilities and individually updating their security measures. However, with a security overlay, a team of security experts of the overlay staff keeps the overlay substantially free of known and new vulnerabilities, thus providing a high level of defense to small and large Internet-based services alike. Owing to the shared nature, the costs are lower and more affordable for content providers.
3. *Advanced Security Features.* The overlay is designed to offer detection and defense against a whole spectrum of attacks, such as SYN and connection flooding attacks, brute force DDoS attacks that generate hundreds of gigabits per second of traffic, malicious request injection attempts, and attempts to hack into

servers. At the network layer, the overlay servers have a hardened networking stack that deals with low level attacks and hacking attempts efficiently, and can mitigate simple attacks from turning into gigabits per second of traffic by denying the attackers requests. The overlay also incorporates a Web application firewall to examine request–response sequences to detect and filter harmful exchanges. With the help of the firewall, several advanced mitigations could be deployed to counter the attacks. For instance, if a certain URL is being used by hackers from a certain part of the world, then a firewall rule could be constructed to block that URL when requested from that part of the world. Such an action could be used to thwart the attackers, while leaving the site open for legitimate users.

4. *Shielding the Origin.* The overlay can offer another simple yet powerful feature for origin protection against hackers. The origin of an Internet-based service is more vulnerable to hacker attacks if its IP addresses are externally known to users. However, when using a security overlay, the origin can be configured so that it receives only traffic known to originate from the servers of the overlay that use a small range of IP addresses. This allows the origin's administrator to entirely firewall-off the origin from users not in the specified IP address range, thus easily identifying and blocking any traffic that arrives at the origin from outside the secure overlay. Additional protective measures such as authenticating all origin-to-overlay communication using a shared secret can further thwart the attackers.
5. *Control Design.* Even though the security overlay is a shared infrastructure, it should have features that allow an individual content provider to maneuver in response to an ongoing attack. In particular, controls must be provided for individual content providers to change the security configurations, firewall rules, filters, and so on. as pertaining to their own site at any time and at short notice after an attack is detected.

16.5.2 Security Benefits

We present some statistics collected during a recent DDoS attack on a content provider who uses a security overlay. Figure 16.9(a) shows a sudden increase in the traffic of the content provider's website due to the DDoS attack. The website that is normally accessed at a rate less than 50 pages per second was accessed at a much higher rate of 9000 pages per second during the attack. Figure 16.9(b) shows the firewall rules getting triggered in response to the attack and denying over 90% of the attackers' requests, and thus protecting the origin from the significant surge of traffic.

16.6 CONCLUSION

In this chapter, we reviewed the rationale for building overlays as a means for providing availability, performance, scalability, and security for Internet-based services. Although overlays are a powerful technique for building a variety of new functionality on top of the Internet, we focused three key types of overlays. The caching and routing overlays

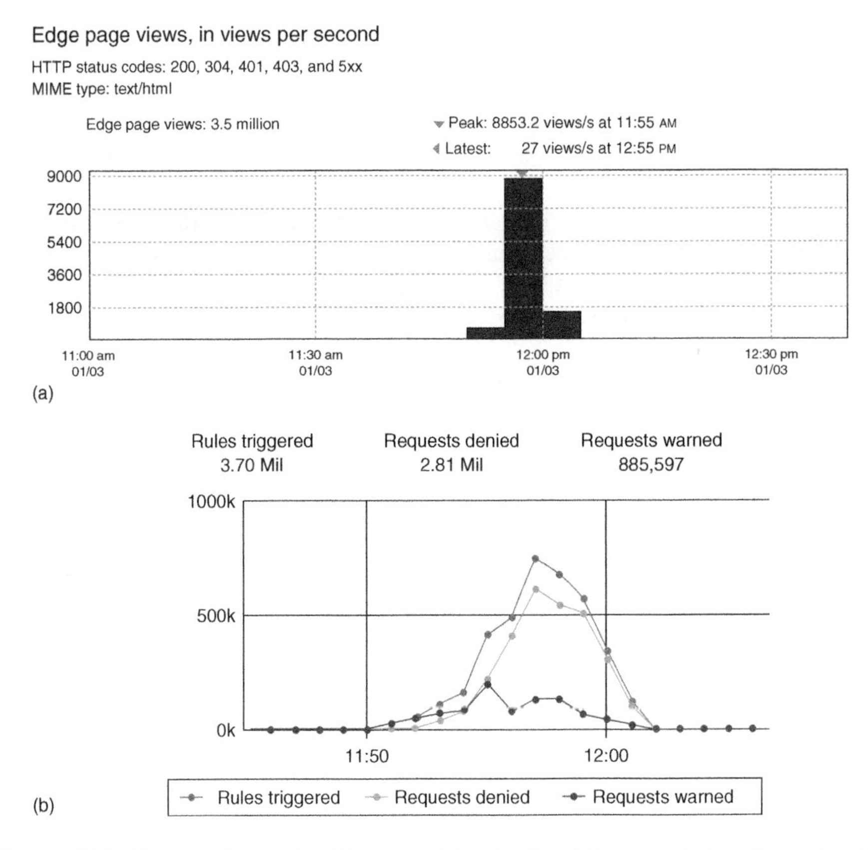

Figure 16.9 The security overlay filters out DDoS traffic at the edge before it reaches the origin. (a) Traffic spike because DDoS attack and (b) rules are triggered at the edge servers to filter out the attack traffic.

are perhaps the most ubiquitous and form a critical part of the Internet infrastructure. The security overlay is a novel and emerging tool to defend against DDoS attacks and other security threats that are rapidly becoming more sizable and more frequent. The future requirements of Internet-based services are hard to predict, even as modern trends like social networking were largely unforeseen during the turn of the century. The promise that overlays hold for the future is their ability to bridge the gap between what the vanilla Internet offers and what future Internet services may need. From that perspective, overlays hold the keys to the rapid evolution of Internet services, even as the underlying Internet architecture is slow to change.

REFERENCES

1. Nygren E, Sitaraman RK, Sun J. The Akamai Network: a platform for high-performance Internet applications. ACM SIGOPS Oper Syst Rev 2010;44(3):2–19.

2. 2013. SE-ME-WE 4 disruptions. Available at http://en.wikipedia.org/wiki/SEA-ME-WE_4. Accessed 10 October 2013.
3. ZDNet. Amazon Web Services suffers partial outage. Available at http://www.zdnet.com/blog/btl/amazon-web-services-suffers-partial-outage/79981. Accessed 2014 May 10.
4. Renesys Blog. 2008. Wrestling with the zombie: sprint depeers cogent, internet partitioned. Available at http://www.renesys.com/blog/2008/10/wrestling-with-the-zombie-spri.shtml. Accessed 2014 May 10.
5. Akella A, Maggs B, Seshan S, Shaikh A, Sitaraman R. A Measurement-Based Analysis of Multihoming. Proceedings of the ACM SIGCOMM Conference on Applications, Technologies, Architectures, and Protocols for Computer Communication (SIGCOMM); August 2003.
6. Amogh D, Luckie M, Huffaker B, Claffy K, Elmokashfi A, Aben E. Measuring the deployment of IPv6: topology, routing and performance. ACM Internet Measurement Conference (IMC); November 2012.
7. eSecurity Planet. 2013. DDoS attacks: growing, but how much? Available at http://www.esecurityplanet.com/network-security/ddos-attacks-growing-but-how-much.html. Accessed 10 September 2013.
8. Eriksson H. The multicast backbone. Commun ACM 1994;37(8):54–60.
9. Koch RR, Leighton FT, Maggs BM, Rao SB, Rosenberg AL, Schwabe EJ. Work-preserving emulations of fixed-connection networks. ACM Symposium on Theory of Computing (STOC); 1989.
10. Cole RJ, Maggs BM, Sitaraman RK. Multi-scale self-simulation: a technique for reconfiguring processor arrays with faults. Proceedings of the 25th Annual ACM Symposium on Theory of Computing (STOC); May 1993. p 561–572.
11. Leighton FT, Maggs BM, Sitaraman R. On the fault tolerance of some popular bounded-degree networks. Proceedings of the 33rd Annual Symposium on Foundations of Computer Science (FOCS); 1992 October. p 542–552.
12. Sitaraman RK. Communication and fault tolerance in parallel computers [PhD thesis]. Princeton University; 1993.
13. Andersen DG, Balakrishnan H, Frans Kaashoek M, Morris R. Resilient overlay networks. Proceedings 18th ACM SOSP; 2001 October; Banff, Canada; 2001.
14. Dilley J, Maggs B, Parikh J, Prokop H, Sitaraman R, Weihl B. Globally distributed content delivery. IEEE Internet Computing 2002;6(5):50–58.
15. Andreev K, Maggs BM, Meyerson A, Sitaraman R. Designing overlay multicast networks for streaming. Proceedings of the 15th Annual ACM Symposium on Parallel Algorithms and Architectures (SPAA); June 2003.
16. Kontothanassis L, Sitaraman R, Wein J, Hong D, Kleinberg R, Mancuso B, Shaw D, Stodolsky D. A transport layer for live streaming in a content delivery network. Proc IEEE 2004;92(9):1408–1419.
17. KaZaa. Available at http://en.wikipedia.org/wiki/Kazaa. Accessed 2014 May 40.
18. Gnutella. Available at http://en.wikipedia.org/wiki/Gnutella. Accessed 2014 May 40.
19. Stoica I, Morris R, Karger D, Kaashoek MF, Balakrishnan H. Chord: a scalable peer-to-peer lookup service for internet applications. ACM SIGCOMM Comput Commun Rev 2001;31(4):149–160.
20. Ratnasamy S, Francis P, Handley M, Karp R, Schenker S. A scalable content-addressable network. ACM SIGCOMM Comput Commun Rev 2001;31(4):161–172.

21. Zhao BY, Huang L, Stribling J, Rhea SC, Joseph AD, Kubiatowicz J. Tapestry: a resilient global-scale overlay for service deployment. IEEE J Sel Areas Commun 2004;22(1):45–53.
22. Rowstron A, Druschel P. Pastry: scalable, distributed object location and routing for large-scale peer-to-peer systems. IFIP/ACM International Conference on Distributed Systems Platforms (Middleware); 2001 November; Heidelberg, Germany. p. 329–350.
23. Akamai. Akamai NetSession Interface (Client Side Delivery) Overview. Available at http://www.akamai.com/client/. Accessed 2014 May 10.
24. Assad AA. Multicommodity network flow—a survey. Networks 1978;8(1):37–91.
25. Konstantin A, Maggs BM, Meyerson A, Saks J, Sitaraman RK. Algorithms for constructing overlay networks for live streaming. arXiv preprint arXiv:1109.4114, 2011.

17

NEXT-GENERATION CDNs: A CoBlitz PERSPECTIVE

Vivek S. Pai

Princeton University, Princeton, NJ, USA

17.1 INTRODUCTION

Previous chapters have discussed the mechanics and operation of pure-play content delivery networks (CDNs), operated by CDN service providers. This chapter focuses on a variety of next-generation CDNs, specifically those addressing markets other than pure-play CDNs. Although pure-play CDNs are still the dominant form of CDN-based traffic, a number of CDN variants have arisen to address different market needs and requirements. Some of these needs may be entirely strategic or operational—that a CDN may be viewed as something so central to the delivery infrastructure of a company that it would like to have more direct control over one. In other cases, organizations may view their CDN needs as so basic that they feel they can gain operational efficiency by running their own rather than negotiating prices with pure-play providers. This chapter discusses a number of next-generation CDNs, such as those operated by carriers, those managed by CDN service providers on behalf of a carrier, CDNs that are federated between different organizations, and finally licensed CDNs where CDN technology is sold as a building block. This chapter concludes with a case study of one such licensed CDN, CoBlitz.

In the original CDN model, now called pure-play CDNs, the CDN service provider runs the CDN as a business and tries to attract content providers to use it. The CDN

Advanced Content Delivery, Streaming, and Cloud Services, First Edition.
Edited by Mukaddim Pathan, Ramesh K. Sitaraman, and Dom Robinson.

service provider in these scenarios is typically a middleman, owning neither the content rights nor the last-mile connections to reach the end-user. Owners of either of these components may find themselves in a position where they may want to enter the CDN market in some manner to exploit the resource they own. In this chapter, we cover some of the next-generation CDN scenarios. The notion of CDNs run by network operators had been considered since the early days of CDNs, but quickly faded with the original dot-com bust of the early 2000s. However, nearly a decade later, with the growth of video traffic, the economic considerations revitalized interest in what became known as operator CDNs.

The original interest in CDNs owned by network operators and hosting organizations was the observation that they already had some of the skills and infrastructure necessary for running a CDN, and the CDN business provided much higher gross margins than their existing lines of business. Many of these organizations also had content providers as customers, so seeing those content providers moving traffic (and revenue) over to CDN service providers was also a driving factor.

Among the problems faced by these operator CDNs was how to provide service outside their own network footprints, since a content provider would be much more likely to pick a CDN provider with a wide reach rather than signing many individual contracts with limited-footprint CDN providers. To this end, some of the companies providing hardware and software for building CDNs also developed plans for federating CDNs, wherein two CDN providers with separate footprints could redirect requests to each other so that each CDN appeared to have a larger service footprint than its actual physical footprint. Cisco created the content alliance for operators using its appliances [1], and Inktomi, America Online, and Adero created the content bridge for operators primarily using Inktomi's software [2]. These efforts would provide peering between CDN services similar to the way that network operators already had bandwidth peering arrangements. These efforts were largely victims of the dot-com bust, when CDN spending became uncertain, and new entrants in the CDN service space faced greater investor scrutiny.

Nearly a decade later, though, the scenario had changed due to the growth of video traffic and the increase of video-over-Internet Protocol (IP) services being used by network providers. Some CDN services that were delivering video were often doing it from centralized locations, rather than within each provider's footprint. This approach caused increased bandwidth at the peering points. However, the growth of video was causing network congestion throughout the operator's network topology, so even if the peering points could be offloaded, the cost within the network was still substantial. Despite the large size of video libraries, some content was still disproportionately popular, opening the way for some form of caching. Even caching just this popular content within the network could help with congestion. At the same time, with IP-based video-on-demand (VoD) services, and more viewing on IP-based devices, network operators had to solve similar video delivery problems themselves.

With these concerns, the idea of operator CDN became feasible again, both to serve VoD traffic but also to cache over-the-top (OTT) traffic from other sources. For large operators, it might be possible to negotiate directly with content providers to deliver their traffic within the operator's footprint. For small operators, who did not have favorable peering arrangements, the ability to own a CDN and cache popular traffic presented an opportunity to reduce their bandwidth costs.

The new entrants in the CDN space cover a range of different approaches, including carrier CDNs, managed CDNs, licensed CDNs, and federated CDNs. Each of these has a CDN at the core, but the origin and operation of the CDN can differ. The rest of this chapter describes some of the motivation and deployment scenarios for each of these next-generation CDNs. Also included is a longer case study of CoBlitz, a research CDN that became a commercial licensed CDN.

17.2 CARRIER CDNs

Some operators will opt to build their own CDNs, using some combination of internal software or open-source projects. Very often these CDNs start as the outgrowth of a research division of the company or an extension of the Web hosting division.

a. At small scale, a set of replicated hosting servers combined with a geography-aware domain name system (DNS) can behave as a CDN. Even the DNS system can be purchased as a global load balancer or can be outsourced to companies that run geo-aware DNS. For extending a Web hosting service where all of the content is already in the Web hosting facility, a remote synchronization system can keep the replicated servers acting as mirrors. This approach is feasible for smaller libraries, and exploits the operational strengths and existing customer relationships of the Web hosting division of these companies.
b. For larger carrier CDNs, a number of open-source projects can be used as starting points. Possible choices include the Squid proxy configured as a Web accelerator, the Varnish reverse proxy, or the Apache Traffic Server, which began life as Inktomi's Traffic Director. Each of these systems has the ability to cache content and pull content on demand, reducing the need for any explicit synchronization of content between machines. Some of these systems, either alone or with extensions, also support the ability to invalidate cached content or force a new copy to be pulled from the source.
c. Finally, some carrier CDN projects start with home-grown components, sometimes the outgrowth of research projects. These can include traditional CDN systems or hybrid approaches that also try to include peer-to-peer (P2P) behavior. These hybrid systems were viewed as being attractive when BitTorrent's share of network traffic was at its peak, since hybrid CDNs could potentially address BitTorrent caching as well as handling regular Web traffic. For these approaches, load balancing, request distribution, and even monitoring and reporting can be augmented with commercially available systems.

Carriers entering the CDN market sometimes view it as a way of addressing the growth of OTT content, defined as content originating and being delivered outside the carrier's footprint. The growth of OTT video content has been problematic for many carriers, since network topologies typically provision much lower sustained bandwidth per user at the peering points versus the last mile. For typical Web surfing, traffic usage is bursty, and this provisioning allows latency in Web page download times while reducing network costs. However, long-form video traffic means that the statistical multiplexing assumptions used to size the capacity of peering points have to be modified, resulting in more peering point traffic.

Carriers can use CDNs to address OTT traffic by moving this OTT traffic into their own network, either through arranged agreements with content providers or via transparent caching. Although transparent caching has existed in Web proxy servers since the late 1990s, the opportunity is different from OTT video traffic. In particular, OTT traffic consists of a small number of sources responsible for a large fraction of bandwidth, so targeting just this traffic is possible. In transparent Web proxies, small objects provide a very low cost/benefit ratio in caching, but OTT video provides the opportunity to improve the cost/benefit ratio by ignoring small objects and just focusing on large streams, possibly only from specific content providers.

At a technical level, carriers could use the CDN as a platform to transparently cache and deliver OTT traffic initially, and then reach cooperative agreements to provide CDN service to those same content providers. The legal considerations, especially in the US market, have been more problematic. The interaction of transparent caching and network neutrality regulation has been an open area of concern, as has the business case for transparent caching. For Tier 1 network providers, who have the cheapest access to bandwidth, transparent caching systems may be more costly than the peering point traffic they save. For smaller network providers, however, transparent caching becomes much more attractive.

Some of the carriers to deploy CDNs at some level have included AT&T, Telecom Italia, Telefonica, France Telecom, Telstra, Verizon, and Comcast [3]. Some of the details of these deployments have been made public, but others can also be inferred. France Telecom and AT&T have announced partnerships with Akamai for CDN services. In the past, when AT&T originally announced their carrier CDN, a Novell caching/CDN spin-off called Volera announced AT&T as one of their customers [4]. Telefonica has made public statements about using a home-grown P2P CDN [5], while Comcast has announced the use of Apache Traffic Server [6], which was originally developed by Inktomi and then open-sourced by Yahoo many years after it acquired Inktomi. Telstra's CDN is home-grown, using a combination of Cisco content networking and routing equipment [7]. Verizon used CDN services from Velocix, which was given as one of the reasons for Velocix's acquisition by Alcatel-Lucent [8]. Velocix originally focused on transparent caching, including BitTorrent traffic, when it was known as CacheLogic.

17.3 MANAGED CDNs

Managed CDN are an outgrowth of traditional CDN service providers, and a way for them to cement relationships with carriers. In this model, a CDN service provider becomes the outsourced managing service for a CDN that operates primarily in the carrier's network footprint [9]. The content on such CDNs is expected to be primarily that of the carrier, such as VoD offerings or IP video services, to enable computers, tablets, and phones to watch the same video available over a cable network. The carrier is often replacing legacy video streaming equipment or growing a new service, but does not want to be in the business of building or operating a CDN. At the same time, a CDN may be seen as desirable for a number of reasons, such as cost or flexibility.

In these cases, a carrier may contract with a CDN service provider such that the CDN service provider installs and operates a CDN, but with ultimate control belonging to the carrier. For example, in this model, the carrier may dictate where servers are placed in the network, enabling CDN server placement to interact with traffic management and network engineering. By choosing where CDN servers are placed, the carrier may be able to offload congested links or delay upgrades in network links. These kinds of placement decisions would clearly not be the primary concern of a traditional CDN service provider, since the service provider's customers are content providers, not carriers. However, in a managed CDN, the carrier is the customer, so placement decisions are more important.

The cost of a managed CDN may be based on several factors, including the usage, capacity, configuration, or level of service being provided to the carrier by the CDN service provider. In general, if any component includes a volume-based cost, such as a per-gigabyte cost, it will be lower than what the CDN service provider charges content providers, since the content being delivered will be the carrier's own content. The managed CDN's value to the carrier primarily derives from the convenience of not having to run the system themselves, rather than as a new source of revenue. As such, its pricing is considered relative to other operational expenses, particularly the cost to train and keep the personnel needed to operate a CDN. Some of the companies offering managed CDN services include Highwinds, Akamai, and Limelight Network's Limelight Deploy service.

17.4 FEDERATED CDNs

Once an operator has such a CDN in place, it may want to use CDN services beyond its own footprint, or it may want to increase the usage of the CDN that it has within its footprint. These desires have led to the resurrection of the federated CDN. However, the difference is that the federated CDN is now largely being viewed as an outgrowth of other CDN use cases, rather than a means for smaller CDN service providers to appear larger. Some of these use cases are discussed as follows:

a. Off-network access to content—in cases where a carrier CDN is being used for VoD content, some portion of users may want to access those videos even when they are away from home. Particularly with the growth of mobile devices, it is desirable to watch videos while traveling. In these use cases, the carrier may still want authenticated access to its CDN-based VoD service, but may want to deliver content anywhere. It has the option, of course, to start delivering content from its own footprint, and use the public Internet to reach its remote customer. However, for all of the latency-related reasons CDN tries to place nodes locally, but this approach may not result in a good remote viewing experience. Instead, and especially if, the carrier CDN is a managed CDN provided by a global CDN service provider, the carrier may federate with the global CDN service provider to handle off-network delivery of its content. In this case, the remote user would still get access to the carrier's content, but it would be delivered via the global CDN provider. The general Internet population at large, however, would not get access to this content via the global CDN provider.

b. Higher utilization of the managed CDN—when a carrier contracts with a CDN provider to run a managed CDN service, the primary goal of the managed CDN is to deliver what the carrier wants. However, the CDN equipment and software may be identical to what the CDN service provider already runs, so, in some sense, it is a fungible resource. Carriers often have to overprovision services at installation time based on forward plans, and if they find themselves with excess capacity, one way of monetizing it is to federate their local CDN with the CDN service provider. The service provider would get to use some portion of the managed CDN's capacity to deliver content on behalf of the service provider's other customers. The compensation for this usage may be in the form of reduced management fees, or it may be some revenue sharing of the content delivered. In either case, the carrier sees higher utilization of its CDN, and it recovers some amount of money from the CDN service provider. Note that since this delivery is within the carriers on footprint, the higher utilization of existing network connections is largely not an issue, since they are already a fixed cost. Only in the cases where managed CDN nodes were being used to reduce network traffic, congestion might be a mitigating factor in the desire to federate the CDN.

c. Regional CDN exchange—some national-level CDNs are used to provide a service to a linguistically distinct population, and these markets may not be primary markets for global CDN service providers. For example, local-language content in many countries may have limited appeal outside of those countries. However, with the Internet-connected global diaspora, many pockets of immigrants exist in other countries that want to stay connected to the content popular in their home country. If such content is made available over the public Internet, it may result in large international bandwidth bills. In such cases, regional CDNs may opt to provide service to each other users to address these kinds of issues. They may not be a strong driver for CDN federation by themselves, but may be part of a larger business exchange between two carriers. For example, if two neighboring countries have CDNs and also have peering links, it may be preferable to have CDN federation than to send the CDN content over the peering links.

When CDNs federate across different legal boundaries, care must be taken to observe the appropriate laws and agreements of all interested parties. For example, different countries have different data retention and privacy laws, so federated CDNs may have to observe different logging policies for content delivered to different regions. Even though each CDN may deliver content only within its own footprint, when it delivers logs containing a customer information to another CDN, it may be required that the receiving CDN observe the privacy laws where the customers reside. Likewise, some content may be geographically restricted, based on criteria such as agreements with the rights holders, or due to content produced with taxation-based support. This content may be offered for free within national boundaries, but may require subscriptions in other regions, and the federated CDNs may have to ensure that geo-restricted content does not leak across the federation.

The technical challenges involved in CDN federation are related to the exchange of content and information, as well as all of the additional metadata involved, such as content metadata, logging information, service monitoring, and verification [10]. This task is particularly complicated when different software is used by the entities involved in the federation. For example, CDNs may want to provide real-time traffic monitoring, as well as more detailed log records that are used to provide billing and usage information to content providers. The formats of these systems and the application program interfaces (APIs) or access methods for them may be different for different CDN software, but the content providers do not want to be affected by limitations caused by CDN federation.

For this reason, CDN federation often revolves around multiple providers using common software (or previously, appliances), as was the case around Cisco's and Inktomi's earlier efforts. The use of a single vendor for a federated CDN eliminates the differences in format and metadata and simplifies data exchange. The CDN software vendor can also develop the federation tools and protocols without going through a standardization body, potentially reducing the development time for new features. For the CDN software vendor, a mass of operators using their software in a federation provides a network effect, and may encourage new operators entering the CDN market to choose the vendor's system over others.

Recent efforts in the federated CDN space include CDN interconnection (CDNI) efforts by Cisco, which publicly announced multiple phases of CDN federation pilots, including carriers such as British Telecom, Orange, Swisscom, Korea Telecom, Telstra, and others [11]. The Internet Engineering Task Force (IETF) working group on CDNI has been active in defining various specifications for CDNI, including exchange frameworks, control interfaces, logging, metadata exchange, and others [12,13].

17.5 LICENSED CDNs

Licensed CDNs are an option for carriers who want to own and run a CDN, but do not want to experiment with developing the core of the CDN. In this model, the carrier licenses a CDN from companies that develop CDN software for just this purpose, or from CDN service companies. The licensing involved can involve anything from yearly pricing to perpetual, transferrable licenses. Cost is often a function of capacity by bandwidth, request rate, hardware configuration, and so on.

Licensed CDNs are an option that lies between carrier CDNs and managed CDNs—they give the operator more control than a managed CDN, but not as much control as a home-grown carrier CDN. The attraction for a licensed CDN is that the CDN software essentially becomes a commodity, and the carrier does not have to maintain a large, knowledgeable development staff devoted to the CDN itself. Instead, those developers can be redirected to projects that provide more differentiation. Large companies often purchase their database software in a similar manner, having the ability to write programs that use the database, without being able to modify the internals of the database itself. Similarly, a licensed CDN, with the right set of APIs and other control mechanisms, allows a carrier to develop a CDN ecosystem without having to modify the CDN software itself.

The emergence of the licensed CDN market is a combination of many factors, including the emergence of companies devoted exclusively to developing CDN software, CDN service companies that pivoted, and the extension of CDN service providers through competition.

a. CDN software companies—seeing the interest in many companies to enter the CDN market in some form, several companies emerged to provide CDN software, sometimes tied with consulting and other services. These companies largely had no interest in providing a general CDN service themselves, but were more focused on approaching content providers and carriers who were interested in deploying their own CDN solutions. This approach allowed the separation of CDN software development from the capital-intensive deployment of a CDN service and the marketing-oriented aspects of approaching content providers. The barrier to entry for a company to develop only CDN software was therefore lower from a capital standpoint.
b. CDN service companies that pivoted—with the revival of interest in CDNs in the mid-2000 decade, many companies received venture funding to enter the CDN service market. These companies typically developed their own CDN software and ran it on a geographically distributed set of nodes. However, as competition in this space was aggressive, some of these new CDNs could only compete on price if they did not offer anything beyond quickly commodifying CDN service. With the ongoing costs of maintaining a global CDN and decreasing gross margins on CDN delivery pricing, some of these CDN service companies largely exited the service business. However, having developed a CDN software stack, they could now license that stack as a source of revenue.
c. Extensions of CDN service companies—in some cases, CDN service companies themselves have entered the licensed CDN market. In the case of smaller CDN service companies, the licensed CDN market may present a larger revenue opportunity. In the case of larger CDN service companies, the licensed CDN market may simply be another opportunity within an existing business relationship, or it may be a means of managing competition. Licensed CDNs are not exclusive of managed CDNs or federated CDNs, so any company that grew dominant in the licensed CDN market could conceivably help turn a set of licensed CDNs into a larger, federated CDN service. For this reason alone, CDN service companies in the licensed CDN market may be viewing it more as a means of addressing a competitive threat. The licensed CDNs may also be used as a federated CDN offering or may eventually turn into a managed CDN offering, so it may also provide opportunities for additional revenue or to deepen business relationships with carriers.

Some of the CDN service providers who also have licensed CDN offerings include EdgeCast, Jet-Stream, and Akamai's Aura Network Solutions. EdgeCast has been used at Deutsche Telekom and AT&T, among others [14]. Some content providers have also taken to building custom CDNs, with the largest being Netflix's Open Connect CDN, which was largely built using open-source tools with customization. By some accounts, Netflix alone accounts for 20–30% of peak-hour traffic [15], and many smaller network operators can reduce payments on their peering point traffic by hosting a Netflix CDN appliance [16]. A similar approach is used by YouTube through what is called the Google Global Cache [17].

The future evolution of licensed CDNs (and operator CDNs in general) is largely dependent on the needs of the carriers they serve and the customers who use them, but their evolution may differ from that of pure-play CDNs for a variety of reasons. The first is that these next-generation CDNs are used in very different environments from pure-play CDNs, and certain customer segments may have little interest in running on a licensed CDN. For example, e-commerce companies want their interactions with customers to be accelerated regardless of what network the customer is using, so the chances of most operators of next-generation CDNs acquiring e-commerce companies as content providers is remote. These companies are likely to remain better served by pure-play CDNs. The wider footprint of pure-play CDNs is important to this class of customer, so features relevant only for this class of customer are not likely to be high priorities for licensed CDN providers. Similarly, other customers with lower traffic but more specialized requirements are unlikely to want to deploy their own CDNs, if the features they want are available on pure-play CDNs. Even if the licensed CDNs can be acquired cheaply, maintaining and managing such systems may require more staff specialization than they are willing to develop and retain internally.

In other cases, however, flexibility and strategic importance may provide a greater motivator in deploying operator CDN than pure cost-reduction concerns. Consider, for example, the case of operators that are deploying CDNs to provide VoD services. Historically, these services were often provided by specialized vendors who were more tied into the specialized delivery requirements associated with set-top boxes and video environments. These environments also had very different cost models than Internet-delivered video. For example, some of the VoD is included in the monthly subscriber bill in many cable networks, while other videos are pay-per-view. In comparison, much Internet-based video tends to be advertising-supported, which may have generated one to two orders less revenue for the same amount of content. As a result, subscriber-based video delivery may place a premium on guaranteed service and total control over the system, rather than reduced cost, so these operators may want to bring everything in-house. Even if the cost of running a CDN internally may not be strictly the lowest option, the strategic concerns may dominate the final decision. Although pure-play CDNs have long had arrangements with these operators, the emergence of viable options for licensed CDN and operator CDNs has no doubt led some CDN providers to be more aggressive about developing their managed CDN offerings.

17.6 CASE STUDY: CoBlitz

The CoBlitz CDN [18] was a research CDN that started at Princeton University in 2004 by KyoungSoo Park and Vivek Pai before being spun out as a company, and then ultimately acquired by Akamai to form the Licensed CDN component of its Aura Network Solutions product line.

To understand the video situation in 2004, commercial content was largely delivered via proprietary protocols over the Internet in order to provide digital rights management and to provide two-way interactivity. Delivering video over standard HyperText Transfer Protocol (HTTP) meant that the video was available in one large file, encoded at a single bitrate. This approach was called "progressive video," and typically required the user to select the bitrate to be used. The video could play as it was being downloaded, but if the bandwidth were insufficient, the video would pause periodically as it was buffered. These videos had no effective protection, and the user could save them to disk instead of watching it in real time. In comparison, the proprietary delivery protocols used custom clients that could report progress to the server, allowing the server to drop frames of video as necessary to meet bandwidth constraints. Some of these systems would operate over port 80 in order to interoperate more easily with firewalls, and so on, but they were typically using HTTP as just a bidirectional transport layer instead of delivering content using the HTTP protocol.

In this environment, when CoBlitz was first developed, the primary usage was expected to be software downloads, since the proprietary video formats did not use standard HTTP. Progressive video could be used with CoBlitz, but since it represented such a small fraction of commercial content, it was not likely to be a large driver of traffic. In the noncommercial space, progressive video was being widely distributed using the BitTorrent Protocol, but this was typically performed without consent of the copyright holder.

The CoBlitz CDN originally operated on a collection of servers called PlanetLab, a research testbed that spanned hundreds of sites worldwide, mostly at universities. These sites typically installed two small servers (often with only 512 MB–1 GB of RAM each) and donated a small amount of bandwidth to the system, typically in the range of 2–10 Mbps. The PlanetLab platform hosted many different projects at the same time, so the memory available to each project was less than the total dynamic random access memory (DRAM) in the system, and consuming more than 200 MB of memory per server would typically cause memory pressure and result in that project being shut down on that machine by a supervisory daemon. CoBlitz was not the first CDN to operate on PlanetLab, but the others typically focused on smaller files due to the memory constraints of the PlanetLab platform [19,20].

The main challenges for CoBlitz were to therefore aggregate the resources of multiple machines in order to deliver any content, since the content might be larger than the available memory in any single server. At the same time, the more machines involved in content delivery, the more likely it was to encounter slowdowns or failures during delivery, so CoBlitz also had to be able to adapt to unpredictable and possibly frequent failures with no manual intervention. The goal became to deliver a reliable CDN service over an unreliable collection of machines.

Another desire was to use HTTP to the extent possible, at the protocol level itself, in order to minimize problems with firewalls. The concept of splitting large files into pieces was not new, with systems such as BitTorrent using this approach, and research systems such as FastReplica using it to populate large files on CDNs. The HTTP protocol had included "range request" support with version 1.1, wherein clients could ask for a range of bytes from a file, instead of starting each download at the beginning [21]. This support was being used to restart downloads that had been interrupted, and was also being used by programs called download accelerators, which would request different ranges of a file simultaneously to receive higher bandwidth.

The problem with range request support was that most caching systems were designed to perform what was called "whole-file caching," wherein objects in the cache represented the entire object, rather than just a part of it. These caches did not fully operate with range requests—they could deliver range requests to clients, but typically could not fetch and cache a portion of a file from an origin server. This stemmed from the range request being part of the request headers separate from the uniform resource locator (URL) itself. So, any system that indexed content based on the URL would have to keep track of multiple pieces of content matching one URL. If a single server would see all of the pieces of a given URL, reassembling the original file from all of the pieces, then just indexing the whole file could address the problem. However, since CoBlitz was interested in spreading pieces of a file to multiple servers, it needed some other approach.

The approach taken by CoBlitz was to treat ranges as separate objects, by augmenting the URLs used within the system to include range information. These requests exist only within a network of CoBlitz servers and are converted to regular URL requests with standard range headers when communicating with origin servers.

The operation of the system, shown in Figure 17.1, is as follows:

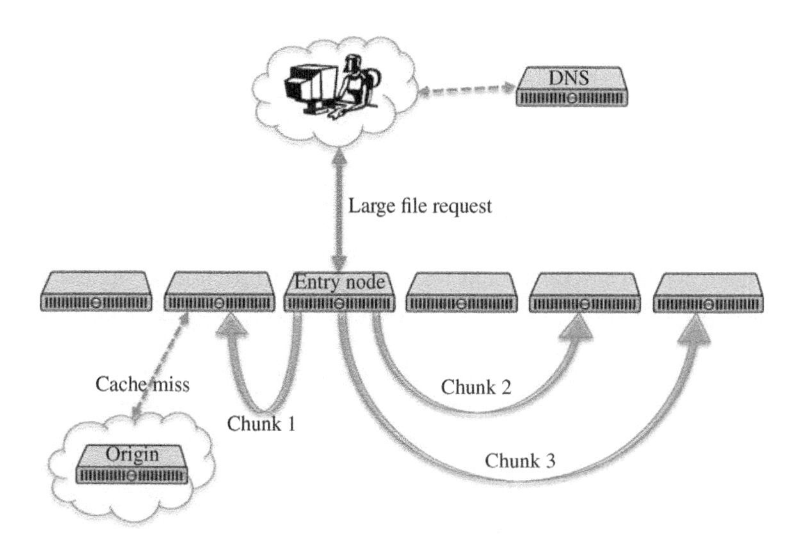

Figure 17.1 The architecture of a CoBlitz cluster. All nodes in the cluster can serve as client-facing nodes as well as cache nodes. A distributed placement algorithm allows nodes to determine where pieces of content reside, without requiring additional signaling traffic.

1. When a client wants a large file, it resolves the domain of the URL, and the CoBlitz DNS server picks an entry node near the client. The entry node is simply a node in a cluster near the client, and is picked in a load-balanced fashion. When the request is received from a client for a large file, the entry node that received the request then generates multiple requests internally that refer to ranges of the file. The number of requests is based on the size of the file—each range is a fixed size, configured typically between 64 KB and 1 MB. The requests generated internally are called chunk requests and have a modified URL that contains the range information.
2. These requests are distributed to other CoBlitz nodes, using a hashing system called highest random weight (HRW) [22]. In this scheme, the URL name is hashed together with the name of each live node in the system, generating the random weights, which are then sorted. The node that has the highest hash value is chosen as the node that is responsible for the URL. The request is sent to that node, which is responsible either for serving the cache hit or for fetching the piece from the origin server.
3. When a node receives a chunk request but does not have that chunk cached, it is responsible for fetching it from upstream, which is either another CoBlitz node in an upper level of the hierarchy or the origin server itself. If the request is being sent to the origin server, the byte-range information is stripped from the URL and inserted into the request header. In this way, the origin servers do not have to be modified to operate with CoBlitz, since they handle only regular HTTP requests.
4. If the entry node notices that any of the nodes handling chunk requests are taking too long, it can send the same request to the next node on the HRW list for that chunk to handle slow nodes or other failures.
5. As the responses flow back from the origin server, the CoBlitz nodes forward them along to the entry node. The entry node assembles the content in order and returns it to the client over the original connection. In this way, the client is unaware of the entire splitting/reassembly process.

This approach has several advantages over both a BitTorrent-style approach and the whole-file approach used in traditional caching.

a. The greatest advantage is that the use of distributed algorithms allows chunk of content to be spread across multiple nodes, allowing aggregation of all the memory, disk storage, disk seeks, and so on, of all of the machines in the system. As more machines are added to the system, the number of entries in the HRW calculation increases, and the new machines will rise to the top of the list for a fraction of the chunks. Most chunks, however, will be unaffected by the addition of one machine to the list, so no mass migration of chunks takes place when new servers are added.
b. Any machine can serve as an entry node, since it will be reassembling content from multiple nodes anyway. Likewise, since the chunks for any piece of content

are spread among multiple machines, no machine becomes a hot spot (or bottleneck) because of handling popular content. The server selection process only has to consider geography/topology and aggregate load, instead of content placement. The elimination of hotspots is particularly important for large files, because the traditional approach to hotspots is to replicate content, which would reduce the aggregate memory storage space significantly in the case of large files [23].

c. No signaling or other coordination is needed to find chunks of content, since the distributed algorithms determine where content should be placed. These algorithms are running on every node in the system, so no explicit coordination is needed to find chunks. The only information that is needed is whether nodes are alive or dead. This information does not have to be updated on a per-chunk basis since, since nodes tend to stay in service for long periods, and any chunk requests being sent to a dead node will timeout from the entry node anyway.

d. Especially for content where users tend to preview just the beginning, the chunk-based approach avoids needlessly fetching and caching the entire file. Large files do not have to be fetched in their entirety in order to be cached, and no special caching algorithms are needed to decide how to split the cache between large files and smaller files [24]. Whole-file caching often has to devise special policies for what to do when a large file that is a cache miss gets requested. For example, if a user starts a download and stops mid-download, should the cache continue the download so that the object can be served from the cache for future accesses? If a client never requests the rest of the object, then that portion of the download was pointless. However, if the cache stops the download and then another client requests the same file, it will again be a cache miss and have to be re-downloaded. How many times should this scenario occur before the cache decides to keep the file? These questions become much easier to handle when content can be partially cached.

When CoBlitz was operating as a research CDN, the content providers that used it were nonprofits and open-source projects, since projects running on PlanetLab were not to be used commercially. Nonetheless, CoBlitz carried terabytes of traffic daily for a number of years. In the research environment, CoBlitz nodes peered with each other across the wide area network (WAN) to fetch chunks, since each location in PlanetLab typically hosted only two servers. The CoBlitz nodes also tried to find peers that were performing well and were geographically close, using networking round trip time as an estimator since network topology was not particularly important. Especially for new releases of operating system distributions, the active dataset being handled by CoBlitz could grow to several gigabytes at a time when most experiments on PlanetLab could only safely allocate 100 MB of memory per server. In this environment, the CoBlitz nodes were self-organized with group sizes ranging from 60 to 120 nodes. Even when group membership differed slightly among nodes, the HRW algorithm used ensured that chunk replication was kept to a minimum. With these larger groups aggregating memory space, and so on, even on large flash crowds, CoBlitz was able to keep content in memory, which could be served much more rapidly than disk-based content.

At the time that CoBlitz became operational, the dominant mechanism used for large file transfer on the Internet was BitTorrent, which accounted for nearly 30% of traffic in many measurements. As such, the tests to gauge CoBlitz performance were largely performed against BitTorrent. The main tests involved having several hundred PlanetLab machines around the world act as clients for both CoBlitz and BitTorrent, and try to download a 50 MB file served from Princeton. This test simulates a flash crowd, i.e., many clients all asking for the same content, and all active in the same time period, and demonstrates a desirable case for BitTorrent. The number of peers for BitTorrent clients was raised to 60 to mimic the peering capacity in CoBlitz, and this increased BitTorrent's performance. Likewise, download times for BitTorrent were taken to include the period when data transfer started, excluding any peering setup time. Any of the clients that finished were also configured to stay connected and continue acting as seeds for other downloads. The results are shown in Figure 17.2 as a cumulative distribution, where higher lines represent better performance. The fastest 20% of the nodes in both systems have similar performance, after which the download times diverge significantly. CoBlitz beats BitTorrent by a factor of 1.5–2.5 in download times on this test. For comparison, the download times for direct download from the origin server are also included and are much worse than either system. The difference between the performance of CoBlitz and BitTorrent in this test stems from several factors, including the time BitTorrent clients have to query each other to find content. In comparison, CoBlitz's use of distributed placement algorithms eliminates such signaling communication. With larger content, the gap between CoBlitz and BitTorrent performance also grows due to memory

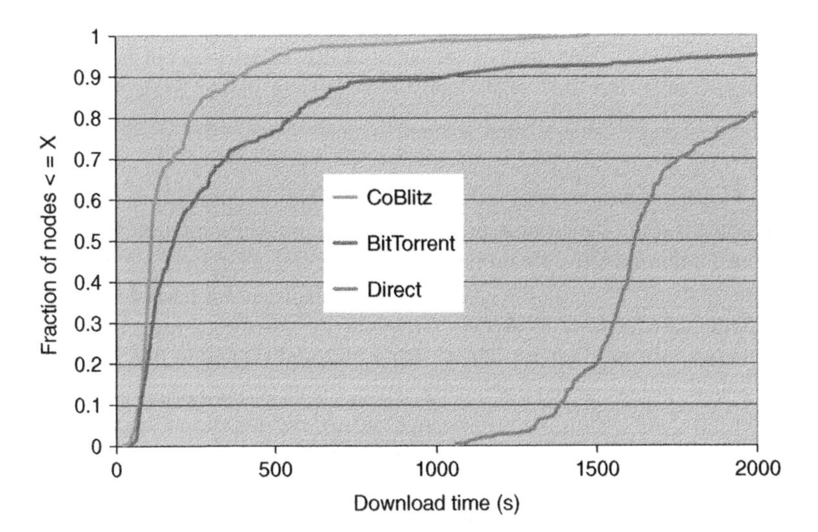

Figure 17.2 In a flash crowd event, CoBlitz performance exceeds that of BitTorrent, and both greatly exceed the performance of having clients directly obtain the content from the origin server. Slower nodes particularly see a significant increase in performance when using CoBlitz, since the elimination of signaling traffic leaves more bandwidth available for downloading content.

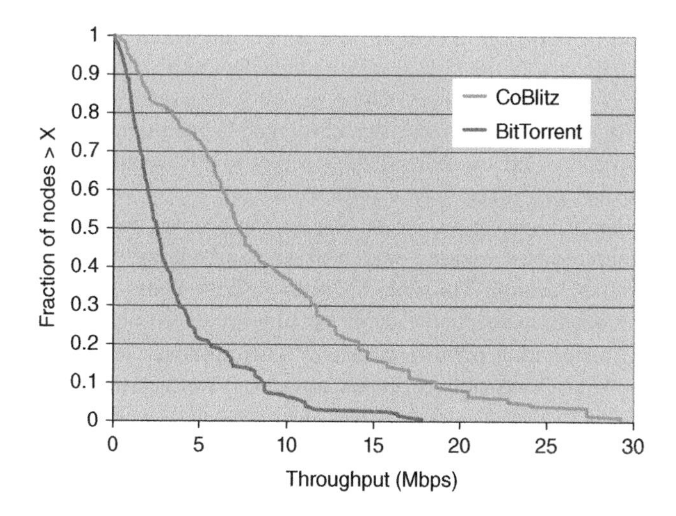

Figure 17.3 After a flash crowd event, clients downloading content from CoBlitz receive significantly higher performance, since they are able to utilize the extra capacity in the CDN. In BitTorrent, the performance improves slightly over the performance seen during the flash crowd.

pressure—since each of the CoBlitz nodes is only responsible for a portion of the total content, more can be kept in memory and served at higher speeds. In BitTorrent, with every node potentially serving every piece of content, the memory pressure can cause more content being served from disk.

In Figure 17.3, the same set of machines re-download the content individually, after the flash crowd test. In this scenario, the clients have no resource competition and the file is cached, so the performance represents a best-case scenario for individual download performance. For BitTorrent, only about half of the clients are able to download the content at speeds exceeding 2.5 Mbps, while the median performance for CoBlitz is a little over 7 Mbps. The differences are similar at slower speeds as well—the 25th percentile for BitTorrent is about 1.3 Mbps and 4.2 Mbps for CoBlitz. In general, the performance of BitTorrent clients in this test is not much greater than their performance on the flash crowd test, which was 2.2 Mbps. In comparison, the CoBlitz clients are able to better utilize the extra available capacity in the system.

17.7 CoBlitz COMMERCIALIZATION

In 2007, CoBlitz was spun out as a commercial entity and entered the licensed CDN market, licensing CDN software by capacity, providing services to install the CDN and operational support.

In comparison to the research deployment of CoBlitz, the commercial environment was much more resource-rich, reliable, and predictable. Commercial servers typically had 8–32 GB of memory, multiple disks, and ample bandwidth. Moreover, the servers were typically dedicated to the task of serving the CDN only, so nodes becoming slow due to resource competition on the machine were not an issue. In commercial deployment, the CoBlitz software, which had been developed in the resource-starved PlanetLab environment, typically used less than 10% of a server's CPU, leaving resources for other value-added services the carrier may want to deploy.

The other big difference between the research and carrier deployments of CoBlitz was that the topology for carriers was well known, relatively stable, and housed multiple servers per location. As such, the need to have nodes peer with each other over the WAN was eliminated, and CoBlitz peering occurred only within the land area network (LAN) inside each point of presence. This approach also made more sense commercially—WAN bandwidth, even for carriers, is a much more limited resource than LAN bandwidth, which can be easily increased. Using the WAN to perform compulsory misses for content makes sense, but using the WAN to reassemble every request makes no sense when storage and random access memory (RAM) resources exist in the cluster. So, all commercial deployments of CoBlitz used the same distributed algorithms, but restricted peering to LAN nodes only. Some of the advantages that CoBlitz provided compared to other options for building a CDN largely stemmed from its distributed design.

a. Since the mechanism for handling large files is to treat them as a series of chunks and then let individual nodes handle the chunks as though they were small files, large files and small files can share the CDN naturally. No special work is needed to handle mixed workloads or to handle changing workload popularity. Especially in environments where the exact popularity of different content types is not known in advance, no provisioning is necessary—all content types can be served from the same hardware, and no special hardware silos need to exist in the CDN.
b. Compared to approaches using DNS for both load balancing and content placement, CoBlitz's approach provides much finer granularity of content placement. Any content can be placed on any node, regardless of the content source. In comparison, DNS-based approaches will want to keep all of the content from a content provider on the same node. With enough content providers, load balancing can be achieved by intelligently partitioning content providers on nodes. With too few content providers, however, replication is needed for load balancing, and replication reduces the aggregate resources of a set of nodes. Especially for carriers who are using a CDN to serve mostly their own licensed content, the notion of multiple content providers is largely meaningless and cannot easily be used for content placement decisions.
c. Another option to provider Layer 7 logic (URL-based decision-making) at the nodes is to use a Layer 7 load balancer, which is typically a specialized hardware device. These devices tend to be expensive, much more than switches of equivalent capacity, so by having all the URL-based decision-making taking place inside the caching software, CoBlitz is able to eliminate the need for specialized hardware. The trade-off with CoBlitz is that the servers needed to be connected to each other over the switch inside each cluster. However, a high-speed LAN switch typically

already existed at each deployment location, and even when one needed to be purchased for the CoBlitz equipment, it was much cheaper than purchasing an L7 load balancer.

d. Since CoBlitz nodes could partially cache content and fetch cache misses from an "upstream" node, they were well suited to being deployed at multiple locations with different number of nodes per location. This approach allowed congestion within a network to be eliminated by adding servers first at larger demand centers, and then gradually expanding to more locations. The smaller locations could use larger locations to handle cache misses, ensuring that more deployments did not result in more traffic to origin servers. This incremental rollout approach also allowed carriers to scale their investment as demand increased, and to grow as their own operational teams became more familiar with the technology. Since carrier CDNs have other parts of the company as the customer, demand growth can be controlled. In comparison, a service CDN asking customers to delay usage of the CDN would be commercially questionable.

The CoBlitz software had other implementation advantages as well, including a custom, user-space file system that used raw device support from the operating system. The user-space filesystem took over all caching and provided semantics that were more suitable for a proxy cache. In particular, it removed directory hierarchies and used a hash-based mechanism to name and index files. This allowed most disk operations, such as reading files, to incur only a single disk seek instead of multiple metadata seeks while traversing a directory hierarchy. It also made file creation and deletion cheap operations that did not require synchronous disk write operations. In practice, the filesystem, implemented in the custom proxy server called Tiger [25], typically outperformed traditional filesystem/proxy combinations by roughly 5x on the workloads of interest. The fact that the filesystem was in user space also meant that it did not need to track changes in the kernel, and that it did not have worry about issues relating to using General Public License (GPL)-only interfaces, which would have raised concerns regarding GPL contamination issues.

17.8 IMPLICATIONS OF HTTP ADAPTIVE STREAMING

As CoBlitz was being commercially deployed, a significant change occurred in the video market that had implications for all CDN solutions. In particular, many major vendors adopted what was called adaptive bitrate video, also called HTTP adaptive streaming (HTTP-AS). This term covers a set of similar approaches that are covered in greater detail in Chapter 2 and Chapter 3; however, the salient points for CoBlitz are summarized below. In this model, a video was broken into multiple fragments, using a configurable fragment length, typically in the range of 2–10 s. Each fragment was encoded at multiple bitrates, typically at least four levels. Each fragment (at each encoding level) had a unique URL, and these fragments were served using standard HTTP protocols. The client-side software was intelligent and would monitor the download time of each fragment. If fragments were arriving too slowly and would result in the video stalling, the client would fetch future fragments at a lower bitrate to ensure smoother video. If the fragments were downloading too quickly, meaning that there was excess unutilized bandwidth, the clients would switch to a higher bitrate for future fragments, improving video quality.

This change eliminated the need to explicitly have the user select a bitrate and could adapt gracefully to changes in network bandwidth. This change eliminated the need to use proprietary video protocols, since fragments could be delivered over standard HTTP. The fragments could also be marked cacheable, enabling the use of standard HTTP CDNs for video delivery without requiring progressive video downloads. At the time, CDNs often charged differently for HTTP and video protocols. The adoption of HTTP-AS made the client experience not only more seamless but also cheaper for content providers to use HTTP-AS instead of proprietary protocols.

The technical challenge for CDNs stemming from HTTP-AS was an explosion in the number of URLs involved in serving video. For example, assume a relatively small video library of 1000 h of video—with 2-s fragments encoded at four different bitrates, this small library would have seven million different URLs. Some of the larger VoD services have in the range of 10,000–50,000 h of content, so this fragment size could yield 70–350 million different URLs. However, different devices have different screen sizes, and content providers often want to encode for a variety of different devices, including computer screens, gaming consoles, phones, tablets, and so on. As a result, videos are often encoded in anywhere from 10 to 100 different sizes [26]. Assuming only 10 different sizes, the larger libraries would have anything from 700 million to 3.5 billion different URLs.

In a CoBlitz system, the location of any URL is easy to determine, independent of the number of URLs—the HRW algorithm is hash-based, so determining which node should handle a particular URL is simply a matter of computing the hashes for the URL. In comparison, some systems that were based on BitTorrent used the notion of trackers, which were centralized repositories tracking the location of each URL in the system. When each URL represented a large video object, all of the interactions of the tracker took place very infrequently. However, with fragment-based systems, trackers became much more of a bottleneck concern.

Another concern in CDNs is the cost of indexing the content on each node. Typically, the index is an in-memory data structure that keeps track of the location of objects on disk. With an explosion in the number of URLs to be indexed, one can also expect a growth in indexing memory. In CoBlitz, this index takes 32–64 bytes per entry, so with 700 million URLs, the total index would be 22–44 GB, which seems quite large. If we consider 3.5 billion URLs and 64 bytes per entry, the total possible index size is 224 GB, which seems infeasible. However, each node would only be responsible for keeping an index on the objects it has cached. If we assume that an average fragment is 500 KB, then a 1 TB disk would be able to hold only two million fragments. A server populated with eight disks would only need to index 16 million URLs, so even at 64 bytes per URL, the per-node index only requires 1 GB of RAM.

For CoBlitz, the rapid switch to HTTP-AS posed no significant problem since the large file support in CoBlitz was built on top of the small-file mechanisms in the CDN. So, the distribution, load balancing, indexing, and delivery were all unaffected. The largest impact on the CDN was in the volume of log records generated for a transfer. With 2-s fragments, a 30-min video generates 900 log entries instead of just one. With log records increasing by three orders of magnitude for the same traffic volume, reporting mechanisms to report this traffic would be affected. For example, reports showing URL frequencies and the most popular URLs would be dominated by reports about fragments, instead of reports about videos. Similarly, the reports going back to content providers would similarly show individual fragments rather than the higher level video. One solution is to combine the log entries for all of the fragments and report the HTTP-AS stream as though it were a single entity instead of a series of separate requests. This approach loses some of the finer information available in the HTTP-AS stream, such as information about changes in bitrate, but is otherwise simple for the content provider to handle. The per-URL transfer information can be kept separately and used for diagnostics or deeper analysis.

17.9 CoBlitz COMMERCIALIZATION LESSONS

Some of the lessons from the commercialization of CoBlitz are broadly similar to that of other research projects, but some are very CDN specific and are worth mentioning.

a. Maintaining HTTP compatibility and on-demand caching, even at the price of performance, significantly reduced the barrier to adoption and made it easier for competitive evaluations. Using HTTP protocols as much as possible, even between caches, made it easier for various parts of a large organization to feel comfortable with an internal CDN. As a widely used protocol, HTTP can be monitored, analyzed, and managed as needed and works nicely with network monitoring products. By requiring no changes to clients or servers, operators can easily decide to test the system and incrementally deploy it. From an academic standpoint, CoBlitz could have gained some performance by having clients contact different nodes directly to fetch pieces of content—having the reassembly take place on the CDN means that content has to be delivered in order, and that extra copies are taking place on the CDN. However, the extra gain in compatibility and transparency easily dominates the small performance cost and opportunity cost.
b. Link asymmetry is real and problematic for upstream transfers. In most research networks, upload bandwidth and download bandwidth are the same, and many protocols designed to exploit this symmetry encounter problems in commercial networks. In particular, protocols that rely on upload bandwidth from end-users will often encounter asymmetric bandwidth at ratios of 5 : 1 or more. In these environments, the download bandwidth cannot be maximized unless multiple customers are serving each active viewer. The network engineers at operators often appreciated that CoBlitz's traffic all flowed downward in their networks, since that did not require any changes to their provisioning assumptions.
c. Silent healing is not desirable. When CoBlitz was a research CDN being operated part-time by two people, any faults in the network or hardware were simply ignored by the system's self-healing properties. This mode of operation made sense since the service continued to run, and any immediate alert would not have been handled anyway. Instead, periodic maintenance was performed to manually resolve

any problems that could be resolved without replacing hardware. Network operators, who wanted immediate alarms on any failures so that they could be addressed as soon as possible, viewed this silent healing property negatively. Even though the service continued to operate, the failures reduce the margin of safety for handling load spikes, and the service contracts they had with their hardware vendors could get any faulty hardware replaced very quickly. By silently healing, CoBlitz was not allowing vendors to take full advantage of their hardware service contracts.

d. Aggressive environmental monitoring is desirable, but aggressive reporting is not. Having been developed in a very unstable hardware environment, the CoBlitz system had numerous environment monitoring systems that checked for misconfigurations, network problems, and abnormal delays. This monitoring could help in identifying environmental problems that may affect the CoBlitz service but were sometimes not directly related to the CoBlitz service itself. They were in effect the networking equivalent of "canaries in a coal mine," providing an advanced warning of possible problems. Unfortunately, the reporting of these problems was merged on the standard system monitoring report and could not be suppressed in the first version of the commercial system. Inside a company, the group responsible for network debugging may be different from the group responsible for running the CDN service, so the aggressive reporting of network problems could cause friction between the groups.

The first two items are ones that were considered important when designing CoBlitz, but were not particularly relevant in academia, where more radical research ideas typically dominate concerns about commercial feasibility or deployability. The latter two items are a reflection of the different personnel concerns between academia and industry—with fewer support personnel in academia, decisions that seem appropriate from a workload standpoint are suboptimal in environments with higher headcount.

17.10 CDN INDUSTRY DIRECTIONS

Over time, the barrier to entry to creating something that behaves as a CDN has dropped, helping commoditize portions of the CDN industry. Hardware and bandwidth prices have dropped dramatically over time, and it is now possible to cheaply acquire server resources and network connectivity from companies that run data centers around the world. Many options now exist for basic CDN software, including open-source software packages and commercially licensable options. Even the DNS logic to appropriately direct requests is available from many sources, ranging from hardware appliances to companies that provide this service. None of these changes ensures that one can profitably start a new CDN company, but the entry of so many new firms into the market has produced competition and has driven down the price and the margins for basic CDN service.

Companies in the CDN space have largely responded by offering a range of services beyond basic content delivery, using CDN as a volume driver and to achieve economies of scale, but not necessarily focus on it for profit. Akamai promotes its global platform and a host of services in addition to content delivery, ranging from on-platform processing to application acceleration services. These additional services are less price-sensitive than basic CDN service, and are also not easily commoditizable. At the other end of the service spectrum would be a company like CloudFlare, one of the newest entrants to the CDN space, which provides basic CDN service as a loss leader, at no cost, using a "freemium" model. It then provides other services for additional fees, such as real-time statistics, site optimization, and protection against distributed denial of service (DDoS) attacks [27]. In addition, CDN service has also emerged as an area of expansion for cloud companies. One notable example is the Web services division of Amazon, which provides on-demand computing and storage resources. Although these resources could (and were) being used to deliver content, customers were using it as they would a hosting provider. Amazon then moved to offer this service itself, creating an explicit CDN service that can deliver content originating from within or outside Amazon.

Over time, this exchange between CDN and cloud services should only continue. CDN companies will likely want to offer more high-margin differentiated services and will become more cloud-like. At the same time, cloud service providers will be able to use their distributed facilities to provide CDN service and will do so to attract or retain customers. Competition will likely prevent any new pure-play CDN-only companies from being more than niche services, since most customers will expect migration paths to more features as their needs grow. At the same time, even if the CDN service is not an overly profitable part of a business, the traffic it drives helps the company obtain better prices on bandwidth that is then useful for reducing the cost of the other services it provides. So, it is unlikely that a company can ever outright abandon basic CDN service as well.

In the broader picture, we can expect this kind of semistandardization of many services related to content delivery. What might start as a fully custom deployment of service on top of a generic cloud infrastructure may develop enough popularity to be duplicated by many providers. Eventually, some form of migration or interoperability arises, and the service becomes commoditized as more entrants drive down the price and margins. Dominant providers may still exist, but cloud providers will be able to make it just another prepackaged service that is offered on their platforms.

ACKNOWLEDGMENTS

The experimental results in this chapter are taken from previous work [18] with KyoungSoo Park.

REFERENCES

1. Jacobs A. Content delivery alliance to set up peering. *Network World*; September 4, 2000.
2. Uimonen T. Inktomi, partners launch content delivery alliance. IDG News Service; August 23, 2000.

3. Wulf J, Zarnekow R, Hau T, Brenner W. Carrier activities in the CDN market-an exploratory analysis and strategic implications. Proceedings of the 14th International Conference on Intelligence in Next Generation Networks (ICIN), 2010; IEEE; 2010. p 1–6.
4. HTRC Group, LLC. Internet and enterprise CDN vendor analysis: Volera's Velocity CDN. Technical report; March 2002.
5. Kishore A. Operator CDNs: Making OTT video pay. Technical report, Heavy Reading; July 2011.
6. Apache Software Foundation. Media alert: The Apache Software Foundation announces Apache Traffic Server v3.2. The Apache Software Foundation Blog; June 2012.
7. Henricks R. Telstra next generation content delivery network. Proceedings of the 11th International Conference on Algorithms and Architectures for Parallel Processing (ICA3PP-11); Melbourne, Australia; 2011.
8. Alcatel-Lucent. Alcatel-Lucent acquires leading content delivery network provider Velocix. Press Release; October 2009.
9. Weller TN, Leiserson CE. Content delivery network service provider (CDNSP)-managed content delivery network (CDN) for network service provider (NSP). US patent 7,149,797. 2006 Dec 12.
10. Day M, Cain B, Tomlinson G, Rzewski P. A model for content internetworking (CDI). IETF RFC 3466; February 2003.
11. Latouche M, Defour J, Renger T, Verspecht T, Le Faucher F. The CDN federation: Solutions for SPs and content providers to scale a great customer experience. Technical report; Cisco Internet Business Solutions Group; 2012.
12. Ma K, Watson G. Use cases for content delivery network interconnection. IETF RFC 6770; November 2012.
13. Niven-Jenkins B, Faucheur FL, Bitar N. Content distribution network interconnection (CDNI) problem statement. IETF RFC 6707; September 2012.
14. Krause R. AT&T uses wireless to build content delivery business. *Investor's Business Daily*; October 11, 2011.
15. Sandvine. Global Internet phenomena report: 2H 2012. Technical report; 2012.
16. Netflix. Open Connect appliance deployment guide v2.4a; May 29, 2012.
17. Axelrod M. Content delivery networks and Google global cache. Proceedings of the 9th African Network Operators Group (AFNOG); Rabat, Morocco; 2008.
18. Park K, Pai VS. Scale and performance in the CoBlitz large-file distribution service. Proceedings of the Third Symposium on Networked Systems Design and Implementation (NSDI 2006); San Jose, CA; May 2006.
19. Michael JF, Freudenthal E, David M. Democratizing content publication with Coral. Proceedings of the 1st USENIX/ACM Symposium on Networked Systems Design and Implementation (NSDI'04); 2004.
20. Wang L, Park K, Pang R, Pai VS, Peterson L. Reliability and security in the CoDeeN content distribution network. Proceedings of the USENIX 2004 Annual Technical Conference; Boston, MA; June 2004.
21. Fielding R, Gettys J, Mogul J, Frystyk H, Masinter L, Leach P, Berners-Lee T. Hypertext transfer protocol–HTTP/1.1. IETF RFC 2616; June 1999.
22. Thaler D, Ravishankar C. Using name-based mappings to increase hit rates. IEEE/ACM Trans Netw, 6, 1; 1998.

23. Fox A, Gribble SD, Chawathe Y, Brewer EA, Gauthier P. Cluster-based scalable network services. Proceedings of Symposium on Operating Systems Principles; St. Malo, France; October 1997.
24. Cao P, Irani S. Cost-aware WWW proxy caching algorithms. Proceedings of the USENIX Symposium on Internet Technologies and Systems (USITS); Monterey, CA; December 1997. p 193–206.
25. Badam A, Park K, Pai VS, Peterson LL. HashCache: Cache storage for the next billion. Proceedings of the USENIX Symposium on Networked Systems Design and Implementation (NSDI); Boston, MA; April 2009.
26. McEntee K. Complexity in the digital supply chain. The Netflix Tech Blog; December 17, 2012.
27. Perlroth N. Preparing for DDoS attacks or just Groundhog Day. *New York Times*; February 17, 2012.

18

CONTENT DELIVERY IN CHINA: A ChinaCache PERSPECTIVE

Michael Talyansky[1], Alexei Tumarkin[1], Hunter Xu[2], and Ken Zhang[2]

[1]*ChinaCache, Sunnyvale, CA, USA*
[2]*ChinaCache, Beijing, China*

18.1 INTRODUCTION

Content delivery industry has a long history in China. ChinaCache began providing content and application delivery services in 2000, and it was the first company (that is not a telecommunications carrier) to obtain a nationwide operating permit from the Ministry of Industry and Information Technology of China to provide content and application delivery services. Chinese Internet landscape is dominated by China Telecom in the South, and by China Unicom formerly known as China Netcom, that "owns" the North. They interconnect in a handful of places, which often experience congestion due to an excessive demand. High transit prices (on the order of \$100–\$200 per Mbps) are a byproduct of this joint monopoly.

As a carrier-neutral service provider, ChinaCache's network in China is interconnected with all telecommunications carriers, major noncarriers, and local Internet service providers in China. It is deployed across networks throughout China, and it uses a private transmission backbone that connects nodes and data centers, thereby optimizing the content and applications delivery performance and reliability (Figure 18.1).

Advanced Content Delivery, Streaming, and Cloud Services, First Edition.
Edited by Mukaddim Pathan, Ramesh K. Sitaraman, and Dom Robinson.

Figure 18.1 ChinaCache data centers and private backbone in Mainland China.

18.1.1 End-User Connectivity in China: Getting Faster but Not Fast Enough

Although the end-user connectivity is still slower than in other well-connected countries, the speeds are on the rise. This means that the importance of benefits provided by content delivery networks (CDNs) is more and more recognized. The following data are based on continuous measurements from the ChinaCache network, which are publicly available from ChinaCache Index [1], the first platform in China that allows Web users to measure their Internet speeds on a real-time basis.

In the fourth quarter of 2012, the average connection speed in China reached about 2.59 Mb/s, compared with 2.31 Mb/s in the third quarter. Shanghai topped the list with an average connection speed of 4 Mb/s, the only region above 4 Mb/s nationwide. Fujian and Zhejiang provinces came in a distant second and third with average speeds of 3.17 and 3.07 Mb/s, respectively. China's western regions fell behind the national average with a connection speed of 2.34 Mb/s, while the Xinjiang Uygur autonomous region lags behind with 1.74 Mb/s.

Comparing China's three major carriers, China Unicom had the slowest average connection speed of 2.3 Mb/s, slightly behind China Mobile's 2.36 Mb/s and China Telecom's 2.63 Mb/s. Web users of China Mobile and China Telecom in South China generally enjoyed faster connection speeds than in the North. Conversely, China Unicom's customers in North China typically had a better experience than those in the South.

18.1.2 Web Hosting—Localized Growth Fuels Need for CDN

A recent analyst report "China Data Center Market Trends 2012–2013" by DCD Intelligence, the research division of DatacenterDynamics, predicts that the Chinese data center market is forecast to grow at a 20% compound annual growth rate (CAGR) for the next 5 years [2].

Beijing has witnessed a 45% growth in "end user white space" in data centers during the period 2011–2012 compared to a lower but still highly significant growth rate of 29.3% across the whole of China. As Web applications are primarily hosted in two major centers—Beijing and Shanghai—the content delivery to the rest of China becomes a necessity for latency-critical solutions (Figure 18.2).

18.1.3 Mobile Explosion

The number of Chinese people accessing the Internet via mobile devices increased to a record high of 420 million at the end of December, and mobile shopping—a quickly expanding mobile application—saw its users reach 55.5 million, according to a report by China Internet Network Information Center issued in January 2013 [3]. A vast majority of mobile users (more than three-quarters) are on slower 2G links, while the 3G sales are finally ramping up.

Mobile Internet users accounted for 74.5% of the nation's total Internet population, which stood at 564 million, amid a slower growth rate since 2010. The number of shoppers on mobile devices grew 136.5% year-on-year in the second half of 2012 to

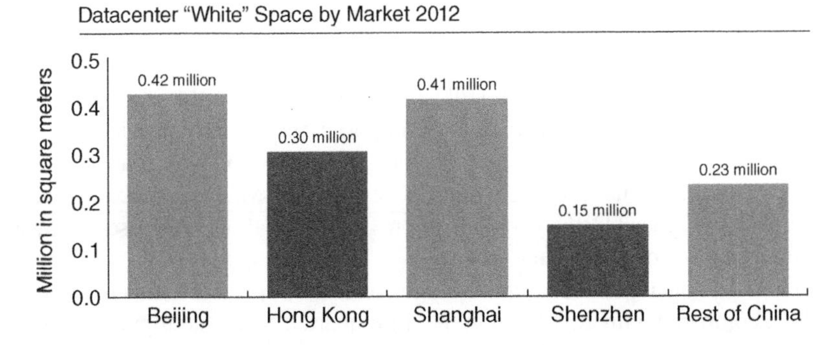

Figure 18.2 Web hosting by city in China (DCD Intelligence).

55.5 million, compared with the nation's 242 million online shoppers in the same period (Figure 18.3).

Although online sales make up a tiny proportion of all retail sales—around 4.32% in 2012—they are becoming the main way to shop in some categories. A research company Roland Berger estimates that around 25% of books were sold online in China in 2011, making it the most popular online category. Online shopping is also expected to account for around 20% of all electronic devices sold, 10% of sportswear, and 6% of mother and infant-care products.

Alibaba Group, which has more than 70% of online retail sales in China, may have 10–20% of its business coming from mobile devices, according to Qiu Lin, an Internet stock analyst at Guosen Securities in Hong Kong, quoted by *China Daily* [4]. According to this article, when Alibaba launched an online shopping promotion through its properties Tmall.com and Taobao.com on November 11, 2012, their sales reached 19.1 billion yuan ($3 billion), with almost one-third of them through mobile devices.

18.2 CONTENT-AWARE NETWORK SERVICES IN CHINA

ChinaCache's content-aware network services (CANS) form a suite of advanced technologies aimed at providing a new level of improvements for a global delivery of content to end-users. Building upon experience of the global CDN industry and ChinaCache's expertise in China, CANS are highly adaptable to different types of applications and to different vertical business segments and supply content providers with analytics related to network connectivity and end-user devices. The three core complementary technological solutions forming the basis of CANS are ICA (Interactive Content Acceleration), TTA (Transparent Transport Acceleration), and ITR (Intelligent Traffic Reduction). ICA and TTA are capable of resolving the network- and distance-related latency problems for well-connected end-users, while ITR is delivering a much-needed relief whenever bandwidth-constrained and congested links are involved.

A common approach to tackling the Web latency problem is to move content close to the end-users, thus reducing the round trip time (RTT) and consequently the total

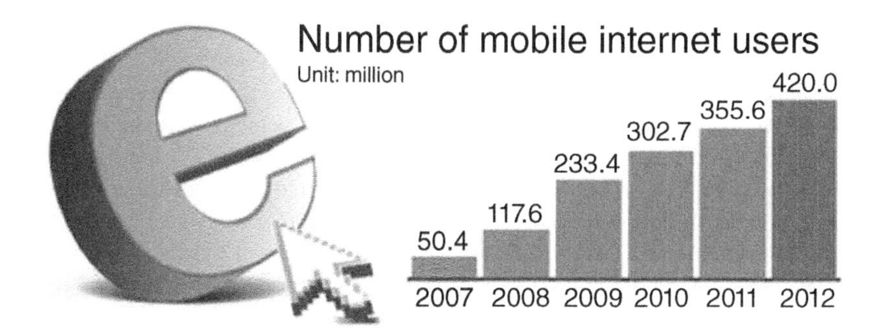

Figure 18.3 Mobile Internet statistics from China Daily [4] (China Internet Network Information Center).

download time. This is how all legacy CDNs dealt with anything that can be cached and served on behalf of the content providers. However, CDNs are facing a challenging problem of inability to store all content requested by users in their caches/servers near those users. Not only the Web content becomes more and more personalized and hence non-cacheable, but also the number of infrequently accessed static objects becomes larger and larger (the so-called long-tail problem). Indeed, according to multiple academic studies, "Online content distribution technologies have witnessed many advancements over the last decade, from large CDNs to P2P technologies, but most of these technologies are inadequate while handling unpopular or long-tailed content. By long-tailed we refer to the popularity of the content, in terms of accesses. CDNs find it economically infeasible to deal with such content – the distribution costs for content that will be consumed by very few people globally is higher than the utility derived from delivering such content" (cited from Reference 5).

A very costly solution to end-perceived latency, which might work only for some Web applications, is to deploy Web servers and back-end systems in data centers all around the Globe to move the content close to end-users. However, there is a more universal and cost-efficient approach pioneered by Netli in the early 2000s. Netli's technology replaced Transmission Control Protocol (TCP) and HyperText Transfer Protocol (HTTP) in the so-called middle mile (where the packets spend most of the RTT) completely transparently from the users. Upon an end-user's demand, the requested content got delivered to Netli's data center closest to this user in a single RTT from the origin server for any object of any size. So instead of paying seven end-to-end RTTs for a 100 KB object, the end-user would deliver this same object from a local server after only one RTT delay. After Akamai bought Netli in 2007 (and most recently Cotendo in 2011), this approach became the most common way of accelerating dynamic content on the Internet.

18.2.1 ICA and TTA

Instead of replicating the proven Netli's solution like many other companies successfully did, ICA and TTA are using a different approach, which preserves compliance to Internet standards. In fact, Netli has received some criticism for deploying nonstandard protocols [6] amid the concerns that such approach is inherently unfair and violates TCP principles. Although almost a decade of experience running such services has not resulted in a collapse of the Internet (or even in any reported congestion event), there is still a considerable value in trying to solve the main problem using a standard compliant approach. This can be achieved through performance-enhancing proxies [7] along with the latest Linux operating system implementing all available Internet Engineering Task Force (IETF) enhancements.

Although ICA utilizes Web proxies with integrated caching and HTTP optimization capabilities, TTA is a pure TCP proxying technique for delivering a near-optimal performance for long-distance TCP-based interactions such as proprietary client server, secure banking, and healthcare applications. Its distinguishing feature is that no information is stored/cached in the ChinaCache's servers. Contrary to ICA, TTA simply relays the TCP payloads, without modifying or inspecting them.

With TTA (similarly to ICA), an end-user's domain name system (DNS) query for an original server is resolved to an Internet Protocol (IP) address of the nearest CDN node. Then, the end-user establishes a TCP connection to this node, which maintains a well-tuned persistent TCP connection to its counterpart located next to the origin server. All TCP payloads are transparently relayed between these two nodes using this persistent TCP link between them without incurring typical TCP overheads for establishing connections, discovering and rediscovering usable bandwidth, unnecessary timeouts due to idle periods, and long retransmission delays. This is achieved using this link for multiple users and by maintaining keep-alive heartbeats between nodes.

For data centers, which are connected by noncongested high bandwidth links, this method gives an optimal one RTT performance. However, from time to time even those data centers can experience a persistent packet loss event. The following simple formula describes the influence of packet loss on maximum achievable TCP throughput. If the packet size is S and the packet loss probability is P, then the average TCP transmission speed is less than $S/(\mathrm{RTT}\sqrt{P})$. Therefore, when packet loss reaches 1% on a link with RTT = 100 ms, the throughput is under 1.2 Mbps (assuming a standard TCP packet size of 1460 bytes). If RTT is 200 ms, then the throughput is less than 600 kbps. Such slowdown will be noticeable for large file downloads by end-users on high speed links. However, most of Web pages will not be significantly affected. Indeed, for a 1% packet loss, the average number of packets that server can send to the client will be 75. Thus, most of transfers of typical Web objects (say, 100 KB or smaller) will be completed in a single RTT (or in two RTTs at most if the server has to resend a lost packet). Under a 4% packet loss (fortunately, such events are very rarely observed in the modern middle-mile environment), the situation becomes much worse. For a 100 KB object, it will take on average four RTTs to deliver it from one server to another even if they maintain a persistent TCP connection with a consistent flow of traffic. In such situations, nonstandard protocols can achieve downloads within two RTTs (or even one RTT if they use data redundancy such as forward error correction).

The most straightforward way of dealing with the problem of an extremely high packet loss is to monitor the links and immediately divert all traffic to another pair of ChinaCache's data centers, which do not experience such congestion. A less universal way is to use jumbo frames, if the whole path allows it. Indeed by increasing the packet size from 1500 to 9000 bytes, the resulting TCP throughput jumps sixfold [8]. Unfortunately, many backbone links still do not support jumbo frames.

Thus, the problem of Web latency induced by protocol inefficiencies can be successfully solved by widely deploying standard compliant solutions. Such solutions work well for low and moderate packet loss between the data centers. In order to tackle severe congestion events, ChinaCache's ICA and TTA use a network of multiple well-connected data centers, so that there are alternative paths that are automatically activated to divert affected traffic from the severely congested links.

An example of performance gains that TTA can seamlessly deliver to an application is depicted in Figure 18.4. The tests were performed by the Gomez Application Performance Monitoring service on a financial application hosted in Europe and downloaded repeatedly from China (both individual measurements and averages are shown).

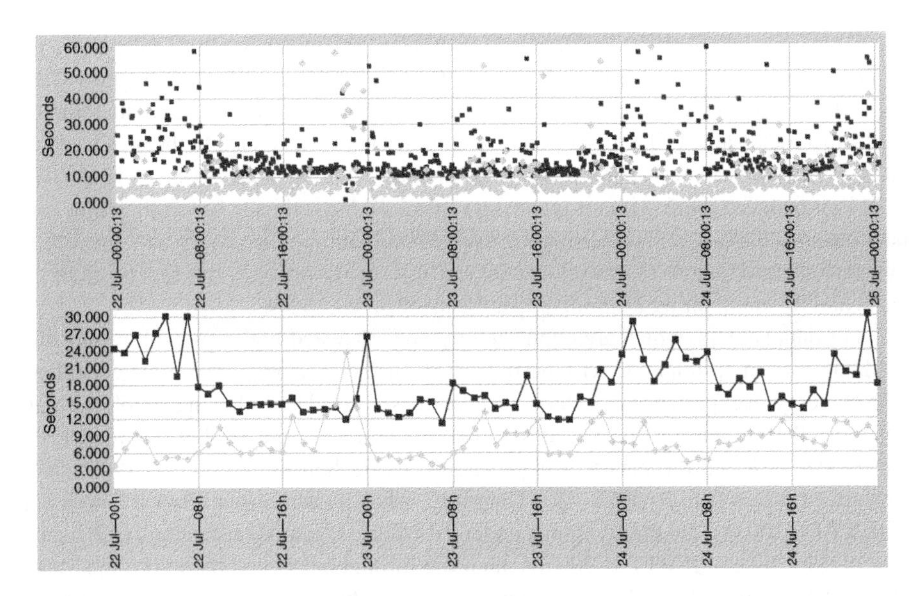

Figure 18.4 Individual and average whole-page download latencies from the origin server (square) and with ICA (diamond) by Gomez/Compuware.

In this case, the observed performance improvements would have been even more striking if all nonsensitive/cacheable objects had been served from a dedicated domain handled by ICA (and, consequently, cached and delivered from edge nodes close to end-users). Such combination of delivery of highly sensitive information via TTA and the rest of content via ICA should provide the best performing and completely secure solution suitable even for the applications that require the highest levels of protection.

18.2.2 ITR

A typical CDN architecture that deals with cache misses and noncacheable/dynamic content is a hierarchical system of Web proxies/servers that constantly relay requests and responses between themselves, generating a huge amount of internal communications. This is how ICA and TTA deal with interactive and personalized content. As long as the bandwidth is not free, the increased costs associated with such communications are a major factor of implementing bandwidth saving solutions within such CDNs. Another important use case is the need to improve internode performance on bandwidth-constrained public links. Indeed, even the most advanced Internet countries such as South Korea (while having the best end-user connectivity) have an insufficient capacity when connecting to the rest of the world. And finally, reducing the size of the objects delivered to bandwidth-constrained end-users (such as mobile phones on 2G or even 2.5G cell networks) is the only way to deal with the poor performance. ChinaCache's ITR was designed to solve these problems with a maximum efficiency within the constraints imposed by the overlay network architecture and by computational resources of its servers and end-user devices.

A well-known approach to bandwidth savings is wide area network (WAN) optimization. Although WAN optimization solutions are designed to provide many benefits for end-to-end traffic, they are costly and cannot be fully integrated with the CDN network elements (in particular they cannot use any feedback mechanisms from the CDN systems). Another major distinction is that content distribution is using a many-to-many architecture where each network element is communicating to many peers, which in turn are communicating to many other nodes. A typical WAN optimization scenario is one-to-one or one-to-many (e.g., when one device is installed in company's headquarters and other devices are in satellite offices). The main difficulty with a many-to-many architecture is the need to have synchronized dictionaries for all pairs of peers.

A common technique used in WAN optimization is called data redundancy elimination (DRE), also known as redundant traffic elimination (RTE) [9]. Under different brand names, it is employed by many products on the market, such as Cisco WAAS [10] and riverbed WAN optimization [11].

An underlying idea of such products is to reduce the number of payload bytes transmitted on a network link by creating a mutable dictionary or cache of data seen on the link, on both sides of the link, and then detect repeating patterns in the network payload; thus, avoid sending the already seen data chunks to the peer, if it is known that peer has these previously seen chunks stored in its dictionary.

There are many variations of this technique, such as object-based deduplication (this is what Web caches also do, to some extent): a smart deduplicator will apply differential encoding algorithms to objects and send only the differences to its peer, provided it knows which version of the object the peer has.

One of the most challenging technical problems common to traffic reduction solutions is the trade-off between the size of the dictionary (which is the main factor on achieving higher compression rates) and the latency caused by longer searches for matches in the larger dictionary and, consequently, higher central processing unit (CPU) loads. With many peers, the size of the dictionary (or of the set of dictionaries corresponding to all pairs of communicating nodes) can grow dramatically, and can result in unacceptable performance and latency levels.

ChinaCache has implemented an innovative approach to reduce the internode traffic within its content distribution network, which does not require significant computer resources and which does not add any perceived latency to Web transactions. Unlike the existing products, a DRE encoder within ITR operates above the TCP stack, compressing a stream of content, so it is implemented as a software module fully integrated with the ChinaCache proxying technologies. Therefore, it allows access to high level (such as HTTP) protocol data.

A key to successful data reduction using DRE algorithm is its dictionary, which contains previously seen chunks of data that are likely to be seen again in the data stream. Encoder and decoder usually start operating with an empty dictionary, populating it from live traffic as they go along. However, this approach cannot scale to a massive CDN-type network deployment and to millions of end-user devices. Instead, ITR runs DRE inside a proxy or a cache with a fixed warm dictionary, which is bootstrapped from a file and is read-only, so multiple peers and even end-users can use it without having to worry who talks to whom. This might slightly reduce the compression ratios, but greatly simplifies the dictionary maintenance, saves memory/disk space, and so on. ChinaCache's real-life experiments show that there is a fair amount of repeated data on the same link, day-to-day, so if we build an optimized dictionary based on analyzing traffic on day N and use it to encode data on day $N + 1$, we can expect only a slight degradation of the compression ratio. ITR efficiently uses a subset of servers for a live dictionary generator (with a larger dictionary than on the rest of the machines). This generated dictionary is then further reduced by picking the best performing content samples (for end-user devices, it is reduced even more to satisfy the memory constraints and to reduce the processing on the device) and is periodically distributed to the rest of the servers (and, subsequently, to end-user devices running a specialized software).

ChinaCache's latest activities are aimed at adapting ITR for end-user devices to improve last-mile performance on bandwidth-constrained and congested wireless, cellular, and fixed-line links. With the increased popularity of mobile communications, this activity becomes even more important and valuable to our customers and their users. Figure 18.5 illustrates savings that can be achieved using this DRE approach to compressing traffic to mobile clients. We show a next-day performance of a fixed dictionary

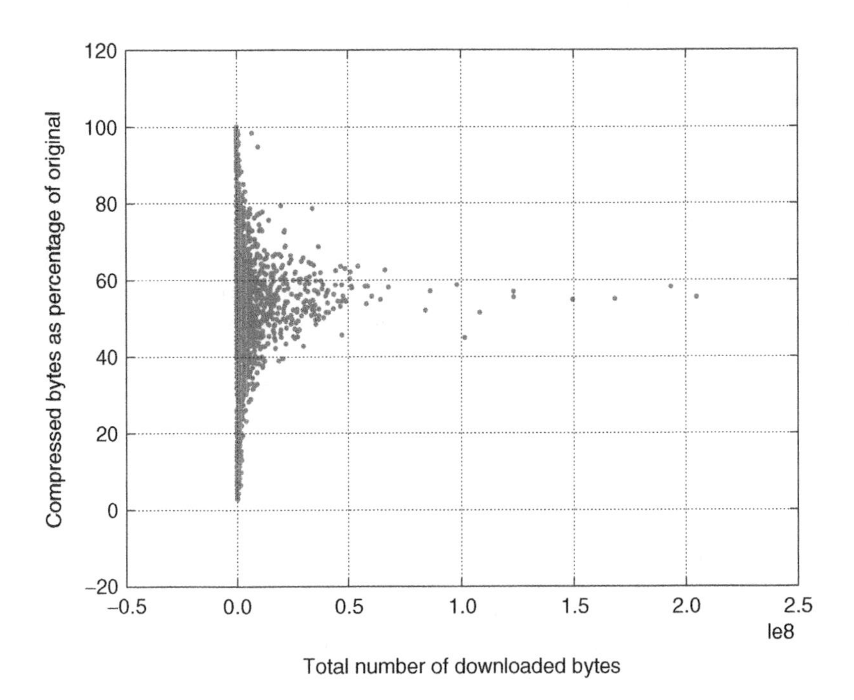

Figure 18.5 Results of DRE compression with a previous-day dictionary for 2978 mobile clients (each data point represents an individual client IP address).

compiled from the previous-day traffic to compress traffic to 2978 clients from the same IP block. The average savings amount to 46%. We can see that some clients have very high compression ratios, while others have very little commonality in day-to-day access patterns. Most of the clients would experience the average compression ratios.

18.2.3 Anti-DDoS Protection

Distributed denial-of-service (DDoS) attacks are a major source of concern for Web properties in China and elsewhere. DDoS usually refers to an attempt to prevent or hinder access to a server by its regular users by sending an excessive number of requests to exhaust the server's resources such as bandwidth and/or processing power. Since more and more efficient DDoS defense mechanisms and tools are proposed and installed on routers and firewalls, the traditional network layer DDoS attacks (such as SYN flooding, ping of death, and Smurf) are much easier to be detected and defended against. Increasingly, they are giving way to sophisticated application-layer attacks.

Application-layer DDoS attack is a DDoS attack that sends out requests following the communication protocol, and thus these requests are indistinguishable from legitimate requests in the network layer. An application-layer DDoS attack may be of one or a combination of the following types: (i) session flooding attack sends session connection requests at a rate higher than legitimate users; (ii) request flooding attack sends sessions that contain more requests than normal sessions; and (iii) asymmetric attack sends sessions with more high workload requests (see Reference 12).

Here is an example of a DDoS attack on ChinaCache's DNS servers that happened in September 2012. Figure 18.6 depicts an onset of the attack, which resulted in an increase in the activity measured in queries per second (QPS) from less than a 1,000 QPS to more than 10,000 QPS.

An ideal defense mechanism against such DDoS would try to minimize the fraction of the rejection of requests from legitimate users over the total number of requests from legitimate users (called the false rejection rate or FRR). Similarly, a false acceptance rate (FAR) should be as small as possible. Although an anti-DDoS algorithm should reduce both FRR and FAR, reducing FRR is more important for both the business and public relations reasons. That is, a server would rather maximally accommodate the legitimate user sessions, even if a small number of attacker sessions were processed normally.

An efficient system for detecting and mitigating such attacks can be deployed at various layers of the network stack (and, optimally, it would tie all of them together). Accordingly, ChinaCache's antiattack (AA) functionality is implemented both in the kernel and in the user space, for example, a successful DNS attack detection algorithm needs to examine the DNS protocol requests, extract the domain, and request type. It also needs to maintain computed heuristic values per top-level domain and whitelist entry, compute hash functions, and perform lookup in hash tables. This processing goes beyond the traditional low touch packet inspections, conventionally performed by kernel-space firewalls. Moreover, a large portion of this work needs to be done by the DNS server anyway to serve a legitimate request, so if such checks are performed in the kernel, they will need to be done a second time in the application, thus adversely affecting performance of the normal operation. The premise of this design is that computational cost

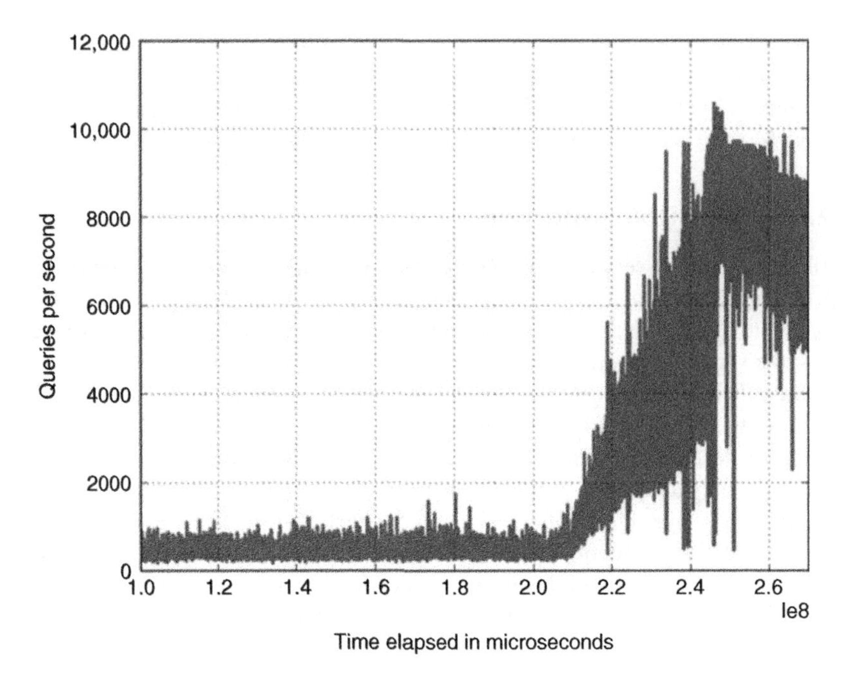

Figure 18.6 DNS DDoS attack in September 2012.

of such checks is comparable with the cost of waking up a user process and reading a User Datagram Protocol (UDP) packet into application space (~50–100 μs needs to be formally profiled), so performing this operation in user space (and not doing it in the kernel) will not result in performance degradation.

Other obvious benefits of doing application-style anti-DDOS protection in the user space include (but are not limited to) the following:

- Safer programming environment
- Much easier to debug, deploy, and maintain
- Availability of application-specific functionality (parsing DNS records, etc.)
- Availability of much wider variety of programming.

The main distinguishing feature of ChinaCache's AA technology is a patent-pending DDoS detection mechanism, which is based on comparing estimates of the current and long-term background activities calculated by different recursive averaging formulas. These formulas are specially tuned for the specific applications and attack patterns, and are optimized for a minimal memory and processing overhead.

All ChinaCache application servers host a detect agent that enables the server's attack-aware functionality. Consequently, the server knows when it is under attack and enables its AA agents at the IP level, TCP level, and HTTP level. The AA has closed-loop controls in detection–action–validation on all network levels, which are very effective in handling different types of DDoS attacks. In addition, ChinaCache has deployed a wide range of topologies that can enhance the DDoS defenses, such as server groupings that can scale for the load and fault tolerance, domain server matching that can help identify which domain are attacked, and IP-based controlling of network interface card (NIC).

18.3 DIRECTIONS FOR FUTURE CDN RESEARCH AND TRENDS IN CHINA

New key Internet technologies such as HTTP 2.0, HTML5, and server-assisted P2P are designed to improve speed and efficiency, and will require upgrading existing CDN platforms. At the same time, ChinaCache will keep optimizing and adapting popular protocols (HTTP, TCP, and UDP) to improve end-to-end communications over bad links. In order to optimize delivery to different browsers and applications under various network conditions, CDN servers can add special "labels" to pass additional information to requester and responder. For example, end-user's attributes might include location, carrier, access type, Wi-Fi (3G, 4G, etc.), real-time link quality, and edge server information. Both requester and responder can use these data to offer a better service. A special client application can be installed on the end-user device to provide even more details.

As cloud computing becomes more prevalent, its users find it increasingly difficult to serve the whole world by one central cloud installation. Poor network reach will be a huge problem, and security threats are another big challenge. ChinaCache is working on a simple, easy service that enables cloud providers to reach/serve global customers, despite the unstable and slow Internet connections. At the same time, CDN's vast flash crowd and content inspection capacities can assure the cloud's safety. ChinaCache will further develop its system to support interactive Web applications as well as non-Web TCP/UDP applications.

As enterprises move into Internet cloud-enabled business, there are many challenges that their IT teams are facing. CDNs will help enterprises to integrate their intranet communications onto Internet environments, enable them to offer more types of Internet information to their staff and their customers, empower cloud access, and protect their business in the public Internet. Technologies such as quality-assured VPN, protocol optimization, content deduplication, data mining, cloud extension, and security enhancements will be targeted to support enterprise Internet business.

As mobile Internet access becomes prevalent and more and more types of applications are becoming essential for our everyday life, ChinaCache will remain focused on mobile Internet service to ensure that mobile devices get the same quality of experience (QoE) as the current well-connected, fixed-line users. This will require adapting content to real-time network conditions and mobile device's screen size and resolution. In addition to that, ChinaCache will also look for ways to better serve location-based services (LBS), maps, geotargeting, and so on, by putting resources and computation near the place where an application runs. ChinaCache will build a smart layer of service and introduce this layer to application developers by either a general service or SDK toolkit. ChinaCache will also work with mobile carriers to improve users' experience and reduce carriers' investments. For example, CDNs can help carriers manage quality of service (QoS)/QoE and do traffic shaping.

18.4 CONCLUSION

Despite all challenges, the content delivery industry in China is expanding rapidly. While historically it was following technological trends of global industry leaders such as Akamai, the growing demands of Chinese market make the next few years an exciting era of new technical and business possibilities and innovations.

China is serious about improving its Internet ecosystem. A recently created China Broadband Development Alliance (CBDA) was founded by the Chinese government in cooperation with telecom operators, major Internet companies, manufacturers, and related scientific research organizations. CBDA will be focused on studying and reporting of the current state of Internet connectivity; researching broadband policies, technologies, and industry; as well as improving the broadband public service. Serving as a bridge linking the government, the enterprise, and the general public, CBDA will promote cooperation and exchange between various parties involved in the broadband industry chain, creating a good environment for a healthy and rapid development of China's broadband industry, and building the next generation of national information infrastructure that should feature adequate bandwidth, high integration, security, and ubiquity, supporting China's and global economic, technological and social development.

REFERENCES

1. ChinaCache Index. 2013. Available at http://ccindex.cn. Accessed 2013 Mar.
2. 2013. Available at http://www.turtleconsulting.com/blog/dcdintelligence-beijing-leads-china/. Accessed 2013 Mar.
3. China Internet Network Information Center. 2013. Available at http://www1.cnnic.cn/IDR/BasicData/. Accessed 2013 Mar.
4. 2013. Number of mobile Internet users rises. *China Daily*. Available at http://usa.chinadaily.com.cn/business/2013-01/16/content_16124925.htm. Accessed 2013 Mar.
5. Traverso S, Huguenin K, Trestian I, Erramilli V, Laoutaris N, Papagiannaki K. TailGate: Handling long-tail content with a little help from friends. WWW'12: Proceedings of the 21st International Conference on the World Wide Web; Lyon, France; April 2012. p 151–160.
6. 2013. Letters to the Editor. http://www.networkworld.com/columnists/2003/0526let.html. Accessed 2013 Mar.
7. Hassan M, Jain R. *High Performance TCP/IP Networking*. NJ: Prentice-Hall; November 2003. 408 pp.
8. Mathis M, Semke J, Mahdavi J, Ott TJ. The macroscopic behavior of the TCP congestion avoidance algorithm. ACM SIGCOMM Comput Commun Rev 1997;27:67–82.
9. Halepovic E, Williamson C, Ghaderi M. Enhancing redundant network traffic elimination. Comput NetwElsevier February 2012;56(1):795–809.

10. Christner J, Seils Z, Jin N. *Deploying Cisco Wide Area Application Services.* 2nd ed. Cisco Press; 2010. 648 pp.
11. Zhang Y, Ansari N, Wu M, Yu H. On wide area network optimization. IEEE Commun Surveys Tuts 2011. DOI: 10.1109/SURV.2011.092311.00071.
12. Yu J, Fang C, Lu L, Li Z. A lightweight mechanism to mitigate application layer DDoS attacks. Infoscale; June 2009. p 175–191.

19

PlatonTV: A SCIENTIFIC HIGH DEFINITION CONTENT DELIVERY PLATFORM

Mirosław Czyrnek[1], Jędrzej Jajor[1], Jerzy Jamroży[1], Ewa Kuśmierek[1], Cezary Mazurek[1], Maciej Stroiński[1], and Jan Węglarz[1]

[1]*Poznan Supercomputing and Networking Center, Poznań, Poland*

19.1 INTRODUCTION

Platon Scientific high definition (HD) interactive TV, PlatonTV in short, is an interactive Internet Protocol (IP) television service built to provide HD digital content to foster education, facilitate knowledge sharing, and popularize science. It is one of several services offered to the research community by the Platon project through deployment of dedicated infrastructure and applications in PIONIER [1], Polish countrywide broadband optical network. PlatonTV platform comprises a variety of components supporting various stages of content processing, from content production, through content packaging to the delivery to end-users, and connects them to provide a complete processing path as shown in Figure 19.1. Content delivery network (CDN) is one of the key platform components, and in this chapter we will concentrate on the issues related to content delivery. However, in order to justify the design choices made for the CDN, we explain its relation to other components and present the assumptions that guided the design process of the entire platform.

Advanced Content Delivery, Streaming, and Cloud Services, First Edition.
Edited by Mukaddim Pathan, Ramesh K. Sitaraman, and Dom Robinson.

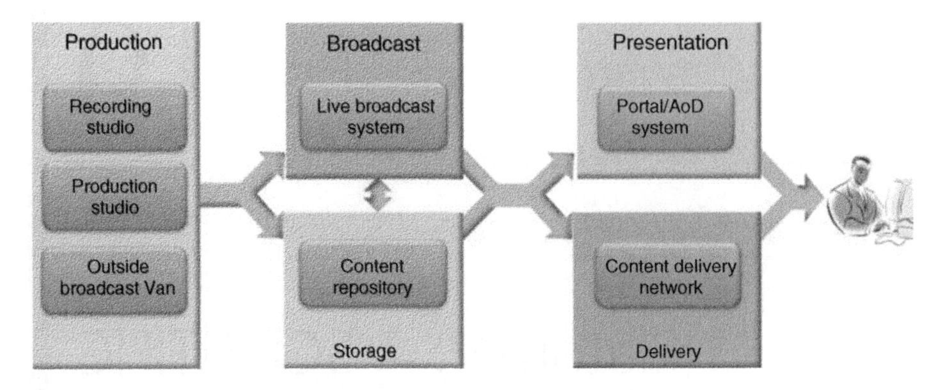

Figure 19.1 PlatonTV overview.

The platform addresses several needs of the research and academic community. Typically, in such a community, there exist many independent rich multimedia content repositories made available by various providers. The fact that they operate separately means that the users need to be aware of many sources and the ways to access their content. On the other hand, there is a need of a number of the universities to have their own TV channels for event reporting to the community. Moreover, international and national projects need a medium to disseminate their results for promotional and educational purposes. PlatonTV provides such medium and integrates content from various sources increasing their accessibility to the community.

PlatonTV was designed with specific goals in mind as far as content and service purpose are concerned. Although the service is offered to the scientific and research community, as stated earlier, the end-user group is not limited to this community but can be characterized as a general audience. Still, entertainment is not the main focus of this service. At the same time, the service design process was guided by the newest multimedia content delivery trends such as multiscreen service with high quality content and various access modes (live, video on demand (VoD), linear), to address users' expectations. All of this means that the platform is capable of delivering high quality content to multiple device types and with multiple protocols, including the major streaming video technologies. Consequently, content is offered in various formats at various bitrates and resolutions. Given a variety of end-users' interests and devices used for content playout, service personalization and contextualization are important issues, as well as interactive access, which was one of the goals set for the platform.

PlatonTV does not use "pure-play" dedicated CDN but rather deploys its own delivery network over PIONIER. The CDN spans 22 data centers, each connected to several Internet service providers (ISPs). The CDN traffic is isolated from the public Internet for internal content distribution and content provisioning to end-users. Hence, the delivery service can be characterized as an over-the-top (OTT) solution deployed over a managed core network. The overall approach in content delivery is to distribute content close to end-users in order to save bandwidth and enhance quality of service (QoS). In the following sections, we provide details on live and VoD content delivery.

In this chapter, we concentrate on PlatonTV content delivery and the related issues. Section 19.2 presents similar systems designed to support education and science popularization with the use of video streaming to provide background for PlatonTV presentation. The platform architecture is introduced in Section 19.3 with the CDN relation to other components explained. Steps leading to and including content delivery, such as content ingest, content distribution, and management within the CDN, and content delivery to end-users are described in Sections 19.4–19.6, respectively. Section 19.7 is devoted to the platform availability and reliability issues and presents our approach to address these issues on multiple levels. We share our visionary thoughts in Section 19.8 and describe possible future research directions in Section 19.9. The platform presentation is concluded in Section 19.10.

19.2 BACKGROUND AND RELATED WORK

In the recent years, there have been a number of initiatives to support education and science popularization through the use of video content. Services developed as a result of these initiatives share certain content and end-user group characteristics. They usually make content available to any user, and thus should be able to deliver content efficiently to clients around the world. The scope of operation may however be limited by factors such as language of the video, for example. Nonetheless, they usually operate on a large scale. Different organizations deal with content distribution in different ways. Some use commercial CDN services from providers such as Akamai. Some others develop new solutions such as OpenCDN [2], relying on a network of participating organizations willing to share their resources in order to bring their content closer to their users. Some organizations use distribution channels such as YouTube or iTunes or a combination of several options. Finally, some organizations partner with linear TV providers, for example, cable or satellite, to make content available in their networks.

Systems developed to provide educational and scientific content often encompass content production service (e.g., MIT OpenCourseWare) as well as content aggregation (e.g., VideoLectures.net). Owing to the specific nature of content needed and promoted by these services, there is a need to guarantee the accuracy and originality of the materials, which is why content is either produced by organization running the service or received only from the participating organizations. Educational and scientific content is usually freely and openly available for noncommercial educational purposes. Typically, it can be used, copied, distributed, and modified, subject to additional optional restrictions. Hence, digital rights management is not a priority for these services.

Research channel based at the University of Washington in Seattle, one of the pioneers, was an educational television operated by a consortium of leading research and academic institutions, which contributed science-related programming. Content was available to users in the United States and in other countries via satellite (DirecTV) and cable television. The project was discontinued in 2010 because of lack of funding.

Another of the education-oriented initiatives is OpenCourseWare that makes educational materials from university undergraduate- and graduate-level courses available online. It was first started at the University of Tübingen in Germany but popularized by the Massachusetts Institute of Technology. Educational materials include not only supplemental content such as lecture notes, reading lists, homework problems, exams, or even textbooks, but more importantly streaming video lectures. The MIT Video website [3] aggregates not only courseware but also content produced by MIT laboratories, centers, and administration including feature and editorial videos, event recordings, and other academic content. Currently, it is a collection of more than 10,000 videos organized into more than 150 channels, where a channel is a group of materials on the same subject. The OpenCourseWare channel is the largest of the channels with more than 2300 lectures. MIT media production group assists in recoding lectures and other course materials, which are encoded into H.264 MP4s. MIT OpenCourseWare publishing infrastructure consists of planning tools, a content management system (CMS), and content distribution infrastructure utilizing Akamai's EdgeSuite platform [4]. Content available from OpenCourseWare website may be used, copied, distributed, translated, and modified for noncommercial educational purposes and made freely available to others. All content is reviewed to determine the correct ownership of the material and obtain the appropriate licenses to make the material openly available on the Web.

VideoLectures.net [5] provides free and open Internet access to high quality scientific content such as open lecture, seminars, and discussions from a number of contributing organizations, MIT OpenCourseWare among others. In addition to content aggregation, it supports also content production. Educational materials are often accompanied by supplemental content such as documents, slides, information, and links. The end-users group includes scientific, research, and general public all over the world with lectures available in a number of different languages. VideoLectures.net participated in the development of an open content distribution channel OpenCDN, which was abandoned in 2008. Some of its content are also available through YouTube and iTunes.

19.3 PlatonTV ARCHITECTURE

Although PlatonTV content delivery system is the main focus of this chapter, we also briefly present other components of the integrated platform and explain the influence they have on the CDN design and implementation.

The CDN is the key component responsible for delivery of content provided by a number of independent content providers, which use media production infrastructure to create VoD and live programs. The VoD content is first ingested to the content repository, which plays the origin server role for the CDN. Live content is processed by the Live Broadcast System before being handed to the CDN. VoD content as well as archived live content can undergo further processing in order to add supplemental material and manage the way it is presented to end-users before they are ingested by the repository. Finally, a set of interactive applications provide access to the available services and content. Figure 19.2 presents PlatonTV platform structure with content and metadata flow among its components. In the following, we briefly describe each component, leaving the CDN to be introduced in details.

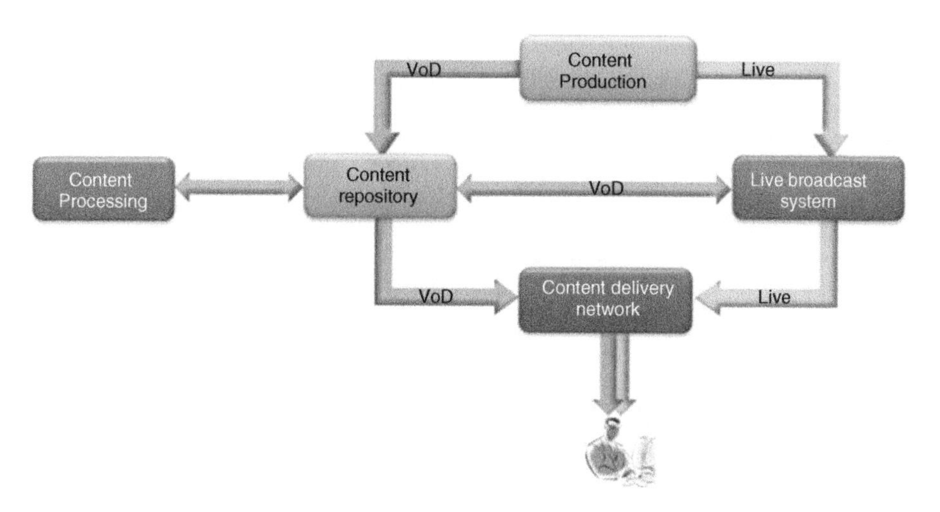

Figure 19.2 PlatonTV components.

19.3.1 Content Production

PlatonTV media production infrastructure consists of 6 production studios, 15 recording studios, and a mobile studio—an Outside Broadcasting Van—that have been deployed and across the country on the participating partners' premises. The production infrastructure has been built with broadcast quality HD equipment including cameras, audio and video mixing devices, live graphics, lightning, and postproduction facilities that enable high quality media production using fully tapeless workflow. Each studio is considered to be a content provider for the platform. The providers can operate independently from one another in different geographical regions and in different scientific areas as they are associated with various research and educational institutions. The content produced includes lectures, recording of events such as conferences, seminars, workshops, and various academic events, as well as presentation of the research results and experiments using unique infrastructures. The platform can also ingest content from external providers, not directly associated with Platon project that have their own production facilities and/or media resources to be published.

19.3.2 Content Repository

The repository does not only ingest and manage content but also adjusts the content to match the required formats and bitrates to the playout devices. The multiscreen service must support a wide variety of the device types from TV set, through PC to hand-held devices, and consequently variety of content format types, bitrates, resolutions, and streaming protocols. PlatonTV content repository transcodes ingested content to a number of preconfigured formats and stores them in the repository alongside the original format. Formats for transcoding and their parameters are selected to cover a set of devices typically used to access content. Since these formats are chosen for delivery of a given content to end-users most of the time, they are generated prior to making content available to end-users, instead of on-the-fly transcoding performed on demand. We consider the latter approach to be suitable for formats that are rarely needed. Therefore, the repository is the primary source of multiple formats of each content for the CDN.

19.3.3 Live Broadcast System

The live streams (feeds) from production studios' encoders are managed through the live broadcast system, which allows one to define live broadcast schedule in the form of a virtual channel. By live program, we mean linear video transmitted synchronously as explained in Chapter 2. The virtual channel can comprise live content and on-demand media available from the content repository. From a user viewpoint, such channel is equivalent to a linear TV transmission. Chapter 2 provides explanation of the nature of live streaming that is more precisely defined as linear streaming transmitted synchronously. From the CDN's viewpoint, channel is a single live media stream provided for distribution. With the live broadcast system, it is possible to record the live transmission, perform simple postprocessing, and export results to the repository. Such approach shortens the time needed to publish the recording of a live transmission as on-demand asset after the broadcast is finished.

19.3.4 Content Processing

As PlatonTV users may vary in terms of the area of interest, the purpose of using the platform, experience, age, and a set of tools, is needed to prepare content for presentation for different user groups. For example, the same scientific experiment will be presented in a different way to a researcher working in a given area and to students of a given subject. It will also need a different supplemental material in both cases. In order to satisfy this requirement, content processing system provides CMS and a number of interactive applications, which we collectively refer to as application on demand (AoD). The former is designed to create content offer, that is, to manage the content and the way it is presented to end-users by access portals. The latter contains a set of tools for adding supplemental content, such as presentations, graphical elements, audio, and text, for generating new content from the source material.

19.3.5 Access Portals

A set of interactive applications provide access to the available services and content. An interactive portal is the primary application providing access to content, content browsing, and searching tools. In addition, the portal provides communication services for its users and between users and content producers. Registered users have an opportunity to comment on the particular video he/she is watching and to provide its evaluation.

19.4 CONTENT INGEST

Ingest of content available through PlatonTV is performed by the content repository. On-demand content is produced in the studio facilities and processed using nonlinear editing systems. The final asset is exported using high quality encoding profile to the content repository, where content metadata is generated. The content repository provides storage for media assets, mechanisms for transcoding content to lower bitrate formats, and for managing content metadata and access policy. In other words, the repository performs tasks necessary to make content ready for delivery.

Format provided by a production studio is referred to as a source format for a given content. Typically, such a format, as a high bitrate format, is not intended for distribution. It is transcoded to a number of so-called derived formats selected to enable content streaming to various types of devices. Currently PlatonTV supports Mac OS®, iOS®, Android, Windows®, and Linux clients, and also experimentally some Digital Living Network Alliance (DLNA) [6,7] and DLNA Media Renderer (DMR) compliant devices. Mac and Windows devices use dedicated Silverlight application for content playback. For Linux, the same application is running using Moonlight. Both of them can play smooth streaming (SS) VoD and live content. IOS and Android can use MP4 h264 AAC media files. DMR devices have very wide spectrum of supported containers and codecs. Unfortunately, it is hard to find a common HD format that would work on all DMRs. Older DMRs cannot play MP4 or Windows Media Video (WMV) files, but accept AVC M2TS streams. The set of derived VoD formats selected to cover most needs is presented in Table 19.1.

Formats for live content are determined by content provider based on the content and the available uplink. The set of live content formats typically used is presented in Table 19.2.

At the time when content is ingested into the repository, its metadata is generated. The metadata includes descriptive data and technical data. Descriptive data is provided by the content provider and characterizes content semantically. It applies to the content as an abstract entity. Technical metadata, on the other hand, characterizes content syntactically and therefore applies to a specific content format. Users select and request content based on the descriptive metadata. Technical metadata is used for the selection of the most suitable format for content presentation.

TABLE 19.1 VoD Formats

Type	Video Codec	Video Streams	Audio Codec
Smooth streaming	H.264	1920 × 1080 at 6000 kb/s, HiP	AAC-LC at 160 kb/s
		1280 × 720 at 2800 kb/s, HiP	
		852 × 480 at 1300 kb/s, HiP	
		640 × 360 at 500 kb/s, HiP	
MP4	H.264	1280 × 720 at 3000 kb/s, HiP	AAC-LC at 96 kb/s
		480 × 272 at 1300 kb/s, BP	
M2TS	H.264	1280 × 720 at 4000 kb/s, HiP, 4.1 level limit	AC3 at 120 kb/s
WMV	VC-1	1280 × 720 at 3500 kb/s	WMA9 at 120 kb/s

TABLE 19.2 Live Content Formats

Type	Video Codec	Video Streams	Audio Codec
Live smooth streaming, HTTP live streaming	H.264	1280 × 720 at 2800 kb/s	AAC-LC at 96 kb/s
		852 × 480 at 1300 kb/s	
		640 × 360 at 500 kb/s	
WMV live streaming	VC-1	1280 × 720 at 2800 kb/s	WMA at 128 kb/s
		640 × 360 at 500 kb/s	

Next to descriptive and technical metadata, there is content access control information that includes permission to distribute content and access channel that defines content accessibility for end-users. The access channel mechanism is very flexible and makes it possible to restrict access based on geographical location of the end-user, the user's Internet service provider, the user group, or the type of the user's end device. User's geographical location is obtained from the external GeoIP database, while the association between the user's IP, ISP name, and network is kept in the internal database and updated with data from public Whois servers, as well as from Border Gateway Protocol (BGP) sessions. A detailed description of PlatonTV content access control mechanisms is provided in Reference 8.

19.5 CONTENT DISTRIBUTION AND MANAGEMENT

Content is distributed and managed by two PlatonTV components: content repository and the CDN. As these two components play different roles, they use different rules for distributing and managing content.

19.5.1 Content Repository

Content repository is the primary source of content for other platform components including the CDN. Therefore, the repository must assure high availability of all contents. Although the repository performs the origin server role, it is in fact a distributed system with a number of storage nodes distributed geographically and high content availability is achieved through replication and archiving. The degree of content replication can be set on a per-content group basis, where a group membership is determined based on the content metadata. For example, a group may contain all contents from a given content provider or all contents belonging to the same series. Content grouping allows the operator to differentiate replication strategy among various content groups. Replication strategy is applied to all contents in a given group and in practice to all their formats.

As the repository is a distributed system with storage nodes placed in various geographical locations, it is possible and beneficial to use "close to the user" principle in the content placement strategy. For the repository, it means placing content in various geographical locations so that CDN nodes can obtain content from the closest location. We define storage node grouping to implement this strategy. Node grouping mechanism is quite flexible and can be based on criteria other than just node location. The operator defines groups, assigns storage nodes to these groups, and configures replication strategy, which maps content groups to node groups. Typically, content is replicated evenly among all location-based node groups. However, it is possible to "regionalize" content location given information on the location of the potential end-users, for example.

19.5.2 Content Delivery Network

PlatonTV CDN consists of two-level mesh of content servers as shown in Figure 19.3. Higher level is built of five regional data centers (RDCs), which group higher level content servers and are responsible for delivering content from the repository or the live broadcast system to the lower level edge servers (ESs). Each RDC is assigned a separate group of ESs for which it is a content source. ESs deliver content to the end-users. Content servers are basically uniform and their role in the CDN can be easily configured to

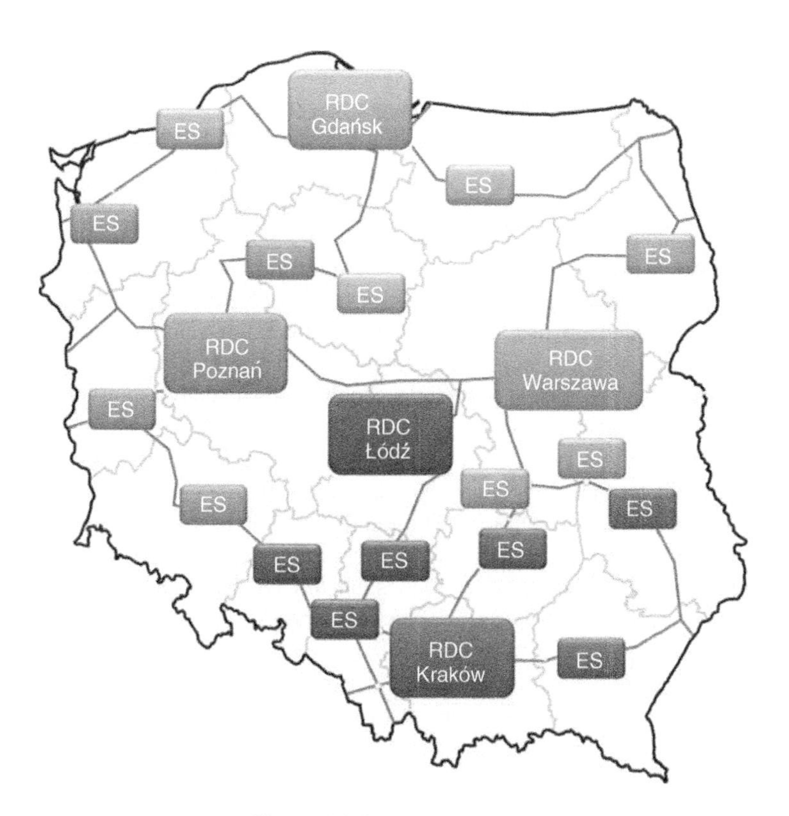

Figure 19.3 PlatonTV CDN.

be either an RDC content server or an ES. Currently, all content servers in RDCs cache exclusively live streams.

PlatonTV provides service primarily to the Polish academic institutions. However, all contents are also available worldwide. Therefore, two groups of ESs are distinguished within our CDN: default group and dedicated group. The default group ESs are located in data centers with direct connection to tier 1 network providers and, most importantly, with connections to the major Polish ISPs. The dedicated group ESs are located at metropolitan area data centers distributed all over the country, with direct connections to the biggest Polish academic centers. The latter group of ESs is dedicated to serve content only to users from the selected *a priori* known networks. The list of the networks, which are recognized by the system, is built semiautomatically based on the BGP data and the policy statements of the institutions that host these networks. The rest of end-users are served by the ESs from the default group. This group is used also as a backup in case of a planned service or a malfunction of the dedicated ESs. There are four data centers that host the default ESs. They are located in Poznań, Warszawa, Łódź, and Kraków. All RDCs and ESs are connected via PIONIER.

Content servers use Microsoft® Internet Information Services (IIS) and Application Request Routing (ARR) to cache content files. Live content is provided by IIS and Windows Media Services (WMS). ESs run IIS and WMS modules for authorization of user requests and user activity logging. The CDN does not have to cache entire multimedia files. According to our workload reports in many cases, end-users do not watch the entire content but rather tend to search through longer contents for interesting fragments. Moreover, multimedia players often use HyperText Transfer Protocol (HTTP) range header[1] to access some fragments of content file prior to playback or seek through the file. As HD multimedia files are heavy objects, often exceeding 1 GB in size, it is not rational to always cache the entire file. ARR supports fragments caching, which shortens response time in the case of cache-miss handling for a request that uses range header, and lowers network bandwidth usage within the CDN core network. Bandwidth requirements and cache space needed are also lowered thanks to streaming technologies used: SS and HTTP live streaming (HLS). Both technologies use content fragmenting concept, with each content fragment requested separately. Some issues related to multimedia content caching are discussed in Reference 9.

In order to avoid an unreasonable replication degree at the ESs, the CDN gives priority to the servers that already cache at least some fragments of the requested content file. Choice of the servers for distribution depends on runtime server and network state monitoring data and on the CDN network topology data collected and stored in the configuration database. The topology defines which CDN nodes are connected and the weight (preference) of each connection. The CDN operator can alter preferred transmission routes by changing the static weights.

There are three policies for choosing source server for content distribution: PULL, PUSH, and LIVE. In PULL case, content distribution is triggered by a user request.

[1]HTTP range header is a HTTP GET request header, which enables requests for fragments of content from a web server. It is commonly used by media players and download managers.

ES selected to process the request uses an RDC content server or the content origin (content repository) directly to obtain content, if it does not have it in its cache. The nearest RDC is preferred as content source, if it is not available there, than other RDCs are checked. Content origin is used in this scenario as the last resort. In case of PUSH policy, content is distributed to a set of ESs selected by the operator. The CDN uses Dijkstra algorithm over cost function calculated from static weights and the current content servers and network load to find the best path for content transfer. LIVE policy is very similar to PUSH. The main difference is that for live streams the CDN makes additional checking for available bandwidth on all network interfaces that would take part in the content transmission.

19.6 CONTENT DELIVERY

Delivery of the content selected by the users involves several steps performed prior to content transmission of the user's end device including content format selection and ES selection.

Content format choice is determined by the end device capabilities. The PlatonTV platform provides customized players for embedding in various browsers and operating systems. Moreover, two more applications are under development: a Home Gateway application and a dedicated Android application enable the use of DLNA-compliant consumer electronic devices such as SmartTVs. Some issues related to delivering content to home ecosystem are presented in Chapter 6. Each of these players has different requirements for content format and transmission protocol. The selection of a proper container and encoding is done by the player application and sent to the CDN. The CDN selects the best material format that does not exceed the given constraints and locates the best ESs for content delivery. ES selection is done upon user's IP address, current ES load, and cache content.

From the CDN side, requests are handled by three instances of a PlayerAccess (PA) Web service located at different RDCs. Each PA comprises a number of application servers (App Server) configured as a net balanced Web farm. PlayerAccess uses configuration and content databases, which include several database engines as shown in Figure 19.4. Primary database site utilizes DB mirroring. Remaining databases are located at different RDCs and work as updatable replicas. User requests are directed to PA farms using domain name system (DNS) round robin with a short time to live (TTL). User's request-handling routes are shown in Figure 19.4 with solid lines while dashed lines represent alternative request routes. Content is delivered by an ES using adaptive streaming (SS, HLS, or WMS) for live content and HTTP progressive download for VoD content. The nature of various content delivery methods, such as adaptive streaming or progressive download, is explained in Chapter 2.

Microsoft WMS use Microsoft's proprietary container Advanced Systems Format (ASF), formerly known as Advanced Streaming Format. One of the advertised features of WMS and ASF is an automatic stream bitrate selection for the best possible user experience. To achieve this, ASF files have to contain multiple audio and video streams encoded at different bitrates/qualities. ASF header defines mutual exclusiveness among

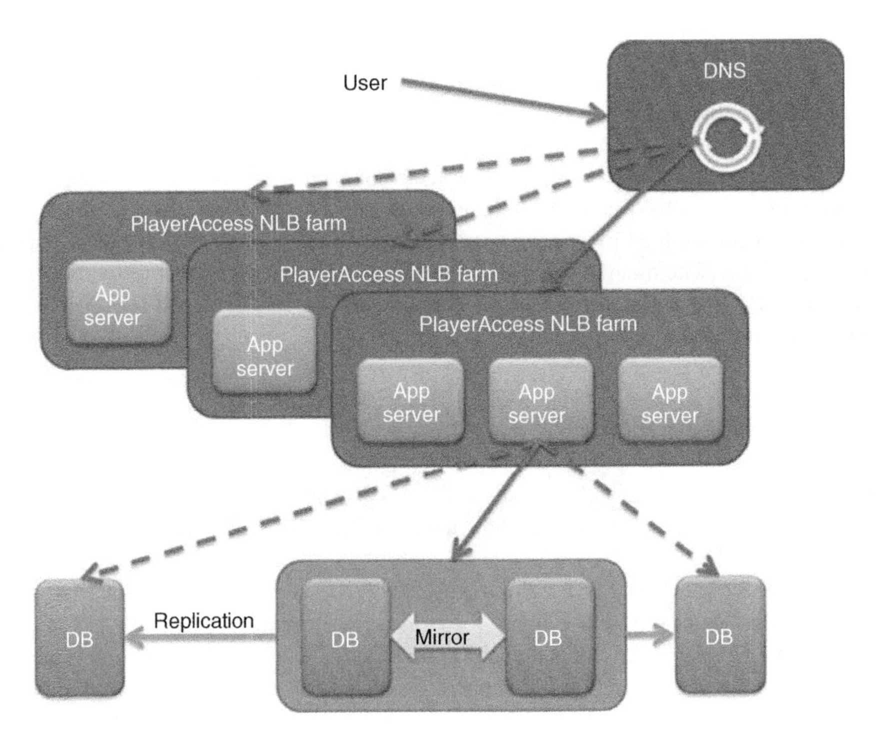

Figure 19.4 User's request-handling process.

streams defining groups of streams that should be delivered to the user. All the streams are written in one file in an interleaved manner. The idea works just fine, but has a serious disadvantage when one server contains a big library of files and users are accessing different files. Even if content hard drive (or redundant array of inexpensive disk (RAID) volume of any kind) is not fragmented, files containing streams are fragmented because of ASF design. HDD are performing much better if data is read in large, continuous blocks. In case of multistream ASF files, content server even under moderate load quickly becomes unresponsive because of long input–output (IO) queues: disk subsystems are unable to execute the excessive number of IO operations. More recent solutions such as SS and HLS use separate files for each bitrate/quality and that approach lightens disk subsystem load.

PlatonTV content can also be accessed from a home area network (HAN) and played on a DMR compliant device. Two applications were developed for home network-based access: an Android smartphone application and Home Gateway PC application. Smartphone application acts as DLNA media Controller (DMC). It allows users to select media content from PlatonTV directory and controls its playback on a DMR using DLNA protocol suite. DMR accesses the content directly from an ES. This approach requires ESs to mimic DLNA media server at a basic level. This scenario has some limitations: the DMR has to be capable of playing at least one of content formats provided by the CDN, because no transcoding can be done. The Home Gateway works in a slightly different manner. It is a gateway between HAN–DLNA and the CDN and acts as a fully featured DLNA media server. It allows browsing and searching for content and provides original media files to DMRs as well as to some transcoding options.

19.7 AVAILABILITY AND RELIABILITY

There are a number of mechanisms used in PlatonTV to ensure high service availability and reliability. The CDN architecture with geographically distributed caches, positioned in peering points of the ISP's and CDN is a good reliability design strategy that prevents congestion and latency issues. As mentioned earlier, the internal topology of the CDN is a two-level mesh of Live Cache and ESs and allows groups of content servers, as well as individual servers within each group, to communicate with one another, giving multiple possible paths for content distribution. The map of all possible data paths is correlated with the underlying network topology and information from the monitoring systems to ensure selection of the best subset of connections.

CDNs often use separate network isolated from the public Internet for internal content distribution and for content provisioning to end-users. PlatonTV also uses this approach with a Level-3 Virtual Private Network, which is one of the network layer services provided by the PIONIER MPLS backbone, built with the multi-10 gigabit DWDM technology (multiple 10 gigabit lambdas on network links). The virtual routed network, isolated from other users of the backbone, connects all 22 data centers housing PlatonTV equipment. While keeping the traffic isolated from other clients, this architecture also provides automatic best path selection and path failover in case of an underlying link failure.

Owing to the video content nature and the streaming protocols requirements, it is difficult to limit the influence of the interruptions in network and system accessibility on the video quality. Infrastructure problems—ranging from congested links, high packet loss, and packet or frame reordering to heavily loaded servers, machine failures, and application issues—affect the overall users' experience, giving ragged video, freezing or missing frames, or even stream playback interruptions. Therefore, it is essential to design a redundant architecture in all aspects—hardware and software. All PlatonTV servers use redundant physical components—power supplies from two different power lines, hard disks with RAID-1 or RAID-50 mirroring, and multiple aggregated network adapters. These measures limit the negative effect of the most common failures that may happen in data centers. The network infrastructure consists of redundant switches in each of the data centers, with multiple uplinks to the core network. Multiple network connections use the 802.3ad dynamic Link Aggregation Control Protocol (LACP) to provide both bandwidth scalability and link redundancy. The best path and failover link selection between the Platon routers and the core network is done by the Open Shortest Path First (OSPF) dynamic routing protocol. The core network uses the Multiprotocol Label Switching technique to provide fast alternative path selection.

Functional components of the system are also multiplied to ensure that at each level of the distribution there is no single point of failure. End-users interact with the CDN in two points, one of them is an ES responsible for provisioning the content, and the other one is the access service, which redirects user request for a particular multimedia content to the designated ES. Naturally, the access service is a critical component in such architecture. The PlatonTV CDN has multiple access services, deployed in three regional data centers, supported by load balancing and high availability clusters of three nodes in each of these locations. This configuration gives both high reliability and high transaction throughput of this service. Content delivery mesh uses multiple ESs in multiple data centers to reduce the impact of possible outage. As mentioned before, the repository—a key component and primary source of content—is also a geographically distributed system to ensure high availability with content replication between two distinct localizations.

Second important aspect, essential in reducing the impact of any problems on system's availability and performance, is live monitoring by a dedicated Network Operations Center (NOC). NOC manages the incidents and supervises the Content Distribution Network 24 h a day, 7 days a week, providing a single point of contact for all issues regarding the system. Aside from the servers used to provide strictly services for the CDN, there are also several servers with specialized software, deployed to monitor hardware infrastructure and network. These tools measure a number of key performance parameters and network statistics, for example, connectivity between nodes, capacity and network traffic on the links, round trip time and packet loss between nodes, jitter on links as well as operating systems' parameters: CPU and memory utilization, storage utilization, throughput, and queue length. There are also tools that automatically test the quality of a network stream from each of the ESs and assess whether a node meets appropriate service level.

Maximal reliability of the system is also achieved by mandatory security audits and periodical patch installations on all components of the system. The access to the system for its administrators is performed over an IPSec Virtual Private Network (VPN) through a firewall, which is a gateway between the internal private network and the Internet.

19.8 VISIONARY THOUGHTS FOR PRACTITIONERS

The volume of media content delivered over the Internet is growing rapidly, forcing the ISPs to cope with the demand for bandwidth and QoS provisioning. The changing behavior of end-users, which expect "anytime-anywhere" access to high quality multimedia, challenges the ISPs to serve more and more traffic in an optimal and cost-effective way. It is obvious that ISPs will always lose race for required bandwidth driven by content providers and OTT service providers. Users would like to have more and more streaming services accessible with not only PCs but also TVs that provide some kind of VoD applications, smartphones, or tablets. While there are enterprises such as Akamai that try to satisfy these requirements, small ISPs are always in a bad situation: unicast streams increase their external network connections load. Moreover, all ISPs do overbook their available bandwidth. Such ISPs may consider using transparent proxies, but it will always have disadvantages. A good solution for them would be to have CDN provider's point of presence in the core network, but it is only a partial solution. A much better one would be to have a common protocol for a CDN interconnection, which is currently developed within the Internet Engineering Task Force (IETF) group [10] (more information on this effort is presented in Chapter 4) and for establishing universal point of presence in ISPs' networks or even in large end-user networks. Such a solution would have to address financial aspects of course, but it would allow one to build a federated CDN capable of providing large-scale streaming worldwide more effectively. In addition, it would improve user's experience because of lower latencies and higher available bandwidth.

19.9 FUTURE RESEARCH DIRECTIONS

PlatonTV provides tools for interactive access to the media assets and enables their use as parts of complex presentation assets (e.g., learning resources). It also allows one to annotate the media to access the indexed fragments in different scenarios. The content production is conducted in professional studios, where people may participate in the program not only in person but also using videoconferencing infrastructure. One of the challenges to be solved is to extend this scenario to enable viewers to take part in the program interactively using video communication solutions that may be lightweight and suitable for larger audiences. Another one is to provide them with tools enabling in-person-like experience and interaction during the program recording. This may be extended with presentation of content and interaction in the augmented reality. Consequently, the participatory TV experience could be achieved in a completely new way. We expect that this will redefine the requirements for service quality, content production, and delivery and will have to be managed in the future.

19.10 CONCLUSION

The PlatonTV platform has been built to present scientific challenges; promote research teams and their results as well as to cover academic lectures, conference, and other events; and made them available to the general audience, through the media assets. The deployed platform enables production and seamless delivery of high quality multimedia over Internet, and provides tools for interactive use of the published contents. The platform architecture has been designed to support many content providers and enable deployment of many access portals, which along with interactive tools, enable content use in different scenarios, including presentation of learning resources. The delivery of the content, made available through dedicated CDN deployed in metropolitan area networks spread over the country, is done effectively and assures high quality of experience for the end-users using different types of end devices.

ACKNOWLEDGMENTS

The PlatonTV platform (available at http://tv.pionier.net.pl), presented in this chapter, has been developed and deployed within the Platon project supported by the European

Regional Development Fund—Innovative Economy programme under the grant number POIG. 02.03.00-00-028/08.

REFERENCES

1. Rychlewski J, Węglarz J, Starzak S, Stroiński M. PIONIER—Polish optical Internet. Proceedings of European Conference on Research and Development for Information Society ISThmus; Poznan; 2000.
2. Falaschi A, Open content delivery network short overview. Proceedings of the TERENA Networking Conference; June 7–10; Rhodes, Greece; 2004.
3. MIT. 2013. MIT OpenCourseWare website. Available at http://ocw.mit.edu. Accessed at 2013 Mar 23.
4. Akamai Inc. Turbo-charging dynamic web sites with Akamai EdgeSuite, White paper; Akamai; 2002.
5. VideoLecture.net. 2013. VideoLecture.net website. Available at http://videolectures.net. Accessed 2013 Mar 23.
6. Digital Living Network Alliance. 2007. DLNA overview and vision whitepaper. Available at http://www.homeelectronics.jp/whitepaper/3.pdf. Accessed 2013 Mar 23.
7. Digital Living Network Alliance. DLNA networked device interoperability guidelines; 2009.
8. Czyrnek M, Jajor J, Kuśmierek E, Mazurek C, Stroiński M, Weglarz J, Platon scientific HD TV platform in PIONIER network. Proceedings of the 5th Conference on Multimedia Communications, Services and Security; Cracow, Poland; May 31–June 1; 2012.
9. Czyrnek M, Kuśmierek E, Mazurek C, Stroiński M, Węglarz J. CDN for live and on-demand video services over IP. In: Buyya R, Pathan M, Vakali A, editors. *Content Delivery Networks*. Springer; 2008.
10. Niven-Jenkins B, Le Faucheur F, Bitar N. Content distribution network interconnection (CDNI) problem statement, RFC 6707. Internet Engineering Task Force (IETF); September 2012.

20

CacheCast: A SINGLE-SOURCE MULTIPLE-DESTINATION CACHING MECHANISM

Piotr Srebrny[1], Dag H.L. Sørbø[1], Thomas Plagemann[1], Vera Goebel[1], and Andreas Mauthe[2]

[1] *University of Oslo, Oslo, Norway*
[2] *InfoLab 21, Lancaster University, Lancaster, UK*

20.1 INTRODUCTION

The original design of the Internet architecture did not consider the single-source multiple-destination datagram transmission, namely *multicast*. However, already in the middle 1980s, it was recognized that this functionality is necessary for applications such as distributed databases, distributed computation, or teleconferencing [1,2]. Applications of this type require efficient multiple-destination data delivery. With content delivery over the Internet being more prominent nowadays, multicast is increasingly becoming a necessary feature. Unfortunately, the Internet still does not support multicast services, and applications must use unicast.

The development of multicast technology has been carried out along two lines. The first direction was set by Aguilar in Reference 1. He proposed to extend the Internet Protocol (IP) header destination field. The extended destination field accommodates a list of data destination addresses. When a source sends a packet to multiple destinations, it lists all destination addresses in the header. This approach does not scale well with the

Advanced Content Delivery, Streaming, and Cloud Services, First Edition.
Edited by Mukaddim Pathan, Ramesh K. Sitaraman, and Dom Robinson.

increasing receiver groups size, since with the growing group size there is less space for data. Ultimately, the group size is limited by the maximum size of a packet, which in the present Internet is 1500 B. Initially the Aguilar proposal was rejected by the networking community, but it has recently been revised and has been found useful as a complement to IP multicast for small groups [3].

The second direction was set by Cheriton and Deering in Reference 2, and it is known as IP multicast. The IP multicast model assumes that a source does not know the receivers of a datagram, rather it sends a packet to a group of hosts identified by an IP multicast address. Therefore, the network is responsible for the delivery of a datagram to all group members; hence, it is also responsible for group management. However, as pointed out by Diot et al. [4], this model of multicast is prone to malicious attacks; it requires global address allocation and reveals serious scalability issues due to group management in the network. To address these issues, the first IP multicast architecture was strongly constrained over the years. Nonetheless, at present, it still lacks vital services such as receiver authorization, authentication, and accounting, to be accepted by content providers and Internet service providers (ISPs). Moreover, IP multicast requires per-group state on forwarding nodes, which poses heavy burden on backbone routers. Another issue is the difficulty to construct multicast congestion control [5]. The solutions for handling heterogeneous receivers are based on layered transmission, which, in turn, requires allocation of expensive multicast groups. Altogether these issues account for the lack of IP multicast services in the Internet. According to AmericaFree.TV from 2007, the IP multicast penetration to the Internet was 2.2% measured as a ratio of the number of IP multicast-enabled systems to the total number of autonomous systems.

The aforementioned issues related to network layer multicast, and difficulties in the deployment of this class of services led researchers to address the problem at the application layer. The application layer multicast (ALM) quickly became very popular not only as a research topic but also as a commercial product. But even though ALM can provide users with full multicast functionality, it "introduces duplicate packets on physical links and incurs larger end-to-end delays than IP multicast" as pointed out Chu et al. [6]. ALM distributes the burden of data transmission across data receivers; however, it does not address the root cause of the problem, which is the lack of the network layer mechanism for single-source multiple-destination data transfers.

At present, the only way to transmit the same data to multiple destinations scattered across many autonomous systems is to use multiple unicast connections. However, this approach results in numerous redundant transfers of the same data over the first few hops from a source. These redundant transfers consume unnecessarily the Internet resources and the waste is particularly severe near to sources sending popular content such as Internet Pay TV (IPTV) servers. This chapter presents CacheCast—a caching system that removes this type of redundant data transfers from network links. The redundancy elimination (RE) is achieved using small caches operating on link endpoints. The caches are deployed in the network hop-by-hop starting from sources of redundant data. A single-link cache operates independently from other link caches; thus, a failure of one link cache does not affect the others. The CacheCast resource requirements are modest. The link cache size is an order of magnitude smaller than the storage space provided by the link queue. This reduction in resource requirements is achieved for the cost of an additional complexity on a source. A source sending the same data to multiple destinations must mark it and transmit it in a batch. Such transmission achieves nearly the same bandwidth utilization as multicast transmission, which is shown in this chapter.

20.2 RELATED WORK

The CacheCast architecture consists of two elements: a distributed architecture of link caches and a server support element. Both aspects have been previously studied in the context of RE on point-to-point channels. The first system introducing the idea of RE on the packet-level granularity has been presented by Spring and Wetherall in Reference 7. To find redundancy in a sequence of packets, it uses a modified algorithm for finding similar files in a file system. Before a packet is transmitted over a channel, the algorithm searches for repetitions in the packet by comparing its content chunk by chunk with the content of previously transmitted packets. The identified repetitions are replaced with short summaries that enable the receiving end to reconstruct the packet. This RE technique is cost efficient when applied to channels with low-to-moderate capacity. However, with the growing channel capacity, the storage and computational costs quickly outweigh the benefits of RE. At present, systems based on this idea are part of wide area network (WAN) optimizers and are often applied to eliminate redundant traffic transmitted between branches of a distributed enterprise.

Anand et al. in Reference 8 propose to use the technique of Spring and Wetherall to achieve universal RE. The key idea is to apply the RE on each link connecting two routers in the Internet. In this case, all redundancy in network traffic is immediately eliminated. The authors also show that the system can yield even greater benefits when the routing algorithms are aware of the RE. The conducted measurements indicate that the RE system with redundancy-aware routing algorithms can reduce the traffic volume inside an ISP by 10–50%. Nonetheless, these are theoretical estimations that do not take into account the architectural limitations of Internet routers and implementation costs. The authors recognize this problem in the follow-up work [9] and address it with an architecture where a single-cache operation is distributed over multiple links. This approach reduces computational effort necessary to detect redundancy. However, the reduction is achieved through an increased management complexity; in consequence the system robustness is decreased.

The key challenge when designing an RE system for the Internet is to keep computational and storage requirements low while maximizing the system efficiency. Two techniques have been proposed to address this challenge. The first technique is to identify packets that are likely to carry redundant data and subject only these packets to RE processing. The remaining packets bypass the RE system unaffected; thus, they do not consume computational and storage resources of the RE system. As shown by Halepovic et al. [10], a simple classification algorithm that distinguishes packet based on the packet size and the content type can increase the RE efficiency by 50%.

The second technique to reduce the RE resource requirements is to model traffic at a source in such a way that link cache processing is simplified. This technique has been studied by Li et al. in Reference 11. A source has methods to transmit data in cache-friendly packet flows. A cache-friendly flow consists of packets where payloads carry either cacheable content or unique content. This significantly simplifies link cache processing, since the redundant part of a packet is always stored at the end of a packet and it forms a single continuous section that can be easily removed and reconstructed.

The two techniques are also part of the CacheCast system. They are implemented by the CacheCast server support element. The server support creates the cache-friendly flows and annotates cacheable packets with a tag that enables the link caches to quickly identify and process the packets. This enables the CacheCast system to achieve high efficiency at low costs.

RE systems are also proposed to improve throughput in wireless networks [12,13]. However, as pointed out by Le et al. in Reference 14, packet corruption, packet loss, and packet reordering events that occur frequently in the wireless networks demand fundamentally different RE architectures. Therefore, the wireless and wired RE systems share only basic principles that are mainly related to redundancy identification techniques.

20.3 CacheCast OVERVIEW

Caching is a well-known mechanism in computer systems to avoid multiple transfers of the same element over the same channel. For example, central processing unit (CPU) caches avoid multiple transfers of the same data/instruction from main memory to the CPU; page caches avoid redundant transfers of the same pages from hard disk to main memory; and Web browser caches avoid multiple transfers of the same elements from Web pages to the same client. The CacheCast system uses the caching technique to avoid multiple transfers of the same payload over the same network link.

Figure 20.1 illustrates the basic idea of eliminating redundant payload transfers over a link. In this example, two consecutive packets originate from the same source, carry the same content, and traverse a few hops over a common path before they are forwarded over disjoint paths to their respective destinations. The first packet traverses the path hop-by-hop. At each hop, the packet payload is cached and the output link for the packet is recorded (steps (1) and (2)). When the second packet is forwarded on this path, the first hop looks up the output link for the packet (step (3)) and identifies that the same payload has been sent with the first packet over that output link. Thus, the payload is already in the next hop cache, and only the packet header needs to be forwarded (step (4)). This procedure is performed on each hop until the last hop of the joint path is reached. The look up operation at the last hop reveals that the output link for the second packet is a different path than for the first packet. Therefore, the payload is not present in the next hop cache and the last hop attaches the payload from its cache to the second packet header, forwards the entire packet (i.e., packet header and payload), and records the output link.

In order to create such packet caching system, two core issues must be addressed. Firstly, it is necessary to identify on a given hop the payloads that are stored in the next hop cache, taking into account that the cache size is finite and that packet reordering or loss can occur. Secondly, it is necessary to determine the cache size that provides

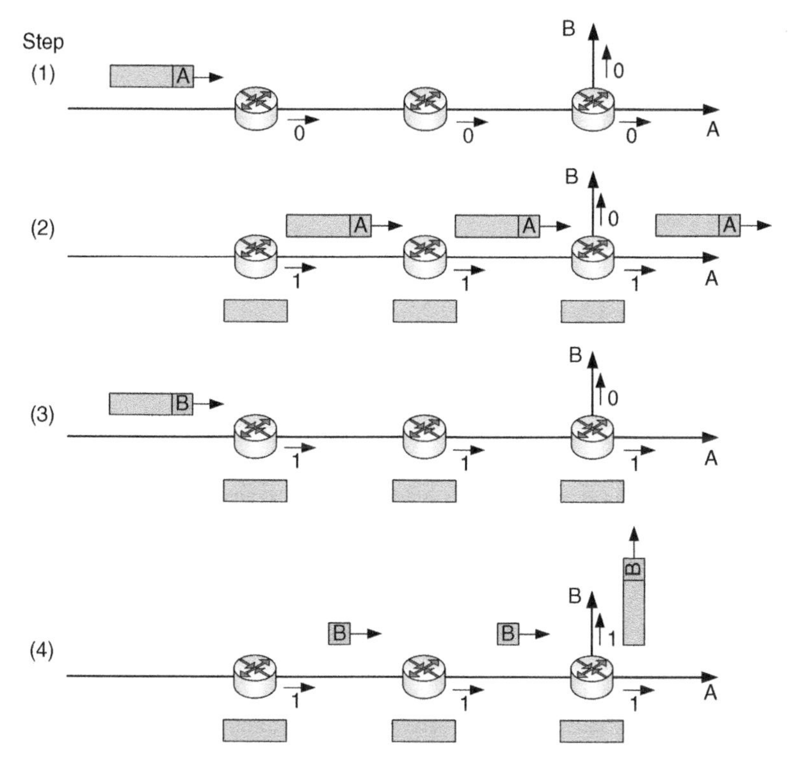

Figure 20.1 General idea for a packet caching system.

maximum efficiency at minimum costs. This is not trivial, because the larger the caches the more expensive they are to build and to manage. However, the larger caches provide the higher hit rate. This trade-off is resolved with explicit support from servers that transmit the same data to multiple destinations. This server support minimizes the required cache size and ensures high link caches efficiency at the same time.

20.4 BACKGROUND ON MULTIDESTINATION TRAFFIC

At the top level, the Internet can be viewed as a graph composed of routers, links, and end-hosts. A network of links and routers enable communication between end-hosts. The role of routers is to build and maintain a local view of the network and to forward incoming packets toward their destinations. The role of links is to transport packets between routers or end-host and router. In order to design a caching system, it is necessary to understand characteristics of these elements and trade-offs that are involved in packet processing and transport.

A link is a communication channel over which packets are transported. It has two endpoints identified with unique addresses. A link can be divided into two directed links that transport data in opposite direction. These two directed links have the same endpoint addresses; however, the destination address is interchanged with the source address. Since a link is a logical structure, it must be implemented using a certain network technology such as the Ethernet or the asynchronous transfer mode (ATM) network. Often, an underlying network technology provides connectivity to multiple nodes (e.g., Ethernet over coaxial cable). In such networks, it is possible to identify individual directed links using endpoint addresses.

The link properties depend on the underlying technology. Surveying technology of links that build the present Internet we identify three general link properties that are essential for link cache design:

Link Capacity. The link capacity is limited by the bit transmission rate. Thus, link can transport many more small packets than large packets within a time unit. This property is a prerequisite for building a link cache, since the link cache improves the link throughput by reducing the packet size. When the link capacity is limited by the packet transmission rate, which does not depend on the packet size, the reduction of the packet size provided by the link cache does not improve the link throughput.

Sequence of Packets. Although the IP specification does not require that the packet order is preserved during the transfer over a link, most technologies implementing links provide ordered transport of packets. Internet backbone links are mainly built using dedicated links or circuit-switched networks, where packet reordering does not occur. In case of local area networks, which are dominated by switched Ethernet, the order of packets is preserved due to single path routing inside a network.

Packet Loss. Links are characterized by very low probability of packet loss during transfer. The transfer reliability is related to robustness of data transmission over wires and fibers, which are additionally protected with erasure codes. The high packet loss occurs primarily in wireless networks; however, the CacheCast system is not designed to operate in this environment.

The two last properties account for the deterministic behavior of packets on links and greatly simplify the CacheCast design.

The second element necessary to build an IP network is a router. A router is a node that connects at least two links. Its basic task is to switch incoming packets according to their destination IP addresses between the links. The first generation of routers uses shared memory architecture to implement the switch fabric. Upon a packet arrival on a link interface, it is immediately moved to the shared memory for further processing. When the router determines the output link, the packet is moved to an interface of this link. Therefore, the router forwarding rate is mainly limited by the memory bus transfer speed and per packet CPU processing. Second- and third-generation routers are based on a distributed architecture where each link interface has its own forwarding engine, and switching is done via a shared bus or a switch fabric. While the first generation routers are sufficient to build small- and medium-size networks, the second- and the third-generation routers are required to handle traffic on the Internet backbone.

The focus of this chapter is on shared memory routers that are used in small-to-medium-size autonomous systems. It is worth noting that for small-to-medium-size networks, the link cost is mainly associated with a physical infrastructure of copper wires or fibers and it greatly outweighs the cost of routers which are cost-efficient boxes. Therefore, the bandwidth savings that can be gained in these networks by caching are important.

20.5 CacheCast DESIGN

20.5.1 Placement of Cache Elements

Packets are cached per link to reduce the number of bits that need to be transferred over it. Each link cache comprises two components that are installed on the edges of a directed link, that is, the link cache management unit (CMU) at the link entry and the cache store unit (CSU) at the link exit (see Figure 20.2). The CMU processes each packet that is marked as the CacheCast packet immediately before it enters the link. If the packet payload is already present in the link CSU on the link exit, the CMU truncates the packet to the header size. On the link exit, the CSU adds those payloads that were removed by the CMU. The CMU has a description of the CSU content in order to manage the CSU. Packet loss on the link can lead to an inconsistent description, but this type of error occurs very rarely on modern wired links, because the major cause of packet loss in a network is congestion. In case of congestion, packets are dropped at the input queue to a link, that is, before they enter the CMU. Therefore, these losses do not impact the link cache.

By removing redundant payload, the CMU reduces the size of a packet with redundant payload to the size of the packet header. Therefore, links can transfer more packets per time unit. The CSU reconstructs the packet at the link exit before it passes it to router for further processing. This enables the router to handle the packet in the standard way (see Figure 20.2).

20.5.2 Caching-Aware Source

A link cache should not introduce delay in packet transfer over a link, since this reduces link utilization and eventually can render caching useless. Therefore, in order to reduce

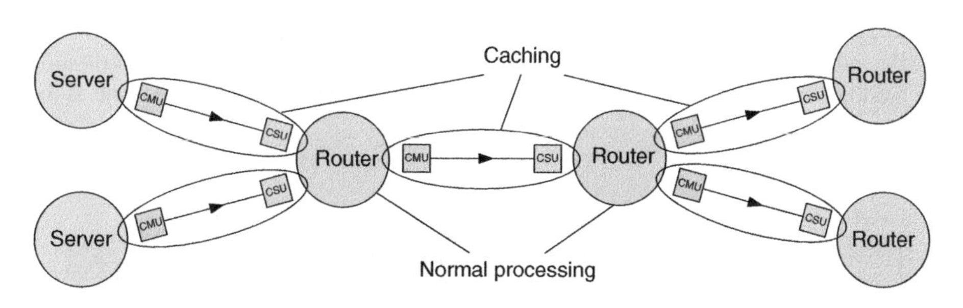

Figure 20.2 Placement of the link cache elements.

the burden of a link cache processing and the cache size, it is necessary to place part of the burden on a source. With this approach, a source transmitting the same data to multiple destinations must ensure that packets carrying the same payload are transmitted within the minimum time interval. In addition, the source provides three key information elements about the packet that simplify caching. This information is stored in an extension header called CacheCast header, which is depicted in Figure 20.3.

1. The source marks packets that carry the same payload as cacheable packets by inserting a CacheCast header between a link header and a network header. Packets with unique payload are not marked and not considered for caching in the network. This increases cache efficiency, because caches handle only redundant payload.
2. The payload part of a packet is of variable size. The source provides the information about the payload size to link cache using the CacheCast header P_SIZE field. Thus, link caches know instantly which part of a packet to cache.
3. The source creates a network-wide unique identifier for each payload. Thus, link caches match payloads based on the payload IDs instead of comparing payloads byte by byte. The payload ID is stored in the P_ID field.

The described support from the server significantly decreases the necessary requirements for the cache memory size and the processing capacity. It is implemented using the presented CacheCast header. The header is further used to convey information from the CMU to the CSU using an additional INDEX field, which is described in Section 20.5.3.

As an alternative, caching could be performed transparently to the source. This would mean that the link caches have to handle all packets, and a substantial amount of storage space would be wasted for nonredundant payloads. Therefore, caches would require larger storage space to achieve a comparable efficiency. Caches would need to determine the payload part and to compare the entire payload in case of a cache hit. The core benefit of transparent caching is that sources do not require any changes and that link caches can identify redundancy in any type of traffic.

20.5.3 Detailed Link Cache Operation

Caching is performed per link and is separated into management and store unit, that is, CMU and CSU. The CMU manages the CSU. It has a table where it keeps information on payloads stored in the CSU. The number of the CMU table entries is the same as the

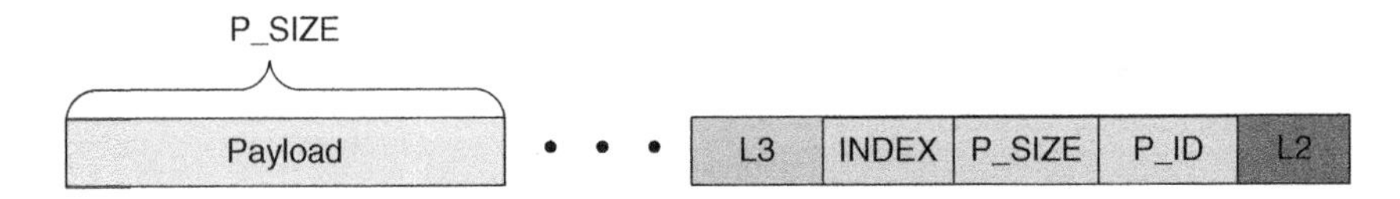

Figure 20.3 CacheCast header.

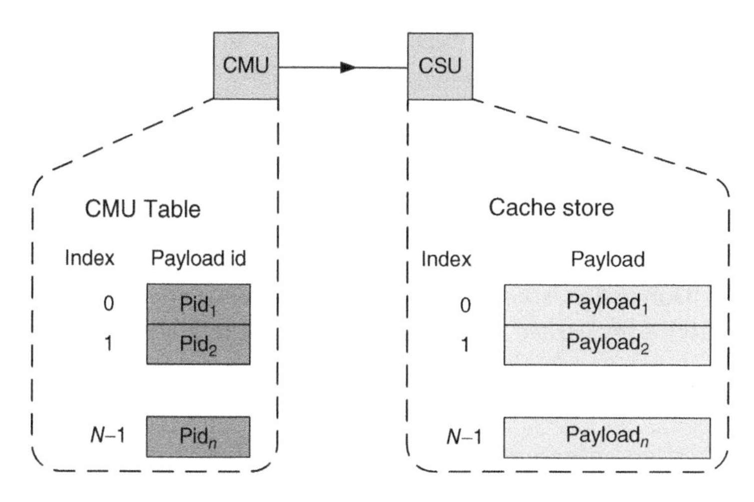

Figure 20.4 CMU and CSU relationship.

number of the CSU slots (see Figure 20.4). The link cache handles the two events "cache hit" and "cache miss" in the following way:

Cache hit occurs when a payload identifier of a packet entering a link is found in the CMU table. This means that the packet payload is in the CSU on the exit side of the link, and there is no need to transmit it again. Thus, the CMU first puts the index of this payload identifier into the INDEX field in the CacheCast header. Next, it removes the payload and transmits only the header part of the packet. When the header arrives at the link exit, the payload with the INDEX in the CSU is attached to it. Then, the whole packet is handed over for further processing to the router.

Cache miss occurs when a payload identifier of a packet entering a link is not found in the CMU table. This means that the packet payload is not present in the CSU on the exit side. The CMU handles the cache miss in the following way. Firstly, it removes one entry from the table according to the selected cache replacement policy, inserting instead the payload identifier of the packet that caused the cache miss. Secondly, it inserts the index where the payload identifier was inserted in the CMU table into the INDEX field of the packet CacheCast header. Finally, the packet is transmitted over the link. Upon arrival at the link exit, the payload is stored at the location pointed to by the INDEX. The previously stored payload can be safely overwritten since it has been evicted from the cache by the CMU. The payload identifier in the table and the corresponding payload in the CSU have the same indexes.

The defined operations of a link cache are simple to implement and execute. The operations do not handle problems related to the inconsistent state of a cache, that is, when the description in the CMU table does not reflect the actual content stored in the CSU. The link cache can enter this temporal state in two cases: (i) a CacheCast packet that carries payload is lost during transfer on a link; or (ii) two CacheCast packets whereby one packet carries payload are reorder during transfer on a link. As we have discussed, these events are exceptionally rare in wired networks where CacheCast operates. Therefore, the link cache mechanism does not attempt to correct inconsistent mappings but only provides a protection against errors that could arise in this state. The inconsistent mappings are evicted after a very short period of time by the constant arrival of new CacheCast packets.

Link cache inconsistency can only result in an erroneous operation of the CSU whereby the CSU attaches incorrect payload from its cache store to a received truncated packet. To protect against this type of error, it is sufficient to store in the CSU payloads together with their IDs. Therefore, before the CSU reconstructs a packet, it compares the payload ID from the CacheCast header with the local ID of payload pointed to by the INDEX. If the two payload IDs differ, then the packet is dropped.

20.5.4 CacheCast Server Support

The CacheCast system requires support from the application to reduce link cache complexity and storage requirements. An application that delivers the same content to multiple clients should send the individual content chunks synchronously to all clients. In the result of such transmission, all redundant transfers of the content chunk over network links occur in a very short period of time. Hence, they can be easily removed by link caches. In this aspect, live streaming applications send individual content chunks synchronously *ipso facto*, since they transmit a new data chunk to all clients immediately after it is generated. Considering applications that deliver content of type download-before-consume, the synchronous transmission can be achieved with an additional effort, which we demonstrate later in this chapter.

To enable an application to execute a synchronous transmission, CacheCast extends the operating system (OS) with a new send function that transmits the same data chunk to multiple clients at the same time. The function ensures that packets carrying the same data are transmitted from the server within the minimum time interval. In addition, the function performs the CMU-related tasks, that is, it removes redundancy from the packets before transmission. The send operation results in a tight sequence of packets carrying the data chunk, as is depicted in Figure 20.5.

20.5.5 Link Cache Size

The key assumption CacheCast takes is that the link cache store size is very small. To achieve a high utilization of these caches, a source of redundant data must transmit the same data to multiple destinations within a minimum time interval. This condition is supported by the server support. In order to determine the optimal size of a link cache, we use the concept of packet train. A packet train is a sequence of packets where only the first packet carries a payload and the remaining packets are truncated to the header size (see Figure 20.6). This structure of packets is generated by a CacheCast server that sends the same data to multiple clients. The duration time of a packet train is the time interval measured from when the first packet is serialized on a link until the last packet is transmitted. While traversing a network, a packet train is divided into smaller packet trains that carry subsets of the original packets. We can assume that to a large extend, the duration time of the subtrains is shorter than the original train.

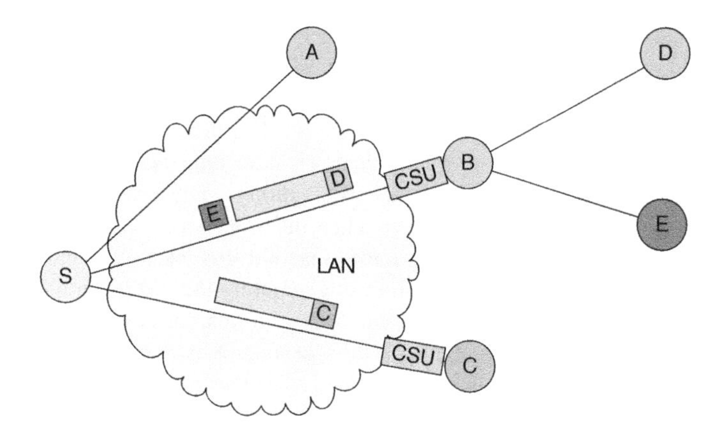

Figure 20.5 Server sending the same data to clients C, D, and E.

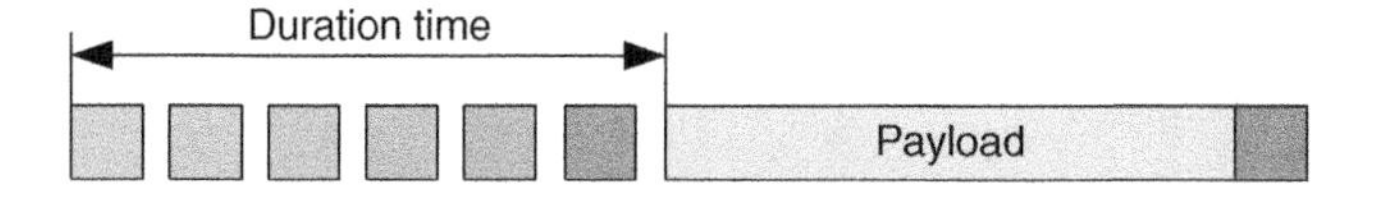

Figure 20.6 Packet train.

It should be observed that a cache can evict a packet payload without any loss in efficiency after a packet train has passed the cache. Therefore, the maximum cache size is related to the maximum duration time of a packet train. The packet train duration time depends on the source uplink speed and the number of addressed destinations. Table 20.1 shows examples of sources with different uplink speeds and size of a packet train that can be sent within three various time intervals. For instance, scaling a link cache store to accommodate 10 ms of link traffic provides a good trade-off between the link cache size and its efficiency. Link caches of this size can remove most of the redundant transfers generated by slow sources sending data to several destinations or by fast sources sending data to hundreds of destinations. In networks where only fast sources are present, it may be beneficial to use even smaller link caches.

TABLE 20.1 Number of Packets a Source Can Send Within a Given Amount of Time

	Time Interval		
Source Uplink Speed	2 ms	10 ms	50 ms
512 Kbps	2	8	40
1 Mbps	4	16	79
10 Mbps	32	157	781
100 Mbps	313	1561	7802

An alternative approach to scale the link cache size is derived from the CacheCast server transmission scheme. According to the CacheCast server support, all cacheable data sent by a server are transmitted in the form of a packet train. Furthermore, packet trains transmitted by different applications that send at the same time are serialized. Therefore, it is sufficient for a first hop cache to store only one payload in order to remove all redundant transfers. We use this observation to state a rule for scaling caches on the first few hop links as follows: given a link that transports traffic aggregated from a certain number of sources, it is sufficient to install a link cache that has the number of cache slots equal to the number of sources generating the traffic. In other words, a link transporting traffic from N sources should have a link cache with N slots. Such a link cache can suppress all redundant transfers generated by these sources when each slot is assigned to each source on a one-to-one basis. This requires that the link cache replacement policy assigns cache slots based on source ID.

Applying the second approach to scale the link cache size can lead to very small caches over a first few hops. However, as the number of aggregated sources increases, it is more beneficial to use the first approach to scale link caches.

20.6 CacheCast EFFICIENCY

The purpose of CacheCast is to remove redundant data transfers from links. Therefore, the measure of CacheCast efficiency is the amount of redundant data transfers that have been suppressed through link caches. To quantify the efficiency, it is sufficient to focus on single-source multiple-destination data transfers. Other type of redundant data transfers are not suppressed by caches as per design. Hence, the CacheCast efficiency should be compared with multicast efficiency.

The upper bound of the CacheCast efficiency is set by a fictitious "perfect" multicast scheme. The perfect multicast slightly differs from the IP multicast. It does not require any additional signaling to establish a multicast tree and to deliver a datagram to receivers. CacheCast cannot achieve the perfect multicast efficiency, since it eliminates only redundant payload transfers while packet headers are still transported per destination. In order to compare the efficiencies, we use the metric proposed by Chalmers and Almeroth [15]. It is expressed by the ratio of the total number of multicast links to the total number of unicast links that are traversed by datagrams during the delivery of the same data to all receivers in a group. It can be denoted as

$$\delta = 1 - \frac{L_{\mathrm{m}}}{L_{\mathrm{u}}} \tag{20.1}$$

When the number of multicast links is similar to that of unicast links, the efficiency is approximately zero, which means that there is no benefit from using multicast. On the contrary, as the efficiency approaches 1, large benefit can be obtained from multicast. The metric expresses the reduction in the total traffic when using multicast instead of unicast. For a simple example of the metric usage, let us consider a tree topology as shown in Figure 20.7. When the server S sends the same data to the hosts A, B, and C using unicast, it must transmit three datagrams each traversing three links; thus, in total,

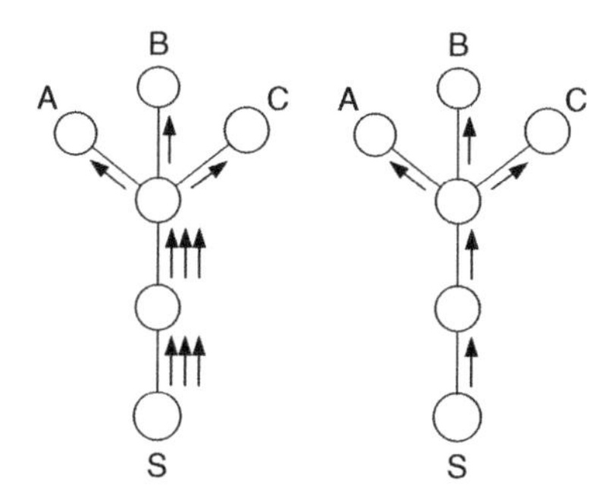

Figure 20.7 Data transmission using unicast (a) and multicast (b).

there are nine transmissions in the network and $L_u = 9$. When the server S uses multicast to send the same data to all hosts, only one datagram traverses the first two hops and is replicated at the branching point; thus, there are only five transmissions in the network and $L_m = 5$. The resulting efficiency is $\delta = 1 - 5/9 \approx 0.44$, which means that multicast reduces the total network traffic by 44%.

20.6.1 Transfer Efficiency

The first reduction in the efficiency of CacheCast when compared to the perfect multicast is related to the transmission of a unique header per the destination. We denote the size of the header part of a packet by s_h and the payload part by s_p. Since the header part of a packet traverses the unicast amount of links, the CacheCast efficiency is

$$\delta_C = 1 - \frac{s_h L_u + s_p L_m}{(s_h + s_p) L_u} \tag{20.2}$$

Denoting the ratio of the header size to the payload size by $r = s_h/s_p$, we can further express the CacheCast efficiency using the perfect multicast efficiency 20.1 and the ratio r as

$$\delta_C = \frac{1}{1+r}\delta \tag{20.3}$$

The factor $1/1 + r$ indicates the difference between the CacheCast efficiency and the perfect multicast efficiency. When the header to payload ratio r decreases, the efficiency of the link layer caching approaches the perfect multicast efficiency. The maximum efficiency is obtained when the packets are of the maximum size while the header part of the packets is of the minimum size. In the Internet, the maximum packet size is limited by the standard maximum transfer unit (MTU) which is, at time of writing, 1500 B. The minimum header consists of the link layer header, the CacheCast header, the IP header, and the transport header, which is approximately the same size as the minimum packet size in the Ethernet network, that is, 64 B. Therefore, at a maximum, CacheCast can achieve approximately 96% of the perfect multicast efficiency in the present Internet.

20.6.2 Incremental Deployment

CacheCast is incrementally deployable. It reduces bandwidth consumption in a network from the very beginning of the deployment. This property is ensured by the link cache architecture. A cacheable packet that exits a link is reconstructed before processing on a router. Therefore, it can be switched to a standard link by the router.

The maximum CacheCast efficiency is achieved when link caches are deployed on all links in a network. The dependency between the deployment range and the CacheCast efficiency is depicted in Figure 20.8, where the deployment range is expressed as the number of hops from a source that are covered with link caches. Figure 20.8 shows that it is sufficient to deploy link caches over first few hops in order to achieve half of the maximum efficiency. Considering a source sending the same data to a small group (here represented by 10 receivers), deploying link caches over the first six hops removes already approximately 70% of redundant transfers. However, in order to achieve this percentage of efficiency for transmission to large group sizes (represented by 100 and 1000 receivers), it is necessary to install link caches on the first nine hops.

The presented results depict the efficiency as a fraction of the maximum efficiency, that is, the efficiency that is obtained by cache deployment in the whole Internet. However, ISPs are mainly interested in their own gains. Thus, the relevant question is what are the direct benefits of deploying caches inside a single ISP? These can be summarized in two points:

- The traffic that is confined to a single ISP will achieve near multicast bandwidth utilization.

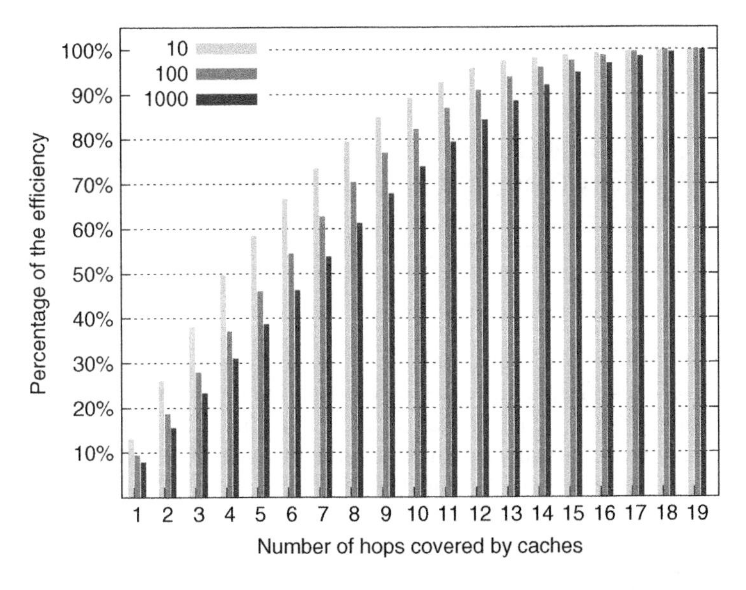

Figure 20.8 Efficiency of the partially deployed CacheCast system.

- The traffic that originates from the ISP and traverses other ISPs will be cached on the way between a streaming source and a gateway inside this ISP. Thus, there will be more available capacity on the ISP's links.

20.7 CacheCast APPLICATIONS

Virtually all applications that deliver content to multiple destinations can benefit from CacheCast. However, to use CacheCast, it is necessary that an application sends the same content chunk to multiple clients synchronously. This requirement may reduce the degree of benefits that can be obtained by the application. In this context, live streaming applications can achieve the greatest gains from CacheCast with the least effort to adapt them to CacheCast. Applications that distribute software or other download before consume type of content can also benefit from CacheCast. However, it requires more effort to adapt this type of application to CacheCast, since receivers arrive at various points in time and download content with variable speed. In the following sections, we give an example of a live streaming application that operates with CacheCast and present the design of an application that uses CacheCast to distribute downloadable content.

20.7.1 Live-Streaming with CacheCast

A live streaming application can serve multiple streams at various rates. However, there is a high likelihood that a single stream is listened to by a group of clients. A standard behavior of a live streaming application is to transmit a new sample to all clients in the group immediately as it is generated. Therefore, the CacheCast requirement to send the same data to multiple destinations within the minimum time interval is fulfilled at the application layer. However, since most applications use Transmission Control Protocol (TCP) connections, there is no guarantee that the application layer send request will result in immediate packet transmission. The transmission may be delayed when there are old samples waiting in the transmission queue. Therefore, in order to guarantee the minimum time interval between packets during the network transmission, applications must use the CacheCast server support to send the samples.

The CacheCast server support is implemented as a send system call that handles the OS data transmission part. The system call invocation results in transmission of a tight packet train. This can be achieved only with protocols that preserve message boundaries such as User Datagram Protocol (UDP) or Datagram Congestion Control Protocol (DCCP). Live streaming applications that deliver content using these protocols can be easily adapted to CacheCast, since the adaptation process requires only replacement of the standard OS send routines with the CacheCast specific. Considering other applications, the adaptation effort may be significant. However, this depends on the specifics of an application.

An example of a live streaming application that delivers data using DCCP is a *paraslash* audio streaming software. We use this software as an example to show how the CacheCast technology can improve efficiency. Let us consider a small network depicted in Figure 20.9. The network consists of a server S, a router R, and two machines A and B

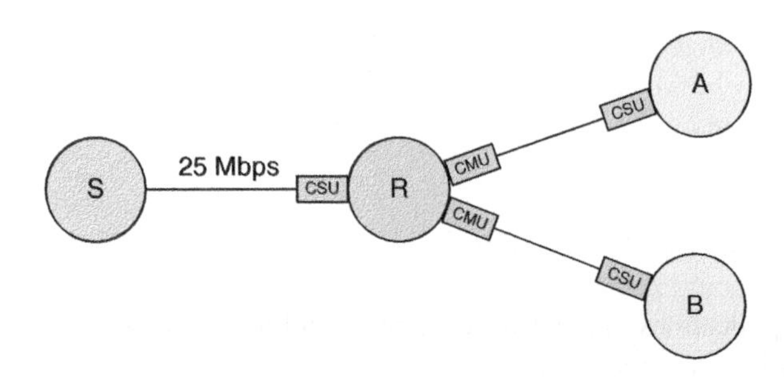

Figure 20.9 CacheCast live streaming application—network setup.

hosting clients. Link caches are installed on all links. The S–R link capacity is limited to 25 Mbps to create a bandwidth constrained environment. The server S uses the *paraslash* server to stream the audio file and the CacheCast support to transmit audio chunks in the CacheCast manner. The router is based on the Click Modular Router software [16]. The client machines (A and B) use the *paraslash* receiver software to obtain an audio stream from the server S, and the Click CSU element is installed on each machine. The client software does not require any modification to communicate with the server.

In the described network, the server S streams an mp3 file at the rate of 320 Kbps using 1024 B data chunks. New clients gradually join the audio stream until the bottleneck link S–R is saturated and no more clients can be served. At the saturation point, the system achieves its maximum utilization serving the maximum number of clients. Table 20.2 depicts the number of clients handled by the original *paraslash* server that does not use CacheCast and the modified server that uses CacheCast. The modified server uses different configuration of the chunk size to show the impact of the size of the transport unit on the system efficiency.

In the bandwidth constraint network, the original server implementation can provide the audio stream for up to 74 clients. Using the same data chunk size, the CacheCast server handles an order of magnitude more clients. The server performance can be further improved by using larger chunks. The reason is that transferring audio stream using larger chunks requires fewer packets per time unit and consequently less packet headers. Since almost all packets carrying the audio chunks are truncated to the packet header,

TABLE 20.2 Audio Streaming Test-Bed Results

Server Type	Chunk Size	Clients	CPU Load
Original	1024	74	2.71 ± 1.1
CacheCast	1024	1020	44.21 ± 1.6
CacheCast	2066	2066	59.63 ± 1.7
CacheCast	4184	4097	58.30 ± 1.0
CacheCast	8364	8001	87.17 ± 1.8
CacheCast	15,674	12,454	91.35 ± 3.049

the consumed bandwidth is directly proportional to the amount of packets transmitted per time unit.

20.7.2 Downloading with CacheCast

The design of a CacheCast content delivery system is fundamentally different from the design of a CacheCast live streaming application. The discrepancies are related to different requirements explained in the following. In order to transfer content through a network, the content is divided into smaller chunks also referred to as blocks. A client downloading content has to obtain all blocks before the content can be viewed. As a consequence of this, the order in which the blocks are delivered to the client is irrelevant. Regarding data integrity, content downloading has stricter requirements than streaming. In a streaming scenario, a moderate loss of content chunks in a network is usually acceptable, since it does not affect a video or voice stream substantially. The lost part of the content is not retransmitted. It could take too long to retransmit the data such that it is obsolete when it arrives. When downloading content, data loss is not acceptable. A single lost block might corrupt the entire content. Therefore, it is crucial that all blocks are received.

The second challenge in content downloading is related to the arrival time of the receivers. While the clients in a live streaming scenario are synchronized *per se*, the clients in a downloading scenario may request content at any time. In order to achieve high utilization of the CacheCast technique, a server must transmit the same chunk of the content to all clients synchronously. To achieve the synchronous transmission, we use the previously discussed fact about the irrelevance of the order of the received blocks. The synchronization is not based on sequential access as with streaming, but on missing blocks. To illustrate, when a new client requests a file, it is missing all blocks and thus it may receive any block. It is a task of the CacheCast server to schedule when this client should receive these blocks. This means that the synchronization is on a block-to-block basis when using content downloading, not on a stream basis as with live streaming.

The third challenge in content downloading is the diversity of the download speed among clients. Streaming applications solve this problem by sending the same content with different rates. Thus, clients can choose the rate that suits them best. Since we want to maximize the CacheCast utilization, sending the same content in multiple streams with predefined rates is suboptimal. The CacheCast server sends only one stream with a transmission rate that is adjusted to serve clients with diverse download speeds. We have chosen to adapt the file transmission rate to the speed of the fastest client. The rationale here is to make sure that the fastest client receives data at its maximum speed. As for the slower clients, end-to-end congestion control will adjust the rate to each client, respectively. All clients, regardless of download speed, can therefore be satisfied.

20.7.2.1 CacheCast File Server Architecture A CacheCast-enabled file server differs from a File Transfer Protocol (FTP) server [17] in a few important aspects. An FTP server handles file selection; however, file transfer is completely delegated to the TCP protocol, which provides reliability and flow control. The CacheCast file server handles file selection and a part of the file transfer. It provides reliability and determines the transmission rate of a file to a group of clients. However, the end-to-end congestion control between the server and each client is delegated to DCCP.

Figure 20.10 presents an overview of the CacheCast file delivery system, including a high level illustration of the reliability and congestion control. The reliability is achieved using fountain codes [18,19] that also greatly simplify the problem of blocks' scheduling. The general idea of fountain codes is to create a stream of encoded packets that is sent toward a client. Each packet is a linear combination of the blocks of the original file. When the client has collected enough packets to decode the whole original content, it notifies the server to stop transmission. Since each encoded block carries new information regardless of what clients have already received, the server can transmit an encoded block to all clients that are ready to receive. The clients' end-to-end flow control is controlled by DCCP. In the following two sections, we elaborate on the transmission rate adaption and the reliability mechanism in the CacheCast file server.

20.7.2.2 Transmission Rate Adaption Similar to a standard FTP server, a CacheCast file server provides multiple files that can be requested by clients at any point in time. The clients that download the same file constitute a group. The transmission of a file to a group is called a file session. For instance, in Figure 20.11, Group 1 is comprises Client 1 and Client 2.

The download speed of clients within the same group is diverse. However, the server transmits only one stream of data for each file session. In order to assure good performance for the whole group of clients, the transmission rate r_j of file j is adjusted to match the capacity of the fastest client. However, finding the fastest client capacity is a nontrivial task, since it can be only determined by sending data to the client. This issue is solved by the rate adaptor element in the following way: The rate adaptor always assumes that there is a spare capacity between the server and the fastest client, and it increases the transmission rate r_j each time it transmits a new data block. When the file transmission rate is higher than the fastest client can receive at, the DCCP instance controlling the client connection notifies the rate adaptor of the congestion event. (In Figure 20.10, the feedback from DCCP is visualized as dashed arrows.) In turn, the rate adaptor rapidly

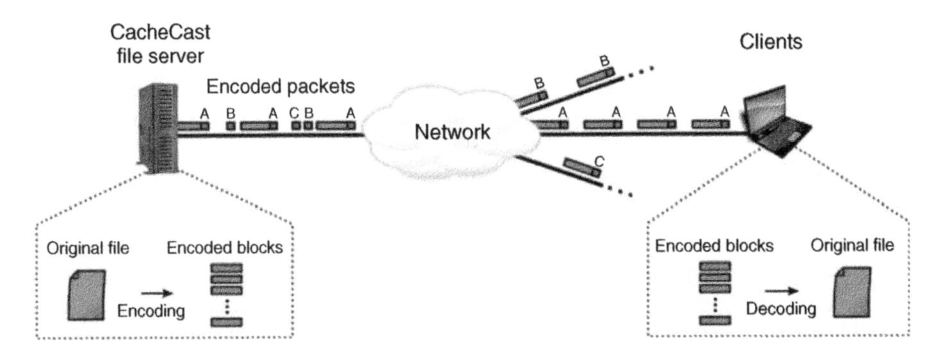

Figure 20.10 Packet flow between the CacheCast file server and clients.

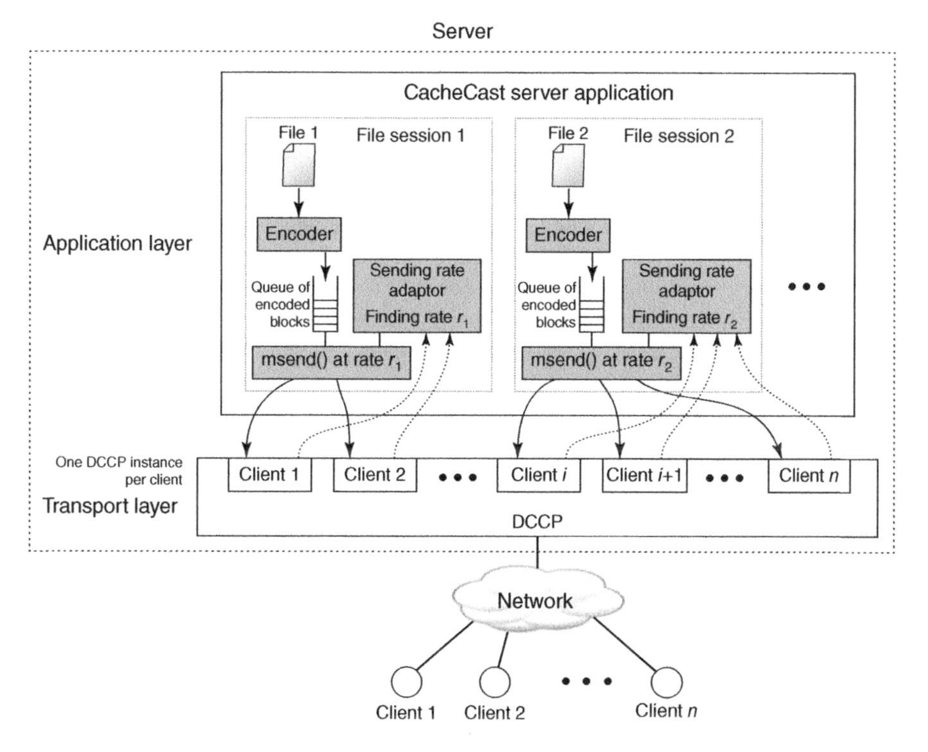

Figure 20.11 CacheCast file server architecture.

decreases the rate and soon after begins to increase it again. The rate adaptation algorithm can be formalized with the following formula:

$$r_j = \begin{cases} r_j + I, & \text{successful transmission} \\ \dfrac{r_j}{D}, & \text{congestion occurred} \end{cases} \tag{20.4}$$

The transmission rate is increased by the constant I when the transmission attempt was successful or divided by the constant D when the DCCP congestion control prohibited the transmission.

Another issue when considering concurrent transmission of multiple files from the server is fairness. When two groups of clients are downloading two different files from the server, the server uplink should be fairly shared among the two downloads. Since we use DCCP to control throughput of individual connections between a server and a client, the fairness in the network is already preserved. Furthermore, the file transmission rate is governed by the rate adaption mechanism that follows the fastest client speed. Consequently, fairness between the concurrent file downloads is based on contention among the fastest clients in the groups. This behavior is further investigated in the evaluation part.

20.7.2.3 Reliable Transfer The second component of the CacheCast file server is the reliable transfer mechanism. As previously discussed, content downloading has strict requirements regarding data integrity, so it is essential that all data are successfully received by the client. To achieve reliability, the CacheCast file server uses fountain codes and transmits a linear combination of the original blocks. A client can decode the original file after it has obtained the sufficient amount of encoded blocks. The server does not need to monitor which blocks have been delivered to which clients. This leads to the fact that all clients require the encoded block and the server can transmit the block to all clients synchronously.

The sending procedure of a CacheCast file server is therefore to (i) create a new encoded block and (ii) send this block to the connected clients. These two tasks can run in parallel. When the current block is being sent, the next block is being encoded. The encoder rate should match the file transmission rate determined by the sending rate adaptor. However, to account for variations in the transmission rate, we insert a queue between the encoder and the transmitter (see Figure 20.11). The same parallelization applies to the client. When a block is received, the previously received block is decoded. Shojania and Li show in Reference 20 that the encoding and decoding operations can be performed even on moderate hardware. With such parallelization, there is no significant overhead caused by the encoding process.

20.7.2.4 File Server Performance The CacheCast file server utilizes significantly less network resources than the standard FTP server. Therefore, it can provide shorter download times for clients. Particularly, the server substantially reduces download times when a network capacity is a bottleneck of the system. We present this with a simple scenario in a small network depicted in Figure 20.12.

The network connects a server and 100 clients. The clients downlink speed is limited according to Table 20.3, which reflects the available DSL and cable offers [21]. The capacity of the first hop link is 10 Mbps; thus, all clients cannot concurrently download files with the maximum downlink speed. Eight clients per second connect to the server and request the same 18 MB large file. After all clients have requested the file, no further clients arrive. The scenario is simulated in the *ns-3* discrete network simulator framework.

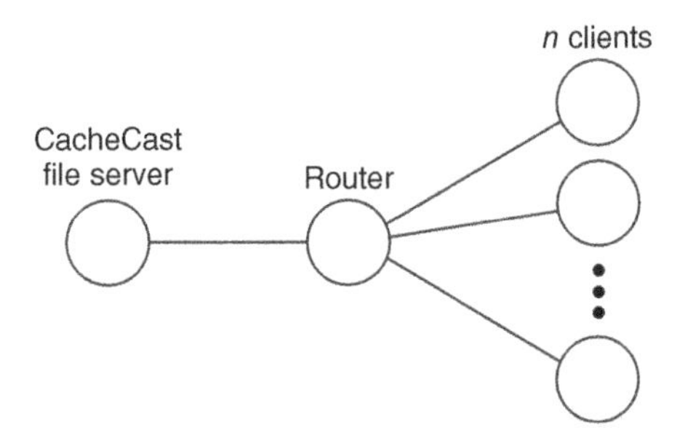

Figure 20.12 CacheCast file server simulation topology.

TABLE 20.3 Downlink Capacity Distribution

Downlink capacity (kbps)	64	256	1500	3000	5000
Clients with this capacity (%)	2.8	14.3	23.3	18.0	37.3

Figure 20.13 shows the cumulative distribution function of the download time for each client in the simulation and the corresponding server uplink traffic. For ease of comparison, the presented download time is normalized. The download time of a client is divided by the time it takes this client to download the file when it is alone in the system. As expected, the results show that the file download time from the CacheCast server is significantly shorter than from the FTP server. This is also reflected in a decrease in the uplink traffic volume. The traffic volume generated by the CacheCast sever is almost 10 times smaller compared to the traffic volume generated by the standard FTP server. It is important to note that the magnitude of these gains depends on the number of concurrently connected clients and their downlink speed. More specifically, it depends on the utilization of the CacheCast mechanism. In the downloading scenario, there are 100 clients with an arrival rate to the server of 8 clients per second. However, the performance decreases when fewer clients per second request the file. In our simple scenario

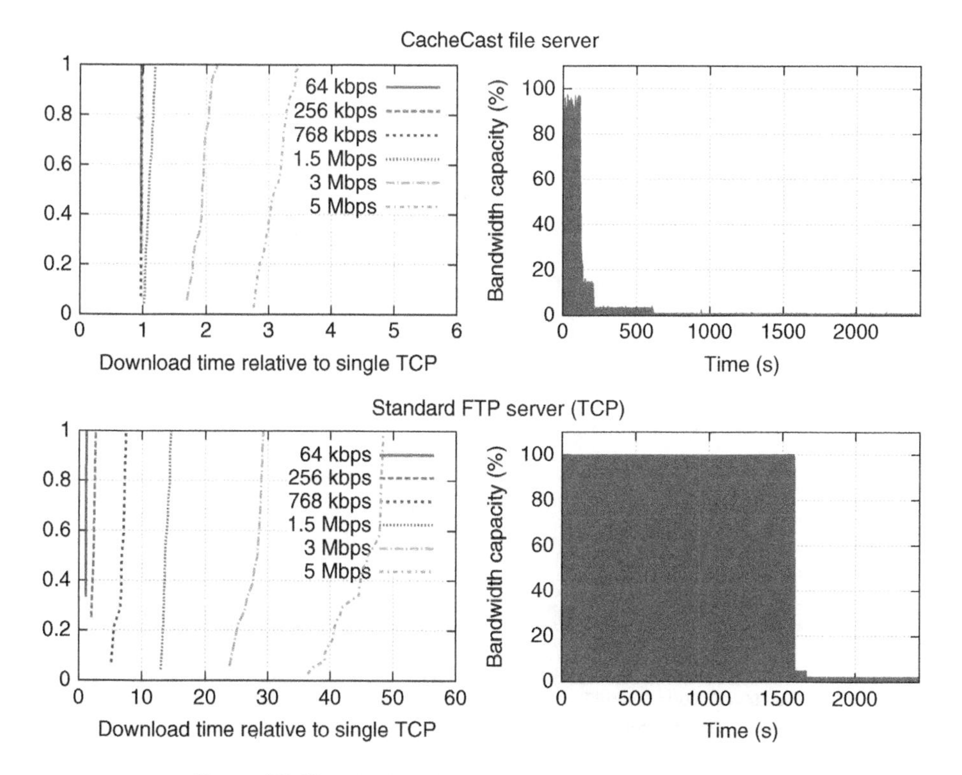

Figure 20.13 File download time and server uplink traffic.

the performance gains are not optimal, since no more clients arrive to the system after all clients has arrived. In a real world, new clients connect to the server constantly, which increases the number of connected clients and the performance gains.

We study the fairness between two groups of clients downloading different files from the server. Two scenarios are investigated. In the first scenario, the total number of clients in the system is small, that is, 10 clients, and in the second scenario a large number of clients is used, that is, 200 clients. The download speed of all clients in both scenarios is 10 Mbps.

Figure 20.14 presents the results of these simulations. When the ratio between the group sizes changes, the larger group gets more bandwidth capacity. However, the ratio of bandwidth share does not fully reflect the group size ratio, especially when considering the scenario with the small number of clients. To explain this behavior, we use an observation about the CacheCast packet train structure. Owing to the removal of redundant payloads, the ratio between the number of packets sent and the number of bytes sent is not equal. Therefore, for short packet trains, the bandwidth share is not proportional to the group size ratio. However, when the packet train length increases, the group size ratio approaches the bandwidth share ratio as in the simulation with 200 clients. Even for different group size ratios, these simulations show that the clients' end-to-end

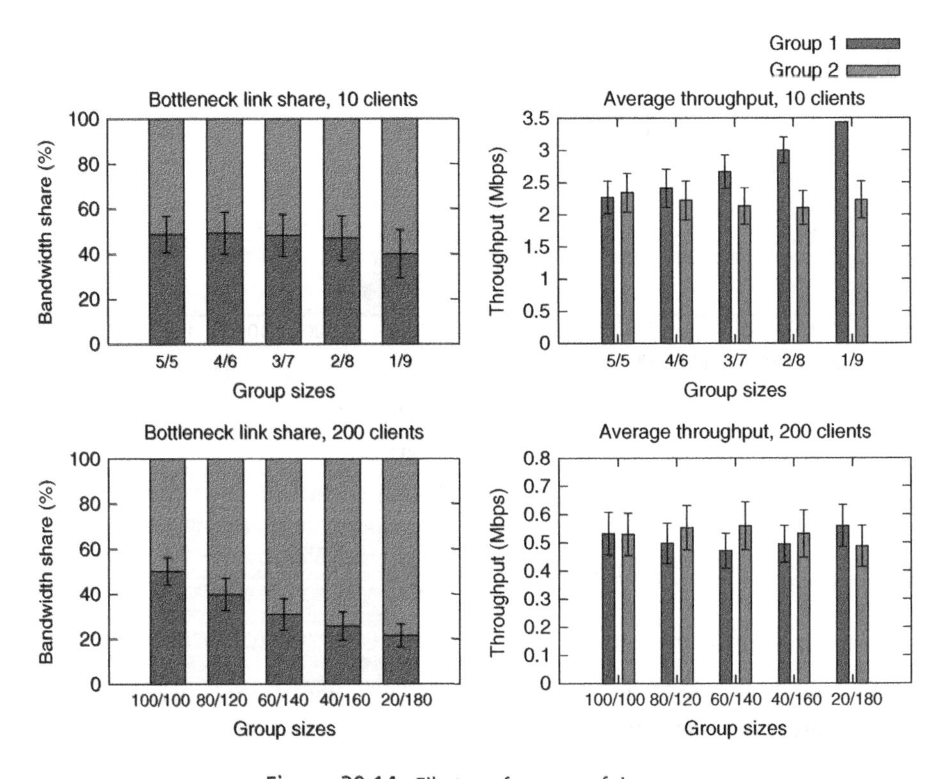

Figure 20.14 File transfer group fairness.

throughput is fairly distributed. This indicates that the CacheCast file server preserves a fair share of network resources between clients, both inside the same group and across different groups.

The presented evaluations show that the content downloading benefits substantially from the CacheCast technique. Given only a moderate number of client requests for the same content, a CacheCast server significantly outperforms a standard FTP server in terms of content download time and resource consumption.

20.8 VISIONARY THOUGHTS FOR PRACTITIONERS

According to the Ipoque studies [22], the major part of today's Internet traffic carries audio/video type of content. This type of content is usually requested by many clients, and the demand is proportional to the content value. The higher the value of the content is, the larger the number of clients wanting it. For example, the newest movie of a famous director is often distributed in millions of copies around the world. Similarly, a top league football match is streamed to hundreds of thousands households. Therefore, in the present Internet, there is a significant amount of redundant transfers that are related to single-source multiple-destination content delivery. These redundant transfers can be suppressed using the CacheCast system, which in turn will improve utilization of the Internet resources.

CacheCast caches are small and can hold data only for several milliseconds. Thus, the system can suppress only redundant transfers that occur within a very short period of time. This type of transfers is characteristic in live streaming applications. Typically, such applications send a new data chunk to all clients immediately after it is ready for transmission. Therefore, the primary use of the CacheCast system is in IPTV services. As we have shown in the previous section, it is relatively easy to adapt an existing live streaming application to the CacheCast system, and it should not pose an additional effort to design new streaming applications that adhere to the CacheCast requirements.

In order to suppress redundant transfers that are related to delivery of downloadable content, CacheCast requires application support. The challenge here is that clients request content at various points in time and download it with variable speed. Thus, an application must organize content delivery in such a way that the same content chunks are transferred synchronously to all clients currently downloading the content. In the previous section, we have discussed a technique that enables an application to achieve this transmission pattern. However, to enable widespread redundant transfer elimination, it is necessary to create a standard transfer protocol that implements CacheCast file transmission—a CacheCast file transport protocol (CFTP).

The CFTP should be available as a library that implements the server and the client part of the protocol. The standard CFTP library would simplify the use of CacheCast in new content delivery applications. The CFTP server could be used to deliver any type of downloadable content. It would ensure multicast-like bandwidth utilization of network capacity. The most promising application of the CFTP server is to accompany the HyperText Transfer Protocol (HTTP) server in content delivery. In such a setup, the HTTP server delegates all large content transfers to the CFTP server. Therefore, the

Web page elements such as images or animations are delivered in an efficient manner. Such support further improves the HTTP server robustness, since the CFTP server can efficiently handle requests for the same data. The compound server architecture is more responsive to the rapid increase in the requests for the same data often called as flash crowd effect.

The efficiency of the CacheCast content delivery depends primarily on the request pattern and the content size. If the client arrival rate is low, there will be a few clients that download content at the same time. Hence, the system efficiency is also low, since it mainly depends on the number of clients concurrently downloading the content. Assuming the same request pattern, the CacheCast system achieves higher efficiency when delivering large content elements than when delivering small content elements. The content size impact on the efficiency is related to the content download time. The number of clients concurrently downloading large content elements is higher, since the time to download a large content element is longer and more clients are aggregated during this time.

The dependence between the file size and the efficiency of the CacheCast system directs to the issue of aggregation of content elements into larger data units. In the example of the HTTP server supported by the CFTP server, we pointed out that the CFTP server could deliver pictures and animations presented on a Web page. To improve the CFTP server efficiency, these content elements should be grouped into a single file and delivered to all clients. This approach significantly increases the system efficiency. In general, to achieve higher CacheCast efficiency, it is beneficial to group smaller content elements into larger units, which we propose to call *content containers*. In such a setup, the CacheCast FTP provides reliability on the content container basis. Thus, a client application cannot consume content unless the complete container has been obtained. This approach has an obvious downside, since a client cannot consume individual content elements as they arrive. In the case of the HTTP server example, pictures and animations cannot be sequentially presented by a client Web browser until all of the container elements are obtained.

An application that uses CFTP to transport content operates on the container granularity. In this aspect, the concept of content container is similar to that of Application Data Unit (ADU) proposed by Clark et al. in Reference 23. An ADU is an aggregate of application data that lower layers of a network stack regard as a unit. The rationale for ADU is that ADUs can be processed out of order and error handling is performed on per-ADU basis. In this sense, a content container is an ADU. However, the rationale for the content container is different, that is, to improve the CacheCast system efficiency.

The more redundancy can be removed the larger is the gain that can be achieved through CacheCast. As such, large-scale content delivery is the obvious application domain. However, removing redundancy is beneficial in any case in which costs in terms of resources or monetary costs play a role. Relating this to the current trend to move from the classical (large-scale) content producers and consumers to prosumers, that is, individuals that produce and consume content, RE might proof also very useful in small-scale applications, for example, streaming live video from a spectacular downhill bicycle ride from a head mounted action camera via a 3G network to a set of friends. First of all, the bandwidth in 3G is limited such that RE would enable to substantially increase the number of live streams that could be supported. Furthermore, the reduced amount of data to be transmitted could reduce costs for the data traffic.

20.9 FUTURE RESEARCH DIRECTIONS

CacheCast operates on logical point-to-point links. For example, three hosts A, B, and C connected by a broadcast medium (such as Ethernet) are connected with separate links A–B, A–C, and B–C from the CacheCast perspective. Therefore, when host A transmits the same data to the hosts B and C, CacheCast does not suppress the second transfer even though host C has already received the data. Similarly, CacheCast does not suppress redundant transfers in link layer switched networks (such as switched Ethernet), since from the CacheCast perspective all network hosts are connected with separate links. Considering the growth of local area networks, it would be valuable to extend the CacheCast approach to remove redundancy from link layer transfers.

In the first generation of routers, packets are switched between input and output interfaces using shared memory. The router CPU reads a packet from an input interface directly to the shared memory where the CacheCast CSU element restores the redundant data that has been previously removed. When the packet output port is determined, the CacheCast CMU element removes from the packet the data that is present in the next hop cache and then the packet is transferred to the output interface. Therefore, packets carrying redundant data are not transferred between the main memory and input/output (I/O) interfaces. However, in the second and third generation of routes, packets are switched using a shared bus or a switch fabric. When a packet arrives at an input interface, the CSU element reconstructs the packet. Then, the output port is determined and the packet carrying redundant data is transferred over a switch fabric to the output interface where the CMU element removes the data that are present in the next hop cache. Since, in the second and third generation of routers, packets carrying redundant data are transferred over a switch fabric, it may create congestion in the switch fabric. Therefore, these routers require additional mechanisms to mitigate this effect.

20.10 CONCLUSION

This chapter presented CacheCast—a link layer caching system for single-source multiple-destination data transfers. CacheCast assumes no multicast support. Hence, in the CacheCast system, a server transfers data to multiple destinations in the standard way by sending the data subsequently to each destination. At the link layer, this sequence of transfers creates a burst of highly redundant traffic that CacheCast eliminates using small caches on links. Therefore, the efficiency of multidestination transfer with CacheCast support achieves close to the multicast transfer efficiency. To minimize the complexity and storage requirements of link caches, CacheCast requires a server support. The server should batch request for the same data and transmit the data within a minimum time interval.

CacheCast preserves the Internet end-to-end relationship between a server and a client. Thus, the CacheCast server can perform authorization, authentication, accounting, and congestion control on a per-client basis, which is necessary for content delivery. The only requirement CacheCast imposes on an application is synchronous transmission of the same data to clients. This requirement is fulfilled *per se* by live streaming application. Systems that deliver downloadable content can also use CacheCast. This, however, requires implementing additional mechanisms such as fountain codes to achieve the synchronous transmission.

The major part of the Internet traffic carries audio/video content that is primarily delivered to multiple clients. In this context, CacheCast can immediately improve the efficiency of popular IPTV services based on unicast transmission. However, to obtain the maximum benefit of CacheCast, it is necessary to implement and widely adapt a CacheCast file transfer system. We expect CacheCast to significantly reduce the traffic volume in the Future Internet.

ACKNOWLEDGMENTS

This chapter has been partially derived from the authors' past work [24,25].

REFERENCES

1. Aguilar L. Datagram routing for internet multicasting. ACM SIGCOMM Comp Commun Rev 1984;14(2):58–63.
2. Cheriton DR, Deering SE. Host groups: A multicast extension for datagram internetworks. SIGCOMM Comput Commun Rev 1985;15(4):172–179.
3. Myung-Ki S, Yong-Jin K, Ki-Shik P and Sang-Ha K. Explicit multicast extension (Xcast+) for efficient multicast packet delivery. ETRI J 2001;23(4):202–204.
4. Diot C, Levine BN, Lyles B, Kassem H and Balensiefen D. Deployment issues for the IP multicast service and architecture. Netw IEEE 2000;14(1):78–88.
5. Yang YR, Lam SS. Internet multicast congestion control: A survey. Proceedings of ICT; Acapulco, Mexico; 2000.
6. Chu Y, Rao SG, Zhang H. A case for end system multicast (keynote address). ACM SIGMETRICS Performance Evaluation Review; 2000.
7. Spring NT, Wetherall D. A protocol-independent technique for eliminating redundant network traffic. ACM SIGCOMM Computer Communication Review; 2000.
8. Ashok A, Archit G, Aditya A, Srinivasan S and Scott S. Packet caches on routers: The implications of universal redundant traffic elimination. Proceedings of the ACM SIGCOMM 2008 Conference on Data Communication; Seattle, WA, USA; 2008. p. 219–230.
9. Anand A, Sekar V, Akella A. SmartRE: An architecture for coordinated network-wide redundancy elimination. ACM SIGCOMM Computer Communication Review; 2009.
10. Halepovic E, Williamson C, Ghaderi M. Enhancing redundant network traffic elimination. Comput Netw 2011;56(2):795–809.

11. Li X, Salyers D, Striegel A. Improving packet caching scalability through the concept of an explicit end of data marker. Hot Topics in Web Systems and Technologies, 2006. HOTWEB'06. 1st IEEE Workshop on; 2006.
12. Fahad RD, Amar P, Himabindu P, Olatunji R and David GA. Ditto: A system for opportunistic caching in multi-hop wireless networks. Proceedings of the 14th ACM International Conference on Mobile Computing and Networking; 2008.
13. Afanasyev M, Andersen DG, Snoeren AC. Efficiency through eavesdropping: Link-layer packet caching. Proceedings of 5th USENIX NSDI; San Francisco, CA; 2008.
14. Le F, Srivatsa M, Iyengar AK. Byte caching in wireless networks. Distributed Computing Systems (ICDCS), 2012 IEEE 32nd International Conference on; 2012.
15. Chalmers RC, Almeroth KC. Developing a multicast metric. Global Telecommunications Conference. GLOBECOM'00. IEEE; 2000.
16. Robert M, Eddie K, John J and M. Frans K. The Click modular router. ACM Trans Comput Syst 2000;18(3):263–297.
17. Postel J, Reynolds J. File transfer protocol. RFC 959; 1985.
18. MacKay DJ. Fountain codes. Communications, IEE Proceedings; 2005.
19. John WB, Michael L, Michael M and Ashutosh R. A digital fountain approach to reliable distribution of bulk data. ACM SIGCOMM Comput Commun Rev 1998;28(4):56–67.
20. Shojania H, Li B. Tenor: Making coding practical from servers to smartphones. Proceedings of the International Conference on Multimedia; 2010.
21. Huang C, Li J, Ross KW. Can Internet video-on-demand be profitable? ACM SIGCOMM Comput Commun Rev 2007;37(4):133–144.
22. Schulze H, Mochalski K. Internet study 2008/2009. IPOQUE Report; 2009.
23. Clark DD, Tennenhouse DL. Architectural considerations for a new generation of protocols. ACM SIGCOMM Comput Commun Rev 1990;20(4):200–208.
24. Srebrny P, Plagemann T, Goebel V and Mauthe A. Cachecast: Eliminating redundant link traffic for single source multiple destination transfers. Distributed Computing Systems (ICDCS), 2010 IEEE 30th International Conference on; 2010.
25. Srebrny PH. CacheCast: A system for efficient single source multiple destination data transfer. University of Oslo; 2011.

21

CONTENT REPLICATION AND DELIVERY IN INFORMATION-CENTRIC NETWORKS

Vasilis Sourlas[1], Paris Flegkas[1], Dimitrios Katsaros[1], and Leandros Tassiulas[1]

[1]*University of Thessaly, Oktovriou, Volos, Greece*

21.1 INTRODUCTION

Information-centric networking (ICN) is emerging as the main future networking environment, given that the vast majority of Internet activities are related to information access and delivery. In ICN, information is explicitly labeled so that anybody who has relevant information can potentially participate in the fulfillment of requests for said information. Publications are issued by clients (publishers) when they have a new information item to publish in the network, while subscriptions are issued by clients (subscribers) to subscribe the items they are interested in. Given the information-centric nature of the distribution utilizing information that is replicated across almost ubiquitously, available storage devices are an almost natural thought. Optimized dissemination of information within transient communication relationships of endpoints is the main promise of such efforts, and efficient replication of information is key to delivering on this promise.

While packet-level in-network opportunistic caching is one of the salient characteristics of ICN architectures, proper cache placement and replica assignment still have an important role to play. Content delivery network (CDN)-like replication distributes a

Advanced Content Delivery, Streaming, and Cloud Services, First Edition.
Edited by Mukaddim Pathan, Ramesh K. Sitaraman, and Dom Robinson.

site's content across multiple mirror servers. When a client is interested in a particular piece of information, his/her request is redirected to one of the existing replication points rather than requiring retrieval from the original publisher. Replication is used to increase availability and fault tolerance, while it has as side-effect load-balancing and enhanced publisher subscriber proximity. Particularly, CDN providers strategically place surrogate servers and connect them to Internet service provider (ISP) network edges so that content can be closer to clients. Given the significant impact that content delivery has on the utilization of an ISP network, some work has recently started to investigate new models and frameworks to support the interaction between ISPs and CDNs.

CDNs and ICN are designed to fulfill the necessity of efficient content delivery. Their main difference is that CDNs build up the end-to-end content delivery in the Internet at the application layer, while ICN is a clean slate proposal for an alternative approach to the core architecture of the network. In this chapter, we present a three-phase framework as a contribution to the problem of information replication in an ICN environment, through the synergy of ICN with CDN techniques. Moreover, we believe that this synergy will enable CDNs to incorporate, through the ICN functional components, dynamic network information on replica selection to determine the best paths over which transfer of content will take place. The objective of the presented framework is to minimize the total traffic load in the network subject to installing a predefined number of replication devices, and given that each device has storage limitations. The presented framework comprises three phases, namely, the *Planning*, the *Offline Assignment*, and the *Online Replacement* phases, which manage the content and the location of each replication device in the network.

In the planning phase, the presented framework selects those nodes of the network to place the replication devices (CDN server) while in the offline assignment phase each information item is assigned, based on its popularity, at a subset of the selected replication points so that the targeted objective is satisfied. Finally, the online replacement/reassignment phase dynamically reassigns information items in the replication devices based on the observed items' changing request patterns. In order to support the presented framework, the three phases should be provided by relevant functional components. Regarding the Planning and the Assignment phases, these components reside outside the network and run offline algorithms at two different but long-medium timescales, while the replacement phase is residing in components installed at each replication device of the network and run in real-time scale. The three-phase replication framework is generic so that it can apply in almost every ICN proposed architecture.

The rest of this chapter is organized as follows. In Section 21.2, a brief related work on ICN architectures and replication is given. In Section 21.3, we present the three-phase replication framework, whereas in Section 21.4, we shortly evaluate the presented algorithms. Section 21.5 is devoted in future research directions, whereas in Section 21.6 we conclude this chapter.

21.2 RELATED WORK

ICN is a flexible communication model that meets the requirements of the information distribution in the Internet, since information is addressed by semantic attributes rather than origin and destination identities. In recent research efforts, among others named data networking/content-centric networking (NDN/CCN) [1] and Publish Subscribe Internet Technology (PURSUIT) [2] aim to switch from host-oriented to content-oriented networking by naming data/content instead of naming hosts in order to achieve scalability, security, and performance.

NDN [1] proposes a name-based routing system for locating and delivering named data packets. The fundamental entities in NDN are Interest and Data packets. When a user wishes to receive data, he/she issues an Interest that contains the data name. The network propagates the Interests to the nearest data source (anycast), and then the requested item is delivered back to the user in the form of a Data packet. NDN uses names to identify content objects only; there is no notion of host name, point of attachment, or path identifier. Content names follow a hierarchical form similar to URLs or file system paths, and by definition, Interest and Data paths are symmetric.

In PURSUIT [2], the design paradigm involves three separate elements and three separate functions: publishers, subscribers, and the REndezvous NEtwork (RENE) on the one hand, and the functions of rendezvous, topology management/formation, and forwarding on the other hand, respectively. While the first three elements also exist in other candidate architectures under different names, the design principle of PURSUIT is to clearly distinguish the latter three functions. In more detail, publishers in PURSUIT advertise the availability of information by issuing a publication message to the RENE. Similarly, subscribers are entities interested in consuming information who express their desire by issuing a subscription message to the RENE for a specific piece of information. The RENE is responsible to match publications to subscriptions through the rendezvous function and choose the best route (through the topology management and the forwarding functions) for the delivery of the requested information.

In the area of replication in ICN in Reference 3, a historic data retrieval publish/subscribe system is proposed, where databases are connected to various network nodes, each associated with a set of items to store. In Reference 3, every information item is stored only once and no placement strategies have been examined. In Reference 4, a set of offline storage planning and replica assignment algorithms for the ICN paradigm are presented, while in References 5, 6 an online approach for the reassignment of items within the replication points is presented.

In the traditional context of CDNs, the placement problem is a thoroughly investigated problem. Particularly in References 7, 8, authors approached the placement problem with the assumption that the underlying network topologies are trees. This simple approach allows the authors to develop optimal algorithms, but they consider the problem of placing replicas for only one origin server. The placement problem is in fact a numeristic polynomial (NP)-hard problem [9] when striving for optimality, but there are a number of studies [10–15] where an approximate solution is pursued. Their work is also known as network location or partitioning and involves the optimal placement of k service facilities in a network of N nodes targeting the minimization of a given objective. In some cases, it can be shown that this problem reduces to the well-known *k-median* problem.

The authors of Reference 16 model replica assignment as a distributed selfish replication (DSR) game in the context of distributed replication groups (DRGs). Under the

DRG abstraction, nodes utilize their storage to replicate information items and make them available to local and remote users. The pairwise distance of the nodes is assumed to be equal, while our framework considers the generic case of arbitrary distances. In the context of DRG and under the same distance assumption, a two-approximation cache management algorithm is presented in Reference 17. Finally, in Reference 18, authors develop a cache management algorithm aimed at maximizing the traffic volume served from the caches and minimizing the bandwidth cost. They focus on a cluster of distributed storages, either connected directly or via a parent node, and formulate the content placement problem as a linear program in order to benchmark the globally optimal performance.

More placement algorithms have been presented in Reference 9. Particularly, authors formulate the problem as a combinatorial optimization problem and show that the best results are obtained with heuristics that have all the stores cooperating in making the replication decisions. Moreover, in Reference 19, authors introduce a framework for evaluating placement algorithms. They classify and qualitatively compare placement algorithms using a generic set of primitives that capture several objectives and near-optimal solutions. In most of the above approaches, a similar cost function (optimize bandwidth and/or storage usage costs for a given demand pattern) is considered. Less attention has been given though to network constraints (limited storage capacity) and the possibility of reassigning items between the replication points as popularity and locality of users demand change.

Finally, in the research area of investigating new models and frameworks to support the interaction between ISPs and CDNs, in Reference 20, the authors highlight that CDN providers and ISPs can indirectly influence each other, by performing server selection and traffic engineering operations, respectively, and they investigate different models of cooperation between the two entities. In Reference 21, the authors propose a framework to support joint decisions between a CDN and an ISP with respect to the server selection process. This framework allows the ISP and the CDN to collaborate by exchanging some local information (network utilization from the ISP side and server conditions from the CDN side), so that it can result in better control of the resources. An ISP-supported CDN service has been presented in References 22, 23, whereby content is stored and served from within ISP domains. This solution, however, can incur high operational costs, given that ISPs will have to maintain large storage capacities, and may thus be economically unviable.

21.3 FRAMEWORK FOR INFORMATION REPLICATION IN ICN

In this section, we present separately the three phases of the framework for the management of the information replication in ICN.

21.3.1 The Planning Phase

The planning phase takes the number of available replication devices an operator wishes to install as an input, the network topology and a long-term prediction of subscriptions in the network. It can run periodically deciding the optimal placement of the replication points at a long-term timescale (e.g., once a year) or whenever the current location of them leads to an inefficient deployment due to significant subscriptions' changes not successfully predicted. Performing and enforcing the decision of the planning component usually involve high level business decisions as there is a high cost associated with moving a replication device to a different physical location or extending their number. In Reference 24, an ICN-oriented planning algorithm is presented for the selection of the replication points in the network based on the local demand for each item and the storage limitations of each replication device.

21.3.2 The Offline Assignment Phase

The offline assignment phase also runs periodically but at a medium- or long-term scale. It takes as input the outcome of the planning phase regarding the locations of the replication devices installed in the network, the physical network topology, and the medium- or long-term forecast. Replicas' relocation can be enforced by instructing the replication devices to subscribe to a different set of information items. The instruction itself is realized as a publication of an information item to which replication devices are subscribed. In general, the replication points act both as publishers and subscribers for the information items they are instructed to store. They subscribe in order to receive new versions of the items, while they act as publishers for the same items to interested subscribers. This way, when a client subscribes to a specific piece of information, one or more publishers/replication points are enabled, based on the operator's policy, to publish the relevant data. Figure 21.1 illustrates the basic modules of the planning and the offline assignment phases.

Before the presentation of the online replacement/reassignment phase, we present below an ICN-oriented planning algorithm for the selection of the replication points in the network based on the local demand for each item and the storage limitations of each replication device. We also present a mechanism for the offline assignment of the replicas of each item in the selected replication points.

21.3.2.1 Modified Greedy Algorithm We use algorithms presented in the context of CDN networks as the base of our planning and offline replica assignment scheme. Particularly in References 9, 10, authors developed several placement algorithms that use workload information, such as latency (distance from the storage points) and request rates, to make the placement decision. Their main conclusion is that the so-called greedy algorithm that places replication devices in the network based on both a distance metric and request load performs the best and is very close to the optimal solution.

The traditional greedy algorithm assumes that there exists only one class of content in the system, or equivalently there is no distinction in the content. We let r_i be the demand (in requests/s) from clients attached to node i. We also let p_{ij} be the percentage of the overall request demand accessing the target server j (traditional placement algorithms replicate a specific origin server) that passes through node i. Also, we denote the propagation delay (hops) from node i to the target server j as d_{ij}. If a replica is placed at

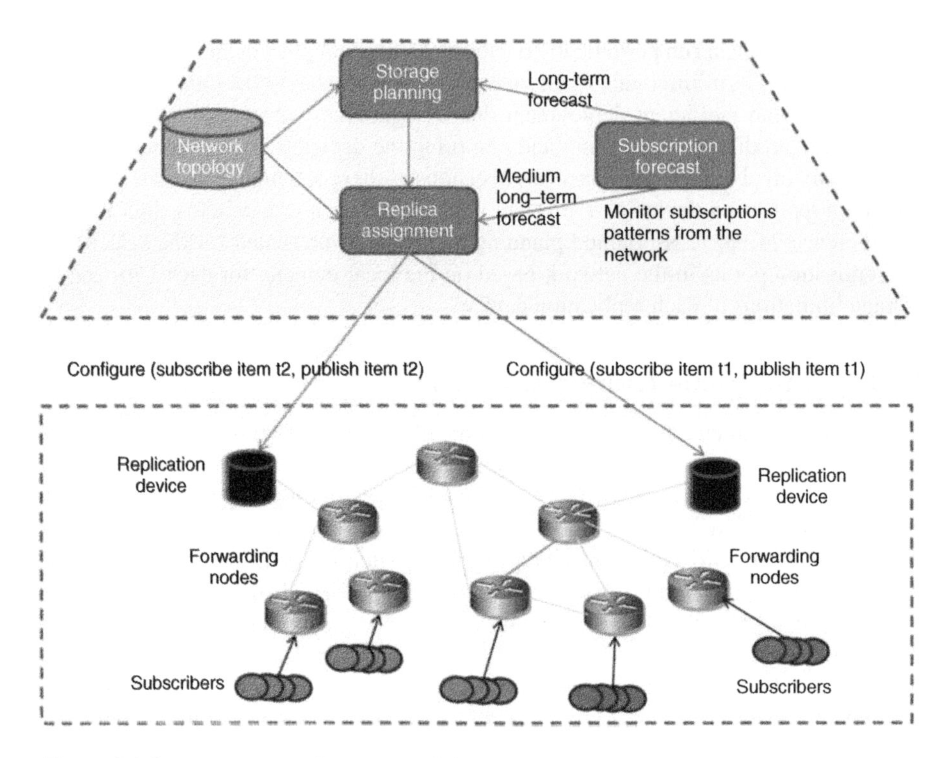

Figure 21.1 Architectural illustration of the planning and the offline assignment phases.

node i, we define the gain to be $g_{ij} = p_{ij} \cdot d_{ij}$. This means that p_{ij} percentage of the traffic would not need to traverse the distance from node i to server j.

The greedy algorithm chooses one replication point at a time. In the first round, it evaluates each node of the network to determine its suitability to become a replication point of the origin server. It computes the gain associated with each node and selects the one that maximizes the gain. In the second round, searches for a second replication point which in conjunction with the one already picked yields the highest gain. The greedy algorithm iterates until all replication points have been chosen to replicate the given server.

In many ICN implementations, the notion of an origin server usually does not exist. Publishers join the network, publish their content, and disappear. So, in order to obtain the location of the replication points, we modify the greedy algorithm. Particularly, we repeat the above-mentioned procedure as many times as the number of the nodes in the network, assuming each time that the targeted server is a different node of the network. We get in that way a set of different possible replication points per iteration. Finally, we select as our final replication points those nodes that appeared more times in the per-element summation of the different sets. The modified greedy algorithm presented here assumes uniform distribution of the probability among the different nodes of the network that publications could occur. Of course, other forms of probability distributions could be used, and each different set should be first weighted with its probability before the per-element summation.

21.3.2.2 Planning and Offline Replica Assignment Algorithm for ICNs

Here, we use the modified greedy algorithm described previously for the case where not one but M different information items in our network exist. The presented algorithm is composed of the following steps:

Step 1. For each item $m \in M$, we execute the modified greedy algorithm and we get M sets of possible replication points S_m.

Step 2. Each vector S_m is weighted by $w_m = \sum_{i=1}^{N} r_i^m / \sum_{m=1}^{M} \sum_{i=1}^{N} r_i^m$, where N is the number of the nodes in the network and r_i^m is the number of requests per second generated at node $i, (i \in N)$ for item $m, (m \in M)$. w_m is the significance of item m (popularity) regarding the traffic demand of each item in the network.

Step 3. We select as our replication points those nodes (as many as the network operator/CDN provider is willing to install) that appeared more times in the per-element weighted summation of S_m vectors. We call that vector the *replication nodes vector* S.

Step 4. For each item m, starting from the most significant (based on the weight), we assign replicas (say k_m) following the procedure below:

For each entry in S_m of item m calculated in Step 1, assign a replica if that entry also appears in S, calculated in Step 3, and only if that replication point has been assigned less than C items (storage capacity of each replication point) until we get k_m replicas (replication degree of item m, which is relative to the weight of each item in the network).

Steps 1–3 of the presented algorithm described previously comprise the planning phase of the algorithm while Step 4 is the assignment phase. Step 4 is also known as the generalized assignment problem, which even in its simplest form is reduced to the NP-complete multiple knapsack problem. For the solution of the assignment problem, we used the heuristic approaches described earlier, while more approaches could be found in the literature (e.g., References 24, 25).

21.3.3 The Online Replacement Phase

As the provisioning periods of the planning and the offline assignment phases can be quite long, the subscription patterns may significantly vary during that period. For this reason, we introduce the online replacement/reassignment phase that enables the replacement of information items to the replication points to take place in real time, based on the changing demand patterns of the users. Distributed components of that phase decide the items every replication point stores by forming a substrate that can be organized either in a hierarchical manner for scalability reasons or in a peer-to-peer organizational structure. Communication of information related to request rates,

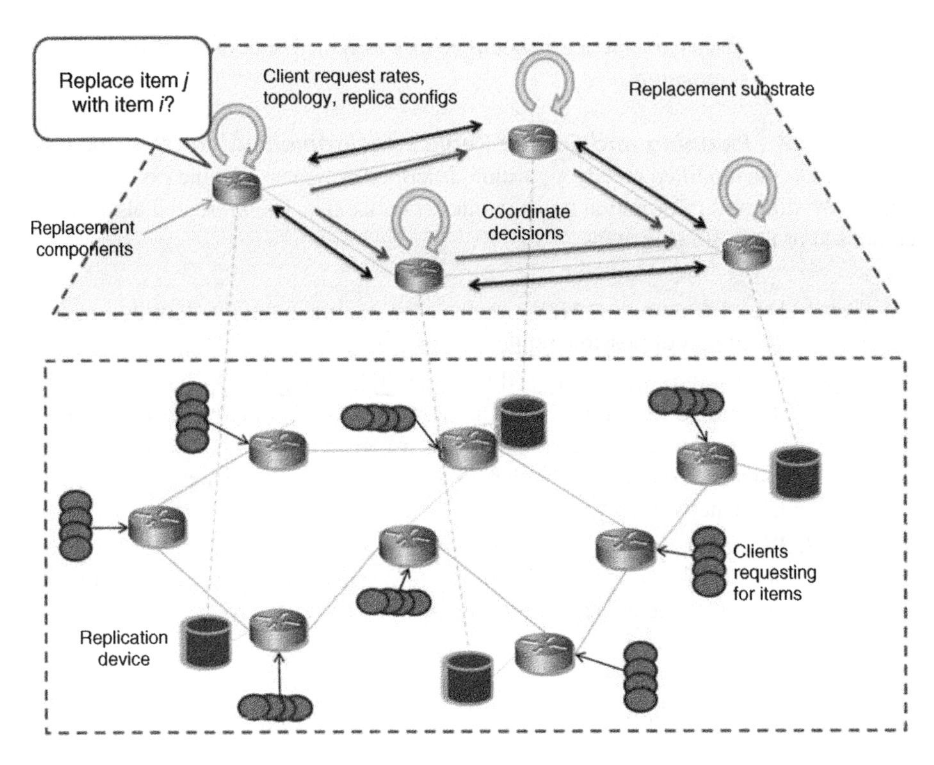

Figure 21.2 Architectural illustration of the online replacement phase.

popularity/locality of information items, and current replication points' configuration takes place between the distributed replacement components through an intelligent substrate.

Every replacement component, as depicted in Figure 21.2, should decide in a coordinated manner with other components whether to store an item. This may require the replacement of an already stored item, depending on the available space. The decision of this replacement of stored items is performed toward maximizing an overall network-wide utility function (e.g., the gain in network traffic), which means every node should calculate the gain the replacement of an item would incur. This approach assumes that every component has a holistic network-wide view of all the replication points' configuration and relevant request patterns, and this information should be exchanged periodically or in an event-based manner when a component changes the configuration of its attached replication device.

Since all the above decisions are made in a distributed manner, uncoordinated decisions could lead to suboptimal and inconsistent configurations. Coordinated decision-making of a distributed solution can be achieved through the substrate mechanisms, by ensuring that components change the configuration of each replication device in an iterative manner, that is, one at a time and not autonomously in a potentially conflicting manner.

Any distributed online mechanism that applies in this third phase should capture the volatile environment under consideration. All of them should be adaptive to popularity and locality changes by fetching new items at a replication device and replacing existing items. We envision two classes of algorithms that could be applied in the online replacement phase that differ in the amount of information that needs to be communicated through the substrate and the required level of coordination among the components. We present them in order of decreasing complexity, in terms of the induced communication overhead.

The first class, henceforth called *cooperative*, aims at minimizing the overall network traffic. This requires that every component needs a holistic network-wide view of the request patterns and the current replication points' configuration. In addition, since each replacement decision affects the whole network, cooperation in the decision-making is required. The cooperative algorithm requires at each iteration each component of the network to compute the relative gain of every possible replacement and through appropriate message exchange to cooperate for the final replacements. In other words, every component participates in the execution of the algorithm at each iteration, but every time only one (the one with the maximum relative gain) performs valid replacements.

The second class, henceforth called *holistic*, also aims at minimizing the overall network traffic and hence requires the same amount of information. However, in the holistic class, there is no need for coordination of the actions of the components, and the required decisions are made in an autonomous manner by each one individually. The holistic class is of similar nature as the cooperative and toward the same objective. Its distinguishing characteristic though is that each component operates in its own performing replacements on the respective replication point. Particularly, only one component at each iteration performs valid and beneficial replacements toward a specific objective. In both cooperative and holistic classes, it can be shown that since any change performed in the replication points' configuration decreases the overall network traffic, the presented algorithms finally converges to a stationary point where no further improvement is possible. The algorithms do not necessarily converge to the optimal assignment but to a local minimum of the objective given the initial configuration. In the presented algorithmic classes, the replacements are based on the real-time observed items' request patterns such as their popularity and locality and not in static offline predictions. In the next section, we present in detail the steps of the cooperative algorithm, and based on them we also describe the functionality of the holistic algorithm.

21.3.3.1 Cooperative and Holistic Online Replacement Algorithms

Since we wish to minimize the overall network traffic cost, we assume that the underlying content delivery mechanism always directs the requests to the closest replication point out of those holding the requested item. Given such an access mechanism, the replacement components have to coordinate their actions toward finding the replication degree and the location where each item should be cached. We assume equal capacity among the replication devices for ease of presentation, and we also assume that each information item is of unit size, which is a typical assumption in the literature (e.g., [16,17]). The presented algorithms are also applicable in the case of different item sizes as well. However, special care needs to be given, since several items may need to be removed from the replication device in order to fit the new item.

At the cooperative algorithm at each iteration, all the replacement components of the network execute the following steps in parallel, given the current storage configuration H and the corresponding total traffic cost $T(H)$.

Step 1. Let C_v denote the set of items that are stored at replication point v and in at least one more replication point in the network. For each item $m \in C_v$ compute the overall performance loss, $l_v^m = T(\overline{H}_v^m) - T(H) \geq 0$, that will be caused if item m is removed from v, leading to a valid new configuration $\overline{H}_v^m$. In this case, all the requests for item m at v will be served by another replication point, which is at least that far.

Step 2. Let P_v denote the set of items that are not stored at v. For each item $m \in P_v$, compute the overall performance gain $g_v^m = T(H) - T(\underline{H}_v^m) \geq 0$ achieved if item m is inserted at replication point v, hence leading to a new configuration $\underline{H}_v^m$. In this case, a certain amount of requests for item m will be served by node v, as the closest replica.

Step 3. Each replacement component v considers as candidate for insertion the item $i \in P_v$ of maximum performance gain and as candidate for replacement the item $j \in C_v$ of minimum performance.

Step 4. Each replacement component at replication point v calculates the maximum local relative gain $b_v = g_i - l_j$ and informs the rest of the replacement components through a report message Rep(b, v, i, j).

Step 5. After receiving the Rep messages, each replacement component calculates the most network-wide beneficial replacement, say Rep $* (b^*, v^*, i^*, j^*)$, the one of maximum relative gain, and updates its configuration matrix H (the matrix that shows where every item is stored in the replication points of the network). At this point, only the configuration matrices of the replacement components are updated. Once the algorithm has converged, these components fetch and cache new information items and replace cached ones (e.g., fetch item i^* and replace item j^*).

Step 6. Repeat Steps 1–5 until no further replacements are beneficial for the network, that is, no positive relative gain exists.

The holistic algorithm is of similar nature and toward the same objective. Its distinguishing characteristic though is that each replacement component operates on its own by performing replacements on the respective replication point. At each iteration, a single component, say v, autonomously decides and executes the following steps. Steps 1–3 are identical to the cooperative algorithm and are omitted:

Step 7. The replacement of maximum relative gain $b = g_i - l_j$ is performed by the component v. The rest of the components are notified through the report message Rep(b, v, i, j).

Step 8. After receiving the Rep message, every component updates its configuration matrix H.

Although the replacements may be applied asynchronously among the replication points, we assume that only a single component may modify the storage configuration at a given time. This is due to the requirement that each component should know the current storage configuration of the network, in order to calculate the gain and loss metrics. Thus, each modification is advertised to the rest replacement components. Relaxing this assumption would lead to a setting where the replacement components make decisions based on outdated information, causing thus some performance degradation and making convergence questionable.

21.4 PERFORMANCE EVALUATION

Here, we evaluate through simulations the performance of the presented framework and the corresponding algorithms. Since no ICN infrastructure has been deployed for commercial use yet, no publicly available datasets exist for performance evaluation. Thus, realistic synthetic workload generators are used instead. The request rate for an item at each node is determined by its popularity. Here, we approximate the popularity of the items by a Zipf law of exponent z_{pop}. Literature provides ample evidence that the file popularity in the Internet follows such a distribution [26–29]. In particular, we consider seven typical values for z_{pop} ranging from -1 to 1. We also assume that in each node of the network, a total of 200 requests per second are generated. Thus, the request rate of each item at each node varies from 0 to 200 requests/s according to its popularity.

We run two sets of experiments: one evaluating both the planning and the offline assignment phases of the presented framework, and the other evaluating only the online replacement phase after an initial planning. We used topologies from the Internet Topology Zoo dataset [30], which contains real network topologies from all over the world.

Figure 21.3 presents the performance regarding the overall network traffic (in responses × hops/s) of the planning and the offline assignment phase. By default, unless explicitly mentioned, we assume that $S = N/8$ replication points are placed in a network of N nodes, and the storage capacity of each replication point is $C = M/4$ information items. We compare the presented offline assignment mechanism with a totally random assignment procedure when we vary the size of the network, the capacity of each replication point, and the number of replication points.

From Figure 21.3, it is obvious that the presented assignment mechanism combined with the presented planning scheme performs on average 10–30% better than the random assignment. For a more detailed evaluation of the planning and the offline assignment phases, refer to References 4, 24, 31, where we observed that in the real world where a storage provider has limitations in the number of replicas that can install, each replication device has storage limitations that the presented scheme is an appropriate solution in almost any scenario.

In Figure 21.4, we investigate the adaptability of the presented online replacement algorithms as the popularity of the demand patterns change. Using as initial storage planning and assignment configuration for a given set of popularity values, we depict the performance of each algorithm as it adapts to the new environmental parameters.

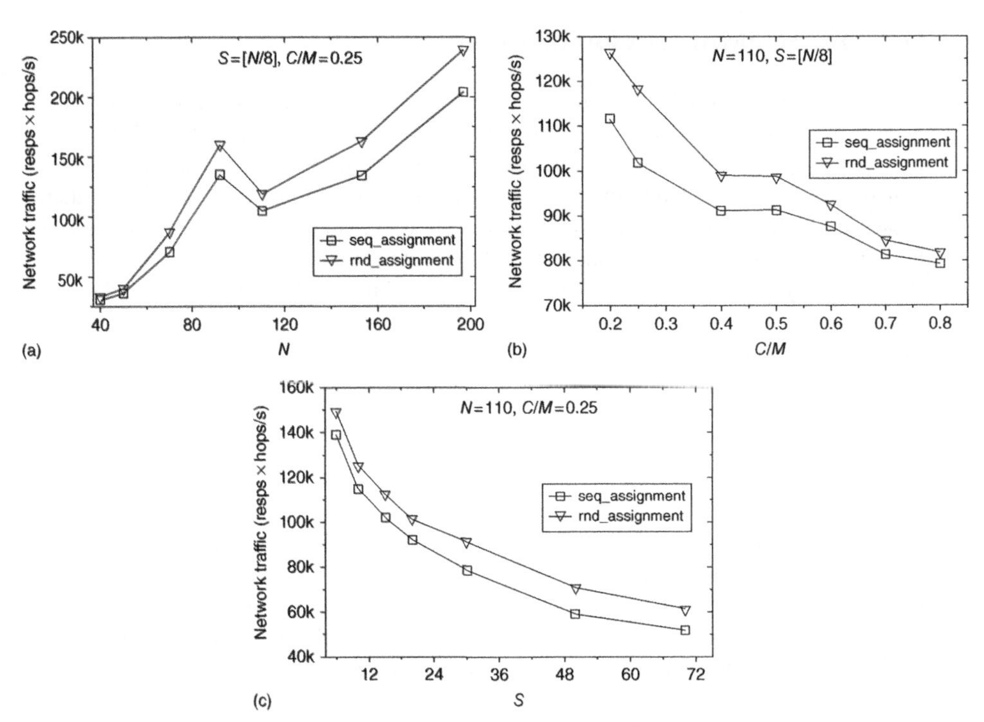

Figure 21.3 Performance of the planning and the offline assignment phase.

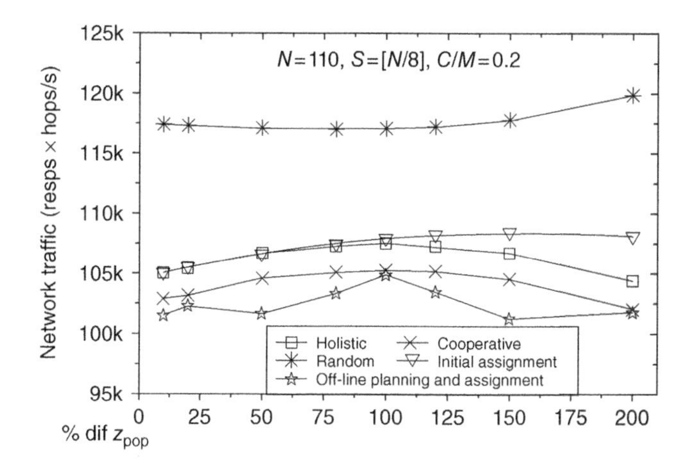

Figure 21.4 Performance of the online replacement algorithms.

Particularly, we initially assume that the popularities assigned to the nodes of the network are given by the vector $Z = (-1, -0.7, -0.5, 0, 0.5, 0.7, 1)$, and at each different experiment (different points in Figure 21.4) this vector changes by a given factor. This factor ranges from 10% to 200%. A change of 10% means that the new vector of popularities is $Z = (-0.9, -0.63, -0.45, 0, 0.45, 0.63, 0.9)$, whereas a change of 100% transforms the vector of popularities to $Z = (0, 0, 0, 0, 0, 0, 0)$, and a change of 200% inverts the vector. We also depict the performance of the initial cache assignment resulting from the initial planning with the new demand pattern as well as the performance of a totally new planning execution.

In Figure 21.4, we observe that the cooperative algorithm performs slightly better than the holistic, whereas the random assignment is at least 25% worse than the two presented algorithms. An interesting finding comes from the comparison of the presented online replacement algorithms with the performance of the initial cache assignment. We observe that when the changing factor of the initial popularities is smaller than 100%, the algorithms with the network-wide knowledge perform only $1 - 3\%$ better than the initial assignment; only when the changing factor is larger than 100% and the popularity vector reverts its sign, we observe a difference is the performance up to 10% regarding the overall network traffic. Finally, the two presented algorithms are performing close to the new planning and offline assignment, which means that a new planning is not needed very often and is useful only under extreme changes in the demand pattern.

Note that Figure 21.4 may also serve as a benchmark for the replacement components in their decision to reassign or not the stored items upon the detection of a change in the popularity pattern. Particularly, the difference between the network traffic cost of the initial storage assignment and the traffic cost after the completion of the algorithms combined with the communication complexity enables the replacement components to perform or skip the content reassignment. For a more detailed evaluation of the online replacement phase, refer to References 5, 6 where our numerical results provide evidence that network-wide knowledge and cooperation give significant performance benefits and reduce the time to convergence at the cost of additional message exchanges and computational effort. In particular, the cooperative algorithm provides the best performance regarding overall network traffic, but requires a high level of cooperation among the managers and hence is of very high computational and communication complexity. On the other hand, the holistic algorithm performs close to the cooperative but converges in a fraction of the iterations required by the cooperative. Thus, the comparison of the presented algorithms may serve as a valuable tool for the network manager so as to select the most appropriate algorithm for his needs, depending on specific network parameters (e.g., network size, number of information items, and volatility of the request pattern).

21.5 FUTURE RESEARCH DIRECTIONS

The work presented in this chapter can be extended in many ways such as optimizing different objectives to serve different quality of service (QoS) metrics and service-level agreements (SLAs) among the storage providers and the content providers. Also it would be interesting, as future work, to explore enhancements to the presented online replacement algorithms that would also take into consideration the cost of replacing the items at the replication points of the network, as well as the processing load of each replacement component when assigning items to them.

Another interesting extension of the presented work would be the identification and analysis of user and content mobility patterns. Considerable research has been dedicated to understand and model the behavior of mobile users from the social and technological perspectives. Most existing work on predicting mobility patterns, attachment points, and connectivity durations aims to minimize periods of disconnection. With ICN, however, content can "follow" mobile users, thus achieving shorter transaction periods; therefore, this work needs to be reevaluated. The outcome will be monitoring mechanisms, as well as algorithms for predicting the future connectivity points in order to move or migrate content accordingly. The presented replication framework could be exploited to identify strategic replication points at the edges of the network. The identification of these points should be based on (i) route selection and transmission scheduling and (ii) trade-offs between energy, delay, and cost. For example, replicating content closer to the user reduces energy consumption and delivery delay, as seen from the end-users point of view, as well as core-network traffic, as seen from the operator's point of view, but it increases replica deployment costs for the operator. Based on the investigation of such trade-offs, the outcome could be (i) the identification of strategic replication points close to the user, (ii) replication strategies to increase the amount of time content stays in the replication points, and (iii) the corresponding effect to the route selection and transmission scheduling algorithms.

21.6 CONCLUSION

In this chapter, we presented a generic three-phase framework as a contribution to the problem of information replication in ICN. The presented phases apply at different timescales and manage the content and the location of each replication point in the network targeting a specific objective. Moreover, in the newly presented online replacement phase, we presented two algorithms that differ in the amount of information that needs to be communicated and the required level of coordination among the replacement components. The online replacement decisions are based on real-time information, such as the observed popularity of the requests, and not on static offline predictions. Our numerical results provide evidence that network-wide knowledge and cooperation give significant performance benefits and reduce the time to convergence at the cost of additional message exchanges.

ACKNOWLEDGMENTS

V. Sourlas' work is cofinanced by the European Union (European Social Fund ESF) and Greek national funds through the Operational Program "Education and Lifelong Learning" of the National Strategic Reference Framework (NSRF) – Research Funding Program: Heracleitus II—investing in knowledge society through the European Social Fund.

REFERENCES

1. Jacobson V, Smetters DK, Thornton JD, Plass MF, Briggs N, Braynard R. Networking named content. ACM CoNEXT; Rome, Italy; December 2009.
2. PURSUIT project. 2011. Available at http://www.fp7-pursuit.eu. Accessed 10 July 2013.
3. Li G, Cheung A, Hou S, Hu S, Muthusamy V, Sherafat R, Wun A, Jacobsen H, Manovski S. Historic data access in publish/ subscribe. Proceedings of DEBS; Toronto, Canada; 2007. p 80–84.
4. Sourlas V, Flegkas P, Paschos GS, Katsaros D, Tassiulas L. Storage planning and replica assignment in content-centric publish/subscribe networks. In S.I. on Internet-based Content Delivery. Comput Netw 2011;55(18):4021–4032.
5. Sourlas V, Flegkas P, Gkatzikis L, Tassiulas L. Autonomic cache management in information-centric networks. 13th IEEE/IFIP Network Operations and Management Symposium (NOMS 2012); Hawaii, USA; April 2012. p 121–129.
6. Sourlas V, Gkatzikis L, Flegkas P, Tassiulas L. Distributed cache management in information-centric networks IEEE Transactions on Network and Service Management. September 2013;10(3):286–299.
7. Li B, Golin MJ, Ialiano GF, Deng X. On the optimal placement of web proxies in the Internet. Proceedings of INFOCOM; March 1999.
8. Cidon I, Kutten S, Soffer R. Optimal allocation of electronic content. Proceedings of INFOCOM; Anchorage; April 2001.
9. Kangasharju J, Roberts J, Ross K. Object replication strategies in content distribution networks. Comput Commun 2002;25:376–383.
10. Qiu L, Padmanabhan VN, Voelker G. On the placement of web server replicas. Proceedings of IEEE INFOCOM; Anchorage, USA; April 2001.

11. Arya V, Garg N, Khandekar R, Meyerson A, Munagala K, and Pandit V. Local search heuristics for k-median and facility location problems. Proceedings of 33rd ACM Symposium on Theory of Computing; 2001.
12. Charikar M, Guha S. Improved combinatorial algorithms for facility location and k-median problems. Proceedings of the 40th Annual IEEE Symposium on Foundations of Computer Science; 1999. p 378–388.
13. Charikar M, Khuller S, Mount D, Narasimhan G. Facility location with outliers. Proceedings of the 12th Annual ACM-SIAM Symposium on Discrete Algorithms; Washington DC; January 2001.
14. Shmoys DB, Tardos E, Aardal KI. Approximation algorithms for facility location problems. Proceedings of the 29th Annual ACM Symposium on Theory of Computing; 1997. p 265–274.
15. E. Cronin, S. Jamin, C. Jin, T. Kurc, D. Raz and Y. Shavitt, Constrained mirror placement on the Internet, IEEE JSAC, 36(2), 2002.
16. Laoutaris N, Telelis O, Zissimopoulos V, Stavrakakis I. Distributed selfish replication. IEEE Trans Parallel Distrib Syst 2006;17(12):1401–1413.
17. Zaman S, Grosu D. A distributed algorithm for the replica placement problem. IEEE Trans Parallel and Distrib Syst 2011;22(9).
18. Borst S, Gupta V, Walid A. Distributed caching algorithms for content distribution networks. IEEE INOFCOM; San Diego, USA; March 2010.
19. Karlsson M, Karamanolis Ch, Mahalingam M. 2002. A framework for evaluating replica placement algorithms. Available at http://www.hpl.hp.com/techreports/2002/HPL-2002-21. Accessed 10 July 2013.
20. Jiang W, Zhang-Shen R, Rexford J, Chiang M. Cooperative content distribution and traffic engineering in an isp network. Proceedings of the Eleventh International Joint Conference on Measurement and Modeling of Computer Systems, ser. SIGMETRICS '09; 2009. p 239–250.
21. Frank B, Poese I, Smaragdakis G, Uhlig S, Feldmann A. Content-aware traffic engineering. Proceedings of the ACM SIGMET-RICS/PERFORMANCE 2012 Joint International Conference on Measurement and Modeling of Computer Systems; 2012. p 413–414.
22. Kamiyama N, Mori T, Kawahara R, Harada S, Hasegawa H. ISP-operated CDN. Proceedings of the 28th IEEE International Conference on Computer Communications Workshops, ser. INFOCOM'09; 2009. p 49–54.
23. Cho K, Jung H, Lee M, Ko D, Kwon T, Choi Y. How can an ISP merge with a CDN? IEEE Commun Mag 2011;49(10):156–162.
24. Sourlas V, Flegkas P, Paschos GS, Katsaros D, Tassiulas L. Storing and replication in topic-based publish/subscribe networks. IEEE Globecom 2010 Next-Generation Networking and Internet Symposium; Miami, USA; December 2010.
25. Cohen R, Katzir L, Raz D. An efficient approximation for the generalized assignment problem. Inform Process Lett 2006;100:162–166.
26. Breslau L, Cao P, Fan L, Phillips G, Shenker S. Web caching and Zipf-like distributions: Evidence and implications. IEEE INFOCOM, NY; March 1999.
27. Padmanabhan VN, Qiu L. The content and access dynamics of a busy wed site; Stockholm, Sweden: ACM SIGCOMM; August 2000.
28. Newman MEJ. Power laws, pareto distributions and Zipfs law. Contemp Phys 2005;46:323–351.

29. Adamic LA, Huberman BA. Zipfs law and the Internet. Glottometrics 2002;3:143–150.
30. Knight S, Nguyen HX, Falkner N, Bowden R, Roughan M. The Internet topology zoo. IEEE JSAC 2011;29(9).
31. Flegkas P, Sourlas V, Parisis G, Trossen D. Storage replication in information-centric networking. International Conference on Computing, Networking and Communications (ICNC 2013); San Diego, USA; 2013. p 850–855.

22

ROBUST CONTENT BROADCASTING IN VEHICULAR NETWORKS

Giancarlo Fortino[1], Carlos T. Calafate[2], Juan C. Cano[2], and Pietro Manzoni[2]

[1] *University of Calabria, Rende (CS), Italy*
[2] *Universitat Politècnica de València, Valencia, Spain*

22.1 INTRODUCTION

In the past years, there has been a growing interest in exploiting wireless technology to enable interaction among vehicles, and between infrastructures and vehicles, for the purpose of improving road safety, managing the vehicular traffic, and assisting drivers during their travel with entertainment and useful information (e.g., accident warnings, road alerts, and traffic information). Vehicular *ad hoc* networks (VANETs) [1,2] have been specifically introduced to cope with the aforementioned purposes. VANETs are mobile *ad hoc* networks that support the communication among neighbor vehicles and can be extended to support communication between vehicles and infrastructures placed in the surrounding environment.

In particular, VANETs can effectively support in-vehicle infotainment, that is, information-based media content or programming that also includes entertainment content in an effort to enhance popularity with audiences and consumers. In the context of VANETs, infotainment is usually transmitted through broadcast-based systems that deliver contents using unidirectional links from antennas to vehicles in mobility [3].

Advanced Content Delivery, Streaming, and Cloud Services, First Edition.
Edited by Mukaddim Pathan, Ramesh K. Sitaraman, and Dom Robinson.

Such systems are affected by several issues that make them difficult to be designed and implemented: lossy communication unidirectional channel, unreliable communication transport protocols, vehicle mobility, and antenna area coverage.

To deal with such issues, in a previous research, we developed a content delivery system (CDS) [4,5] on broadcast antennas based on IEEE 802.11a, which is able to transmit media files using the File Delivery over Unidirectional Transport (FLUTE) protocol [6] integrated with several FEC (Forward Error Correction) schemes, including XOR, Reed–Solomon (R–S), and Raptor [7,8]. The evaluation of the CDS carried out limitedly in a real local controlled environment, highlighting that the CDS based on the Raptor schema performed relatively better, in terms of file transfer time, than the CDS not based on FEC schemes or based on the XOR and R–S schemes.

However, to provide more general results, the evaluation of the CDS should be carried out either on large-scale real environments or on effective simulation environments. In particular, simulation of VANETs [9] requires the joint exploitation of a traffic simulator and a network simulator. The former provides real simulation models of roads and vehicle mobility, whereas the latter offers a robust simulation environment for vehicle-to-vehicle (V2V) and vehicle-to-infrastructure (V2I) communications.

In this work, we aim at evaluating the CDS in a well-established simulation environment to analyze its main performance indices (file transfer time, maximum vehicle speed for file reception, and CDS efficiency) as a function of the FEC schemes, distance between vehicle and broadcast antenna, vehicle speed, and number of broadcasting antennas. For this purpose, we implemented the CDS into the NS-3 simulation framework [10] integrated with a mobility model, the intelligent driver model (IDM) [11], and the MOBIL lane change model [12].

The rest of this chapter is organized as follows. In Section 22.2, we briefly overview the basic concepts about vehicular networks (VNs). Section 22.3 introduces the FEC algorithms. In Section 22.4, we present our CDS, whereas in Section 22.5 our CDS simulation framework is described. Section 22.6 focuses on the simulation results. Finally, conclusions are drawn and some directions of future work are discussed.

22.2 VEHICULAR NETWORKS

VNs are wireless networks where vehicles play a key role as both communication endpoints and message relays in intelligent transportation systems (ITS). VN-based infotainment and multimedia applications are considered to play a very important role in the future of ITS and vehicular infotainment systems. Being able to address vehicle passenger preferences and deliver multimedia content of their interest would allow to support various value-added services in the vehicles, for example, touristic video guide, news, entertainment, and so on.

In VNs, communications can be grouped into two different categories: V2I and V2V. In V2I communications, elements known as RoadSide Units (RSUs) are deployed to offer infrastructure functionality, such as message relaying, infotainment, and Internet access, and also information about relevant and localized events (accidents, slippery road, roadworks, etc.). In V2V communications, vehicles communicate directly among themselves without requiring any infrastructure support. If vehicles also act as message relays toward other vehicles, a VANET is created, extending the possibilities of vehicular communications. Examples of service offers by VANETs include accident alerts, traffic congestion information, and warnings about different types of dangers [5,13].

22.2.1 VANET Radio Communications

In terms of radio communication bands dedicated to VNs, the Federal Communications Commission (FCC) in the United States has allocated a radio spectrum of 75 MHz in the 5.9 GHz band for dedicated short-range communication (DSRC) [14]. In Europe, the European Telecommunications Standards Institute (ETSI) has allocated a radio spectrum of 30 MHz in that same band for ITS applications.

Currently, the best candidate technology for DSRC is the IEEE 802.11p Wireless Access for Vehicular Environment (WAVE) [15] standard, which is expected to be widely adopted by the car industry in the next years. The 802.11p standard supports 10 MHz and 20 MHz bandwidths. Using a 10 MHz bandwidth, the supported data rates are 3, 4.5, 6, 9, 12, 18, 24, and 27 Mbps, depending on the modulation and coding scheme considered. Owing to the limited bandwidth availability, achieving an efficient channel usage is critical to reduce the number of collisions, especially in the presence of massive broadcast traffic [16].

22.2.2 Content Delivery in VANETs

In the literature, we can find some relevant works that focus on content delivery to moving vehicles.

Infoshare [17] addresses the problem of information caching and delivery in intervehicular networks. It is based on a pull paradigm and application-level routing, and its goal is to achieve maximum spreading of information queries among vehicles.

Cabernet [18] is a solution for improving data delivery to moving vehicles, which distinguishes congestion on the wired portion of the path from losses over the wireless link, achieving a two times throughput improvement over Transmission Control Protocol (TCP).

Huang et al. [19] propose a service-oriented information dissemination scheme to support vehicle infotainment that accounts for heterogeneous vehicular communication. They also develop a framework for delivering real-time services over an IP-based network.

FleaNet [20] is a virtual marketplace where users can carry out buy/sell transactions on a VN. In this solution, sellers will disseminate queries, and resolutions are carried out through the FleaNet protocol suite, which is scalable to thousands of nodes.

22.3 FORWARD ERROR CORRECTION TECHNIQUES

FEC is a term that encompasses those techniques where the transmitter adds error correction codes [21] to identify and possibly correct an error in the transmission without requiring retransmitting data.

The simplest FEC schemes for error handling, like XOR, are designed to ensure protection against the loss of a single packet in a group of packets, being applicable only in those cases where error rates are low [7]. The strategy consists of partitioning each source block using a fixed source symbol length, and then adding redundant symbols built as the XOR sum of all source symbols. This process is called Encoding $(k + 1, k)$, where k is the number of source symbols.

An example of a more sophisticated FEC scheme is R–S [22]. It assumes an $RS(N, K)$ code able to produce N codewords of length N symbols, each storing K symbols of data, that are then sent over a lossy channel. Any combination of K codewords received at the other end is enough to reconstruct all of the N codewords. N is usually $2K$, in which case at least half of all the codewords sent must be received in order to reconstruct the original message. RS is used in different commercial applications such as CDs and DVDs, broadcasting satellite, DSL, WiMax, and digital Terrestrial TV (DVB-T).

More recently, RAPTOR encoding was proposed, being considered the most efficient FEC technique currently available. The RAPTOR acronym stands for Rapid Tornado and represents an evolution of the first types of "Erasure Codes." The technique was invented by Amin Shokrollahi [8] and is based on the concept of Luby Transform, being considered the first method able to efficiently encode a block of n symbols in $O(n)$ time. The high efficiency of the RAPTOR is associated with the multiple levels of encoding (and decoding) used: in the first level, it performs a "pre-coding" on source blocks, thus generating intermediate symbols; for this step, the LDPC (low density parity check) is usually adopted. Afterwards, intermediate symbols are processed using LT (Luby Transform) codes to generate the final encoded symbols (repair symbols). One of the main advantages of this strategy is that, when using the LT, an unlimited number of repair symbols can be generated, thereby offering great flexibility to transmission systems.

22.4 A ROBUST BROADCAST-BASED CONTENT DELIVERY SYSTEM

We developed a robust broadcast-based CDS [4,23] of multimedia files. The CDS is based on the FLUTE protocol [6,24] and employs different FEC schemes (see Section 22.3) to increase the robustness of file transmission.

Specifically, the developed CDS is organized in two parts (see Figure 22.1): *Transmitter* and *Receiver*.

The transmitter performs the following actions:

1. *Opening of the Multimedia File.*
2. *Decomposition in Source Blocks.* In particular, the decomposition of the FLUTE works as follows: the structure of a source block is computed to have source blocks as close as possible. Usually, the first group of source blocks shares the same length, whereas shorter length is associated with the second group.
3. *Decomposition in Encoded Symbols.*

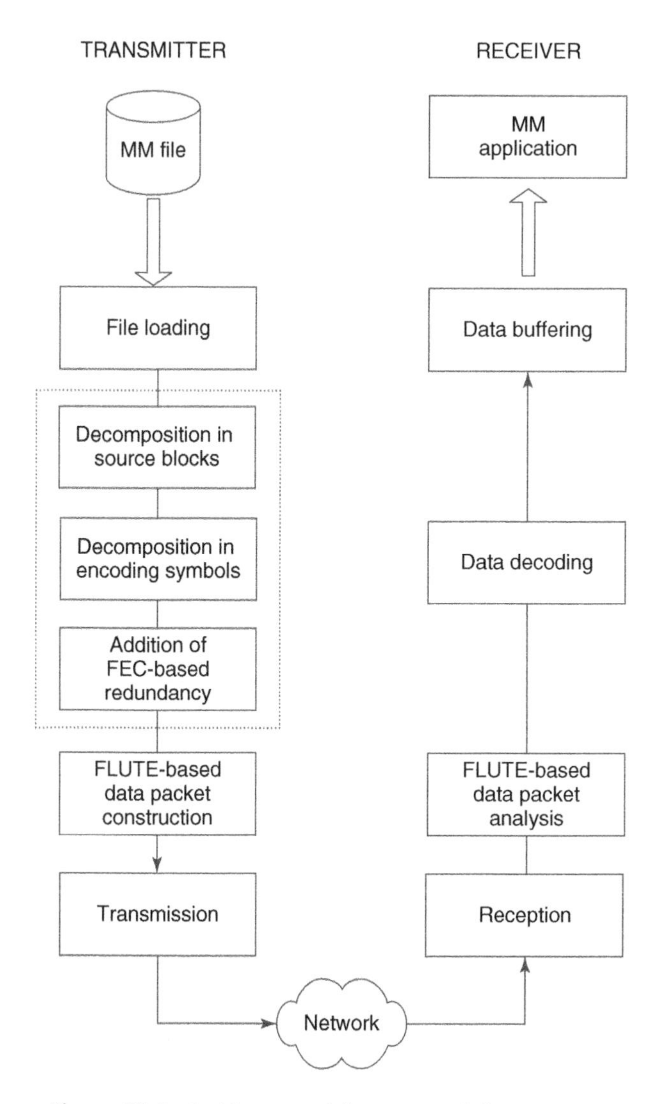

Figure 22.1 Architecture of the content delivery system.

4. *Addition (optionally) of Redundancy According to the Selected FEC Scheme*. The available FEC schemes are XOR, R–S, and RAPTOR. Thus, the *encoded symbols* are the source symbols and the redundancy symbols.
5. *Construction and Transmission of Packets*.

The receiver carries out the following symmetrical actions with respect to the transmitter:

1. *Waiting for incoming packets*
2. *Analysis of the received packet*

3. *Data decoding*
4. *Storing data into a buffer*
5. *Creation of the multimedia file once all data are received*

The CDS is implemented in C atop the Ubuntu operating system. The FEC codes are reused from the library FLUTE mad-fcl 1.7 [24] that provides No-Code, XOR, and R–S FEC schemes. The RAPTOR FEC scheme is included by integrating the coding/decoding RAPTOR library of Digital Fountain [25].

22.5 CDS SIMULATION IN NS-3

Simulation of VANETs is usually based on two simulation environments: *Traffic Simulation* and *Network Simulation*. *Network Simulation* is exploited to evaluate the behavior of network and application protocols under a variety of different conditions, whereas *Traffic Simulation* is used to study transports and traffic engineering. Such simulation environments need to be jointly used to effectively analyze even complex scenarios [17,18]. Two different simulation modes of using the two simulation environments exist: *Federated Simulation* and *Integrated Simulation*.

In the first simulation mode, Network Simulator and Traffic Simulator work in a separated way. A federated simulation is composed of three parts that need to interact, usually through events: the network simulator, the traffic simulator, and the simulation coordinator. Specifically, the Network Simulator is able to use real traffic traces and virtual traffic traces produced by traffic simulators or random vehicle movements. The main advantage of such mode is that it is possible to reuse simulators well established in the network and traffic domain, the disadvantage is the complexity in coordinating the three system simulation parts.

In the second simulation mode, Network Simulator and Traffic Simulator are merged. Such a solution, even though it is more difficult to obtain, allows for a more efficient and direct control of the evaluated scenarios. This also enables its use to users having a superficial knowledge about traffic simulation but that want to build a more realistic simulation environment.

To evaluate the proposed broadcasting technique (see Section 22.3) in VANETs, which include both network communication mechanisms and vehicle mobility, we implemented the CDS in the network simulation framework NS-3 [9,10] purposely integrated with a well-known mobility model component and the IDM [11], and combined with the MOBIL lane change model [12].

NS-3 is a discrete event network simulator written in C++, designed to be used for educational and research purposes and aims to be more efficient and easy to use (specifically with reference to wireless communications) than its predecessor NS-2. The vehicular mobility and the network communications are integrated through events. It is therefore possible to create event handlers able to send network messages or modify the vehicle mobility every time that a network message is received or the vehicle mobility is updated, respectively. To facilitate the creation of simulation scenarios, a rectilinear freeway model, managing the vehicle mobility, has been purposely defined.

22.6 PERFORMANCE EVALUATION

The carried out performance evaluation aims at measuring the file transfer time under different conditions of the transmitter and receiver by varying the used FEC scheme. In particular, the simulation environment consists of a server side (composed of one or two broadcast antennas), which cyclically transmits a file, and of a client side (a vehicle), which receives the file in mobility on a freeway chunk.

The Wi-Fi standard adopted in the simulation is 802.11a; in particular, it is the *WiFi_PHY_STANDARD_80211a* [26] available in NS-3, which uses the OFDM modulation [27] and the broadcast transmission rate of 6 Mbps. The propagation channel adopts a constant model (with speed equal to the light speed) as propagation delay model, whereas the loss propagation model follows the Nakagami distribution [28] (the packet loss occurs for distances greater than 50 m from the transmitting antenna). This configuration was chosen because it models well the VANET environment and it is a stable configuration. Concerning the roadside element, a transmission power of 24 dB is chosen to assure a transmission range of 200–250 m. Finally, according to experimental data, specific parameter value assignments related to the setup of the RAPTOR scheme and of the computation of the packet recovery function were taken. In particular, the tunable parameters of the RAPTOR scheme are set as follows: loss percentage = 20% and additional overhead of redundant packets = 3%.

The first implemented simulation scenario consists of a unidirectional freeway composed of three lanes where a broadcast antenna, which cyclically transmits a file, is added at the roadside (see Figure 22.2). The carried out simulation runs aim at computing the *file transfer time* toward a single vehicle when varying the vehicle speed (20, 40, 60, 80, 115, 130 km/h), the distance between the antenna and the vehicle (10, 50, 55, 60, 90, 100, 125, 140, 150, 175, 185, 200, 210, 230, 240, 250 m), and the file size (15 KB, 50 KB, 100 KB, 220 KB, 500 KB, 1 MB). It is worth noting that the "distance" parameter refers to the distance between the antenna and the vehicle at the moment when the file transmission starts.

In Figure 22.3, the results obtained by setting up the vehicle speed to 130 km/h and the file size to 220 KB are reported.

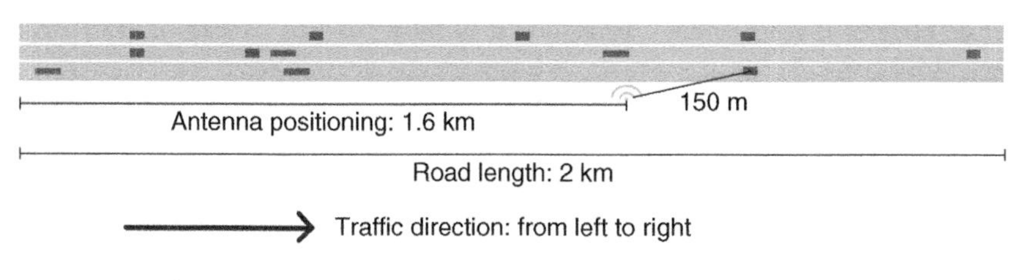

Figure 22.2 Scenario with one roadside broadcast antenna.

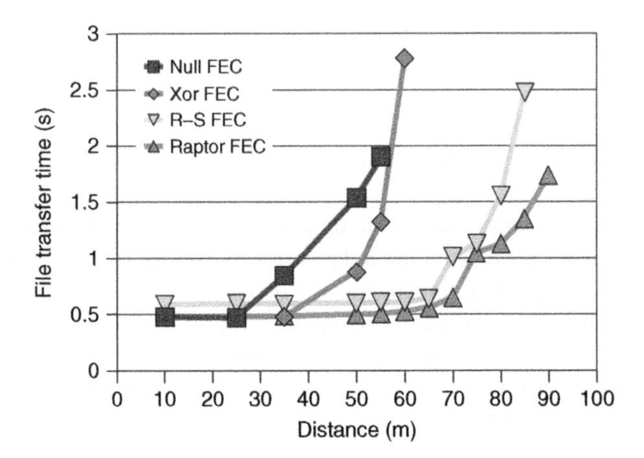

Figure 22.3 File transfer time in function of the distance and varying the FEC scheme, with vehicle speed equal to 130 km/h and file dimension equal to 220 KB.

If the *Distance* is between 10 and 25 m, the highest performance in terms of *file transfer time* is offered by the NULL FEC schema, which does not introduce any redundancy and computational complexity, as no losses occur. The XOR and Raptor FEC schemes provide almost the same performance as the NULL FEC schema. The performance of the R–S FEC schema is slightly lower due to the relatively large quantity of redundancy packet to transmit. After 25 m, the performance worsens as the packet loss rate increases. In such case, it is worth noting that, without FEC schemes, the vehicle needs to wait for a given number of retransmissions to be able to collect the packets needed to reconstruct the file. Also, the performance of the XOR FEC schema starts worsening. This is due to the XOR algorithm that is able to identify the missing packets and fully reconstruct them if the packet losses are limited (maximum one unit per each source block). Differences increase when packet loss rate increases. In such case, the R–S and Raptor FEC schemes are more stable and effective in recover errors than the XOR FEC schema. Between 35 and 55 m, in fact, the Raptor FEC schema outperforms the XOR FEC schema by about 160%. If the distance is between 55 and 90 m, the performance of all the schemes worsens. Specifically, the XOR FEC cannot be used as it does not allow the reception of the file anymore. The performance of R–S and Raptor is stable until 70 m, even though Raptor's performance is higher than R–S. It is evident that if the CDS does not use any FEC schema, it is able to cover only the 60% of the area that, conversely, it is fully covered by exploiting R–S and Raptor.

Although the performance differences between R–S and Raptor seem to be minimal, the Raptor has the capability to produce a minimum quantity of redundant packets; of course, by appositely tuning the Raptor parameters (percentage of loss and code overhead), the quality of produced streams can be increased but the redundancy will increase too.

Considering the file of 220 KB used in the test so far and defining produced packets as the sum of the media file packets and the redundant packets created by the FEC algorithm, the behavior of the used FEC schemes will be evaluated:

- media file packets: 158
- produced packets (transmitted) with no FEC schema: 158
- produced packets (transmitted) with XOR: 161
- produced packets (transmitted) with R–S: 236
- produced packets (transmitted) with Raptor: 205

The efficiency of the CDS can be obtained by computing the ratio between the *efficiency of the FEC algorithm* and the *file transfer time* [5].

$$\text{Efficiency of the FEC schema} = \frac{\text{packages that make up the file}}{\text{produced packets}}$$

$$\text{Efficiency of the CDS} = \frac{\text{Efficiency of the FEC algorithm}}{\text{File transfer time}}$$

As reported in Figure 22.4, the Raptor behavior has higher efficiency than R–S. In particular, in the first 25 m and until a high loss rate is reached, the behavior of the XOR is better than the others as the file transfer time is shorter and the channel is flooded with redundant packets. However, by increasing the distance, the loss rate increases, so that the Raptor allows for the reception of the file in a shorter time than R–S by also producing less redundant packets than the R–S.

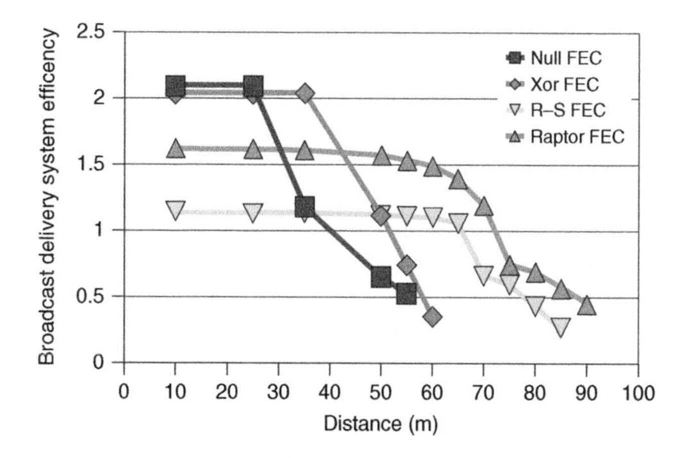

Figure 22.4 Efficiency of the content delivery system in function of the distance and varying the FEC scheme, with vehicle speed equal to 130 km/h and file dimension equal to 220 KB.

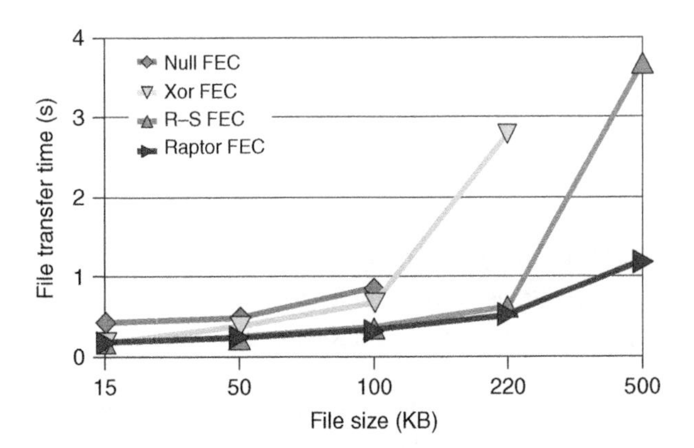

Figure 22.5 File transfer time in function of the *file size* and varying the FEC scheme, with *vehicle speed* equal to 130 km/h and *distance* equal to 60 m.

Figure 22.5 shows how the *file size affects the file transfer time* by varying the FEC schemes. This provides important information about the delivery efficiency of media file of different sizes. If the file has small sizes (10–50 KB), the use of FEC schemes is not so beneficial as the file transfer times are almost similar. If the file size is between 100 and 220 KB, performance of FEC schemes increases. Specifically, without FEC the vehicle is not able to receive the file, with XOR long delays occur, whereas with R–S and Raptor the performance is almost stable. Finally, if file size further increases, the performance of R–S worsens while the Raptor allows for highest performance. Thus, in extreme conditions, where speed is high and file size is large, the loss rate is high and the Raptor exhibits its benefits by producing less redundant packets with higher capability of file reconstruction with less received packets, so allowing the CDS to increase the file transmission frequency.

Finally, we have analyzed the maximum speed of the vehicle, with respect to the distance between the vehicle and the antenna and the FEC schemes, guaranteeing the full reception of the file (size equal to 220 KB) broadcast by the CDS. It is worth noting that R–S and Raptor perform similarly and that the maximum speed of the vehicle, when the distance is 150 m from the antenna, should be very low (30 km/h).

This is partly caused by one of the main limiting factor of VANETs, that is, the reduced antenna transmission range. Such a limit is further worsened by the dynamic nature of the network due to the vehicle mobility that, more or less quickly, exit from the coverage range of the antenna. In the previous (first) scenario, it could be noted that it is possible to deal with such problem, thanks to the FEC schemes. However, to cover wider areas (i.e., all the freeway), it is quite obvious that more antennas need to be introduced and placed on the roadside. Thus, one major problem to deal with is how to distribute such antennas and, specifically, at which distance one from another to place them. The final goal is to maximize the distance between antennas to employ less antennas and reduce the costs, allowing, at the same time, the vehicle to receive files of even large dimensions by maintaining the desired speed (Figure 22.6).

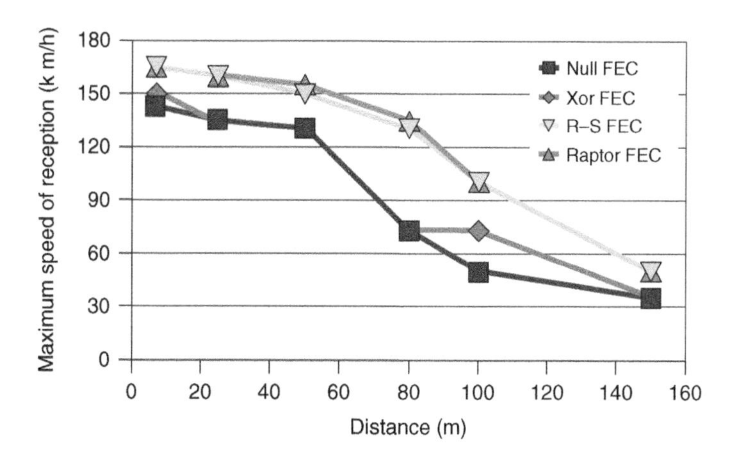

Figure 22.6 Maximum speed of file reception in function of the *distance* and varying the FEC scheme, with *file size* equal to 220 KB.

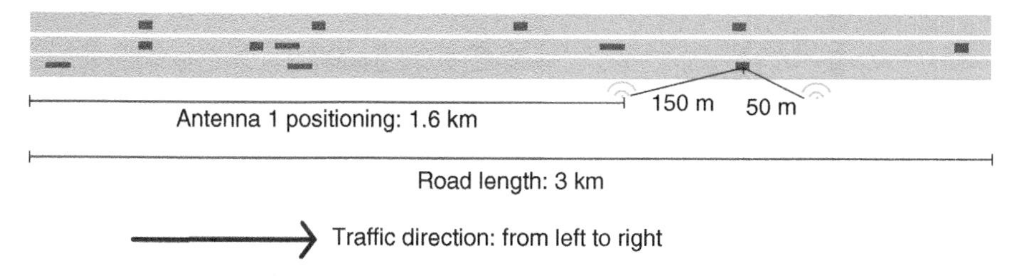

Figure 22.7 Scenario with two antennas on the roadside that are 200 m distant. The file transmission starts when the distance between the first antenna and the vehicle is 150 m, whereas the distance of the vehicle from the second antenna is 50 m.

The second simulated scenario is related to a unidirectional freeway chunk with three lanes to which two broadcast antennas are added to the roadside that both cyclically transmit the same file (see Figure 22.7). Moreover, the second antenna starts the transmission with 0.4 s of delay with respect to the first antenna. The simulation is centered on one vehicle, and the performance of the CDS by varying the vehicle speed, the distance of the vehicle from the first antenna, the file size, and the distance between the two antennas are analyzed.

In this scenario, the performance in terms of *file transfer time* (Figure 22.8) provided by the Raptor are undoubtedly better than those obtainable through the other FEC schemes. With respect to the scenario with one antenna, the exploitation of two antennas increase all the performance in terms of maximum distance of file reception by varying the FEC schemes (Figure 22.9), file transfer time versus FEC schema (Figure 22.10), and maximum distance of file reception by varying the vehicle speed and setting the FEC schema to Raptor (Figure 22.11).

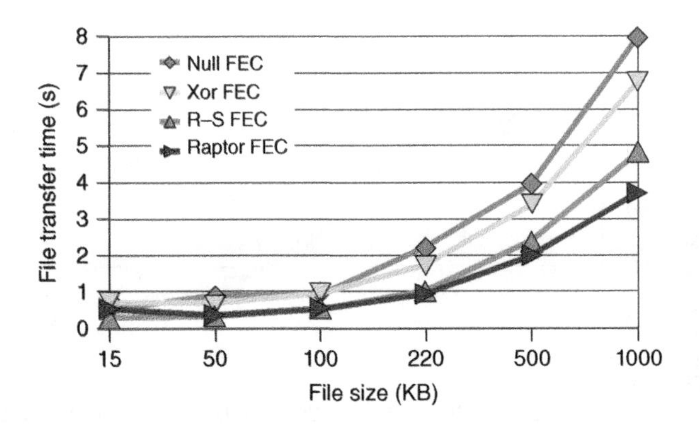

Figure 22.8 File transfer time in function of the *file size* and varying the FEC scheme, with vehicle speed equal to 130 km/h, *distance* between first antenna and the vehicle equal to 150 m, and distance between the two antennas equal to 200 m.

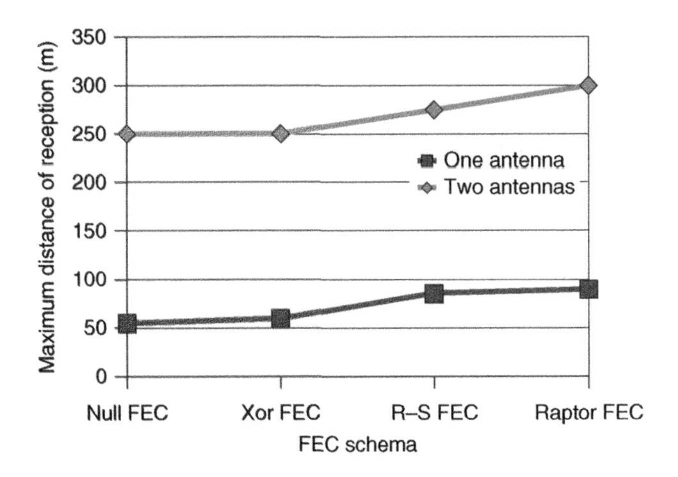

Figure 22.9 Maximum distance of file reception (considering the distance between first antenna and the vehicle) varying the *FEC schema*, with vehicle speed equal to 130 km/h, file size equal to 220 KB, and *distance* between antennas equal to 200 m.

Of course, if we would like to receive files of large sizes and maintain a high speed, the placement of many antennas on the roadside and the use of Raptor are mandatory. In fact, as can be noted in Figure 22.12, displacing the antennas 200 m far away and using the Raptor, although the vehicle moves at 130 km/h, the file transfer time is quite stable even for files of 1 MB.

Finally, we also focused on the dimensioning of the length between the two antennas. As reported in Figure 22.13, if the distance between the two antennas is increased,

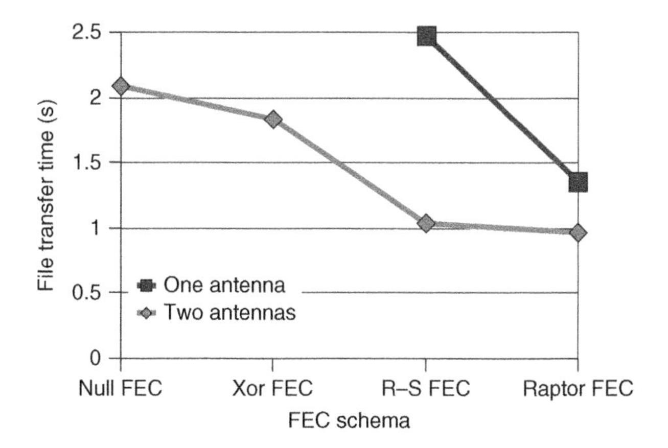

Figure 22.10 File transfer time varying the FEC schema, with distance between first antenna and vehicle equal to 85 m, with vehicle speed equal to 130 km/h, and file size equal to 220 KB.

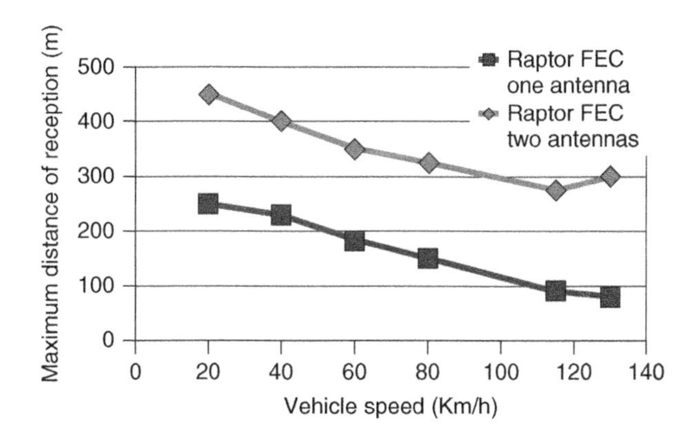

Figure 22.11 Maximum distance of file reception (considering the distance between first antenna and the vehicle) in function of the vehicle speed, with RAPTOR FEC schema and file size equal to 220 KB.

the coverage increases (and the costs decrease as less antennas are needed) but the performance in terms of file transfer time decreases (see Figures 22.14 and 22.15).

With long distances between the two antennas, even the Raptor (which is better than the other FEC schemes) is not able to provide acceptable performance in terms of file transfer time. In such case, if applications require high performance, such as safety and emergency application domains, where the responsiveness of the communication is a fundamental requirement, the only feasible solution is to reduce the distance among the antennas.

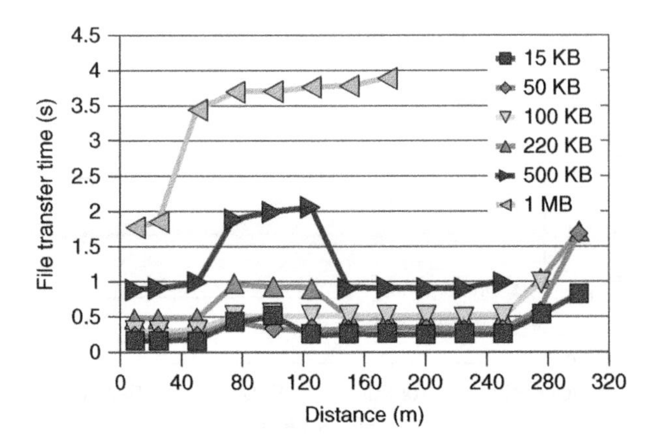

Figure 22.12 File transfer time in function of the distance from the first antenna, with vehicle speed equal to 130 km/h, FEC schema equal to Raptor, and distance between antennas set to 200 m.

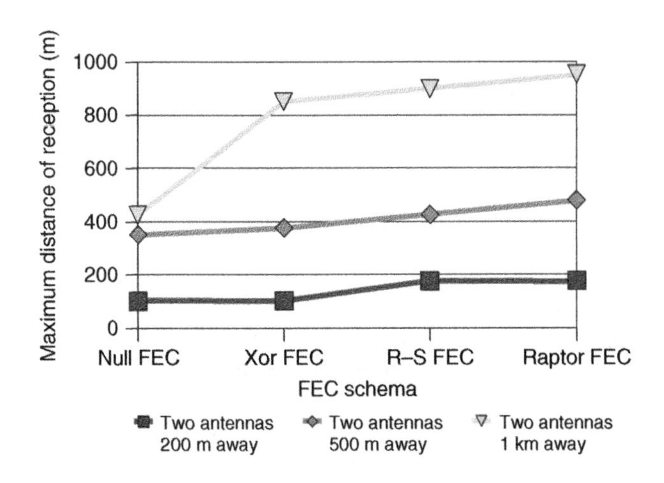

Figure 22.13 Maximum distance of file reception (considering the distance between the first antenna and the vehicle) in function of the FEC schema, with vehicle speed equal to 130 km/h and file size set to 1 MB.

22.7 FUTURE RESEARCH TRENDS

Industry is currently very active in the area of ITS and Smart Cities, and VNs are a key element of these areas. All big IT and communication companies are participating in the development of these sectors.

Presently, in ITS, the focus is mostly oriented toward improving global transportation systems by increasing mobility and improving safety for millions of people in an

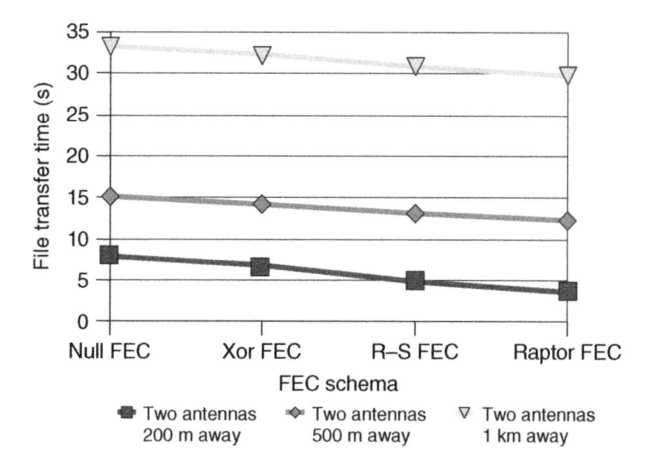

Figure 22.14 File transfer time in function of the FEC schema, with distance between the first antenna and the vehicle of 100 m, vehicle speed of 130 km/h, and file size of 1 MB.

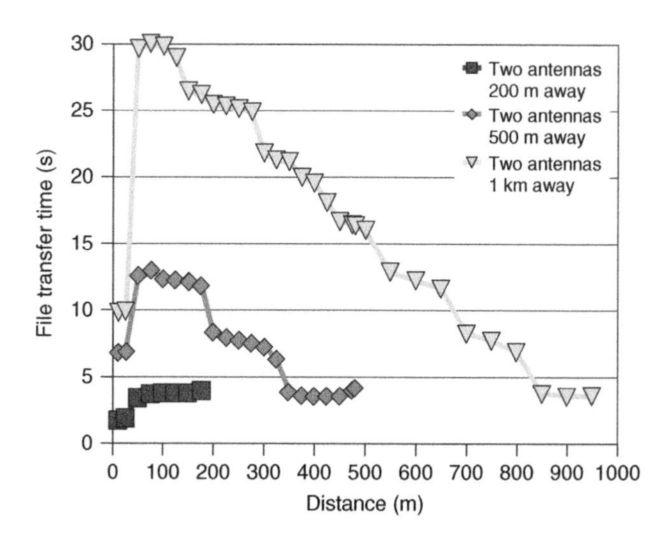

Figure 22.15 File transfer time in function of the distance between the first antenna vehicle and the vehicle with vehicle speed equal to 130 km/h and FEC schema set to Raptor.

environmentally conscious manner. Smart parking, traffic signals that adapt to real-time conditions, and roads that sense the surrounding environment are becoming more commonplace. But, in the close future, the application of advanced content delivery technologies will help cities to better understand, predict, and intelligently respond to patterns of behavior and events.

Leveraging the power of real-world data generated by vehicles and cities' systems of services will require the following: the collection and management of the right kind of data, the integration and analysis of the data, and the optimization of systems to achieve desired systems behavior based on insights gained through advanced data analysis.

So, even though, to the best of our knowledge there is no currently any commercial CDN product, in the next 5 years we will surely assist to the spring of solutions of this type, where the presence of a robust and reliable technology for content broadcasting would be fundamental. In this sense, our work provides a general contribution to CDN solutions since the analysis made and the results obtained are generic enough to allow their adoption in different CDN frameworks. In particular, the FEC analysis emphasizes on the need to protect against losses to optimize content delivery speed and effectiveness. In addition, the antenna/coverage analysis shows that content delivery in high speed environments may require more than one antenna to guarantee successful content delivery, especially when content size is significant.

22.8 SUMMARY AND CONCLUSION

The potential of wireless content delivery infrastructure (e.g., for advertising, entertainment, traffic, and emergency) is significant and can be used both for vehicles and for moving people.

In this chapter, we proposed a system for broadcasting robust multimedia content focusing on highway environments with moving vehicles. The system integrates the algorithm FLUTE and various FEC schemes (e.g., XOR, RS, and RAPTOR) in such a way as to make the distribution of content more robust, reliable, and fast.

The performance evaluation of the system was performed in two different scenarios: the first involved a three-lane unidirectional highway with a single RSU that periodically broadcasted a file; in the second scenario two RSUs were used.

The experiments showed the following two relevant results:

- The use of FEC schemes is critical within a VANET with infrastructure. In particular, RAPTOR coding is quite efficient and rapid in contexts in which the distance from the vehicle to the antenna, the vehicle speed, and the size of the file that you want to receive are high. Moreover, it is quite evident that the use of FEC code, even the simplest, such as XOR FEC, should be adopted as they allow improving the performance of about 66% compared to no use of FEC schemes.
- If the objective is the reception of large files while traveling at the fastest speed possible and in a short time (the latter is a fundamental requirement in application fields such as safety or emergency road), the antennas must be installed not far away from each other.

In future works, we will extend our implementation to more complex environments such as, for example, urban scenarios, where transmissions are altered by the presence of buildings and other obstacles and by other wireless networks that produce congestions and packet collisions. In this sense, there is no specific protection by any protocol (with the possible exception of RTCP [29]); some studies [30] have shown that this is a significant problem within a VANET especially when safety information is to be transmitted regarding the state of the roads and traffic adjacent. These issues may be theoretically solved with the use of RAPTOR codes that restrict this phenomenon.

ACKNOWLEDGMENTS

This work was partially supported by the "*Ministerio de Ciencia e Innovación*," Spain, under Grant TIN2011-27543-C03-01.

REFERENCES

1. Galaviz-Mosqueda GA, Aquino-Santos R, Villarreal-Reyes S, Rivera-Rodriguez R, Villaseñor Gonzalez L, Edwards A. Reliable freestanding position-based routing in highway scenarios. Sensors 2012;12:14262–14291.
2. Gramaglia M, Bernardos CJ, Calderon M. Virtual induction loops based on cooperative vehicular communications. Sensors 2013;13:1467–1476.
3. Salvo P, De Felice M, Cuomo F, Baiocchi A. Infotainment traffic flow dissemination in an urban VANET. Proceedings of Globecom 2012 - Ad Hoc and Sensor Networking Symposium; 2012.
4. Calafate CT, Fortino G, Fritsch S, Monteiro J, Cano J-C, Manzoni P. An efficient and robust content delivery solution for IEEE 802.11p vehicular environments. J Netw Comput Appl 2012;35(2):753–762 Elsevier. DOI: 10.1016/j.jnca.2011.11.008.
5. Fortino G, Calafate C, Manzoni P. A robust broadcast-based multimedia content delivery system for urban environments. Next Generation Content Delivery Infrastructures: Emerging Paradigms and Technologies, IGI Global; June 2012.
6. Paila T, Luby M, Lehtonen R, Roca V, Walsh R FLUTE - File deliver over unidirectional transport, RFC 3926; October 2004.
7. Peltotalo S, Peltotalo J, Roca V. 2004. Simple XOR, Reed-Solomon, and parity check matrix-based FEC schemes. IETF RMT Working Group. Available at draft-peltotalo-rmt-bb-fec-supp-xor-pcm-rs-00.txt. Accessed 5 April 2013.
8. Shokrollahi A. Raptor Codes. IEEE Trans Inform Theory 2006;52:2551–2567.
9. Arbabi H, Weigle MC. Highway mobility and vehicular ad-hoc networks in ns-3. Proceedings of the Winter Simulation Conference; Baltimore, MD; December 2010.
10. 2006. NS-3 network simulator. Available at http://www.nsnam.org. Accessed 5 April 2013.
11. Treiber M. 2006. Intelligent driver model (IDM). Available at trafficsimulation.de/IDM.html. Accessed 5 April 2013.
12. Treiber M. 2006. Minimize overall braking decelerations induced by lane changes (MOBIL). Available at traffic-simulation.de/MOBIL.html. Accessed 5 April 2013.
13. Fiore M, Härri J. The networking shape of vehicular mobility. Proceeding of the 9th ACM International Symposium on Mobile Ad Hoc Networking and Computing; 2008. p 261–272.
14. ASTM. ASTM E2213-03. Standard specification for telecommunications and information exchange between roadside and vehicle systems —5 GHz band dedicated short range communications (DSRC) medium access control (MAC) and physical layer (PHY) specifications; 2003.

15. Eichler S. Performance evaluation of the IEEE 802.11p WAVE communication standard. Proceedings of the Vehicular Technology Conference (VTC-2007 Fall); Baltimore, MD, USA; 2007.
16. Jiang, D, Chen, Q, Delgrossi, L. Optimal data rate selection for vehicle safety communications. Proceedings of the Fifth ACM International Workshop on VehiculAr Inter-NETworking VANET'08; New York, NY, USA; ACM; 2008. p 30–38.
17. Fiore M, Casetti C, Chiasserini CF, Garetto M. Analysis and simulation of a content delivery application for vehicular wireless networks. Perform Eval 2007;64(5):444–63.
18. Eriksson J, Balakrishnan H, Madden S. Cabernet: vehicular content delivery using WiFi. 14th ACM International Conference on Mobile Computing and Networking; 2008.
19. Huang C-J, Chen Y-J, Chen I-F, Wu T-H. An intelligent infotainment dissemination scheme for heterogeneous vehicular networks. Expert Syst Appl 2009;36(10):12472–12479. DOI: 10.1016/j.eswa.2009.04.035.
20. Lee U, Lee J, Park J-S, Gerla M. FleaNet: a virtual market place on vehicular networks. IEEE Trans Veh Technol 2010;59(1):344–55.
21. Luby M, Vicisano L. Compact forward error correction (FEC) schemes. RFC 3695; February 2004.
22. Lacan J, Roca V, Peltotalo J, Peltotalo S. 2009. Reed Solomon error correction scheme. IETF RMT Working Group, RFC 5510 ("Standards Track/Proposed Standard").
23. Fiore M. Vehicular mobility models. In: Olariu S, Weigle MC, editors. *Vehicular Networks: From Theory to Practice*. Boca Raton: CRC Press/Taylor & Francis; 2009.
24. Peltotalo S. 2013. FLUTE MAD project. Tampere University of Technology, Institute of Communications Engineering. Available at http://mad.cs.tut.fi. Accessed 5 April 2013.
25. DF Raptor R11 Encoder/Decoder 2.2.1 Software Development Kit, Digital Fountain; 2010 Feb 2.
26. Lacage M, Henderson TR. Yet another network simulator. 57th Proceedings from the 2006 Workshop on ns-2: The IP Network Simulator; 2006 Oct 10; Pisa, Italy; 2006.
27. Shelswell P. The COFDM modulation system, the heart of digital audio broadcasting, BBC Research and Development Report, BBC RD 1996/8; 1996.
28. Laurenson D, Laurenson DI. Indoor radio channel propagation modelling by ray tracing techniques [PhD thesis]. University of Edinburgh; 1994.
29. Schulzrinne H, Casner S, Frederick R, Jacobsoni V. RTP: A transport protocol for real-time applications; July 2003.
30. Zang Y, Stibor L, Cheng X, Reumerman H-J, Paruzel A, Barroso A. Congestion control in wireless networks for vehicular safety applications. Proceedings of the 8th European Wireless Conference; Paris; April 2007.

23

ON THE IMPACT OF ONLINE SOCIAL NETWORKS IN CONTENT DELIVERY

Irene Kilanioti[1], Chryssis Georgiou[1], and George Pallis[1]

[1]*Department of Computer Science, University of Cyprus, Nicosia, Cyprus*

23.1 INTRODUCTION

As the task of *content delivery networks* (*CDNs*) is the improvement of Internet service quality via replication of the content from the origin to surrogate servers scattered over the Internet, the area of CDNs faces three major issues concerning the maximization of their overall efficiency [1,2]: (i) the best efficient *placement* of surrogate servers with maximum performance and minimum infrastructure cost, (ii) the best *content diffusion* placement either in a global or in a local scale, that is, which content will be copied in the surrogate servers and to which extend, since this requires memory, time, and computational cost, and (iii) the *temporal diffusion* related with the most efficient timing of the content placement.

The increasing popularity of *online social networks* (*OSNs*) [3–5] and the growing popularity of streaming media have been noted as being the primary causes behind the recent increases in HyperText Transfer Protocol (HTTP) traffic observed in measurement studies [6]. The amount of Internet traffic generated everyday by online multimedia streaming providers such as YouTube has reached huge numbers. Although it is difficult to estimate the proportion of traffic generated by OSNs, it is observed that there are

Advanced Content Delivery, Streaming, and Cloud Services, First Edition.
Edited by Mukaddim Pathan, Ramesh K. Sitaraman, and Dom Robinson.

more than 400 tweets per minute with a YouTube link [7]. These providers often rely on CDNs to distribute their content from storage servers to multiple locations over the planet. Toward this direction, we can exploit information diffusion analyzing the user activity extracted from OSNs. Thus, the improvement of user experience through scaling bandwidth-demanding content largely depends on the exploitation of usage patterns found in OSNs, and can be conducted either through the selective prefetching of content, also taking into account timing issues, or through the strategic placement of surrogate servers. Furthermore, the cost of scaling such content in CDNs can be expressed in different ways. For example, it might be the number of replicas needed for a specific source or it may take into account the optimal use of memory and processing time of a social-aware built system. Thus, it is crucial to support social network analysis tasks that accommodate large volumes of data requirements for the improvement of user experience (e.g., through prefetching via a CDN infrastructure).

The goal of this chapter is to present existing approaches that can be leveraged for the scaling of rich media content in CDNs using information from OSNs. Specifically, we present a taxonomy of the relative research (outlined in Figure 23.3 in the next section), taking into account phenomena related with rich media content and its outspread via OSNs, and measurement studies on OSNs that could provide valuable insight into CDN infrastructure decisions for the replication of the content, as well as systems built with the leverage of OSNs' data.

The remainder of this chapter is organized as follows: In Section 23.2, the main concepts of OSNs and social cascades are presented. In Section 23.3, the properties and approaches that characterize the social cascades and affect the CDN performance are described. The performance measurements of associating rich media content diffusion with social networks are given in Section 23.4. Section 23.5 gives the outline of existing works that exploit information extracted from OSNs for the improvement of content delivery. In Section 23.6, some future directions are given, and key areas of interest concerning the diffusion of rich media content over OSNs are explored with some commercial/practical implications for CDNs. Section 23.7 concludes our study.

23.2 ONLINE SOCIAL NETWORKS BACKGROUND

Formally, an OSN is depicted by a directed graph $G = (V, E)$, where V is the set of the vertices of the graph representing the nodes of the network and E are the edges between them, denoting the various relationships among the nodes of the graph [5]. The semantics of these edges are different for different social networks: for Facebook, friendship is usually translated in personal acquaintance, whereas in LinkedIn means business contact. As far as the directionality of the edges of the social graph is concerned, it depends on the OSN the graph depicts. For Facebook, an edge means mutual friendship between the endpoints of a link. For Twitter, if the edge between B and A points at A, B is a follower of A, meanings that A's posts (tweets) appear in B's main Twitter page. The *neighbors* of a node are defined as the nodes that are in a 1-hop away distance from it in the social graph. Figure 23.1 depicts an example of a Twitter social graph. Contrary

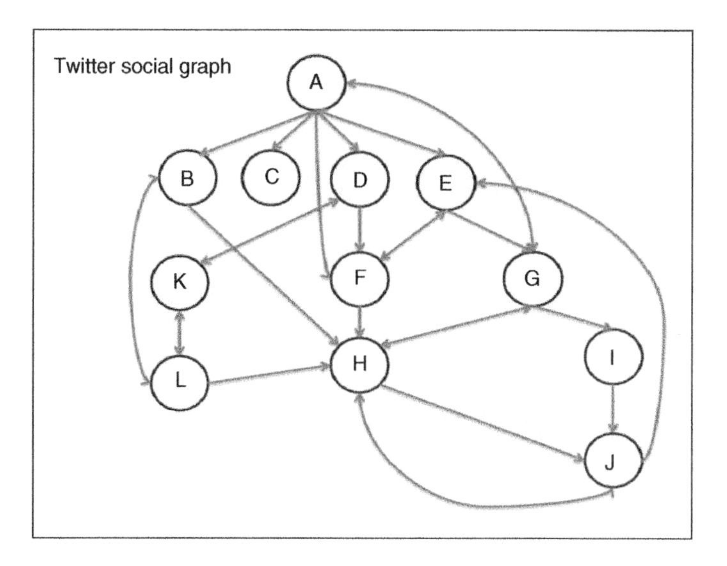

Figure 23.1 An example of a Twitter social graph.

to other OSNs, a Twitter user may follow another user to receive his/her tweets, forming a social network of interest. Furthermore, it is not necessarily the case that two users are mutual followers. Thus, Twitter is represented by a directed graph, where nodes represent the users and a direct link is placed from a user to another user, if the first follows the tweets of the latter. Users A and G are mutual followers, while users A and B are not (A follows B but not vice versa).

A fusion of bandwidth and storage demanding media, which may include text, graphics, audio, video, and animation, is characterized as *Rich Media*. Rich media is currently ubiquitous due to the proliferation of smartphones, video editing software, and cheap broadband connections. The *diffusion* of information in a network is essentially interweaved with whether a piece of information will become eventually popular or its spread will die out quickly. A large proportion of rich media is distributed via OSNs' links (for example, YouTube videos links through retweets in Twitter) that contribute significantly to Internet traffic. Facebook and Twitter users increasingly repost links they have received from others. Thus, they contribute to *social cascades* phenomena [8], a specific case of *information diffusion* that occurs in a social network, when a piece of information is extensively retransmitted after its initial publication from an originator user. Therefore, it would be beneficial to know when such cascades will happen in order to proactively replicate popular items (*prefetching*) via CDN infrastructures. *Content diffusion* placement and *temporal diffusion* could significantly benefit from such "prefetching" policies.

Social cascades can be represented as rooted directed trees where the initiator of the cascade is the root of the tree [8]. The *length* of the cascade is the height of the resulting tree. Each vertex in the cascade tree can have the information of the user, and the identity of the item replicated in the cascade. Figure 23.2 depicts an example of a

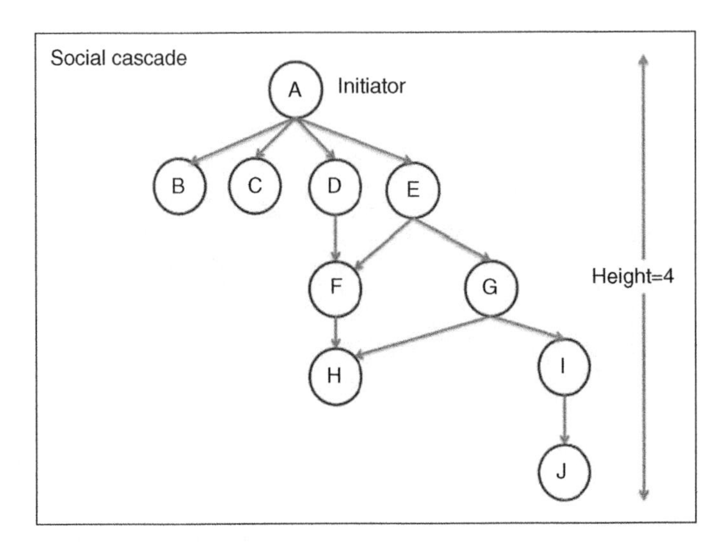

Figure 23.2 An example of a social cascade.

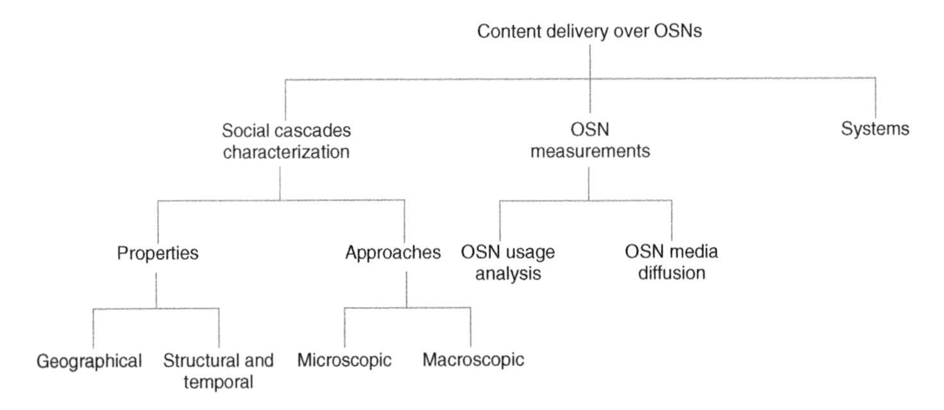

Figure 23.3 A taxonomy of content delivery over OSNs.

cascade, initiated by user A over the social graph of Figure 23.1. Then, the length of the cascade is the height of the resulting tree.

The presented taxonomy of content delivery over OSNs is outlined in Figure 23.3. Our taxonomy presents the various properties and approaches in the literature for the *characterization of cascades*. The branching of topics in our presented taxonomy continues with *OSN measurement* works that focus on phenomena and measurement studies, providing valuable insights into *usage analysis* and *media diffusion*. The last dimension of our taxonomy consists of content delivery *systems* built based on OSNs' data.

23.3 CHARACTERIZATION OF SOCIAL CASCADES

As mentioned earlier, it would be beneficial to characterize the social cascades in order to proactively replicate popular items via CDN infrastructures. The following sections present the key properties of cascades (geographical, structural, and temporal) and the existing approaches (microscopic and macroscopic) for the characterization of the extent a cascade will receive.

23.3.1 Geographical Properties

According to a recent study [1], the geography of the requests influences the performance of CDNs. Therefore, it would be useful to understand whether an item becomes popular on a global scale or just in a local geographic area.

Local cascades affect only a relatively small number of individuals and typically terminate within one or two steps of the initiator. The size of local cascades is determined mostly by the size of an initiator's immediate circle of influence, not by the size of the network as a whole. In global cascades, the opposite happens: they affect many individuals, propagate for many steps, and are ultimately constrained only by the size of the population through which they pass.

A cascade is local if it spreads in a fraction φ of the network lower than a threshold ω or else, we say that the cascade is global. Let the classification function f_1 for a cascade C_z, $z \in \mathbf{N}$, be as follows:

$$f_1(G, C_z) = \begin{Bmatrix} 0 & \text{if } \varphi < \omega \text{ local cascade} \\ 1 & \text{if } \varphi \geq \omega \text{ global cascade} \end{Bmatrix}$$

Formally, a cascade should be characterized as global or local with the maximum accuracy $\alpha = (\sigma/v) \times 100$, where σ is the number of the correctly classified cases and v is the number of all sampled cases, such that the cost of replicating a simple object c is minimized: min $\sum_{C_z} c = f_2(t_p, \varphi, N_{cl}, C_z)$ for all cascades $C_z \in C$, where φ is the fraction of the network that the object is bound to spread, N_{cl}, $cl \in \mathbf{N}$ is the number of clients requesting a specific object, and t_p is the amount of traversing path p between client and the server finally serving the request.

Other geographical properties that have been presented in the literature in order to characterize social cascades are the *geodiversity* and *georange*. Specifically, the geodiversity denotes the geometric mean of the geographic distances between all the pairs of users in the cascade tree, whereas the georange denotes the geometric mean of the geographic distances between the users and the root user of the cascade tree [1].

23.3.2 Structural and Temporal Properties

Social cascades are characterized by the structural properties of size and length. The *size* is the number of participants, including the initiator, and the *length* [8] denotes the height of the cascade tree. Social cascades are also characterized by temporal properties such as the *time delay* between two consecutive steps of the cascade [1], the *time duration*, and the *rate of the cascade* [9]. The latter, for the epidemiological model of [9], is the basic reproductive number $R_0 = \rho_0 \overline{k}^2 / \overline{k}^2$, where $\rho_0 = \beta\gamma\overline{k}$, in which β is the transmission rate, γ is the infection duration, and k is the node degree. With σ_0 the probability that a person will adopt the shared piece of information (under the assumption that duration infection is equal to the timelife of the user, much larger than duration of the cascade, and, thus, the information will be definitely shared among connections), it applies $\rho_0 = \sigma_0\overline{k}$, and σ_0 can empirically be estimated by identifying an infected node and counting the fraction of its connected nodes subsequently becoming infected. Another temporal property related to the *susceptibility* of the network to new items is the time to the first step of the cascade from the infector's point of view and the duration of exposure to an item before infection from the infectee's view [9].

23.3.3 Approaches

Even though a percentage of the occurred information flow in cascades is ascribed to homophily, namely, the tendency of individuals to associate with similar others, as "similarity breeds connection" [10], and research has been conducted for the discrimination between the two cases (homophily or influence) [11], different approaches are presented for the characterization of the extent a cascade will receive. Some of them are related to the extent that nodes are influenced by their neighbors on a *microscopic* level, such as the "vulnerability" that Watts [12] introduces, some to factors that function as obstacles to the spread of a cascade on a *macroscopic* level, such as those that Kleinberg and Easley [5] or Steeg et al. [13] introduce, and others follow an hybrid approach [14]. The discrimination is based on the view of the OSN as a whole or on its study based on user-level properties.

23.3.3.1 Microscopic Approaches In Reference 12, Watts defines a global cascade as a "sufficiently large cascade," covering practically more than a fixed fraction of a large network. Watts introduces a simple model for cascades on random graphs.

- The network comprises of n nodes with threshold distribution $f(\varphi)$, and the degree distribution of the graph is p_k, namely, each node is connected to k neighbors with probability p_k; z is the average node degree ($\overline{k} = z$).
- The initial state of each node is state 0 (inactive) and each node is characterized by a threshold φ. If at least a threshold fraction φ of the node's k neighbors acquire state 1 (active), the node will switch from inactive to active.
- Nodes with $k \leq \lfloor 1/\varphi \rfloor$ are said to be "vulnerable" and will switch state if just one of their neighbors becomes active. Otherwise, nodes are called "stable."

Watts uses *percolation theory* [15], the theory studying how connected clusters behave in a random graph to investigate the conditions under which a small initial set of seed nodes can cause a finite fraction of infinite nodes to switch from inactive to active.

Percolation in this case is interpreted as follows: a global cascade is said to occur when the vulnerable vertices percolate, namely, the largest connected vulnerable cluster of the graph must occupy a finite fraction of the infinite network. For infinite Poisson random graphs, Watts defines a region, inside which a finite fraction of an infinite network would switch from inactive to active state if at least one arbitrarily selected node switched from inactive to active state. Simulations on finite graphs of 10,000 nodes give similar results.

Watts makes the observation that the frequency of global cascades is related to the size of the vulnerable component, with the larger the component, the higher the chance for the cascade to be global. He also states that the average size of a global cascade is governed by the connectivity of the network as a whole. In sparsely connected networks, cascades are limited by the global connectivity of the network, and in dense networks cascades are limited by the stability of individual nodes.

23.3.3.2 Macroscopic Approaches In Reference 5, Kleinberg and Easley claim that clusters are obstacles to cascades, and, moreover, that they are the *only* obstacles to cascades: "Considering a set of initial adopters of behavior A, with a threshold of q for nodes in the remaining network to adopt behavior A: (*i*) If the remaining network contains a cluster of density greater than $1 - q$, then the set of initial adopters will not cause a complete cascade. (A cluster of density p is a set of nodes, so that each node in the set has at least a p fraction of its network neighbors in the set.) (*ii*) Moreover, whenever a set of initial adopters does not cause a complete cascade with threshold q, the remaining network must contain a cluster of density greater than $1 - q$."

In Reference 13, Steeg et al. find two additional factors related with the multiple exposure of users to stories due to the highly clustered nature of Digg, which drastically limit the cascade size in Digg. The reproductive number R_0 of the epidemical model used, which intuitively expresses the average number of people infected by a single infected person, is the product of the average number of *fans* times the *transmissibility*. As far as the first factor is concerned, it is implied that only the number of new fans (those that have not already been exposed to a story) should be taken into account. For the second factor, transmissibility for actual cascades is observed to remain constant until about a number of people have voted, and then begin to decline (may be due to decay of novelty [16] or decrease in visibility [17] as a consequence of new stories being submitted to Digg). From this point of view, cascades are limited.

In Reference 14, Dave et al. combine microscopic and macroscopic level approaches to identify how empirical factors like users' and their neighborhood's influencing ability or a specific action's influencing capability and other user and network characteristics affect the *reach* quantity, and come to the conclusion that action dominates in the prediction of the spread of the action. Specifically, they quantify the *reach* $^{a}(u)$ of a user u as the number of cascades it can reach with a specific action α as

$$reach^{\alpha}(u) = \sum_{u_i \in \overrightarrow{P^{a}(u)}} 1 + \frac{1}{2} \times \sum_{u_j \in \overrightarrow{P^{a}(u_i)}} (reach^{\alpha}(u_j))$$

$$0 \quad \text{otherwise}$$

where a user gets the complete credit for action propagation to his immediate neighbors, or a decaying factor for nonimmediate neighbors. $\overrightarrow{P}^{\alpha}(u)$ is the propagation set of user u, consisting of all his immediate neighbors u_i, such that there was an action propagation from u to u_i.

23.4 ONLINE SOCIAL NETWORK MEASUREMENTS

OSNs can provide information including the location of users, the items shared by users and structural and temporal properties of a social cascade. Finding ways of harnessing the potential of information constantly generated by users of OSNs is a key and promising research area for the networking community [1]. In this section, we investigate whether information extracted from social cascades can effectively be exploited to improve the performance of CDNs. Recently, thanks to the availability of large datasets, many studies have been presented. In the following sections, we present some indicative large-scale analysis and media diffusion measurements that have been conducted in the context of OSNs and their findings have implications in CDN's performance. Section 23.4.1 focuses on OSN usage analysis, whereas Section 23.4.2 focuses on OSN media diffusion.

23.4.1 OSN Usage Analysis

A first large-scale analysis of multiple OSNs data encompassing Flickr, YouTube, LiveJournal, and Orkut, social networks for sharing photos, videos, blogs, and profiles, respectively, by Mislove et al. [18] highlighted the difficulties of crawling a social network and came to the following conclusions: Although node degrees in the studied OSNs varied by orders of magnitude, key findings are the same. The studied OSNs are power-law, small-world, scale-free, the in-degree matches out-degree distribution (due to link symmetry, an observation at odds with the Web graph, that increases OSNs' network connectivity and reduces their diameter), there is a densely connected core of high degree nodes surrounded by small clusters of low degree nodes, the average distances are lower, and clustering coefficients higher than those of the Web graph (studied OSNs clustered 10,000 more times than random graphs, 5–50 times more than random power-law graphs).

Wilson et al. [19] conducted the first large-scale analysis of Facebook, by crawling and use of "networks" (15% of total 10 M users, and 24 M interactions). In Reference 20, Kumar et al. study Flickr and Yahoo!360, finding that they follow power-law degree distributions. Low diameter and high clustering coefficient, as well as power laws for in- and out-degree distributions were confirmed for the Twitter social graph by Java et al. in Reference 21.

In terms of user workloads in OSNs, Benevenuto et al. [22] collected traces from a social network aggregator website in Brazil, enabling connection to multiple social networks with a single authentication, and, thus, studied Orkut, MySpace, Hi5, and Linked. Benevenuto et al. presented a clickstream model to characterize users' interactions, frequency, and duration of connection, as well as frequency of users' transition to activities, such as browsing friends' profiles, and sending messages, with their analysis showing that browsing, which cannot be identified from visible traces, is the most dominant behavior (92%). They also reinforced the social cascade effect, since more than 80% of rich media content such as videos and photos was found through a 1-hop friend.

23.4.2 OSN Media Diffusion

Zhou et al. in Reference 23 explore the popularity of photos in Facebook, noting that the request pattern follows a Zipf distribution, with an exponent $\alpha = 0.44$, significantly lower than that of traditional distributions (ranging from 0.64 to 0.83 [24]). They interprete this as shift of interest from popular items to items in a long tail. In the same context, Yu et al. [25] analyze PowerInfo, a video-on-demand system deployed by China Telecom and note that the top 10% of the videos account for approximately 60% of accesses, and the rest of the videos (the 90% in the tail) account for 40%. Unaccessible via the official distribution channels (television networks or record companies) independent video content generated by the users, denoted as user-generated content (UGC), becomes available to a wide number of viewers via services as YouTube or the US-based Vimeo. Cha et al. [26] investigate the long tail opportunities in the UGC services, such as YouTube video content, taking into account the fluctuation of the viewing patterns due to the volatile nature of the videos (videos may appear and disappear) and the various sources that direct to the content (recommendation services, RSS feeds, Web reviews, blogosphere, etc.)

In Reference 27, Yang and Leskovec examine the temporal variations of Twitter hashtags and quotations in blogs, creating time series of the number of mentions of an item i at time t, thus measuring the popularity given to the item i over time. By grouping together items so that item i is in the same group have a similar shape of the time series x_i with a clustering algorithm, they infer items with similar temporal pattern of popularity and find that temporal variation of popularity of content in online social media can be accurately described by a small set of time series shapes, with most press agency news depicting a very rapid rise and a slow fading.

In a subsequent work [6], Christodoulou et al. confirm the higher impact of the social cascading effect on a more focused set of geographic regions, and, furthermore, study the social cascading effect of YouTube videos over Twitter users in terms of its impact on YouTube video popularity, dependence on users with a large number of followers, the effect of multiple sharing follows, and the distribution of cascade duration. They come to the conclusion that the video retweet likelihood is increased as the number of user's follows who have already shared the same tweet increases, with the increase seeming to be exponential when the same tweet is shared by more than eight follows. This observation is consistent with Reference 3, where it is claimed that the vast majority of YouTube videos do not spread at all, since large cascades are rare, and, finally, that links to videos can quickly spread over social networks, leading to many views in a short period of time. However, it should be noticed that Christodoulou et al. do not take factors such as the recency of the studied videos or their popularity in general into account.

From the above-mentioned measurements, it occurs that OSN content is different from more traditional Web content and affects significantly the navigation behavior of users. Specifically, the above studies have shown that social cascading impacts the diffusion of information. In this context, CDNs can take advantage of the fact that social cascades have high impact on a more focused and less diverse set of geographic regions. Also, the findings regarding the temporal evolution of social cascades are a critical issue that affects the CDN performance.

23.5 SYSTEMS

In this section, we present systems that could provide valuable insights concerning the exploitation of information extracted from OSNs for scaling of content diffused via OSNs. The cost for scaling of content tailored for a small number of users can be expressed in terms of required bandwidth for quick access to the content or required storage capacity for caching of content. Specifically, the long tails observed in Internet can impact system efficiency. For example, Facebook engineers developed an object storage system for Facebook's Photos application with the aim of serving the long tail of requests seen by sharing photos [28]. These optimizations are important, as requests from the long tail accounted for a significant amount of their traffic, and most of these requests are served from the origin photo storage server, rather than by Facebook's CDN.

In the direction of distributing long-tailed content while lowering bandwidth costs and improving QoS, although without considering storage constraints, Traverso et al. in Reference 2 exploit the time differences between sites and the access patterns users follow. Instead of naively pushing UGC immediately, which may not be consumed and contribute unnecessarily to a traffic spike in the upload link, the system can follow a pull-based approach, when the first friend of a user in a point of presence (PoP) asks for the content. Moreover, instead of pushing content as soon as a user uploads, content can be pushed at the local time that is off-peak for the uplink and be downloaded in a subsequent time bin, also off-peak for the downlink, and earlier than the first user in the PoP is bound to ask for it. The larger the difference between the content production bin and the bin in which the content is likely to be read, the better the performance of the system.

In Reference 29, Sastry et al. built a prototype system called Buzztraq, which leverages the information encoded in social network structure to predict users' navigation behavior and may be partly driven by social cascades. The key concept of Buzztraq is to place replicas of items already posted by a user closer to the location of friends, anticipating future requests. The intuition is that social cascades are spread rapidly through population as social epidemics. For instance, friends usually have common interests; consequently, if a user shares a video through his/her network, many of user's friends may find it interesting and share it to their network. Experimental results showed that social cascade prediction can decrease the cost of user access compared to location based placement, improving the performance of CDNs.

Zhou et al. [23] leverage the connection between content exchange and geographic locality (using a Facebook dataset, they identify significant geographic locality not only concerning the connections in the social graph but also the exchange of content) and the observation that an important fraction of content is "created at the edge"

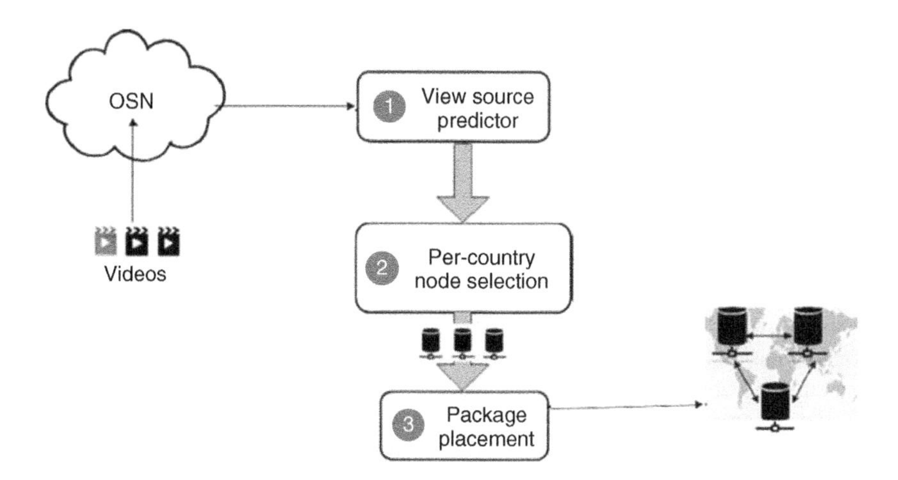

Figure 23.4 Overview of VOD service placement strategy leveraging OSNs.

(*is user-generated*), with a Web-based scheme for caching using the access patterns of friends. Content exchange is kept within the same Internet service provider (ISP) with a drop-in component, which can be deployed by existing Web browsers and is independent of the type of content exchanged. Browsing users online are protected with k-anonymity, where k is the number of users connected to the same proxy and are able to view the content.

Instead of optimizing the performance of UGC services exploiting spatial and temporal locality in access patterns, Huguenin et al. in Reference 30 show on a large (more than 650,000 videos) YouTube dataset that content locality (induced by the related videos feature) and geographic locality are in fact correlated. More specifically, they show how the geographic view distribution of a video can be inferred to a large extent from that of its related videos, proposing a UGC storage system that proactively places videos close to the expected requests. Such an approach could be extended with the leverage of information from OSNs, in the way that Figure 23.4 depicts.

23.6 FUTURE RESEARCH DIRECTIONS

In order to harness the power of social networks diffusion over CDN infrastructures, the key areas of interest that need to be explored include the large-scale datasets, the OSN evolution, and semantic annotation.

23.6.1 Large-Scale Datasets

The amount of information in OSNs is an obstacle, since elaborate manipulation of the data may be needed. An open problem is the efficient handling of graphs with billions of nodes and edges [38]. Facebook, for example, reported that it had one billion monthly active users as of October 2012 and 604 million monthly active users who used Facebook mobile products as of September 2012.

In order to generate aggregations and analyses that have meaning, the Facebook custom built-in data warehouse and analytics infrastructure have to apply *ad hoc* queries and custom MapReduce jobs [31] in a continuous basis on over half a petabyte of new data every 24 h, with the largest cluster containing more than 100 PB of data and the process needs surpassing the 60,000 queries in Hive, the data warehouse system for Hadoop, and Hadoop compatible file systems.

The desired scaling property refers to the fact that the throughput of the presented approaches should remain unchanged with the increase in the data input size, such as the large datasets that social graphs comprise and the social cascades phenomena that amplify the situation. The cost of scaling such content can be expressed in different ways. For instance, in the case of CDNs, it can be the number of replicas needed for a specific source, or it may take into account the optimal use of memory and processing time of a social-aware built system.

23.6.2 OSN Evolution

Existing works examine valuable insights into the dynamic world by posing queries on an evolving sequence of social graphs (e.g., Reference 32) and time-evolving graphs tend to be increasingly used as a paradigm also for the emerging area of OSNs [33]. However, the ability to process queries concerning the information diffusion in a scalable way remains to a great extent unstudied. With the exception of sporadic works on specialized problems, such as that of inference of dynamic networks based on information diffusion data [34], we are not aware of relative studies on the information diffusion through OSNs under the prism of graphs dynamicity.

23.6.3 Semantic Annotation

It would also be interesting to know which social cascades will evolve as global and which of them will evolve as local, possibly making some associations with their content or context features. It is challenging to discover contextual associations among the topics, which are by nature implicit in the UGC exchanged over OSNs and spread via social cascades. In other words, we would like to derive semantic relations. In this way, the identification of a popular topic can be conducted in a higher, more abstract level with the augmentation of a semantic annotation. Although we can explicitly identify the topic of a single information disseminated through an OSN, it is not trivial to identify reliable and effective models for the adoption of topics as time evolves [35,36], characterized with some useful emergent semantics. Such knowledge would improve caching of Web content in CDN infrastructures [1].

To sum up, OSNs create a potentially transformational change in users' behavior. This change will bring a far-reaching impact on traditional industries of content, media, and communications. In this context, the rapid proliferation of OSNs sites is expected to reshape CDN's structure and design [37]. Investigating the geographical, structural, and temporal properties of social cascades, new CDN infrastructures will be built where cache replacement strategies will exploit these properties. Traditional CDN systems support content distribution with specific needs, such as efficient resource discovery, large-scale replication of popular resources that follow zipfian resource-popularity distributions, and simple access-rights. In contrast, the next generation of CDN systems are required to support a variety of social interactions conducted through an open-ended set of distributed applications, going beyond resource discovery and retrieval and involving the following: synchronous and asynchronous messaging; "push" and "pull" modes of information access; finer access control for reading and writing shared resources; advanced mechanisms for data placement, replication and distribution for a large variety of resource types, and media formats.

23.7 CONCLUSION

Understanding the effects of social cascading on content over the Web is of great importance toward improving CDN performance. Given the large amount of available resources, it is often difficult for users to discover interesting content. Relying on the suggestions coming from friends seems to be a popular way to choose what to see. Taking into account the increasing popularity of OSNs and the growing popularity of streaming media, we have presented existing approaches that can be leveraged for the scaling of rich media content in CDNs using information from OSNs.

ACKNOWLEDGMENTS

Irene Kilanioti is a PhD candidate at the University of Cyprus whose work is supported by the Greek State Scholarships Foundation, Lifelong Learning Programme (Customized Evaluation Process).

REFERENCES

1. Scellato S, Mascolo C, Musolesi M, Crowcroft J. Track globally, deliver locally: Improving content delivery networks by tracking geographic social cascades. Proceedings of the 20th International Conference on World Wide Web; Hyderabad, India; 2011.
2. Traverso S, Huguenin K, Triestan I, Erramilli V, Laoutaris N, Papagiannaki K. Tailgate: Handling long-tail content with a little help from friends. Proceedings of the 21st International Conference on World Wide Web; Lyon, France; 2012.
3. Bakshy E, Rosenn I, Marlow C, Adamic L. The role of social networks in information diffusion. Proceedings of the 21st international Conference on World Wide Web; Lyon, France; 2012.
4. Chard K, Caton S, Rana O, Bubendorfer K. Social cloud: Cloud computing in social networks. Proceedings of the 3rd IEEE International Conference on Cloud Computing (CLOUD); Miami, FL, USA; 2010.
5. Easley D, Kleinberg J. *Networks, Crowds, and Markets*. Cambridge University Press; 2010.

6. Christodoulou G, Georgiou C, Pallis G. The role of twitter in youtube videos diffusion. Proceedings of the 13th International Conference on Web Information System Engineering (WISE); Paphos, Cyprus; 2012.
7. Brodersen A, Scellato S, Wattenhofer M. Youtube around the world: Geographic popularity of videos. Proceedings of the 21st international Conference on World Wide Web; Lyon, France; 2012. p 241–250.
8. Bakshy E, Hofman JM, Mason WA, Watts DJ. Everyone's an influencer: Quantifying influence on twitter. Proceedings of the Fourth ACM International Conference on Web Search and Data Mining; Kowloon, Hong Kong; 2011.
9. Cha M, Mislove A, Adams B, Gummadi KP. Characterizing social cascades in flickr. Proceedings of the First Workshop on Online Social Networks; Seattle, WA, USA; 2008.
10. McPherson M, Smith-Lovin L, Cook JM. Birds of a feather: Homophily in social networks. Annu Rev Sociol 2001;27:415–444.
11. La Fond T, Neville J. Randomization tests for distinguishing social influence and homophily effects. Proceedings of the 19th International Conference on World Wide Web; Raleigh, NC, USA; 2010.
12. Watts DJ. A simple model of global cascades on random networks. Proc Natl Acad Sci 2002;99(9):5766–5771.
13. Steeg GV, Ghosh R, Lerman K. What stops social epidemics? Proceedings of the 5th International Conference on Weblogs and Social Media (ICWSM); Barcelona, Spain; 2011.
14. Dave KS, Bhatt R, Varma V. Modelling action cascades in social networks. Proceedings of the 5th AAAI International Conference on Weblogs and Social Media; Barcelona, Spain; 2011.
15. Stauffer D, Aharony A. *Introduction to Percolation Theory*. CRC: ; 1994.
16. Wu F, Huberman BA. Novelty and collective attention. Proc Natl Acad Sci 2007;104(45): 17599–17601.
17. Hogg T, Lerman K. Stochastic models of user-contributory web sites. Proceedings of 3rd international AAAI Conference on Weblogs and Social Media; San Jose, CA, USA; 2009.
18. Mislove A, Marcon M, Gummadi KP, Druschel P, Bhattacharjee B. Measurement and analysis of online social networks. Proceedings of the 7th ACM SIGCOMM Conference on Internet Measurement; San Diego, CA, USA; 2007.
19. Wilson C, Boe B, Sala A, Puttaswamy KP, Zhao BY. User interactions in social networks and their implications. Proceedings of the 4th ACM European Conference on Computer Systems; Nuremberg, Germany; 2009.
20. Kumar R, Novak J, Tomkins A. Structure and evolution of online social networks. In: *Link Mining: Models, Algorithms, and Applications*. 2010. p 337–357.
21. Java A, Song X, Finin T, Tseng B. Why we twitter: Understanding microblogging usage and communities. Proceedings of the 9th ACM WebKDD and 1st SNA-KDD 2007 Workshop on Web Mining and Social Network Analysis; San Jose, CA, USA; 2007.
22. Benevenuto F, Rodrigues T, Cha M, Almeida V. Characterizing user behavior in online social networks. Proceedings of the 9th ACM SIGCOMM Conference on Internet Measurement; Chicago, IL, USA; 2009.
23. Zhou F, Zhang L, Franco E, Mislove A, Revis R, Sundaram R. Webcloud: Recruiting social network users to assist in content distribution. Proceedings of IEEE International Symposium on Network Computing and Applications; Cambridge, MA, USA; 2012.

24. Breslau L, Cao P, Fan L, Phillips G, Shenker S. Web caching and zipf-like distributions: Evidence and implications. INFOCOM'99. Eighteenth Annual Joint Conference of the IEEE Computer and Communications Societies. Proceedings; New York, NY, USA: IEEE; 1999.
25. Yu H, Zheng D, Zhao BY, Zheng W. Understanding user behavior in large-scale video-on-demand systems. ACM SIGOPS Operat Syst Rev 2006;40:333–344.
26. Cha M, Kwak H, Rodriguez P, Ahn Y-Y, Moon S. I tube, you tube, everybody tubes: Analyzing the world's largest user generated content video system. Proceedings of the 7th ACM SIGCOMM Conference on Internet measurement; San Diego, CA, USA; 2007.
27. Yang J, Leskovec J. Patterns of temporal variation in online media. Proceedings of the Fourth ACM International Conference on Web Search and Data Mining; Kowloon, Hong Kong; 2011.
28. Beaver D, Kumar S, Li HC, Sobel J, Vajgel P. Finding a needle in haystack: Facebook's photo storage. Proceedings of OSDI 10. 9th USENIX Symposium on Operating Systems Design and Implementation, Proceedings Usenix; Vancouver, BC, Canada; 2010; Vol. 10. p 1–8.
29. Sastry N, Yoneki E, Crowcroft J. Buzztraq: Predicting geographical access patterns of social cascades using social networks. Proceedings of the Second ACM EuroSys Workshop on Social Network Systems; Nuremberg, Germany; 2009.
30. Huguenin K, Kermarrec A-M, Kloudas K, Taäni F. Content and geographical locality in user-generated content sharing systems. Proceedings of 22nd SIGMM International Workshop on Network and Operating Systems Support for Digital Audio and Video (NOSSDAV); Toronto, Canada; 2012.
31. Dean J, Ghemawat S. Mapreduce: Simplified data processing on large clusters. Commun ACM 2008;51(1):107–113.
32. Ren C, Lo E, Kao B, Zhu X, Cheng R. On querying historical evolving graph sequences. Proceedings of the 37th International Conference on Very Large Data Bases (VLDB); 2011.
33. Fard A, Abdolrashidi A, Ramaswamy L, Miller JA. Towards efficient query processing on massive time-evolving graphs. Proceedings of the 8th IEEE International Conference on Collaborative Computing: Networking, Applications and Worksharing (CollaborateCom); Pittsburgh, PA, United States; 2012.
34. Rodriguez MG, Leskovec J, Schölkopf B. Structure and dynamics of information pathways in online media. Proceedings of ACM International Conference on Web Search and Data Mining (WSDM); Rome, Italy; 2013.
35. Garcá-Silva A, Kang J-H, Lerman K, Corcho O. Characterising emergent semantics in twitter lists. Proceedings of the 9th International Conference on the Semantic Web: Research and Applications (ESWC); Heraklion, Greece; 2012.
36. Lin CX, Mei Q, Jiang Y, Han J, Qi S. Inferring the diffusion and evolution of topics in social communities. Proceedings of ACM SIGKDD Workshop on Social Network Mining and Analysis (SNAKDD); San Diego, CA, USA; 2011.
37. Han L, Punceva M, Nath B, Muthukrishnan S, Iftode L. Socialcdn: Caching techniques for distributed social networks. Proceedings of IEEE 12th International Conference on Peer-to-Peer Computing (P2P), 2012. Tarragona, Spain: IEEE; 2012. p 191–202.
38. Leskovec J, McGlohon M, Faloutsos C, Glance N, Hurst M. Patterns of cascading behavior in large blog graphs. Proceedings of SIAM International Conference on Data Mining (SDM) 2007; Minneapolis, Minnesota, USA; 2007.

INDEX

Advanced Content Delivery, Streaming, and Cloud Services, First Edition.
Edited by Mukaddim Pathan, Ramesh K. Sitaraman, and Dom Robinson.

WILEY SERIES ON PARALLEL AND DISTRIBUTED COMPUTING

Series Editor: Albert Y. Zomaya

Parallel and Distributed Simulation Systems / Richard Fujimoto

Mobile Processing in Distributed and Open Environments / Peter Sapaty

Introduction to Parallel Algorithms / C. Xavier and S. S. Iyengar

Solutions to Parallel and Distributed Computing Problems: Lessons from Biological Sciences / Albert Y. Zomaya, Fikret Ercal, and Stephan Olariu (*Editors*)

Parallel and Distributed Computing: A Survey of Models, Paradigms, and Approaches / Claudia Leopold

Fundamentals of Distributed Object Systems: A CORBA Perspective / Zahir Tari and Omran Bukhres

Pipelined Processor Farms: Structured Design for Embedded Parallel Systems / Martin Fleury and Andrew Downton

Handbook of Wireless Networks and Mobile Computing / Ivan Stojmenović *(Editor)*

Internet-Based Workflow Management: Toward a Semantic Web / Dan C. Marinescu

Parallel Computing on Heterogeneous Networks / Alexey L. Lastovetsky

Performance Evaluation and Characterization of Parallel and Distributed Computing Tools / Salim Hariri and Manish Parashar

Distributed Computing: Fundamentals, Simulations, and Advanced Topics, *Second Edition* / Hagit Attiya and Jennifer Welch

Smart Environments: Technology, Protocols, and Applications / Diane Cook and Sajal Das

Fundamentals of Computer Organization and Architecture / Mostafa Abd-El-Barr and Hesham El-Rewini

Advanced Computer Architecture and Parallel Processing / Hesham El-Rewini and Mostafa Abd-El-Barr

UPC: Distributed Shared Memory Programming / Tarek El-Ghazawi, William Carlson, Thomas Sterling, and Katherine Yelick

Handbook of Sensor Networks: Algorithms and Architectures / Ivan Stojmenović *(Editor)*

Parallel Metaheuristics: A New Class of Algorithms / Enrique Alba (*Editor*)

Design and Analysis of Distributed Algorithms / Nicola Santoro

Task Scheduling for Parallel Systems / Oliver Sinnen

Computing for Numerical Methods Using Visual C++ / Shaharuddin Salleh, Albert Y. Zomaya, and Sakhinah A. Bakar

Architecture-Independent Programming for Wireless Sensor Networks / Amol B. Bakshi and Viktor K. Prasanna

High-Performance Parallel Database Processing and Grid Databases / David Taniar, Clement Leung, Wenny Rahayu, and Sushant Goel

Algorithms and Protocols for Wireless and Mobile Ad Hoc Networks / Azzedine Boukerche (*Editor*)

Algorithms and Protocols for Wireless Sensor Networks / Azzedine Boukerche (*Editor*)

Optimization Techniques for Solving Complex Problems / Enrique Alba, Christian Blum, Pedro Isasi, Coromoto León, and Juan Antonio Gómez (*Editors*)

Emerging Wireless LANs, Wireless PANs, and Wireless MANs: IEEE 802.11, IEEE 802.15, IEEE 802.16 Wireless Standard Family / Yang Xiao and Yi Pan (*Editors*)

High-Performance Heterogeneous Computing / Alexey L. Lastovetsky and Jack Dongarra

Mobile Intelligence / Laurence T. Yang, Augustinus Borgy Waluyo, Jianhua Ma, Ling Tan, and Bala Srinivasan (*Editors*)

Research in Mobile Intelligence / Laurence T. Yang *(Editor)*

Advanced Computational Infrastructures for Parallel and Distributed Adaptive Applicatons / Manish Parashar and Xiaolin Li *(Editors)*

Market-Oriented Grid and Utility Computing / Rajkumar Buyya and Kris Bubendorfer *(Editors)*

Cloud Computing Principles and Paradigms / Rajkumar Buyya, James Broberg, and Andrzej Goscinski (*Editors*)

Algorithms and Parallel Computing / Fayez Gebali

Energy-Efficient Distributed Computing Systems / Albert Y. Zomaya and Young Choon Lee *(Editors)*

Scalable Computing and Communications: Theory and Practice / Samee U. Khan, Lizhe Wang, and Albert Y. Zomaya (*Editors*)

The DATA Bonanza: Improving Knowledge Discovery in Science, Engineering, and Business / Malcolm Atkinson, Rob Baxter, Michelle Galea, Mark Parsons, Peter Brezany, Oscar Corcho, Jano van Hemert, and David Snelling *(Editors)*

Large Scale Network-Centric Distributed Systems / Hamid Sarbazi-Azad and Albert Y. Zomaya *(Editors)*

Verification of Communication Protocols in Web Services: Model-Checking Service Compositions / Zahir Tari, Peter Bertok, and Anshuman Mukherjee

High-Performance Computing on Complex Environments / Emmanuel Jeannot and Julius Žilinskas (*Editors*)

Advanced Content Delivery, Streaming, and Cloud Services / Mukaddim Pathan, Ramesh K. Sitaraman, and Dom Robinson *(Editors)*